Principles of Aircraft Engineering

Principles of Aircraft Engineering

Edited by
Natalie Spagner

WILLFORD PRESS

www.willfordpress.com

Published by Willford Press,
118-35 Queens Blvd., Suite 400,
Forest Hills, NY 11375, USA

ISBN: 978-1-68285-581-2

Cataloging-in-Publication Data

Principles of aircraft engineering / edited by Natalie Spagner.
   p. cm.
Includes bibliographical references and index.
ISBN 978-1-68285-581-2
1. Aeronautics. 2. Airplanes. 3. Flight engineering. 4. Aerospace engineering.
I. Spagner, Natalie.
TL545 .P75 2019
629.134--dc21

For information on all Willford Press publications
visit our website at www.willfordpress.com

WILLFORD PRESS

Contents

Preface

Aircraft engineering is studied in the engineering field of aeronautical engineering. It studies the design, manufacture and operation of aircrafts. The science of aerodynamics is central to the development of this field. Three broad areas of incompressible flow, compressible flow and transonic flow are widely studied in the branch of aircraft engineering. This book is a compilation of a range of topics that aim to explain the key concepts and principles related to aircraft engineering while also illustrating the modern technological advancements. From theories to research to practical applications, case studies related to all contemporary topics of relevance to this field have been included in this book. It aims to serve as a reference to a broad spectrum of readers.

This book is a result of research of several months to collate the most relevant data in the field.

When I was approached with the idea of this book and the proposal to edit it, I was overwhelmed. It gave me an opportunity to reach out to all those who share a common interest with me in this field. I had 3 main parameters for editing this text:

1. Accuracy – The data and information provided in this book should be up-to-date and valuable to the readers.

2. Structure – The data must be presented in a structured format for easy understanding and better grasping of the readers.

3. Universal Approach – This book not only targets students but also experts and innovators in the field, thus my aim was to present topics which are of use to all.

Thus, it took me a couple of months to finish the editing of this book.

I would like to make a special mention of my publisher who considered me worthy of this opportunity and also supported me throughout the editing process. I would also like to thank the editing team at the back-end who extended their help whenever required.

Editor

CFD Analysis of Mars Phoenix Capsules at Mach Number 10

Madhunuri Raju*

Department of Aeronautical Engineering in MLR Institute of Technology, Hyderabad, India

Abstract

Numerical simulations are performed for flow over Phoenix entry vehicle at Mars at zero angle of Attack and Mach number 10. Flow field features like bow shock, shear layers, expansion fan and separation bubble will be captured using CFD commercial package FLUENT. The computed wall data of pressure and temperature will be compared with experimental results at Mars atmospheric conditions. This project deals with the study of flow over Phoenix entry vehicle at Mach number 10. The concept of atmospheric entry has applications in various fields. Vehicles that typically undergo this process include exo-orbital trajectories. In this project, the type of entry vehicle considered is entry capsule, which enters Mars atmosphere from an orbit. The primary design consideration of entry capsule requires large spherical nose radius of their fore body that gives high aerodynamic drag and a short body length for reducing the total structural weight and the ballistic coefficient.

Keywords: Phoenix entry vehicle; CFD; Aerodynamic drag; Mach number 10

Introduction

Motivated by the curiosity about the universe origin, the search for the extraterrestrial intelligence, and the interest in the space recourses exploitation, exploration missions to Mars have been carried out several times since 1960s. THE Phoenix Mission was launched on Delta II rocket to mars. Phoenix is designed to study the history of water and search for complex organic molecules in the ice-rich soil of the Martian arctic [1].

In order to know about the planets and the natural satellites present in our solar system, there is a need for us to enter an orbit around that plant or satellite and observe them. For better understanding, we might even have to send rovers onto the surface of the planet or satellite to conduct experiments and if possible, bring them back to Earth. For this to happen, the space probe must pass through their atmosphere, reach the surface intact, conduct experiments there, travel back to Earth and perform another entry phase in order to reach Earth's surface with the data collected. For this to happen, the entry vehicle must be designed accordingly.

Atmospheric entry refers to the movement of human made or natural objects as they enter the atmospheric of a planet from outer space, in the case of Earth from an altitude above the "edge of space" [2]. The vehicle which enters into the atmospheric conditions from the outer space is known as Entry Vehicle. Entry vehicle design addresses the design of controlled entry vehicles which are intended to reach the planetary surface intact. Vehicles that typically undergo this process include ones returning from orbit and ones on exo-orbital trajectories.

The use of CFD methods can valuable knowledge for future spacecrafts such as crew exploration vehicle, especially with the recent call to move back to the old Apollo shaped entry vehicle capsule rather than the usage of space shuttles. Hence, we choose to so analysis on entry capsules. The Phoenix Mars Lander seeks to verify the presence of water and habitable conditions in the Martian arctic.

Study the history of water by examining water-ice below the Martian surface.

- Determine if the Martian article soil could support life.

- The details of design model are indicated in Figure 1 and also

mentioned clear specifications in Table 1.

Objectives of this paper are to find:

- The entry vehicles considered and Martian entry capsules.

- In the present project, CFD analysis is done on two design configurations of Martian entry capsules, PHOENIX using software FLUENT.

- A study is taken up on the flow fields featured around the capsule at Mach number 10.

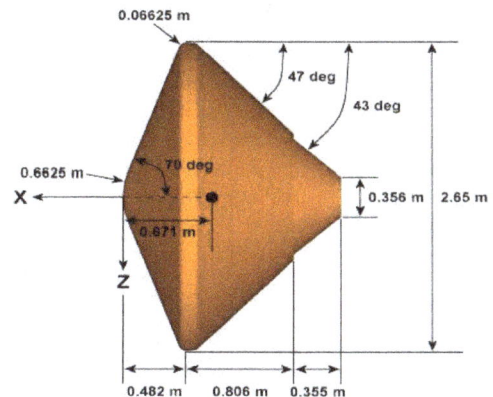

Figure 1: Dimensions of phonex mission.

***Corresponding author:** Madhunuri Raju, Department of Aeronautical Engineering in MLR Institute of Technology, Hyderabad, India
E-mail: raju.madhunuri98@gmail.com

Specifications	Values
Diameter, m	2.65
Entry mass, kg	602
Relative entry velocity (km/s)	5.5
Relative entry FPA (deg)	-13.2
Free stream Mach number at parachute deployment	1.65
Hypersonic α (deg)	0
Control	Non-spinning

Table 1: Phonex specifications.

Methods

Numerical simulation methods

Computational Fluid Dynamics technique was undertaken to predict the aerodynamic force for the Mars entry vehicle. A finite volume approach was used to solve the full Navier– Stokes flow field equations for calorically perfect gas [3]. The code used the van Leer flux-vector splitting method for the inviscid fluxes with the 2nd order correction using the van Leer limiter. For time integration, the non-iterative implicit method was used for rapid convergence.

Constant inflow condition was imposed on the far field boundary, and the extrapolation was used on the outflow boundary in all solutions. Non-slip wall boundary condition was implemented and the thermal state of the surface is radiative equilibrium with a fixed surface emissivity of 0.78.

The flow field is assumed to be steady and laminar. Unsteady effects are weak at these high speeds for long duration entry flight, and moreover, unsteady aerodynamic contribution is negligibly small in the after body. Transition to turbulence could occur but is weak due to low Reynolds number in the relatively thin Martian atmosphere.

where the reference viscosity $\mu_0 = 1.48 \times 10\text{-}5$ kg/(m•s), the reference temperature $T_0 = 293.15$ K, and the Sutherland constant C=240 K. Sutherland's law agrees well with theoretical calculation.

Engineering prediction methods

Surface panel method was used to calculate the surface pressure coefficient for rapid prediction. The pressure coefficient for each individual element is the function of inflow velocity vector, the element area and the surface normal vector of the given element, and has no relation with the neighbor element [4]. The total aerodynamic coefficients are then calculated by integrating the panel pressure coefficient along the total surface.

The main work of the surface panel method is to calculate the surface pressure coefficient. Multiple engineering methods to predict the surface pressure coefficient have been developed so far, with their peculiar applications. The modified Newtonian theory is believed to be quite suitable for hypersonic aerodynamics prediction of blunt body, such as reentry shapes, Mars entry capsules, etc. The modified Newtonian formula can be described as:

$$Cp = Cpmas(V^{-}T/V\text{œ})2$$

It can be seen that the stagnation pressure coefficient is the only function of inflow Mach number and the specific heat ratio. It is noted that the modified Newtonian formula is valid for windward region, and the coefficient in the leeward region is assumed to be zero because the pressure effect here is negligibly weak [5]. The velocity ratio is the cosine of the angle between the inflow velocity and the surface normal vector.

Effective specific heat ratio method

Multi component mixture exists in the Martian atmosphere, including approximately 95% CO_2, 3% N_2, 2% Ar and other negligible species. Since the real values of specific heat ratio markedly decrease at high-Mach-number flight conditions when the gas goes across the shock, effective specific heat ratio was specified to take the non-air and high-temperature real-gas effects into account. The post-shock specific heat ratio, which only depends on the local temperature for thermal perfect gas, is expected to be the effective value γ_{eff}. Considering the vibration excitation, the local specific heat ratio is:

where m is the number of gas species, wj is the mass fraction the species j. For each species, cP and cV are determined by molecular transitional, rotational and vibrational energy, so we

$$cP = cV + R$$

Validation case

The validation model is a scaled 70° sphere-cone configuration with the available numerical and experimental data. The wind tunnel experiment use CO_2 as the test gas with the inflow enthalpy of 1.89 MJ/kg (U∞=1908 m/s, p∞=1010 Pa). Calculated results from numerical simulation and engineering prediction based on the effective specific heat ratio method are compared against the experimental data [6]. The calculation results of the surface pressure from both two mentioned approaches, the DPLR data and the experimental data are all shown in Figure 2 for comparison. The comparison highlights a good agreement between calculated and experimental data, thus confirming reliability of the numerical simulations and engineering prediction for Mars entry capsules.

Design and Analysis

The various materials that can be used for different temperatures are mentioned in Table 2.

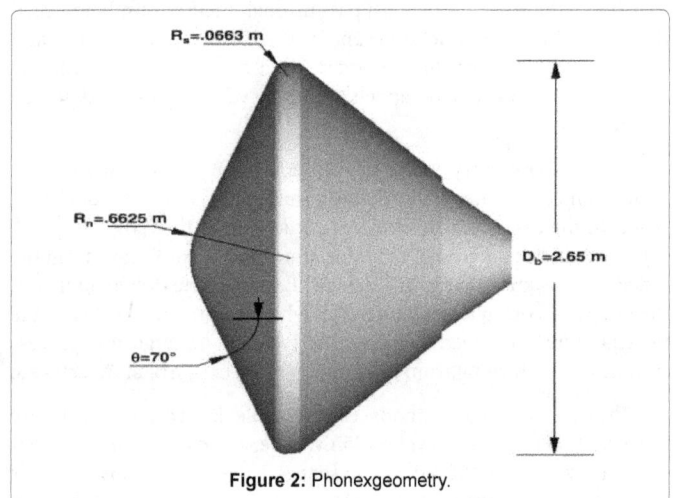

Figure 2: Phonexgeometry.

Materials	Energy to vaporize, Btu/lb	Melting Temperature, R⁰
Tungsten	1870	6500
Titanium	3865	3700
Beryllium oxide	13400	2900
Graphite	28700	6800

Table 2: Materials that can be used for the heat shield.

Here, Btu means the British thermal unit which is a traditional unit of energy equal to about 1,055.05585 joules. It is approximately the amount of energy needed to heat 1 pound (0.454 kg) of water from 276.9 K to 277.4 K.

Spacecraft specifications are Mass: The lift-off mass was 1,350 kg (2,980 lb), including 852 kg (1,878 lb) of propellant Bus: The spacecraft's bus is a modified I-1 K structure and propulsion hardware configuration, similar to Chandrayaan-1, India's lunar orbiter that operated from 2008 to 2009, with specific improvements and upgrades needed for a Mars mission. The satellite structure is constructed of an aluminium and composite fibre reinforced plastic (CFRP). Power: Electric power is generated by three solar array panels of 1.8 m × 1.4 m (5 ft 11 in × 4 ft 7 in) each (7.56 m^2 (81.4 sq ft) total), for a maximum of 840 watts of power generation in Mars orbit. Electricity is stored in a 36 Ah Lithium-ion battery.

Propulsion: A liquid fuel engine with a thrust of 440 newton is used for orbit raising and insertion into Mars orbit. The orbiter also has eight 22-newton thrusters for attitude control. Its propellant mass is 852 kg (1,878 lb).

Atmospheric conditions taken in software to process the flow analysis are mentioned in the below Table 3.

Results and Discussions

Geometry and grid generation

The proposed geometry for aerodynamic performance analysis is the Mars phonex vehicle with sphere-cone fore body and three-sectional-cone after body as described in Figure 2.

Two-dimensional surface grid is required for rapid aerodynamic prediction using the panel method, while full three-dimensional grid is necessary for numerical simulation of overall flow field. The computational grid for phonex vehicle configuration is a multi-block structured grid, mainly growing from the surface grid taking the boundary layer into consideration [7]. To assure accurate prediction, the grid is everywhere orthogonal to the body at the surface. Grid independence analysis shows that grids for the wall cell Reynolds number $Rec = \rho\infty U\infty dw/\mu\infty$ can fulfill the accuracy requirement of aerodynamics analysis, where dw is the normal grid distance close to the wall. Figure 3 shows both the two-dimensional surface grid and the grid on the symmetry plane.

Mars entry flow field

In order to figure out the detailed hypersonic flow field, steady numerical simulations for a specific hypersonic flight state are herein performed. The flight speed is 3300 m/s and the Mach number is 10.

Hypersonic flow field for the phonex entry vehicle is complicated, especially for the vortex flow. Figure 4 shows the main steady structure of the hypersonic flow on the symmetry plane. A strong bow shock wave is detached from the surface and lies very close to the heat shield [8]. The gas across the shock wave is compressed to form a shock layer,

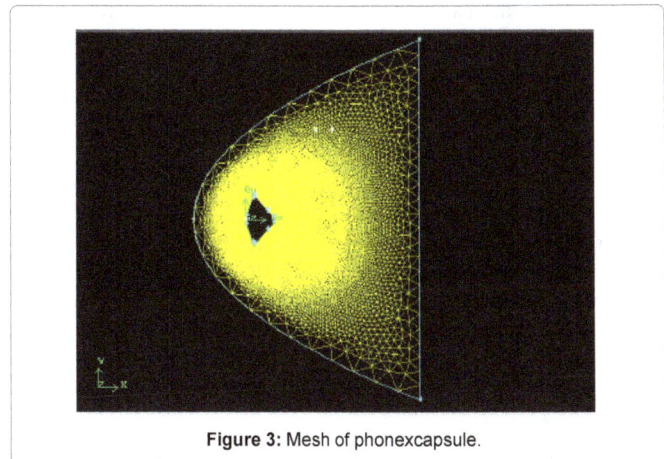

Figure 3: Mesh of phonexcapsule.

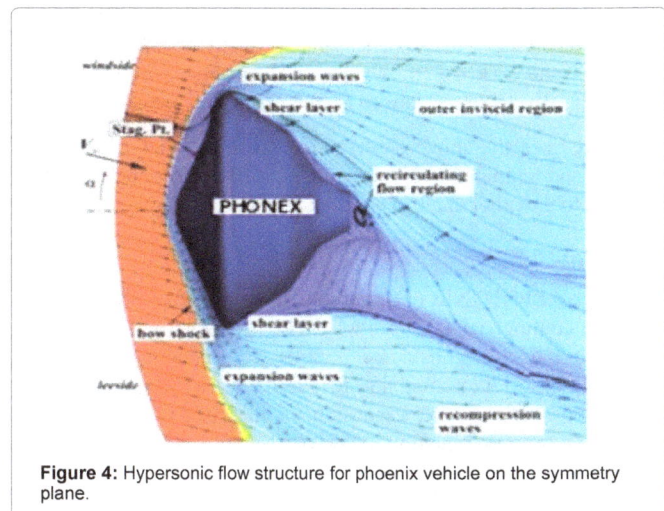

Figure 4: Hypersonic flow structure for phoenix vehicle on the symmetry plane.

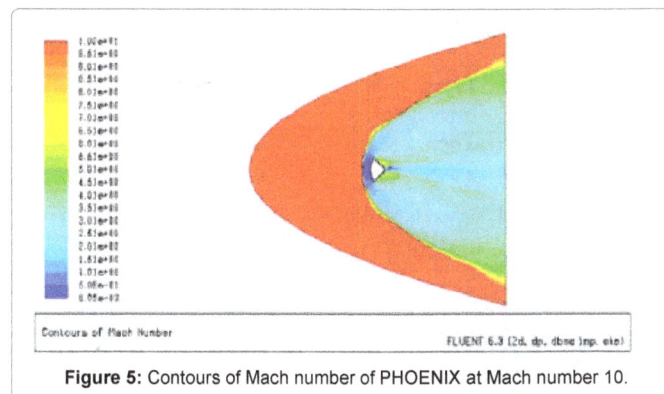

Figure 5: Contours of Mach number of PHOENIX at Mach number 10.

much thinner than that of the air flow. Inflow uniform streamlines are deflected across the shock and travel around the body (Figure 5). Expansion waves occur near the shoulder due to large deflection angle. The gas gets accelerated in this region and merges into the external inviscid flow downstream from the shoulder. The heat shield boundary layer thickness increases from the stagnation point to the shoulder, and the shear layer appears after the shoulder, resulting in local vortex motion. Consequently, flow separation arises on the backshell (Figure 6). A large region of recirculating flow leads to relatively low pressure

Parameters	MARS	EARTH
Surface gravity (g)	3.71 m/s^2	9.8 m/s^2
Surface heat at constant pressure(Cp)	841 J/Kg-K	1006.43 J/Kg-K
Thermal conductivity (k)	0.01633 w/m/k	0.0242 w/m/k
Viscosity	1.51e-05 kg/m-s	1.716e-05 kg/m-s

Table 3: Initial conditions given in FLUENT.

distribution on the backshell, which has a negligible contribution to the total coefficients. The wake flow after the capsule meets and then accumulates to form the recompression waves, which are nearly parallel to the inflow direction, and grow weaker to the outer flow.

Below, in figures are images of the Mach number and pressure coefficient contours over PHOENIX capsule at mach number 10.

We can observe from the above contour that distinct shock wave that forms. The velocity magnitude scalar clearly illustrates the flow of the air coming to a slow halt as it gets closer and closer to the heat shield of the vehicle. In the Figure 5, of the pressure gradient scalar, we can observe a very similar shape for the shock wave; however, we also observe an expansion shock beginning to form from the corners of the craft. This is due to the fact that air cannot turn quickly enough around those corners.

The flow expands around the corners and separates to form a large recirculation region behind the vehicle. The flow in this region is highly unsteady and three dimensional with a large range of length and time scales. The shear layers enclosing the recirculation region come together at the neck region. A recompression shock is formed at this point that turns the outer flow to make it parallel to the vehicle axis.

Apart from the above observation it is found that variations of pressure and temperature is minimum at at 10 degrees angle of attack compare to the other angle of attack at Mach no 10 (Table 4).

Conclusion

Flow over PHOENIX entry capsules is computed at Mach number 10 using commercial CFD software FLUENT for ideal gas model. Turbulent model is considered. Results were obtained for various angles of attack. All essential flow field features are fairly well captured such as bow shock wave, expansion on the corner, recompression shock wave and re-circulation in the base region.

The following observations can be made about the flow characteristics from the results obtained for the configurations considered:

• Sonic line occurs near the junction between the fore body heat shield and the corner fillet of capsule

• The shock wave formed comes closer to the body with increase in Mach number

• At high speeds, even a slight change in the angle of attack can severely alter the activity of the entry capsule including the shock wave which plays a major factor in the fate of the craft

• With increase in angle of attack, the flow region affected increases and effective pressure and temperature decreases

• We have also seen that a change in the capsule shape can alter the wake formation behind the craft, which changes the deceleration.

• The flow characteristics obtained from the present study agree quite well with those obtained by earlier investigators.

Acknowledgment

This research was supported by Dr Satya Narayana Gupta, Department of Aeronautical engineering in MLR Institute of technology, Hyderabad.

References

1. Gnoffo PA, Weilmuenster KJ, Braun RD, Cruz CI (1995) Effects of sonic line transition on aerothermodynamics of the Mars pathfinder probe in Proceedings of AIAA.

2. Prabhu DK, Saunders DA (2002) On heat shield shapes for Mars entry capsules in Proceedings of AIAA, 1221.

3. Viviani G, Pezzella (2009) Aerodynamic analysis of a capsule vehicle for a manned exploration mission to Mars in Proceedings of the 16th AIAA/DLR/DGLR International Space Planes and Hypersonic Systems and Technologies Conference AIAA 7386.

4. Liever PA, Habchi SD, Burnell SI, Lingard JS (2002) CFD prediction of the Beagle 2 aerodynamic database in AIAA 0683.

5. Hirschel EH, Weiland C (2009) Selected aerothermodynamic design problems of hypersonic flight vehicles Springer Press New York.

6. Anderson JD (2006) Hypersonic and high-temperature gas dynamics. J space cr Rockets AIAA.

7. Brauckmann GJ, Paulson Jr, Weilmuenster KJ (1995) Experimental and computational analysis of shuttle orbiter hypersonic trim anomaly. J Spacecr Rockets 32: 758–764.

8. Weilmuenster KJ, Hamilton HH (1982) A comparison of computed space shuttle orbiter surface pressures with flight measurements. AIAA 0937.

AOA	Temparature	Pressure	Velocity
0	5.00e + 003	6.23e + 006	1.311e + 004
10	4.957e + 003	1.595e + 007	3.109e + 003
15	4.951e + 003	1.557e + 007	3.105e + 003

Table 4: Various properties at different angle of attack.

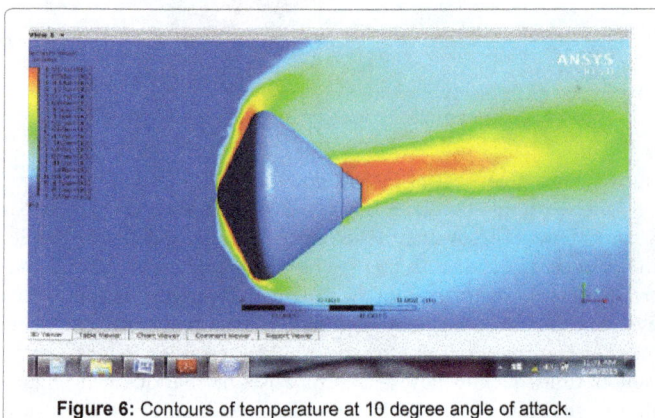

Figure 6: Contours of temperature at 10 degree angle of attack.

2

Visualisation of Vortex Structures Developed on the Upper Surface of Double-Delta Wings

Abene Abderrahmane*

University of Valenciennes and Hainaut Cambrai, ISTV Mount Huy, F-59304 Valenciennes, France

Abstract

A large number of studies of flow visualisations, developed on the upper surface of delta and of gothic wings, and of cones, have been carried out in the wind tunnel of the Valenciennes University's Aerodynamics and Hydrodynamics Laboratory. These studies have provided a better understanding of the development and of the positioning of vortex structures and have enabled, in particular, the preferential nature of intervortex angles, thereby defined, to be determined.

This study of the vortical structures developed on the upper surface of a double-delta wing has revealed that these vortex flows are quite complex and that vortex structures interact with one another. Indeed, it would seem that vortex behaviour has something of a universal nature. An angular conformity between primary and secondary vortex torques and the leading edges of the wing can be expressed by the law of filiation. Intervortex angles evolve with increasing incidence while fragmentation is a function of the apex angle. It would be interesting to recall that this particular spatial organisation of vortex structures, citing the concept of preferential angles, also appears in standard theories on aerodynamics as, for example, in those governing aerodynamic drag. Nevertheless, the link between interior and exterior vortex structures remains to be investigated further. Such studies might even prove the existence of a supplementary torque. In addition, the least resistance of secondary vortices in relation to their fragmentation inevitably calls for experiments to be undertaken with other possible combinations of slender bodies although these areas of research are beyond the scope of this article.

Keywords: Intervortex angles; Vortex flow; Apex angle

Nomenclature

θ_m: preferential angle associated with the whole numbers l and m

ℓ and m :whole numbers such as m>0 and $\ell \geq m$

β: apex angle

C_o: height of wing

h: height of cone

R: radius of the circular base of the cone

i: incidence

V_o: speed of the flow at infinite up stream

Re: Reynolds number

α_1: the main or interior inter vortex angle for wings

α_2: the secondary or exterior intervortex angle for wings

ω_1: the main or intervortex angle for ogives or cones

ω_2: the secondary or exterior intervortex angle for ogives or cones

Introduction

Quite a considerable number of studies have been carried out to date into delta wings, ogival wings, cones and also into more or less simple slender bodies formed from combinations of such components; the findings have dealt as much with the development of approximate theories as with the definition of models specifying vortex lift by unit area.

Visualisations of hyper lifting vortex structures, mainly those carried out by Werle H [1-6], the analysis of pressure and speed fields created by these vortices, with or without breakdown – notably the analysis by Solignac et al. [7] also provide quite outstanding studies that are the standard works in their fields.

Already described fully in such papers as, for example, those by [8-11] and Stahl [12], these findings offer today entirety a thorough knowledge of the properties of various types of slender bodies.

However, given that the character of most of the aspects referred to remains empirical and limited to this or that degree of incidence [1-4] or to a numeric range [5], the way lies open, starting out from experimental data and various factors of analysis [6], for new attempts to be undertaken to examine the fundamental problems related to the position of vortices created by such slender bodies.

A large number of photographic visualisations, concerning vortex flows developed on the upper surface of delta or ogival wings and cones, have been carried out at the Valenciennes University (France) laboratory [7] in such a way as to provide a better understanding of the development and positioning of vortex structures at not only low and mean incidence but also at high incidence.

These visualisations have enabled priority to be accorded to the study of examples of the most elementary shaped section, i.e., delta and ogival wings.

*Corresponding author: Dr. Abene Abderrahmane, University of Valenciennes and Hainaut Cambrai, ISTV Mount Huy, F-59304 Valenciennes, France
E-mail: a.abene@yahoo.fr

The results obtained in these two cases, fully described in previous articles and papers [6,8], may be acknowledged to have remarkable simplicity and consequently convey the fundamental nature of these studies.

The intervortex angles have been found, under experimental conditions, to have a preferential nature thereby underlining a simple angular characterisation of the relative positions of single or double vortex torques.

Vortex Structures of a Double-Delta Wing

Geometrical description

The profile under investigation in the wind tunnel is a double-delta wing having an apex angle $\beta=29°$ and a chord Co=240 mm. It is 1 mm thick (Figure 1).

Analysis of the results

The evolution of the vortex phenomena was traced in terms of that parameter which exerts the greatest influence on them, namely the angle of incidence of the configuration in relation to the flow. The visualisations were carried out at an upstream speed of flow of 3 m/s.

A reminder of the main phenomena taking place at low and mean incidences

• **i=0°**: the flow is uniform on the upper side of the wing. The boundary layer is observed but there is no flow separation as yet.

• **i=2°, i=5°**: the upstream flow skirts around the leading edges of the profile. Three zones become organised into a central zone and two external ones. The hyper lifting vortices resulting from the separation of the boundary layer begin to appear as increasingly organised structures.

• **i=8°**: the main and secondary vortices are clearly detected and have now become individualised, concentrated and separated from boundary layer. The central zone is visualized and fading out.

• **i=10°, i=15°, i=20°**: the vortices increase in strength. Both the vortex flow and the direction of the rotation of the vortices are clearly seen. Then central zone has disappeared. The presence of tertiary vortices is to be noted although they are extremely difficult to visualise. There is no breakdown as yet.

• **i=25°**: the breakdown phenomenon makes its appearance. The main vortices are breaking down a long way downstream from the

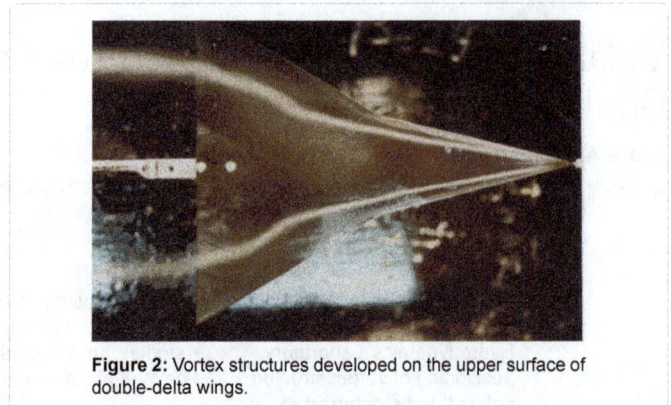

Figure 2: Vortex structures developed on the upper surface of double-delta wings.

profile: in fact, a more diffused mass of smoke is observed in this zone. The secondary vortices have broken down upstream from the trailing edge; as for the tertiary ones, which are difficult to observe because of their positioning at the edge of the boundary layer, it seems that they break down in the area close to the apex and coil around the main and secondary vortices. Once the secondary vortices have broken down, they also coil around the main ones. The asymmetry of the breakdown point of the main and secondary vortices is to be noted.

• **i=30°**: the tertiary vortices have now completely disappeared. The secondary ones are breaking down in the area of the trailing edge of the profile.

• **i=40°**: the secondary vortices break down near the apex whereas the point of breakdown of the main ones has advanced to a third of the way along the chord. A sudden expansion at the core of the main vortices is still discernible, followed by an unstable zone showing quite considerable turbulence.

Not that the position of the breakdown point is estimated on the basis of a main reading of the respective breakdown points of the right-hand and left-hand main vortices.

• **i=45°**: the main vortices break down at the fore quarter of the chord; the secondary ones are absorbed by the main vortices at the apex and no longer visible.

• **i=50°**: a total breakdown of the vortices takes place at the apex. Intense turbulence is observed at about the trailing edge (Figures 2-4).

Interpretation

Comments: the breakdown of the secondary vortices is not as impressive as that of the main ones. In fact, the vortex cores of the latter, after having maintained a cone-shaped form increasing in diameter towards the downstream side, undergo a sudden expansion into a brush-shaped form (the rate of expansion is approximately three to four times the diameter of the vortex core upstream from the point of breakdown). The point of breakdown is immediately followed by a zone in which the flow circulates again and the by an area of intense turbulence. The thin streams of smoke indicating the main vortices seem to change direction at the point of breakdown and to follow a spiral trajectory downstream.

• The pulsation phenomenon of the vortex breakdown has become highly conspicuous from now on the breakdown points undergo quirt considerable positional fluctuations. The broken down vortices seem "to dance".

• On placing an object in the axis of the broken down vortices, it can

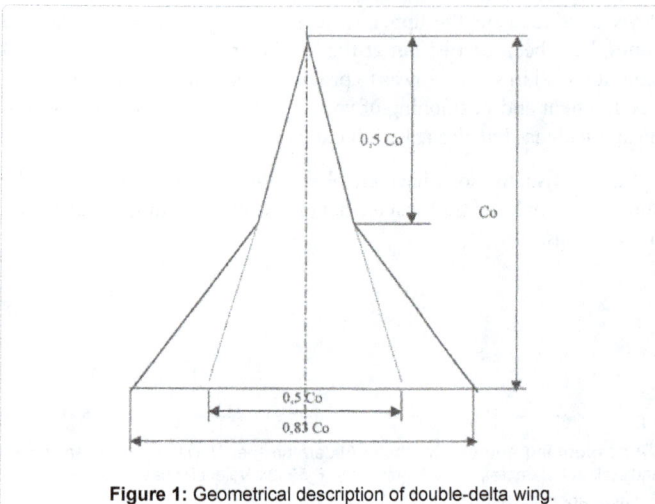

Figure 1: Geometrical description of double-delta wing.

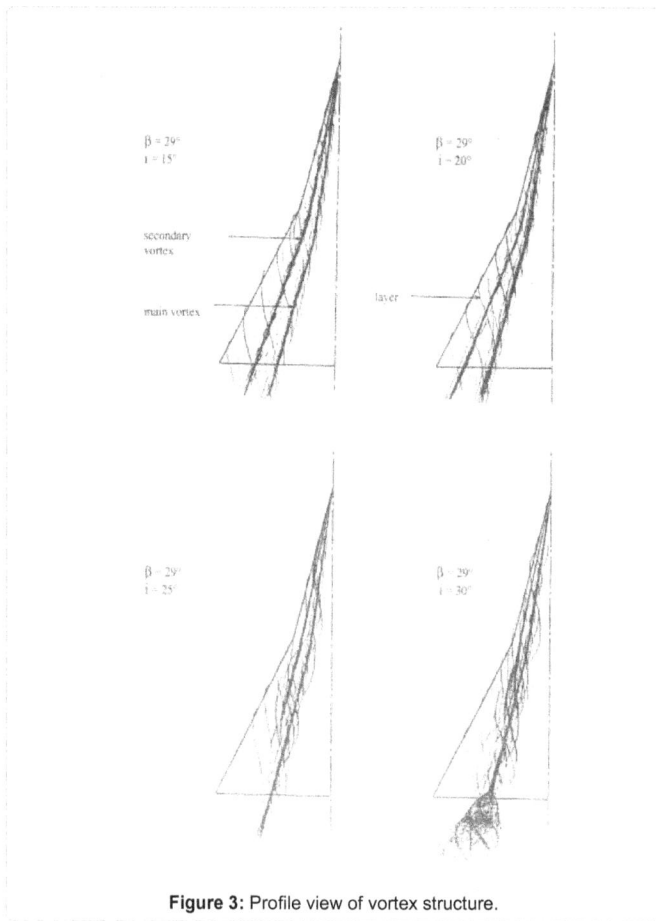

Figure 3: Profile view of vortex structure.

be seen that the breakdown points "rise again" in an upstream direction but it is also noticed that the breakdown point of the secondary vortices is still located upstream from that of main vortices.

The presence of preferential intervortex angles is observed.

Intervortex angles on the wings:

$$10° \leq i \leq 15° \Rightarrow \alpha_1 = 20.7°$$
$$15° \leq i \leq 26° \Rightarrow \alpha_1 = 19.5°$$
$$26° \leq i \leq 45° \Rightarrow \alpha_1 = 18.4°$$
$$\alpha_2 = 26.6°$$

The experiment has shown: first, a layer on the upper surface of the wing as regards low incidences and the start of the formation of vortex structures with steadily flowing thin streams becoming separated from each other and heading towards the leading edges. Furthermore, the formation of two vortices since all wings under investigation has a non-preferential apex angle. Moreover, for i>45° the two vortices overlap each other. Therefore, this leads to the appearance of the torch phenomenon (spiralling effect) constituting the disappearance of the intervortex zone.

The exterior secondary system is less significant than and not as dense as the main system and is the first one to deteriorate.

The law of filiations is examined between the two systems: the angle of the secondary system begets that of the main system.

The breakdown evolves in relation to the incidence. It is a function

of the apex angle: the greater the apex angle, the sooner the breakdown takes place (Figures 5 and 6).

As regards moderated angles of incidence, the vortex structures of the aerofoil wing and of the canard surface remain separated.

When the angle of attack increases, the vortices coming from the canard plan evolve closer to the surface of the rear section of the aerofoil wing and towards its exterior.

As regard high incidence, the merging process of the two vortex structures begins at about the aerofoil wing: this leads to the fact that no more than one sole vortex structure exists on the surface of the aerofoil wing being the result of the joining of the other two structures.

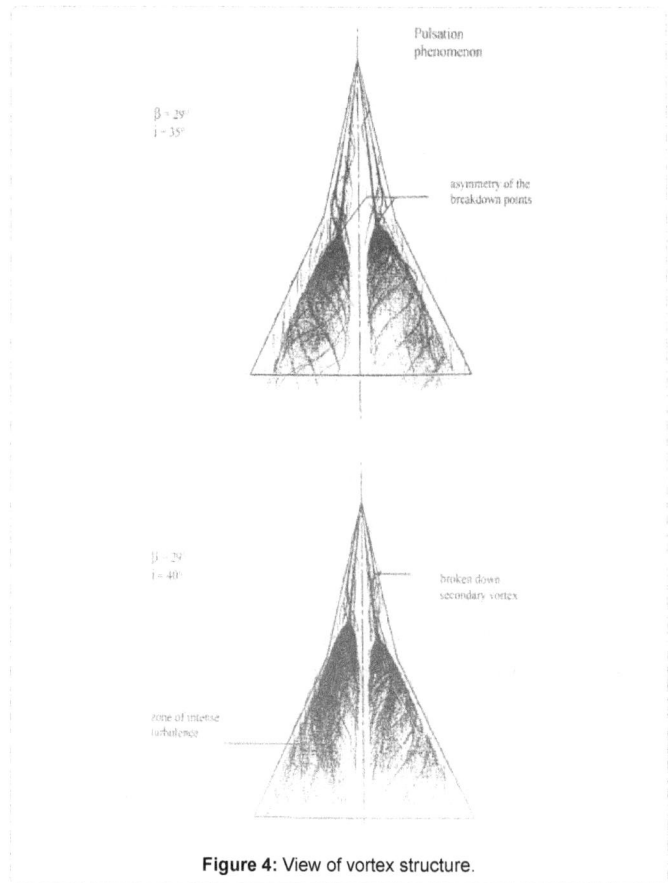

Figure 4: View of vortex structure.

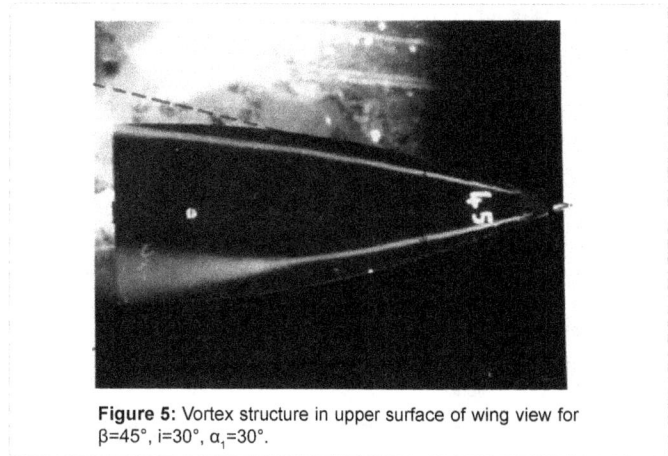

Figure 5: Vortex structure in upper surface of wing view for β=45°, i=30°, α₁=30°.

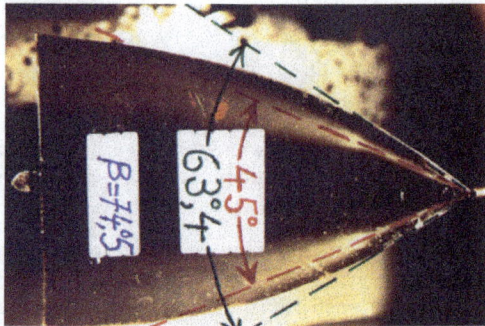

Figure 6: Vortex structure in upper surface of wing view for β=74.5°, i=30°, α_1=45°, α_2=63.4°.

Conclusion

Those vortex flows which are developed on such double-delta or delta-canard-wing-body configurations are quite complex : vortex flows can, of course, be observed on the leading edges of the airfoil wing, but they also appear on the leading edges of the canard-wing. These two vortex structures interact with one another [9].

Giving prominence to the preferential nature of those intervortex angles present on the upper surface of cones constitutes, moreover, a generalisation–as regards the basic three-dimensional cone–formed from observations previously made at valenciennes on delta and ogival wings; it has thereby been revealed that vortex behaviour has something of a universal nature.

At the present time, no complete theoretical approach seems to be sufficiently adequate to provide a straightforward explanation of the remarkable simplicity of these findings [10].

In other respects, it would appear interesting to recall that those particular spatial organisations of vortex structures citing the concept of preferential angles also appear in stand and theories on hydrodynamics and aerodynamics as in, for example, those governing aerodynamic drag. In such slipstreams, the crests of the waves, in the form of curvilinear triangles, will in fact each disappear at two counter-flow points, the alignment of which, along two right-hand sides constituting the raised edges of the slipstream, reveals a double angle of 19.4° in relation to the axis of the same slipstream [11-28].

A large number of questions remain to be investigated; further research can be undertaken in particular into the link between interior and exterior vortex structures. In addition, the least resistance of secondary vortices in relation to the breakdown inevitably calls for experimenting with other possible elementary combinations of slender bodies that would enable the appearance of the breakdown to be delayed. Somme studies would even show the existence of a supplementary vortex torque.

References

1. Werle H (1986) Visualization of unsteady flows in water tunnels of ONERA using quantitative methods to facilitate their exploitation. ONERA, Note technique.

2. Werle H (1986) Delamination structures on cylindrical wings. La Recherche Aérospatiale 3: 13-19.

3. Werle H (1987) Transition and turbulence (hydrodynamic visualizations). ONERA Note technique.

4. Werle H (1965) Vortices very slender thin wings. La Recherche Aérospatiale 109: 3-12.

5. Werle H (1989) Vortex interactions on delta wings fixed or oscillating (hydrodynamic visualizations). La Recherche Aérospatiale 2: 43-48.

6. Werle H, Gallon M (1976) Study by viewing various hydrodynamic processes control of separated flows. La Recherche Aérospatiale 2: 75-94.

7. Werle H (1962) Peeling on the body of revolution at low speed. La Recherche Aéronautique 90: 1-14.

8. Solignac JL, Pagan D, Molton P (1983) Review of some properties of the flow on the upper surface of a delta wing. Rapport technique.

9. Solignac JL, Pagan D, Molton P (1982) Fundamental study on the formation and flow of vortex structures, basic experience and modeling. ONERA Direction Aérodynamique.

10. Solignac JL, Pagan D, Molton P (1989) Experimental study of flow in the extraction of a delta wing in incompressible regime. La Recherche Aérospatiale 6: 47-65.

11. Delery J, Pagan D, Solignac JL (1987) The bursting of the vortex generated by the delta wing. Baden 6.

12. Stahl WH (1993) Experimental investigations of asymmetric vortex flows behind elliptic cones at incidence. AIAA Journal 31: 966-968.

13. Stahl W (1990) Suppression of vortex asymmetry behind circular cones. AIAA Journal 28: 1138-1140.

14. Stahl WH, Mahmood, Asghar A (1990) Experimental Investigations of the Vortex Flow on Very Slender, Sharp-edged Delta Wings at High Incidence. DLR-Interner Bericht.

15. Stahl WH, Mahmood, Asghar A (1992) Experimental investigations of the vortex flow on delta wings at high incidence. AIAA Journal 30: 1027-1032.

16. Stahl WH, Hartmann K (1990) Development to testing and a Nasengeometric for missiles with large anstelling. German Aerospace Research Establishment, Cologne, Germany.

17. Ayoub A, Mclachlan BG (1987) Slender delta wing at high angles of attack - A flow visualization study. NASA Technical Reports Server (NTRS).

18. Rediniotis OK, Stapountzis H, Telionis DP (1993) Periodic vortex shedding over delta wings. AIAA Journal 31: 1555-1562.

19. Morteveille A, Tournier L (1985) Behavior of slender cones at high incidence. Rapport Université de Valenciennes.

20. Pagan D, Benay R (1988) Numerical study of vortex breakdown subject to pressure gradients. La Recherche Aérospatiale.

21. Leray M, Deroyon MJ, Deroyon JP, Minair C (1985) Angular stability criteria of a helical vortex or a pair of vortices rectilinear angles privileged role in the optimization of wings, sails, hulls of aircraft and ships. Bulletin de l'ATMA, 85: 511-529.

22. Minair C (1987) Preferred angles, great invariants and universals: an approach based on fluid dynamics, aesthetics and physiological biology. Université de Valenciennes France.

23. Abene A (1988) Documents videographic and photographic graduate (vortices formed on the upper surface of various slender body (ribbed wings, cones, cylinders cones). University of Valenciennes France.

24. Abene A (1990) Systematic study of the positions and the stability of vortex structures above ribbed wings and cones. University of Valenciennes France.

25. Abene A, Dubois V (2012) Fundamental aspects of the vortex flow on cones. Aerosp Sci Technol 22: 1-8.

26. Benkir M (1990) Persistence and destruction of vortices concentrated or partial over delta wings. University of Valenciennes France.

27. Bergmam A, Hummel D, Gelker HCH (1990) Vortex formation over a close-coupled canard-wing-body configuration in unsymmetrical flow. AGARD-CP-49414.

28. Ericsson LE, Reding JP, Guenther RA (1971) Analytic Difficulties in Predicting Dynamic Effects of Separated Flow. J Spacecraft Rockets 8: 872-878.

Night Vision Imaging Systems Development, Integration and Verification in Military Fighter Aircraft

Sabatini R[1]*, Richardson MA[2], Cantiello M[3], Toscano M[3], Fiorini P[3], Zammit-Mangion D[1] and Jia H[1]

[1]Department of Aerospace Engineering, Cranfield University, Cranfield, Bedford, UK
[2]Defence Academy of the United Kingdom, Shrivenham, Swindon, Cranfield University, UK
[3]Italian Ministry of Defense, Air Staff, Rome, Italy

Abstract

This paper describes the research and experimental flight testactivities conducted by the Italian Air Force Official Test Centre (RSV) in collaboration with Alenia S.p.A. and Cranfiled University, in order to confer the Night Vision Imaging Systems (NVIS) capability to the Italian TORNADO IDS (Interdiction and Strike) and ECR (Electronic Combat and Reconnaissance) aircraft. The activities included various Design, Development, Test and Evaluation (DDT&E) activities, including Night Vision Goggles (NVG) integration, cockpit instruments and external lighting modifications, as well as various ground test sessions and a total of eighteen flight test sorties. RSV and Litton Precision Products were responsible of coordinating and conducting the installation activities of the internal and external lights. Particularly, an iterative process was established, allowing an in-site rapid correction of the major deficiencies encountered during the ground and flight test sessions. Both single-ship (day/night) and formation (night) flights were performed, shared between the Test Crews involved in the activities, allowing for a redundant examination of the various test items by all participants. An innovative test matrix was developed and implemented by RSV for assessing the operational suitability and effectiveness of the various modifications implemented. Also important was definition of test criteria for Pilot and Weapon Systems Officer (WSO) workload assessment during the accomplishment of various operational tasks during NVG missions. Furthermore, the specific technical and operational elements required for evaluating the modified helmets were identified, allowing an exhaustive comparative evaluation of the two proposed solutions (i.e., HGU-55P and HGU-55G modified helmets). The results of the activities were very satisfactory. The initial compatibility problems encountered were progressively mitigated by incorporating modifications both in the front and rear cockpits at the various stages of the test campaign. This process allowed a considerable enhancement of the TORNADO NVIS configuration, giving a good medium-high level NVG operational capability to the aircraft. Further developments also include the internal/external lighting for the Italian TORNADO "Mid Life Update" (MLU) and other programs, such as the AM-X aircraft internal/external lights modification/testing and the activities addressing low-altitude NVG operations with fast jets (e.g., TORNADO, AM-X, MB-339CD), a major issue being the safe ejection of aircrew with NVG and NVG modified helmets. Two options have been identified for solving this problem: namely the modification of the current Gentex HGU-55 helmets and the design of a new helmet incorporating a reliable NVG connection/disconnection device (i.e., a mechanical system fully integrated in the helmet frame), with embedded automatic disconnection capability in case of ejection. Other relevant issues to be accounted for in these new developments are the helmet dimensions and weight, the NVG usable FOV as a function of eye-relief distance, and helmet centre of gravity (moment arms) with and without NVG (impact on aircrew fatigue during training and real operational missions).

Keywords: Night vision imaging systems; Night vision goggles; NVG compatibility; Military avionics systems

Introduction

In recent years, the Italian Air Force (ITAF) set the requirements for Night Vision Imaging Systems (NVIS) to be integrated on TORNADO-IDS (Interdiction and Strike version) and ECR (Electronic Combat and Reconnaissance version) aircraft for operational missions at medium and high altitudes.

The initial operational capability (operational certification for employment in peace-keeping operations) was achieved by RSV after a ground and flight test campaign (three ground sessions and six flight test sorties) conducted on modified aircraft interior and external lighting configurations, using the AN/AVS/9 (F4949) NVG manufactured by ITT-Night Vision. Successively, the full technical/formal process of avionics certification was undertaken under the direction of the Italian Ministry of Defense Aeronautical Armaments Certification Authority (Armaereo). The related flight test activities were conducted by the Italian Official Flight Test Centre with participation of the AleniaS.p.A. Flight Test Department. During the activity, Cranfield University provided technical advice regarding the mathematical models and analytical tools required for NVIS performance prediction and evaluation. The specific objectives of the TORNADO ground and flight test activities were the following:

• Internal and external lighting day and night evaluation with and without N/AVS/9 NVG (F4949)

• Workload assessment in single-ship and formation flights

Corresponding author: Roberto Sabatini, Department of Aerospace Engineering, Cranfield University, Cranfield, Bedford, UK
E-mail: r.sabatini@cranfield.ac.uk

• Ergonomic and operational evaluation of the HGU-55P and HGU-55G modified helmets

• N/AVS/9 NVG (F4949) cockpit stowage evaluation

• Determination of the TORNADO-NVIS combination resolution characteristics

• Determination, by ground tests and analysis, of the TORNADO-NVIS range performance

After brief overview of NVIS technology, this paper described the DDT&E activities performed, with a special focus on cockpit design and ground/flight test methods developed and progressively refined throughout the activity.

NVIS Technology Overview

The Image Intensifier (I^2) is the core element of NVIS systems. I^2 devices are electro-optic systems used to detect and intensify reflected energy in the visible and near infrared regions of the electromagnetic spectrum. They require some external illumination in order to operate because the image quality is a function of the reflective contrast. The performances of I^2 devices are also dependant on atmospheric and environmental conditions. Particularly, penetration through moisture can be quite effective (especially when compared to other Electro-Optic (EO) devices, like FLIR systems), while smoke, haze and dust can significantly reduce I^2 performance. Signal-to-noise ratio (SNR) is the parameter commonly used to characterize I^2 systems performance.

Generation I (GEN I) NVG's were introduced into service in the mid 1960's during the Vietnam War. They used starlight scopes based on electron acceleration (i.e., no micro channel plates). Therefore, they were characterized by high power requirements and tube gains between 40,000 and 60,000. Multiple staging, required to increase gain, often determined an increase of image distortion, and the overall systems were large/heavy (i.e., not suitable for head mount). Furthermore, GEN I systems were very susceptible to booming and the MTBF of a typical GEN I NVG was in the order of about 10,000 hours.

Generation II (GEN-II) NVG's were introduced in the late 1960's and they were small enough to be head mounted. They used electron multiplication (i.e., micro channel plate-MCP), with increased tubes gain, reduced power requirements, and reduced size/weight. Furthermore, the new I^2 technology reduced distortion and blooming (confined to specific MCP tubules halos). Typical GEN-II systems were the AN/PVS-5 ground system, and the AN/AVS-5A system modified for aircraft usage. The MTBF of typical GEN-II systems was in the order of about 2000-4000 hours (worst than GEN I), the tube gain was approximately 10,000, and there was no inherent resolution improvement with respect to GEN I systems.

Improved photocathode performance, obtained by Gallium Arsenide (GaAs) components, determined a substantial improvement in spectral response with Generation III (GEN-III) systems. GEN-III matches night sky radiation better than GEN I and GEN-II systems, and can operate also in the absence of moon (starlight capability). Improved MCP performance were obtained by Aluminium Oxide coating, which decreases ion hits and increases MTBF (>10,000 hours). Today, GEN-III systems are widely used on most ground and in aircraft applications. Figure 1 shows the relative responses of the GEN-II/GEN-III NVG systems and the human eye, together with the average night sky radiation [1,2]. The improvement obtained with GEN-III NVG systems is evident.

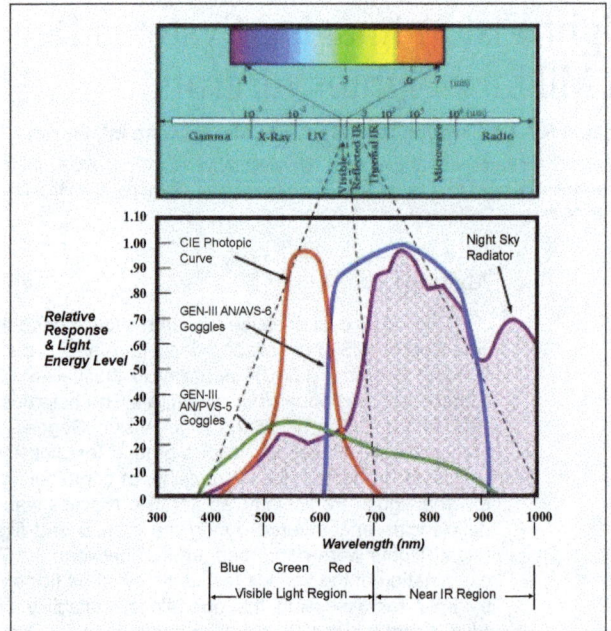

Figure 1: Relative responses of NVGs and the human eye.

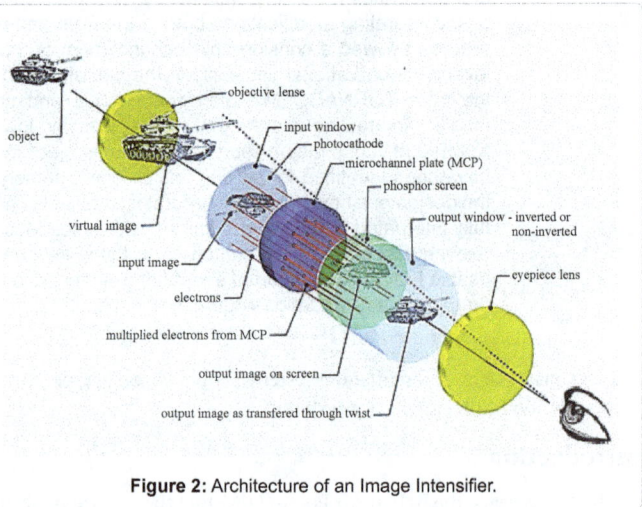

Figure 2: Architecture of an Image Intensifier.

As illustrated in figure 2, an I^2 device is typically composed by the following elements:

• Objective Lens

• Minus Blue Filter

• Photocathode

• Ion Barrier Film

• Microchannel Plate

• Phosphor Screen

• Image Inverter

• Eyepiece Lens

The Objective Lens combines the optical elements and focuses incoming photons onto the photocathode (inverted image. In most airborne NVG's, the Objective Lens is coated with a "minus blue" filter (necessary for compatible cockpit lighting). It focuses from several inches to infinity (depending on NVG). Particularly, in airborne applications, infinity focusing is used in order to obtain:

- NVG external viewing

- Look Under/Around NVG for cockpit and instrument viewing

In airborne NVG's a "minus-blue" filter is coated inside the objective lens. Its purpose is to reject visible light and to prevent other specific wavelengths from entering the image intensifier. Therefore, it allows the use of properly emitting/filtered lighting to illuminate the cockpit for viewing underneath the goggles. There are three different classes of NVG objective lens filters:

- Class A: blocks below 625 nm (blue/green)

- Class B: blocks below 665 nm (blue/green/reduced red)-allows use of color displays

- Class C (leaky green)-incorporates notch cut-out to permit viewing of specific wavelength

The Photocathode (PC) converts light energy (photon) to electrical energy (electrons). The PC Inner surface is coated with a photosensitive material. Particularly, we list the following materials used in GEN-I/II and GEN-III systems:

- GEN-I/II: S-20 multi-alkali compound, sensitive between 400 and 850 nm (peak sensitivity at 500-600 nm);

- GEN-III: Gallium Arsenide (GaAs), sensitive from 600-900 nm (impact of photons cause release of electrons).

Typical PC responsivity figures are 250-550 μA/lm for GEN-II systems and 1,000-1,800 μA/lm for GEN-III systems. As illustrated in figure 3, GEN-III I² tubes are currently fabricated with a so called Ion Barrier (IB) film. This film extends tube life (protects the PC) but reduces the system performance (i.e., degrades signal-to-noise ratio).

The Microchannnel Plate (MCP) is a thin wafer (about 1mm)

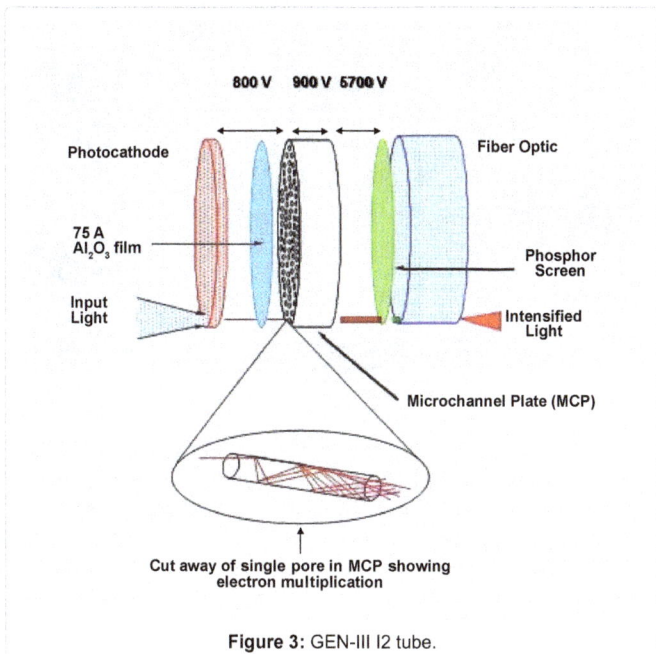

Figure 4: MCP working principle.

containing various millions of glass tubes or channels (typically 4-6 millions). Electrons from the PC enter the MCP tube (tube walls coated with lead compound rich in electrons) which is tilted (about 5 degrees) to ensure the impact of the electrons with the wall (Figure 4). When an electron impacts the tube wall, more electrons are released resulting in a cascade process. Electrons are then accelerated towards the phosphor by an electrical potential differential (positive pole at phosphor). The ultimate output is number of electrons and their velocity. Resolution is a function of number of MCP tubes.

The Phosphor Screen (PS) is a thin layer of phosphor at the output of the MCP. Phosphor emits light energy when struck by electrons (electro-luminescence). Light emitted by phosphor creates a visible (green) image.

The Image Inverter (INV) is a bundle of millions of light transmitting fibers. The bundle rotates 180 degrees to reorient the image (fiber optic twist). It also collimates image for correct positioning at the viewer's eye. Problems in INV manufacturing and installation result in adverse image effects, such as distortion and honeycomb appearance. Some NVG designs do not incorporate a fiber optic twist for reorienting the image.

The Eyepiece Lens (EL) is the final optical component of the NVG. It focuses the visible image on the retina of the viewer and, generally, a limited diopter adjustment is allowed to permit some correction for individual vision variations. In general, corrective lenses must still be worn by users (the system does not correct for astigmatism). Most GEN-II systems have a 15 mm eye-relief and a nominal 40° FOV. GEN-III systems typically have 25 mm nominal eye-relief which also provides the 40° FOV but enhances the ability to look under/around the NVG.

Signal to Noise Ratio (SNR) is a measure of image intensifier performance (resultant of the image intensification process). SNR for a NVG is defined as the ratio of electrons produced by ambient light (signal) to stray electrons (noise). Improved performance (larger SNR's) is produced by increasing the ambient light and/or improving the I² (e.g., increasing PC sensitivity and decreasing the space between the elements).

NVIS Compatibility Issues

Intensified imagery of the outside scene is of primary importance to the aircrew. Incompatible light from cockpit sources and external lights are detected by the NVG and intensified, thus reducing the NVG gain. The resulting degraded image quality may not be readily apparent to the aircrew.

NVG compatible lighting results in instruments and displays being easily read with the unaided eye at night. However, all instruments must

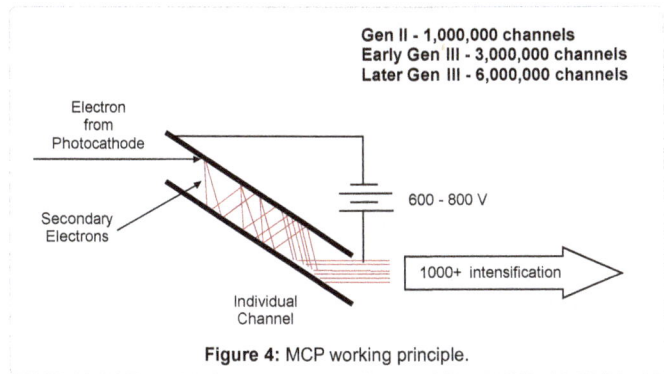

Figure 3: GEN-III I2 tube.

still be readable during day. NVG compatible lighting is often invisible to the NVG, while "friendly" lighting may be visible to the goggles, but without changing the gain state of the goggle. Typically, NVG compatible instruments and displays only emit wavelengths to which the eye is most responsive (i.e., little red and no near-IR emission).

There are basically two different implementation methods which can be adopted for integrating NVG compatible lighting in the cockpit. These methods are the following

• Permanent lighting: Including integral instrument/display lighting, post and bezel lighting, food lighting using existing aircraft light fixtures or LED based light sources

• Temporary lighting: Including chemical light sticks and Light Emitting Diodes (LED) wiring harness

Also NVG compatible external lights have can be used in order to increase mission effectiveness, increase flight safety and decrease aircraft vulnerability (IR covert mode). Also in this case, there are basically two different approaches possible

• Introducing new equipment: Including conventional/filtered, electro-luminescent and LED technologies

• Retrofitting existing lights: Including filtering and modifying the existing light source

Another important aspect to be considered with NVIS compatible aircraft developments is the NVG-helmet integration. Particularly, the following are the main goals to be achieved:

• Reduce the NVG-helmet moment arms

• Reduce the weight

• Maximize usage of the available FOV (considering eye relief, exit pupil, etc.)

• Allow use of various types of visors (including laser protection visors)

Description of Test Articles

The test activities were carried out using the NVG mod. AN/AVS/9 F4949G (P/N 264359-8) produced by ITT-Night Vision (Figure 5). This is a GEN-III NVG, with class B filter and 40° nominal Field-of-View (FOV).

The goggles were installed on both the Gentex HGU-55/G and HGU-55/P standard helmets, using the ITT Night Vision helmet modification kit NSN 5340-01-442-641 as illustrated in the figures 6 and 7.

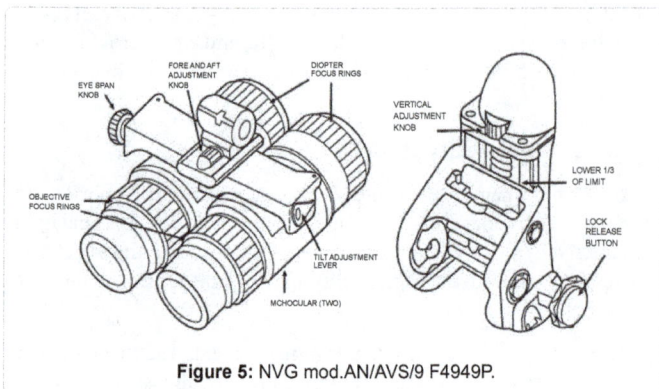

Figure 6: Modified HGU-55/P helmet with NVG installed.

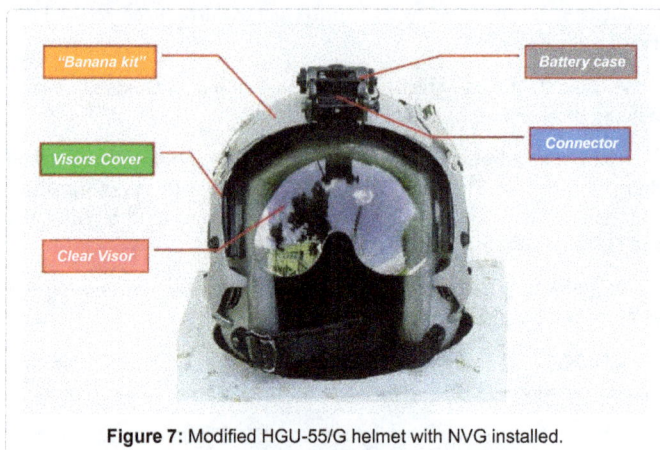

Figure 7: Modified HGU-55/G helmet with NVG installed.

Control Panel Setting				VIS Emission			IR Emission		
ON/OFF	BRIGHT/DIM	VIS/IR	CODE	Tail light	Wing tip	Intake	Tail light	Wing tip	Intake
ON	BRGT	VIS	C	PUNG	PUNG	PUNG	OFF	OFF	OFF
ON	BRGT	VIS	1,2,3,4	1,2,3,4	1,2,3,4	PUNG	OFF	OFF	OFF
ON	DIM	VIS	C	Steady (DIM)	Steady (DIM)	PUNG (DIM)	OFF	OFF	OFF
ON	BRGT	IR	C	OFF	OFF	OFF	PUNG	PUNG	OFF
ON	BRGT	IR	1,2,3,4	OFF	OFF	OFF	1,2,3,4	1,2,3,4	OFF
ON	DIM	IR	C	OFF	OFF	OFF	Steady (DIM)	Steady (DIM)	OFF

Figure 8: External lighting system functions.

Figure 5: NVG mod.AN/AVS/9 F4949P.

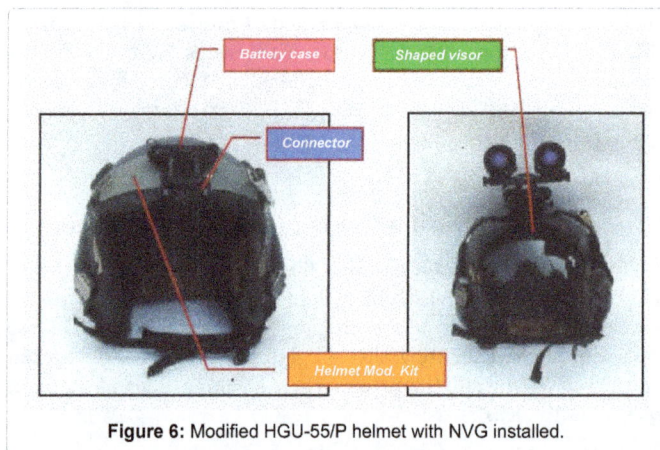

The great majority of the TORNADO IDS/ECR cockpit displays, control panels and lights were modified by filtering or substituting the existing light sources, in order to obtain NVG compatible emissions. Also the aircraft external lights were modified, introducing an NVG friendly (IR emission) functional mode, and adding new functionalities in to the already existing visible lights. The new functionalities incorporated into the aircraft external lighting system are described in figure 8.

Particularly, a new control box was installed in the cockpit allowing the pilot selection of the various external lights functional modes. Five different codes, all square wave in nature (codes 1, 2, 3, 4 and C in figure 4), were programmable in the control box (using an EPROM). One of these codes was programmed with equal on and off times, while the other codes were programmed according to aircrew requirements, selecting code sequences with flash repetition frequencies and flash durations well discernible in flight.

During the flight test activities, after introducing a large number of modification into the TORNADO IDS/ECR front and rear cockpits, it was observed that certain areas of the front/rear main instrument panels and of the front/rear left and right consoles were not sufficiently illuminated by self-contained and/or general purpose cockpit lighting. Therefore, it was decided to test a 'finger light' both in the front and in the rear cockpits. The finger light FINGERSTAR (P/N 4790-NF-01A) used in the trials had both IR and visible emissions available, selectable by the operator using a finger-switch located on an adjustable (left/right hand) switching rail.

Test and Analysis Methods

An innovative test matrix was used for assessing the operational suitability and effectiveness of the various modifications implemented in the cockpit (Figure 9). Particularly, both flight safety and operational effectiveness/suitability of the NVIS configuration were considered in the test matrix, allowing a direct correlation between the flight test rating criteria and the standard evaluation rating scale used by RSV. This approach was applied both to the single modified items under test (displays, lights, panels, etc.), and to the overall cockpit NVIS configuration.

Modified aircraft external lights (both VIS and IR modes) were tested in formation flights (chase aircraft), performing the following tasks:

- Tactical Rejoin
- Fighting Wing
- Close and Battle Formation
- Air-to-air Refueling

Also important was definition of criteria for Pilot and WSO workload

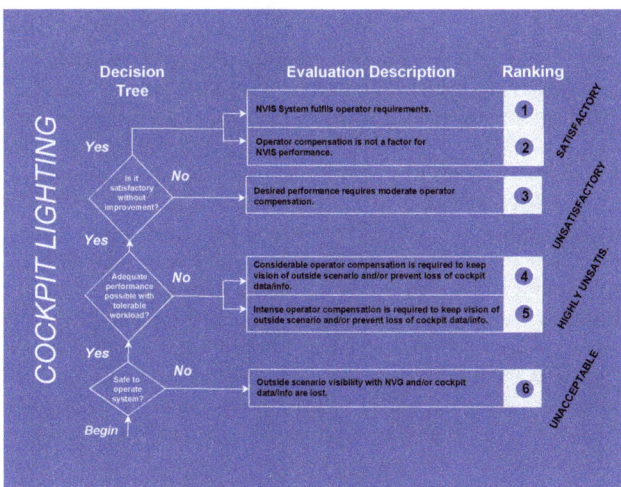

Figure 10: Workload evaluation matrix.

assessment during the accomplishment of various operational tasks during NVG missions (Figure 10). Particularly, a workload evaluation matrix was implemented that allowed identification of the workload levels associated with the various Pilot and WSO operational tasks during real missions. These included ferry flights, attack, formation flights and tactical evasive/escape manoeuvres. The operational tasks considered were the following:

- Navigation
- Automatic Flight Director System (AFDS) operation and monitoring
- Engine/airplane systems operation and monitoring
- Manual flight path control
- Communications
- Command decisions
- Collision avoidance

For each of the above tasks performed on the TORNADO NVG configuration, the levels of mental effort and physical difficulty, together with time required for the specific tasks and the understanding of horizontal/vertical position (spatial orientation) during execution of the tasks, were compared with the respective levels/values found for the standard TORNADO aircraft. Furthermore, the specific technical and operational elements required for evaluating the modified helmets were identified, allowing an exhaustive comparative evaluation of the two proposed solutions (i.e., HGU-55P and HGU-55G modified helmets). These elements included: measurement of the available FOV and calculation of the Projected FOV Area Reduction (PFAR), weight/balance, comfort and stability, crew fatigue in low and high dynamics flights. Furthermore, the NVG connection/disconnection devices were tested performing high dynamics manoeuvres (with NVG both in the up-locked and down-locked positions).

In order to assess the operational suitability of the modified HGU-55/P and HGU-55/G helmets, the related test activities focused on the following aspects:

- Measurement of the available Field-of-View (FOV) with minimum eye-relief

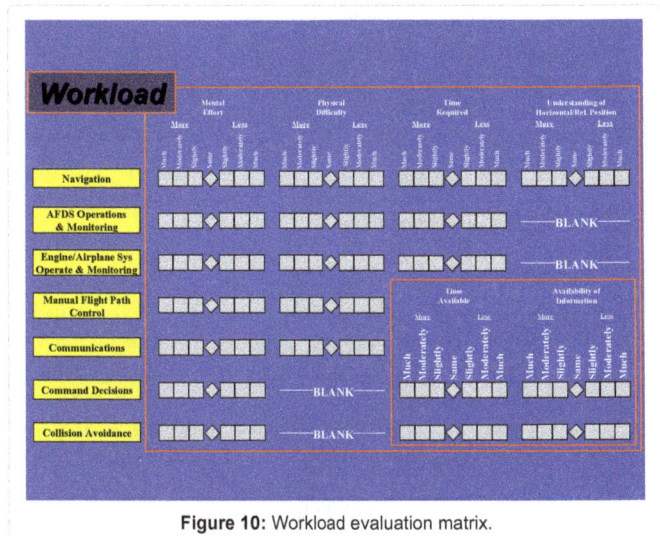

Figure 9: Cockpit evaluation test matrix.

• Determination of the minimum Projected Area FOV reduction (P-FOV)

• NVG helmets fitting and stability

• Clearance with a/c structure (NVG up-locked and down-locked)

• Fatigue in low dynamics flight

• Fatigue in maneuvering flight

• Possible use of protection visors

The spatial resolutions obtainable with the F4949 visors in the various sectors of the TORNADO canopy (normal sectors for external clearing), were also measured. This was done by using the USNTPS 20/20–20/70 resolution bars method [3]. Particularly, a resolution table was prepared (Figure 11), composed of 16 groups of bars with dimensions and spacing corresponding to visual acuities between 20/70 and 20/20. The resolution bars table was illuminated with an artificial light source reproducing typical night illumination conditions.

During a ground test, using the bars target shown in figure 13, together with the low illumination lamp, the spatial frequencies (cycle/mrad) corresponding to various 2-D discrimination levels were determined for the F4949 system used on TORNADO, in the various sectors of the aircraft canopy. Using these experimental data it was possible to calculate the detection, recognition and identification ranges of the NVG system, for targets of given aspect dimensions located in certain regions of the Pilot and WSO external clearing scanning patterns.

Before executing the on-board ground tests, a preliminary session were performed by the same aircrews (with NVG) positioned on the ground at a distance of 20 feet from the resolution table (illuminated by the low illumination lamp). In this condition, the groups of bars resolved were annotated. Also during successive the on-board session, the distance between the Pilot/WSO Reference Eye Positions (REP's) and the bars target was maintained to exactly 20 feet, and the resolution table was rotated about the REP's as shown in figure 12. Particularly, the following Pilot/WSO sectors were considered:

• Max Rear (Field-of-Regard limit)

• Lateral Sector 90°

• Lateral Sector 60°

• Lateral Sector 15°-30°

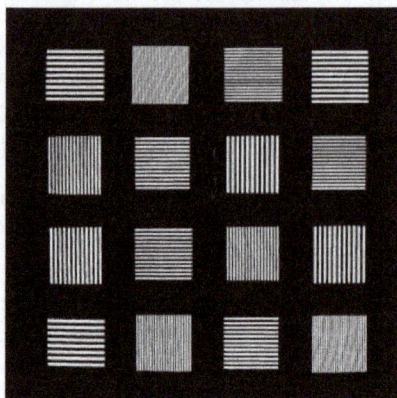

Figure 11: Resolution table (20/70–20/20).

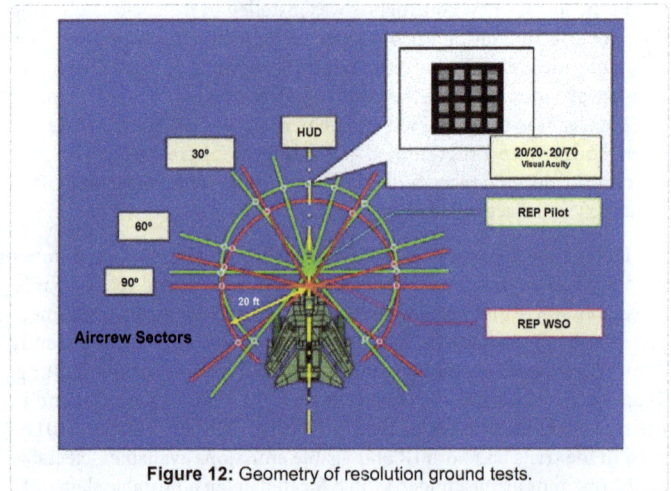

Figure 12: Geometry of resolution ground tests.

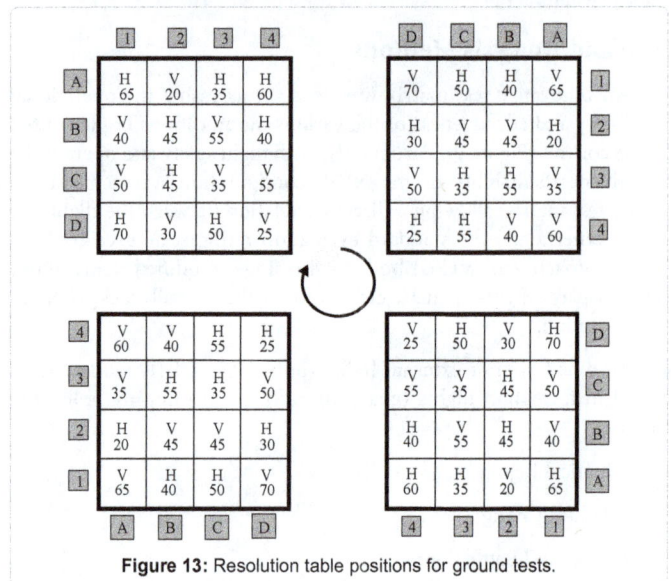

Figure 13: Resolution table positions for ground tests.

• Pilot HUD (0°-15°)

In each relevant position, the resolution target was rotated in four different positions as shown in figure 13. In each case, the Pilot/WSO abilities to resolve the various groups of bars were recorded.

NVG range performance predictions require a mathematical model that describes the eye/brain image interpretation process. Unlike the response of an electronic circuit, the response of a human observer cannot be directly measured but only can be inferred by many visual psychological experiments. The lowest level of discrimination is a distinction between something and nothing. The final level is the precise identification and description of a particular object. Between these two extremes lay a continuum of discrimination levels. In the late fifties, Johnson studied image intensifiers discrimination performance at the US Army Engineering and Research Laboratories [1]. He arbitrarily divided visual discrimination into four categories: detection, orientation, recognition, and identification. Johnson's results allowed correlating detectability with the sensor threshold bar pattern resolution (Table 1). In Johnson's work, the (angular) spatial frequency (SF) is defined as:

$$SF = \frac{R_T}{W_{1c}} \qquad (1)$$

Discrimination level	Meaning	Cycles across minimum dimension
Detection	An object is present (object versus noise)	1.0 ± 0.025
Orientation	The object is approximately symmetrical or unsymmetrical and its orientation may be discerned (side view versus front view)	1.4 ± 0.35
Recognition	The class to which the object belongs (e.g., tank, truck, man)	4.0 ± 0.80
Identification	The object is discerned with sufficient clarity to specify the type (e.g., T-52 tank, friendly jeep)	6.4 ± 1.50

Table 1: Summary of Johnson's experimental results.

Discrimination level	Meaning	Cycles across min. dimension (N_{50})
Detection	An object is present	1.0
Recognition	The class to which the object belongs	4.0
Identification	The object is discerned with sufficient clarity to specify the type	8.0

Table 2: Current industry criterion for 1-D discrimination (50% probability level).

Probability of discrimination	Multipler F_m
1.00	3.0
0.95	2.0
0.80	1.5
0.50	1.0
0.30	0.75
0.10	0.50
0.02	0.25
0	0

Table 3: Discrimination cumulative probability.

Where:

R_T=sensor-to-target range;

W_{1c}=width of one cycle of target,

And the 'cycle' is defined as the sum of one bar and one space on the reference target. Johnson applied the number of cycles across the target minimum dimension, without regard to the orientation of the minimum dimension (his image intensifier imagery was radially symmetrical and therefore it was reasonable for him to ignore the bar orientation). Johnson's approach, known as the equivalent bar pattern approach, became the foundation for the discrimination methodology used today.

Successive studies and tests performed at the US Army Night Vision Laboratories and by industry suggested modifications to the values originally found by Johnson. Table 2 provides the current industry standard for one-dimensional target discrimination [2]. Orientation is a less popular discrimination level. Because current standards are based upon Johnson's work, they are labelled as the Johnson criterion though they are not the precise values found by him.

The Johnson criterion provides an approximate measure of the 50% probability of discrimination. Results of several tests provided the cumulative probability of discrimination or target transfer probability function (TTPF). The TTPF can be used for all discrimination tasks by simply multiplying the 50% probability of performing the task (N_{50} in table 2) by the appropriate TTPF multiplier in table 3 [2].

For instance, the probability of 95% recognition is $2N_{50}=2(4)=8$ cycles across the target minimum dimension. Similarly, the cycles

required for detection, recognition and identification with a probability level of 80% are 1.5, 6 and 12 respectively. An empirical fit to the data provides [3]:

$$P(N) = \frac{\left(\dfrac{N}{N_{50}}\right)^E}{1+\left(\dfrac{N}{N_{50}}\right)^E} \tag{2}$$

Where:

$$E = 2.7 + 0.7 \cdot \left(\frac{N}{N_{50}}\right) \tag{3}$$

Visual psychophysical experiments suggest that the eye response follow a log-normal distribution [4]. The probability density function follows:

$$p(N) = \frac{1}{\sqrt{2\pi} \cdot \log(\sigma)} \cdot e^{-\frac{1}{2}\left[\frac{\log(N)-\log(N_{50})}{\log(\sigma)}\right]^2} \tag{4}$$

Where $\log(\sigma) = 0.198$. The cumulative probability is:

$$P(N) = \int_0^{\log N} p(N) d\log(N) \tag{5}$$

Both the empirical fit of eq (3) and the log-normal approach (based upon a physically plausible foundation) of eq (5) provide similar numerical results. As clutter increases, the ability to discern a target decreases. To account for this reduced capability, N_{50} must increase. Most studies have broadly categorized clutter into high, moderate and low regions, and defined the signal-to-clutter ratio (SCR) as:

$$SCR = \frac{\max \ target \ value \ - \ background \ mean}{\sigma_{clutter}} \tag{6}$$

Where:

$$\sigma_{clutter} = \sqrt{\frac{1}{N}\sum_{i=1}^{N}\sigma_i^2} \tag{7}$$

And σ_i is the rms value of the pixel values in a square cell that has side dimensions of approximately twice the target minimum dimension. The scene is composed of N adjoining cells. The use of adjoining cells introduces a spatial weighting factor that is similar to the spatial integration performed by the eye/brain process. Clutter sizes that are equal to the object size weigh more heavily in this calculation.

The results are presented in table 4 [5].

Field experiments demonstrated that the Johnson detection criterion applies to a "general medium to low clutter" environment. Therefore, the 50% probability of detection in table 4 where normalized in moderate clutter to one cycle. These experimental findings roughly follow the empirical TTPF of eq (2). It is convenient to use 0.5, 1.0 and 2.5 as a multiplier (F_d) to N_{50} for low, moderate, and high clutter environments respectively.

In order to obtain the two-dimensional discrimination levels required in a 2-D performance prediction model, each value in the

Probability of detection	Multiplier F_d		
	Low Clutter SCR>10	Moder. Clutter 1<SCR<10	High Clutter SCR<1
1.0	1.7	2.8	**
0.95	1.0	1.9	**
0.90	0.90	1.7	7.0*
0.80	0.75	1.3	5.0
0.50	0.50	1.0	2.5
0.30	0.30	0.75	2.0
0.10	0.15	0.35	1.4
0.02	0.05	0.1	1.0
0	0.0	0.0	0.0

** No data available. * Estimated

Table 4: TTPF when clutter is present.

Discrimination level	Meaning	Cycles across minimum dimension ($N_{50\text{-}2D}$)
Detection	An object is present	0.75
Recognition	The class to which the object belongs	3.00
Identification	The object is discerned with sufficient clarity to specify the type	6.00

Table 5: Discrimination levels for the 2-D model (50% probability level).

one-dimensional criteria (Table 5) is multiplied by 0.75. The results are presented in table 5.

The US Night Vision Laboratory Static Performance Model [6] uses the minimum dimension (1-D), whereas most 2-D models refer to the object critical dimension [7]:

$$h_c = \sqrt{W_{TGT} \times H_{TGT}} \qquad (8)$$

Where W_{TGT} and H_{TGT} are the horizontal and vertical object dimensions. In this case, the number of cycles used for range performance calculations is that associated to the critical dimension h_c.

In conclusion, our 2-D range performance prediction model is summarized by the following equations:

$$R = \frac{h_c}{\left(N_{50-2D} \times F_d\right)} \times SF \quad \text{for detection} \qquad (9)$$

$$R = \frac{h_c}{\left(N_{50-2D} \times F_m\right)} \times SF \quad \text{for recognition and identification} \qquad (10)$$

where

R=predicted slant range;

h_c=target critical dimension;

S_F=measured spatial frequency;

$N_{50\text{-}2D}$=cycles required for detection, recognition and identification;

F_m, F_d=multipliers for the various discrimination levels.

Test Results

The activities on TORNADO IDS and ECR both included various ground test sessions and a total of eighteen flight test sorties (7 night flights and 2 day flights for each aircraft type). RSV and Litton Precision Products were responsible of coordinating and conducting the installation activities of the internal and external lights. Particularly, an iterative process was established, allowing an in-site rapid correction

of the major deficiencies encountered during the ground and flight test sessions. Both single-ship (day/night) and formation (night) flights were performed, shared between the Test Crews involved in the activities (Test Pilots/WSOs), allowing for a redundant examination of the various test items by all participants.

The technical results of the activity were very satisfactory. Particularly, the internal lighting compatibility problems were progressively mitigated by incorporating modifications both in the front and rear cockpits at the various stages of the development test program. This process allowed a considerable enhancement of the TORNADO cockpits NVIS configurations, giving a good medium-high level NVG operational capability to the aircraft. The Air Force Operational Certifications for both the IDS and ECR aircraft configurations were achieved by 2002. Figure 14 shows the initial and final results of the overall cockpit evaluation.

All external lighting modifications incorporated into the aircraft where satisfactory. Particularly, all medium-high level flight tasks required were performed successfully, after an adequate level of aircrew training. Close formation flights were indeed some of the most demanding tasks during NVG operations, requiring an appropriate level of aircrew training in order to estimate other aircraft distance, attitude and speed (dept/distance perception is severely degraded by NVG).

The workload assessment also gave encouraging results, demonstrating that the modifications of the aircraft interior and exterior lighting increased the levels of Pilot/WSO situational awareness and therefore their ability to perform operational tasks in night conditions. Particularly, medium-high level navigation and communications tasks where performed without a significant increase of aircrew workload, while the increase of workload experienced in AFDS/Engine/Airplane Systems operation and monitoring was counterbalanced by the substantial reduction of workload experienced in manual flight path control, command decisions, and collision avoidance tasks (e.g., formation flights). Again, it was readily apparent during the tests, that aircrew training was the key to increase flight safety and operational effectiveness in NVG operations.

The results of the NVG-helmets ergonomic evaluation are summarized in figure 15. The modified HGU-55/G helmet was heavier

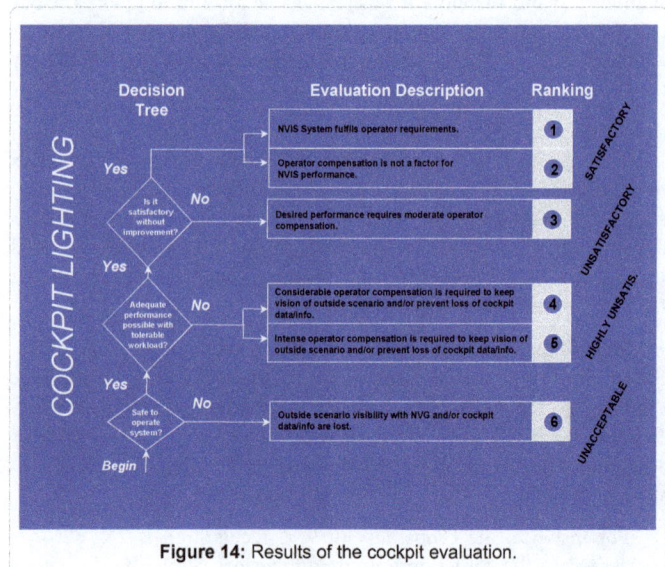

Figure 14: Results of the cockpit evaluation.

Figure 15: Results of helmets ergonomic evaluation.

FOV		Diff. FOV	PFAR		Diff. PFAR
HGU-55P	HGU-55G	1.98	HGU-55P	HGU-55G	10.14%
39.19	37.21		4.30%	14.44%	

Table 6: FOV e PFAR measurements.

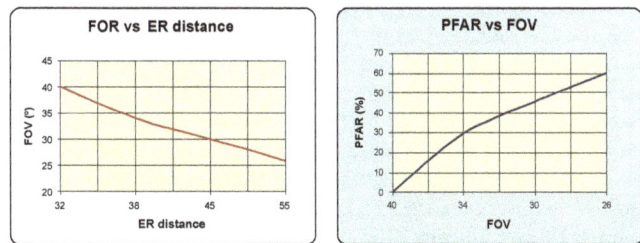

Figure 16: FOV vs. ERD and PFAR vs. FOV curves.

and less stable/balanced than the HGU-55/P helmet, and also gave a reduced NVG FOV due to increased eye-relief. However, the HGU-55/P helmet was not suitable for operational use, due to difficulties in installing and removing the clear/laser protection visors during night operations with NVG (flying with protection visors is required on TORNADO to protect the aircrew, in case of ejection, against windblast and canopy fragmentation).

Table 6 shows the experimental data relative to the NVG FOV and PFAR, obtained with the HGU-55/G and HGU-55/P modified helmets, used by an operator with average percentiles, wearing a medium size helmet and a medium size oxygen mask (similar results were obtained with operators having different percentiles).

Compared to the 40° nominal FOV of the F43949 system, it is evident that there was a decrease in FOV of about 0.8° for the HGU-55/P helmet, and of 2.8° for the HGU-55/G helmet (i.e., the HGU-55/P helmet gives a 2° increase of FOV due to a reduced eye-relief). With the same operator, the PFAR (i.e., reduction of imaged scene area covered by the NVG), was about 4% for the HGU-55/P and about 14% for the HGU-55/G. Therefore, there was a difference of about 10% in the area covered by the NVG between the two helmets.

Based on the F4949 design data (provided by ITT Night Vision),

figure 16 shows the FOV calculated as a function of the eye-relief distance and the PFAR vs. FOV curve.

The experimental PFAR data (Figure 17) were essentially coherent with the theoretical calculations. It is worth to underline that an ERD increase of 1 mm determines a 1° reduction in FOV, and an increase of the PFAR of about 5%. Compared to the ideal case of FOV=40°, this would equate to a 20% reduction of the area covered by the NVG for the HGU-55/G helmet, and of about 10% for the HGU-55/P helmet

Based on visual acuity measurements results, the NVG detection, recognition and identification range performances were calculated using equations (9) and (10), for different types of targets. Particularly, the detection/recognition/identification range performances were calculated with 80%, 90% and 100% probability levels. Furthermore, the detection performances (80%, 90% and 100% probability) were also calculated in low, medium and high clutter conditions [8]. Examples of the results obtained are shown in figure 18.

Lessons Learned

The Human Factors risks in NVG operations are directly related to the quality of the interior and exterior aircraft lighting, the quality of aircrew training and the ability to detect and quantify under NVG. The most important technical and operational lessons learned during the TORNADO NVG flight test activities were the following:

• Unaided readability is just as important as NVG compatibility. NVG flight can be regarded as 'visually aided' instrument flight.

• A poor installation can spoil a good modification design (e.g., incompatible light leaks).

• Daylight readability may be more difficult after NVG modifications. Suppression of warning/caution indicators within the NVG FOV has to be avoided.

• The same design rationale for standard lighting applies to NVIS lighting.

• Standard lighting cannot be turned down enough to be NVG compatible.

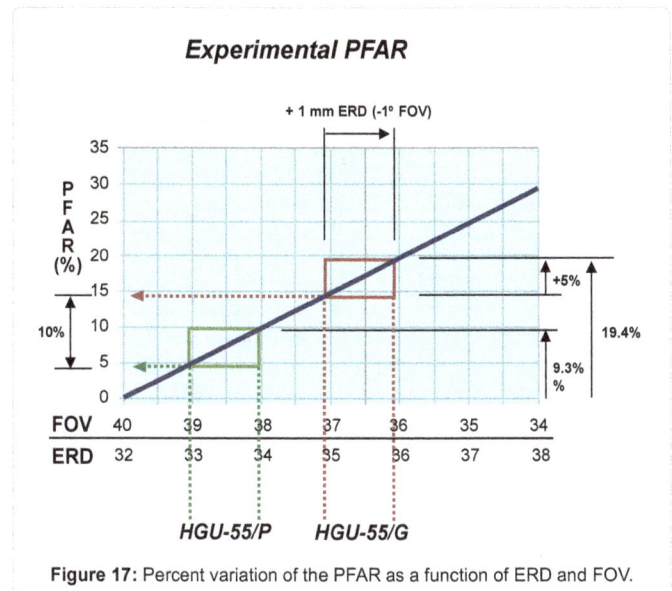

Figure 17: Percent variation of the PFAR as a function of ERD and FOV.

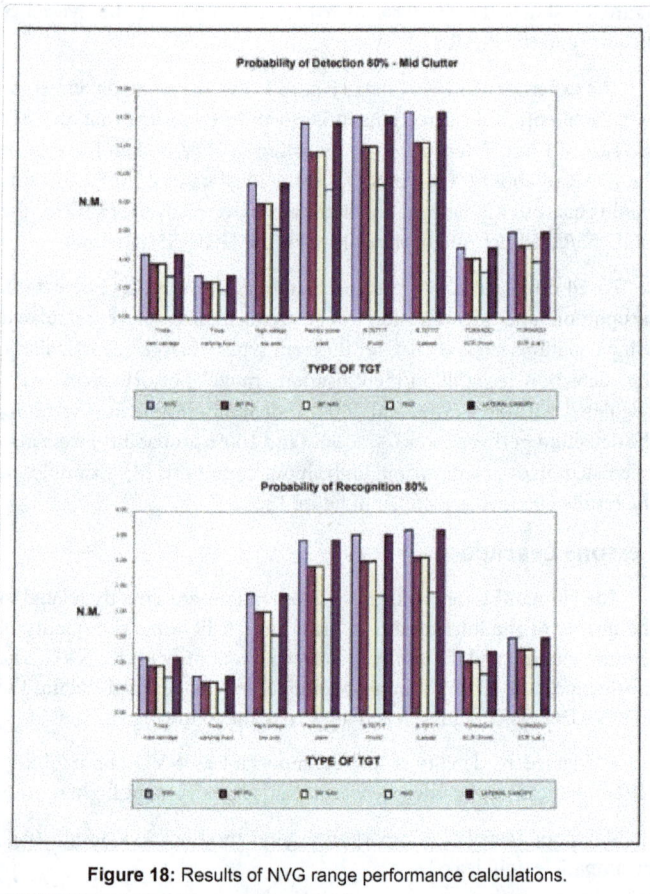

Figure 18: Results of NVG range performance calculations.

• Partial modification is usually not successful. NVGs used for long periods may result in increased workload for aircrew.

• Properly designed NVIS lighting is usually superior to lighting it replaces. Particularly, it reduces reflections on canopies; it makes instruments easier to read at lower brightness levels, and reduces eye fatigue.

Conclusions and Further Developments

In this paper we have described the development and testing activities conducted on the Italian TORNADO IDS/ECR in order to confer a medium-high level NVG operational capability to the aircraft. The TORNADO development activities, addressing the aircraft interior/exterior lighting and the helmet modifications (NVG integration), were conducted by RSV and supported by industry (Litton Presion Products). Also the ground and flight test activities were conducted by RSV, with participation of industry to the test flights (Alenia).

Particularly important for RSV was the clear identification of the technological alternatives available for aircraft modifications, as well as the definition of suitable test methods for both internal and external lighting evaluation. Also very important was the adoption of appropriate NVG performance analysis models, which leaded to the development of a standard PC based data analysis tool.

The technical results of the TORNADO NVG activities were very satisfactory. Particularly, the internal lighting compatibility problems were progressively mitigated by incorporating modifications both in the front and rear cockpits at the various stages of the development test program. This process allowed a considerable enhancement of the

TORNADO cockpits NVIS configurations, giving a good medium-high level NVG operational capability to the aircraft.

The workload assessment also gave encouraging results, demonstrating that the modifications of the aircraft interior and exterior lighting increased the levels of Pilot/WSO situational awareness and therefore their ability to perform operational tasks in night conditions. However, it was readily apparent during the tests, that aircrew training was the key to increase flight safety and operational effectiveness in NVG operations.

The NVG-helmets tests allowed a comprehensive verification of the ergonomic and technical elements in favor or against each of the proposed solutions (i.e., modified HGU-55/G and HGU-55/P helmets). Overall, the HGU-55/P helmet was rejected due to difficulties in installing and removing the clear/laser protection visors during night operations, while the modified HGU-55/G was selected for TORDADO IDS/ECR operations (although not fully satisfactory).

In conclusion, a considerable experience was gained during the TORNADO NVG activities and further developments were launched in this area, taking advantage of the technical and operational lessons learned, to increase the ITAF aircraft operational capability and safety. Further developments include the Alenia internal/external lighting design for the Italian TORNADO "Mid Life Update" (MLU) and various other Air Force programs, such as the AM-X aicraft internal/external lights modification/testing and other activities addressing low-altitude NVG operations with fast jets (e.g., TORNADO, AM-X, MB-339CD). A major issue encountered is the safe ejection of aircrew with NVG and NVG modified helmets. Two options have been identified for solving this problem: modification of the current HGU-55 helmets and the design of a new helmet incorporating a reliable NVG connection/disconnection device (i.e., a mechanical system fully integrated in the helmet frame), with embedded automatic disconnection capability in case of ejection. Other relevant issues to be accounted for in these new developments are the helmet dimensions and weight, the NVG usable FOV as a function of eye-relief distance, and helmet centre of gravity (moment arms) with and without NVG (impact on aircrew fatigue during training and real operational missions). A pictorial representation of the system initially proposed by Gentex and ITT Night Vision in order to match the Italian and German Air Forces TORNADO helmet requirements is shown in figure 19.

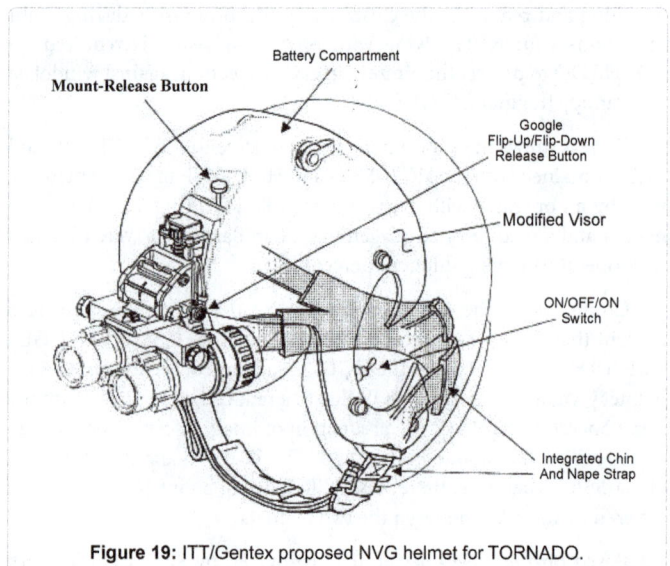

Figure 19: ITT/Gentex proposed NVG helmet for TORNADO.

The ITAF requirements for a new helmet allowing a safe and practical usage of the F4949P NVG were established so that no restrictions were applied to the aircraft operational flight envelopes due to use of the NVG system. In order to achieve this, the new development should address the following main issues:

• maximise the operator's usage of the NVG performance

• maximise the balancing, stability and comfort of the new helmet

• maximise the level of safety (normal use and ejection)

The overall goals to be achieved in the development are the following:

• No modifications of the existing F4949P NVG system

• NVG usable in "up-locked" and "down-locked" positions

• Practical and safe connection/disconnection of the NVG/Adapter

• Maximum usage of the available NVG FOV

• No protrusions on the helmet

• No Helmet weight increase

• NVG-Adapter moment arm minimisation

• Maximum comfort and stability also under g's

• Use of helmet visors (inner clear/laser visor for NVG operations and dark outer visor for operations without NVG)

• Availability of documentation required for Helmet/Adaptor Qualification and Certification (i.e., System Performance Specification, System Design Documentation, and Development Test Reports)

The new developments shall not include modifications of the existing F4949P NVG system. Furthermore, the NVG should be usable both in 'up-locked' and 'down-locked' positions, without possibility of NVG disconnection in these positions from the Adapter-Helmet. Manual disconnection of the NVG from the Adapter-Helmet should be possible only in a dedicated 'intermediate' position. Self-disconnection during ejection should be guaranteed independently from the NVG position.

Connection and disconnection of the F4949P NVG, of the Helmet Adapter and of the NVG-Adapter block should be possible for the operator with a single action and using a single hand. Particularly, the entire NVG-Adapter block should be removable as one section (e.g., before ejection), the F4949P NVG should be separately removable from the Adapter-Helmet (e.g., for normal stowing of the NVG), and the adapter should be also separately removable from the Helmet (using the same device available for removal of the NVG-Adapter block). Additional detailed requirements are:

• During the initial phase of a seat-ejection (i.e., acceleration phase) the NVG-Adapter block should fall off the helmet without any action of the crew

• The modified Helmet-Adapter should allow usage of the maximum Field Of View (FOV) provided by F4949P NVG

• The Helmet should be free from significant protrusions. The Adapter block should be designed to minimise protrusions, as to allow a smooth surface of the Helmet-Adapter combination

• All efforts should be placed in order to minimise the weight of the modified Helmet. Particularly, it is desirable that the weight of the

new helmet does not increase with respect to the current helmets and, if feasible, it should be reduced

• The moment arm of the NVG-Adapter block should be minimised, in order to obtain a balanced Helmet and to maximise the Helmet stability and fitting comfort

• The inner part of the helmet should be modified in order to enhance the helmet stability (also under g's) by using combined Chin-Nap Straps or other stability enhancing features

• The helmet should be equipped with two visors: an inner visor (i.e., clear visor or laser visor) and an outer visor (i.e., dark visor). The F4949P NVG system will be used with the inner visor down

Recent studies conducted by ITT-Night Vision and Gentex, in collaborationwith ITAF and the Italian MoDhave leaded to the NVG-helmet solutions shown in figures 19 and 20. Particularly, two different technical options were identified: one which is based on the HGU-55/G helmet (Figure 20) and another based on the HGU-55/P helmet (Figure 21).

Acknowledgements

The authors would like to acknowledge the valuable contribution given by Alenia and Litton Precison Products, during the TORNADO IDS/ECR NVG development and flight test activities. Great thanks go to the staff of RSV for strongly supporting NVIS programs. Last, but not least, the author wishes to thank the aircrews and

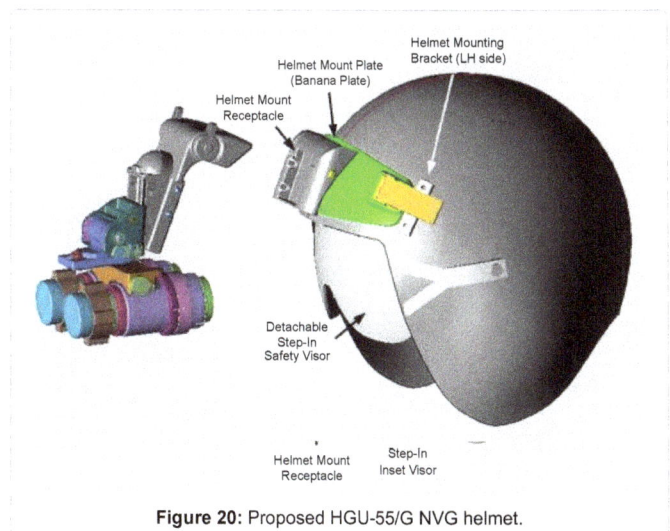

Figure 20: Proposed HGU-55/G NVG helmet.

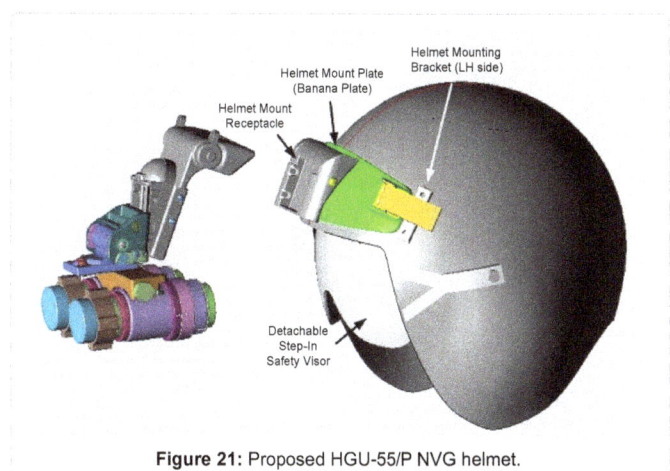

Figure 21: Proposed HGU-55/P NVG helmet.

technical personnel of the Alenia Flight Test Department. Thanks also go to all Air Force, Alenia, Litton, ITT and Gentex personnel not explicitly mentioned here, which supported in different ways the TORNADO NVG development programs. A shorter version of this paper was presented at the SPIE Photonics Europe 2012 Conference, held in Brussels in March 2012.

References

1. Johnson J (1985) Analysis of Image Forming Systems. Proceedings of SPIE, USA.

2. Ratches JA (1976) Static Performance Model for Thermal Imaging Systems. Opt Eng 15: 525-530.

3. Howe JD (1993) Electro-Optical Imaging System Performance Prediction. The Infrared & Electro-Optical Systems Handbook. SPIE Press 57-116.

4. Gerald HC (1992) Applying the log-normal distribution to target detection. Proc SPIE 1689: 213-216.

5. Schmieder DE (1983) Detection Performance in Clutter with Variable Resolution. IEEE Transactions on Aerospace and Electronic Systems. AES-19: 622-630.

6. James RA, Walter LR, Luanne OP, Richard BJ, Thomas CW (1975) Night Vision Laboratory Static Performance Model for Thermal Viewing Systems. Army Electronics Command Fort Monmouth NJ.

7. Holst GC (2003) Electro-optical imaging system performance SPIE Press monograph. JCD Pub.

8. Roberto S, Mark RA, Maurizio C, Mario T, Pietro F, et al. (2012) Night vision imaging systems design, integration, and verification in military fighter aircraft. Proceedings of the SPIE, USA.

4

Transonic Flow Simulation Around the Pitching Airfoil with Accurate Pressure-Based Algorithm

Djavareshkian MH* and Faghihi AR

Faculty of Engineering, Department of Mechanical Engineering, Ferdowsi University of Mashad, Mashad, Iran

Abstract

A new pressure based implicit procedure to solve the Euler and Navier-Stokes equations is developed to predict transonic viscous and inviscid flowsaround the pitching airfoil with high resolution scheme. In this process, nonorthogonal and non moving mesh with collocated finite volume formulation areused. In order to simulate pitching airfoil, oscillation of flow boundary condition is applied. The boundedness criteria for this procedure are determined from Normalized Variable Diagram (NVD) scheme. The procedure incorporates the k-ε eddy-viscosity turbulence model. In the new algorithm, the computation time is considerably reduced. This process is tested for inviscid and turbulent transonic aerodynamic flows around pitching airfoil.The results are compared with other existing numerical solutions and with experiment data. The comparisons show that the resolution quality of the developed algorithm is considerable.

Keywords: Pitching; Transonic; Inviscid; Viscous; Boundary condition

Nomenclature

A, $\tilde{D}$, D	: Finite difference coefficients
C_1, C_2, C_μ	: empirical coefficients
c	: chord length
C_l	: airfoil lift coefficient
f	: physical frequency
K	: a factor in SBIC scheme to determine a special scheme
M_∞	: free stream Mach number
u, v	: Mean (time-average) velocity components in x and y directions
V	: Velocity vector
Γ	: Diffusivity coefficient
κ	: reduced frequency=$\omega c/(2U\infty)$
α	: angle of attack
ε	: Volumetric rate of dissipation
$\tilde{a}$	: Cell face cell
b	: one-half of the chord length
C_m	: airfoil moment coefficient about quarter chord
F	: Flux
I	: Flux
K	: Kinetic energy of turbulence
T	: stress tensor
U_∞	: free stream velocity
Γ_φ^t	: Turbulent diffusivity coefficients
$\delta\upsilon$	: Cell volume
t	: Time

α_m	: angle of attack in mean position
ω_a	: circular frequency, $2\pi f$
δ_{ij}	: Kronecker delta
G	: Generation of turbulence kinetic energy
μ	: Dynamic viscosity
P	: Pressure
σ_ε	: Turbulent Prandtl numbers for dissipation rate
θ	: angle between mean and moment chords
ρ	: Density
R_e	: Reynolds number
μ_t	: Turbulent viscosity
σ_k	: Turbulent Prandtl numbers for turbulent kinetic energy
ϕ	: Scalar quantity

Introduction

In the field of Computational Fluid Dynamics (CFD), there are two categories of numerical methods for simulating moving boundary flow problems. One is the moving grid method [1], which constantly updates the grid according to the position of object. The major limitation of moving grid method is the regeneration of mesh at every time step, which may consume much time and reduce computational efficiency. To overcome this drawback, a pseudo grid-deformation

---

*Corresponding author: Djavareshkian MH, Faculty of Engineering, Department of Mechanical Engineering, Ferdowsi University of Mashad, Mashad, Iran E-mail: javareshkian@um.ac.ir

approach was developed [2]. This approach calculates the grid speed through analytical expression of grid movement. The method is feasible to simulate rotational motion of the object. However, to simulate axial motion of the object, the volume change of grid cells should be considered. Another type of approaches for handling moving boundary problems is the field velocity method [3,4] which adopts the grid speed technique to simulate the velocity change of flow field. This method is especially suitable for calculation of step change of airfoil, and has been successfully applied to calculate the gust response of the airfoil/wing [5-8]. The method of conventional field velocity is usually used to calculate the indicial response by incorporating unsteady flow conditions via grid movement in CFD simulations. The main privilege of this method is direct calculation of aerodynamic responses to step changes in flow conditions. An impulsive change in the angle-of-attack can be considered as an impulsive superposition of a uniform velocity field to the free stream. The magnitude of the indicial change for the angle of attack is used for calculation of the magnitude of normal velocity. In this method, the necessity of uniform distribution of time step over the entire flow domain is guaranteed. In addition, the airfoil is not made to pitch. Hence, the influence of pure angle-of-attack and pitch rate are decoupled efficiently. A similar methodology for simulating responses of an airfoil to step changes in pitch rate and interaction with vertical gusts exists. Moreover, the field velocity method is also applied for prediction of the effects of the trailed vortex wake from the other rotor blades in helicopters, compressors or other turbo machineries. A time dependence study illustrates that a smooth and accurate solution in time requires the consistent evaluation of time metrics in order to satisfy the geometric constitutive law Sitaraman et al. [9].

The objective of the present work is to compute unsteady transonic inviscid and viscous flow fields over a pitching NACA0012 airfoil at various angles of the attack.A new pressure based implicit procedure to solve the Euler and Navier-Stokes equations is developed to predict flowsaround the pitching airfoil with high resolution scheme. In this process, nonorthogonal and non moving mesh with collocated finite volume formulation are used. In order to simulate pitching airfoil, oscillation of flow boundary condition is applied. The boundedness criteria for this procedure are determined from Normalized Variable Diagram (NVD) scheme. The procedure incorporates the k-ε eddy-viscosity turbulence model. The algorithm is tested for inviscid and turbulent transonic aerodynamic flows around pitching airfoil.The results are compared with other existing numerical solutions and with experiment data. The comparisons show that the resolution quality of the developed algorithm is considerable.

Governing equations and discretization

The basic equations, which describe conservation of mass, momentum and scalar quantities, can be expressed in Cartesian tensor form as:

$$\frac{\partial}{\partial t}(\rho) + \frac{\partial}{\partial x_i}(\rho u_i) = 0 \tag{1}$$

$$\frac{\partial}{\partial t}(\rho u_i) + \frac{\partial}{\partial x_j}(\rho u_i u_j - T_{ij}) = S_i^u \tag{2}$$

$$\frac{\partial}{\partial t}(\rho \phi) + \frac{\partial}{\partial x_i}(\rho u_i \phi - q_i) = S^\phi \tag{3}$$

The stress tensor and scalar flux vector are usually expressed in terms of basic dependent variable. The stress tensor for a Newtonian fluid is

$$T_{ij} = -\left(P + \frac{2}{3}\mu\frac{\partial u_k}{\partial x_k}\right)\delta_{ij} + \mu\left(\frac{\partial u_i}{\partial x_j} + \frac{\partial u_j}{\partial x_i}\right) \tag{4}$$

The scalar flux vector usually given by the Fourier-type law is

$$q_i = \Gamma_\phi\left(\frac{\partial \phi}{\partial x_i}\right) \tag{5}$$

Turbulence is accounted for by adopting $k - \varepsilon$ turbulence model. The governing

$$\frac{\partial}{\partial t}(\rho \kappa) + \frac{\partial}{\partial x_i}\left(\rho u_i k - \Gamma_k \frac{\partial k}{\partial x_i}\right) = G - \rho\varepsilon + D_{comp} + \Theta_{diff} \text{ equations for} \tag{6}$$

these quantities are

$$\frac{\partial}{\partial t}(\rho \varepsilon) + \frac{\partial}{\partial x_j}(\rho u_j \varepsilon - \Gamma_\varepsilon \frac{\partial \varepsilon}{\partial x_j}) = C_1\frac{\varepsilon}{k}G - C_2\rho\frac{\varepsilon^2}{k} \tag{7}$$

The turbulent viscosity and diffusivity coefficients are defined by

$$\mu_t = C_\mu \rho \frac{k^2}{\varepsilon} \tag{8}$$

$$\Gamma_\phi^t = \left(\frac{\mu_t}{\sigma_\phi^t}\right) \tag{9}$$

and the generation term G in eqs. (6) and (7) is defined by

$$G = \mu_t\left[\left(\frac{\partial u_i}{\partial x_j} + \frac{\partial u_j}{\partial x_i}\right)\frac{\partial u_i}{\partial x_j} - \frac{2}{3}\delta_{ij}\left(\frac{\partial u_m}{\partial x_m} + \rho k\right)\frac{\partial u_i}{\partial x_j}\right] \tag{10}$$

The term D_{comp} and Θ_{diff} are additional $k - \varepsilon$ contributions to the standard model often introduced to account for the effects of compressibility [10,11]. In this work, the models proposed by Yang et al. [10] are adopted, namely,

$$D_{comp} = -\frac{9}{55}\ k\frac{\partial_i}{\partial x_i} - \frac{1}{\rho}\cdot\frac{t}{\rho}\cdot\frac{\partial}{\partial x_i}\frac{\partial}{\partial x_i}$$

$$\Theta_{diff} = 0$$

The latter being appropriate for high Reynolds number flows, as it is the case here. The values of the turbulence model coefficients used in the present work are given in (Table 1) [10].

The discretization of the above differential equations is carried out using a finite-volume approach. First, the solution domain is divided into a finite number of discrete volumes or cells, where all variables are stored at their geometric centers (Figure 1). The equations are then integrated over all the control volumes by using the Gaussian theorem. The development of the discrete expressions to be presented is effected with reference to only one face of the control volume, namely, e, for the sake of brevity.

For any variable ϕ (which may now also stand for the velocity components), the result of the integration yields

$$\frac{\delta \upsilon}{\delta t}[(\rho\phi)_p^{n+1} - (\rho\phi)_p^n] + I_e - I_w + I_n - I_s = S_\phi\delta\upsilon$$

Where $I_{(s)}$ are the combined cell-face convection I^C and diffusion I^D fluxes. The diffusion flux is approximated by central differences and

C_1	C_2	C_μ	σ_k	σ_ε
1.44	1.92	0.09	1.0	1.3

Table 1: Values of emperical coefficients in the standard k-ε turbulence model.

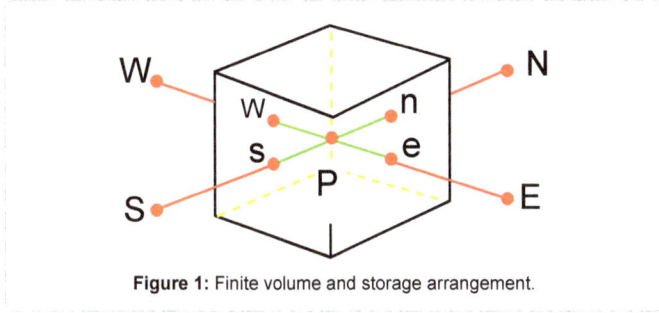

Figure 1: Finite volume and storage arrangement.

can be written for cell-face e of the control volume in Figure 1 as an example as:

$$I_e^D = D_e(\phi_p - \phi_E) - S_e^\phi$$

Where S_e^φ stands for cross derivative arising from mesh nonorthogonality. The discretization of the convective flux, however, requires special attention and is the subject of the various schemes developed. A representation of the convective flux for cell-face e is:

$$I_e^c = (\rho.V.A)_e\phi_e = F_e\phi_e$$

The value of the dependent variable ϕ_e is not known and should be estimated using an interpolation procedure, from the values at neighboring grid points. ϕ_e is determined by the SBIC scheme Djavareshkian [12], that it is based on the NVD technique, used for interpolation from the nodes E, P and W. The expression can be written as

$$\phi_e = \phi_w + (\phi_E - \phi_w).\tilde{\phi}_e$$

The functional relationship used in SBIC scheme for $\tilde{\phi}_e$ is given by

$$\tilde{\phi}_e = \tilde{\phi}_P \quad \text{if} \quad \tilde{\phi}_P \leq 0 \text{ or } \tilde{\phi}_P \geq 1$$

$$\tilde{\phi}_e = -\frac{\tilde{x}_P - \tilde{x}_e}{K(\tilde{x}_P - 1)}\tilde{\phi}_P^2 + \left(1 + \frac{\tilde{x}_P - \tilde{x}_e}{K(\tilde{x}_P - 1)}\right)\tilde{\phi}_P \quad \text{if} \quad 0 < \tilde{\phi}_P < K$$

$$\tilde{\phi}_e = \frac{\tilde{x}_P - \tilde{x}_e}{\tilde{x}_P - 1} + \frac{\tilde{x}_e - 1}{\tilde{x}_P - 1}\tilde{\phi}_P \quad \text{if} \quad K \leq \tilde{\phi}_P < 1 \quad 0 < K \leq 0.5$$

where

$$\tilde{\phi}_e P = \frac{\phi_P - \phi_W}{\phi_E - \phi_W} \quad \tilde{\phi}_e = \frac{\phi_e - \phi_W}{\phi_E - \phi_W} \quad \tilde{x}_e = \frac{x_e - x_W}{x_E - x_W} \quad \tilde{x}_P = \frac{x_P - x_W}{x_E - x_W} \quad (19)$$

The limits on the select each value of K could be determined in the following way. Obviously the lower limit is to keep $K=0$, which would represent switching between upwind and central differencing. This should not be favored because; it is essential to avoid the abrupt switching between the schemes in order to achieve the converged solution. The upper limit of K is 0.5, since it represents the constant gradient and there is no need to use anything else than central differencing in that case. The value of K should be kept as low as possible in order to achieve the maximum resolution of the scheme.

According to Eq. (17), if $\tilde{\phi}_P$ (or $\tilde{\phi}_C$ normalized variable at the central node) does not belong to [0,1], the space discretization is first order, otherwise the SBIC scheme has second order accuracy from point of view space discretization. The details of how the interpolation is made is dealt with [12]; it suffices to say that the discretized equations resulting from each approximations take the form:

$$A_P.\phi_P = \sum_{m=E,W,N,S} A_m.\phi_m + S_\phi'$$

Where A(s) are the convection-diffusion coefficients. The term S_ϕ' in Eq. (19) contains quantities arising from non-orthogonality, numerical dissipation terms, external sources, deferred correction terms, and $(\rho\delta\upsilon/\delta t)\phi_P$ of the old time-step/iteration level. For the momentum equations it is easy to separate out the pressure-gradient source from the convected momentum fluxes.

Solution algorithm

The set of Eq. (19) is solved for the primitive variable (velocity components and energy) together with continuity utilizing pressure-based implicit sequential solution methods. The technique used is the PISO scheme presented herein Issa [13]. In this technique, the methodology has to be adapted to handle the way in which the fluxes are computed in Eqs. (15-18). The adapted PISO scheme consists of a predictor and two corrector sequence of steps at every iteration. The predictor step solves the implicit momentum equation using the old pressure field. Thus, for example, for the component, the momentum predictor stage can be written as

$$u^* = H(u^*) - D\nabla p^o + S_u'$$

Where H contains all terms relating to the surrounding nodes and superscripts * and o denote intermediate and previous iteration values, respectively. Note that the pressure-gradient term is now written out explicitly; it is extruded from the total momentum flux by simple subtraction and addition. The corrector-step equation can be written as

$$u^{**} = H(u^*) - D\nabla p^* + S_u'$$

Hence, from Eqs. (20) and (21)

$$u^{**} - u^* = -D\nabla(p^{**} - p^*) \quad or \quad \delta u = -D\nabla\delta p$$

Now the continuity equation demands that

$$\frac{\delta\rho}{\delta t} + \nabla\left(\rho^* u^{**}\right) = 0$$

For compressible flows it is essential to account for the effect of change of density on the mass flux as the pressure changes. This is accounted for by linearizing the mass fluxes as flows

$$\rho^* u^{**} \approx \rho^o u^* + \rho^o \delta u + u^* \delta\rho$$

Or

$$\rho^* u^{**} \approx \rho^o u^* - \rho^o D\nabla\delta p + u^*(\frac{d\rho}{dp})\delta p$$

Where Eq. (22) is invoked to eliminate δu and $\delta\rho$ is related to δp by the appropriate equation of state. Substitution of Eq. (24) into Eq. (23) yields a pressure-correction equation of the form

$$A_P . \delta p_P^* = A_E . \delta p_E^* + A_W . \delta p_W^* + A_N . \delta p_N^* + A_S . \delta p_S^* + S_P$$

Where S_p is the finite difference analog of $\nabla(\rho^o u^*)$, which vanishes when the solution is converged. The A coefficients in Eq. (25) take the form (the expression for A_E is given as an example)

$$A_E = (\rho^o \tilde{a} D)_e - \lambda_e (\tilde{a} u^*)_e . \left(\frac{d\rho}{dp}\right)_e$$

where λ is a factor whose significance is explained subsequently. The mass flux at a cell face is computed from nodal values of density and velocity, the cell-face values of ρ_e^o and u_e^* in Eq. (26) are not readily available. To compute those values, assumptions concerning the variations of ρ need to be made. In upwinding $\lambda = 1$ when u is positive; otherwise it would be zero. Alternatively, in central difference formula $\lambda = 1/2$.

Such assumptions have no influence whatsoever on the final solution because they affect only the pressure-correction coefficients, and as δp goes to zero at convergence, the solution is, therefore, independent of how those coefficients are formulated; however, they do influence the convergence behavior.

The structure of the coefficients in Eq. (25) simulates the hyperbolic nature of the equation system. Indeed, a closer inspection of expression (26) would reveal an upstream bias of the coefficients (A decreases as u increases), and this bias is proportional to the square of the Mach number. Also, note that the coefficients reduce identically to their incompressible form in the limit of zero Mach number.

In the present work, Crank-Nicolson scheme is applied for discretization of time derivative with second order accuracy. This option seems to be the most obvious as it requires the minimum amount of memory storage of the velocity fields. The system of equation is solved by biconjugate gradient method.

New time advancement algorithm

In this research three time advancement algorithms are used for the simulation. Figure 2 shows the flowchart of them. Algorithm 2(a) has external loop to satisfy convergence criteria for each iteration. An internal loop just for pressure equation is used in Figure 2b. The new time advancement algorithm, Figure 2c, is utilized an internal and external loop for calculation.

Boundary conditions

At the inlet of the domain, only three of the four variables need to be prescribed: the total temperature, the angle of attack, and the total pressure. The pressure is obtained by zeroth order extrapolation from interior points. At outlet, the pressure is fixed. Slip boundary conditions are used on the lower and upper walls. In the case of viscous flow, the

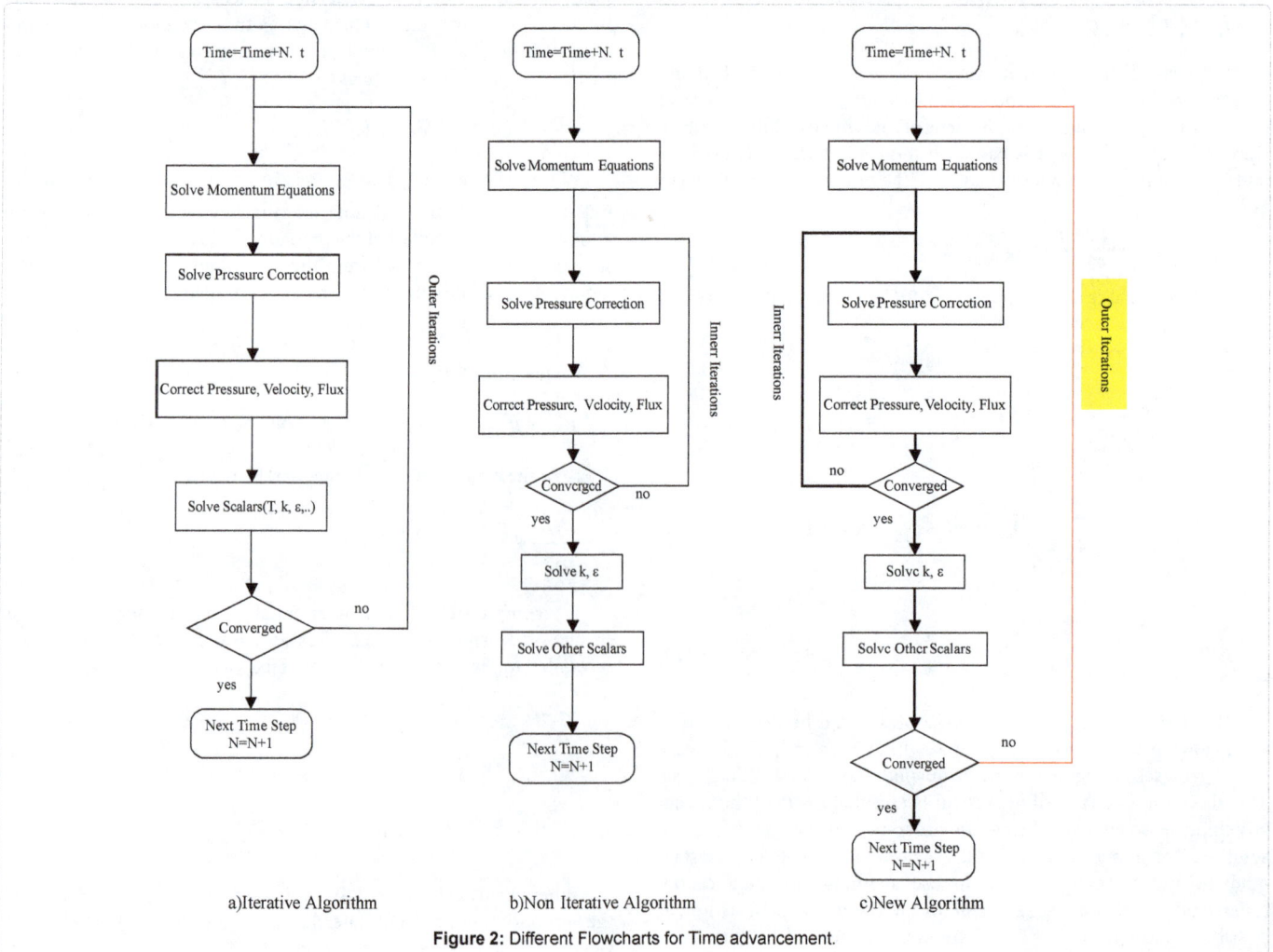

a)Iterative Algorithm   b)Non Iterative Algorithm   c)New Algorithm

Figure 2: Different Flowcharts for Time advancement.

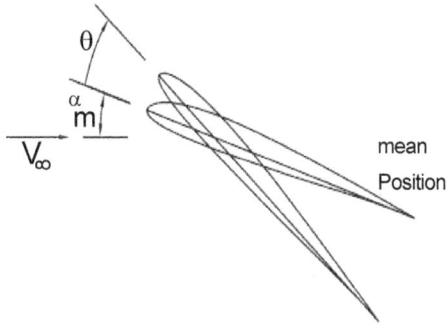

Figure 3: Pure pitch definition.

κ	M_0	α_m (deg.)	α_p (deg.)	c
0.0814	0.755	0.016	2.51	1.0

Table 2: Pure pitch motion parameters.

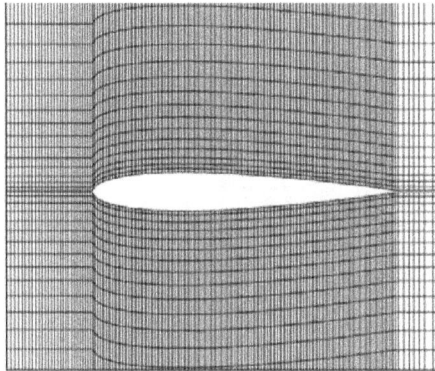

Figure 4: Part of the H Grid.

non-slip condition is applied at the airfoil surfaces. To account for the steep variations in turbulent boundary layers near solid walls, wall functions, which define the velocity profile in the vicinity of no-slip boundaries, are employed. The far-field boundary is set to 30c from the airfoil to minimize its undesired effects on the flow surrounding and is set to slip boundary conditions.

Results and Discussion

In this section, the results of the inviscid and viscous flows over a pitching NACA0012 airfoil along its quarter chord axis are indicated. The simulations are performed at a higher Reynolds number. In particular, we aim to validate the simulation with existing experiment results of a pitching airfoil, and study the lift and drag characteristics of a pitching airfoil. The steady state solutions are used as initial conditions for time-marching calculations. Figure 3 provides an illustration of pure-pitch motion for an airfoil with a mean angle of attack of αm. The parameters of motion and flow field are described in Table 2. The airfoil is forced into an oscillation around an axis located at the quarter-chord. The angle of attack is specified as:

$$\alpha(t) = \alpha_m + \alpha_p \sin(\omega_\alpha t) \tag{27}$$

The free stream velocities for unsteady computations are set to $u_{inlet}=U_\infty \cos(\alpha(t))$ and $v_{inlet}=U_\infty \sin(\alpha(t))$. A H-type mesh is generated to model the airfoil and the surrounding flow. The schematic of this grid which used in the present simulation is shown in Figure 4. The

grid dependence test for Navier-Stokes Equation on the NACA0012 airfoil at M_∞=0.755, α=-1.8° is indicated in Figure 5. Three different mesh sizes were considered: 27680, 57950 and 115960 cells and each simulation emerged from its fully converged solution. Thus the mesh of 57980 cells was selected as a baseline mesh for further analyses. Convergence histories for the inviscid flow are shown in Figures 6 and 7 compare the computed viscous case surface pressure distribution with the experimental data [14]on NACA0012 with M_∞=0.755, α_m=0.016°, α_p=2.51°, k=0.0814 for two angles of attacks. As it is seen from these results, there is quite a good agreement between the present method and the measurement of Landon [14]. These comparisons show that the solutions using oscillating boundary condition method has good prediction.

The computed variation of the lift coefficient versus angle of attack for inviscid and viscous flows during the third cycle is compared with that Landon [14] and Uzun [15] and in Figure 8. The existence of this variation loop is the result of induced velocities, which result in different lift coefficients between the up and down strokes. For presented viscous case, the turbulence quantities were specified at inlet to correspond to 0.008 turbulence intensity and a dissipation length scale of 10% of the airfoil chord. The value of K in SBIC scheme for this case is 0.3. Figure 8a shows the computed variation of lift coefficient versus angle of attack for viscous case which is in close agreement with experimental data. Because the flow around a pitching oscillation airfoil is disturbed and turbulence models can influence the results, the little difference between the numerical prediction and experimental data could be due to turbulence model. Figure 8b shows the C_l versus α for inviscid

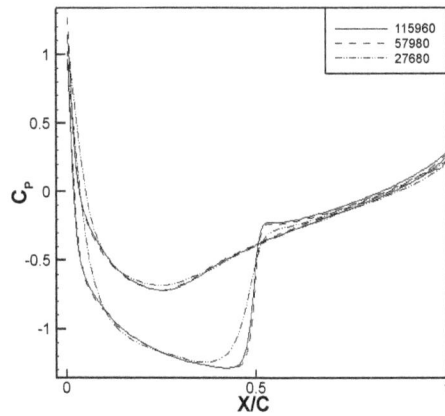

Figure 5: Grid dependency results for NACA0012, M_∞= 0.755, α=-1.8°.

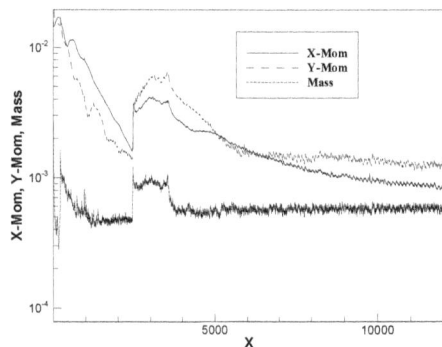

Figure 6: Convergence histories for NACA0012, M_∞= 0.755, α=-1.8°.

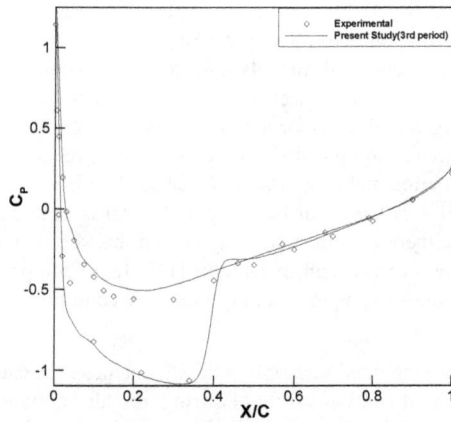

a: Pressure coefficient distribution

$\alpha(t) = -0.54°$.

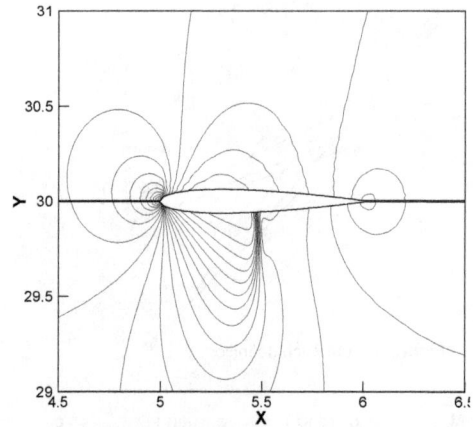

b: Pressure coefficient contours

$\alpha(t) = -0.54°$.

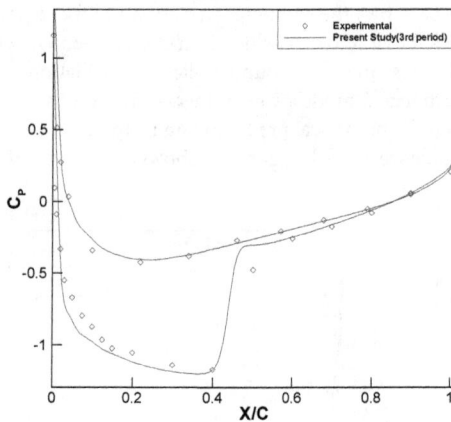

c: Pressure coefficient distribution

$\alpha(t) = 2.34°$.

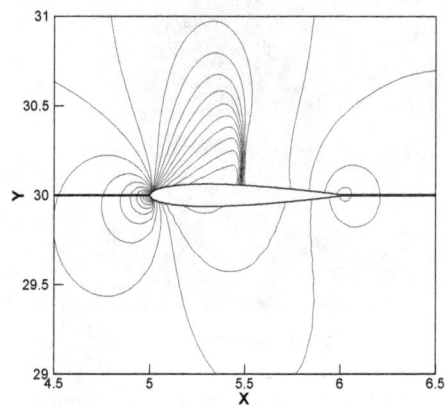

d: Pressure coefficient contours

$\alpha(t) = 2.34°$.

Figure 7: Pressure distribution on NACA0012, $M_\infty = 0.755$, $\alpha_m = 0.016°$, $\alpha_p = 2.51°$, $k = 0.0814$.

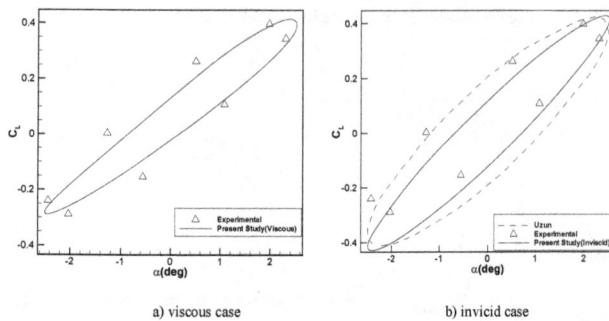

a) viscous case

b) invicid case

Figure 8: Lift coefficient versus angle of attack for $M_\infty = 0.755$, $\alpha_m = 0.016°$, $\alpha_p = 2.51°$, $k = 0.0814$.

Figure 9: Drag coefficient versus angle of attack for viscous case at $M_\infty = 0.755$, $\alpha_m = 0.016°$, $\alpha_p = 2.51°$, $k = 0.0814$.

case. Uzun [15] used a parallel algorithm for the solution of unsteady Euler equation on unstructured reformatting grids while this study

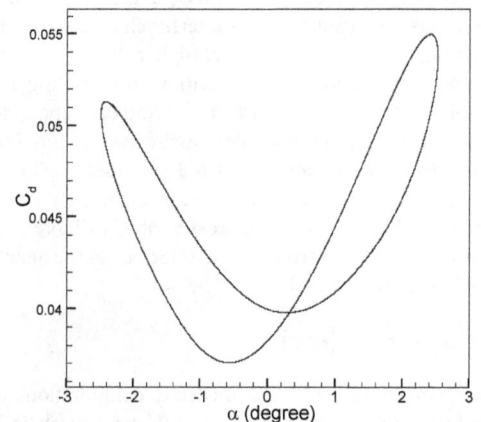

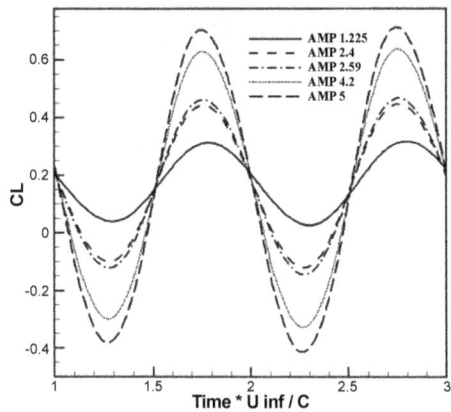

Figure 10: Instantaneous Lift coefficient versus non-dimensional time M_∞=0.755, k=0.0814.

	Iterative Algorithm	Non-Iterative Algorithm	New Algorithm
Internal Loop No.	-	20-30	2-3
External Loop No.	1000		1-2
CPU Time (min)	8600	2800	120

Table 3: CPU Time comparison for different algorithms.

non moving mesh with oscillation of flow boundary condition is applied. It can be seen that both methods are not good agreement with experimental data particularly at the lowest angle of attack. The reason for this difference is caused by the lack of consideration of viscosity. In other words, the viscosity can effect on the separated vortex from the airfoil and aerodynamic coefficients in unsteady flow.

The predicted drag coefficients versus angle of attack are illustrated in Figure 9. The upstroke C_d and C_{dmin} are higher than the down stroke one. In this work, the effect of the airfoil amplitude of oscillation on the simulated lift coefficients is assessed. The instantaneous C_L versus τ where K=0.0184, M=0.755 on NACA0012 is indicated in Figure 10. As illustrated, the maximum lift coefficients increases at higher amplitudes of oscillation, the calculated lift coefficients are periodic and resemble harmonic-like patterns. Furthermore, increasing amplitude endues significant lead in the C_L results that C_{Lmax} is obtained at a lower τ. This can be attributed to the stronger effects of the shed wake and vertical structures on the surrounding fluid at the higher amplitudes. Table 3 indicate CPU Time comparison for different algorithms. The numbers of iterationto satisfy convergence criteria for the external loops of algorithms (a),(b) and (c) are approximately 10,000 and 1-2 respectively. For internal loops, the numbers of iterations of these algorithms are about 0, 20-30 and 2-3 respectively. As a result, the two algorithms (a) and (b) are time consuming and CPU time for new method is considerably decreased.

Conclusions

A pressure based implicit procedure to solve the Euler and Navier-Stokes equations is developed to predict transonic viscous and inviscid flows around the pitching airfoil with high resolution scheme. In order to simulate pitching airfoil, oscillation of flow boundary condition is applied. The boundedness criteria for this procedure are determined from Normalized Variable Diagram (NVD) scheme. The main findings can be summarized as follows: 1-The pitching airfoil simulation with the oscillation of flow boundary condition with fix grid is very simple and has low cost. 2-The grid dependence test with high resolution

scheme indicates that an acceptable solution can be obtained even on fairly coarse 3-The agreement between numerical and experimental data is considerable. 4-The CPU time for new method considerably reduce.

References

1. Feng D, Yizhao W, Xueqiang L (2007) Numerical Simulation of Two-Dimensional Unsteady Viscous Flow Based on Hybrid Dynamic Grids. Journal of Nanjing University of Aeronautics & Astronautics 39: 444-448.

2. Tadghighi H, Liu Z, Ramakrishnan S (2005) A Pseudo Grid-Deformation Approach for Simulation of Unsteady Flow Past a Helicopter in Hover and Forward Flights. 43rd AIAA Aerospace Sciences Meeting and Exhibit 10-13 January Reno Nevada USA.

3. Hao Z, Wei-qi Q (2007) Numerical simulation of gust response for airfoil and wing. Acta Aerodynamica Sinica 25: 531-536.

4. Hao Z, Wei-qi Q (2009) Numerical simulation on gust response of elastic wing. Chinese Journal of Computational Mechanics 26: 270-275.

5. Gopalan H, Povitsky A (2008) A Numerical Study of Gust Suppression by Flapping Airfoils. 26th AIAA Applied Aerodynamics Conference 18-21 August Honolulu, Hawaii.

6. Raveh D (2010) CFD-Based Gust Response Analysis of Free Elastic Aircraft. ASD Journal 2; 23-24.

7. Raveh DE (2011) Gust-Response Analysis of Free Elastic Aircraft in the Transonic Flight Regime. J Aircraft 48: 1204-1211.

8. Raveh DE (2007) CFD-Based Models of Aerodynamic Gust Response. J Aircraft 44: 888-897.

9. Sitaraman J, Baeder J, Iyengar V (2003) On the Field Velocity Approach and Geometric Conservation Law for Unsteady Flow Simulations. 16th AIAA Computational Fluid Dynamics Conference, USA.

10. Yang ZY, Chin SB, Swithenbank J (1991) On the modelling of the k-equation for compressible flow. Proceedings of the 7th International Conference, Stanford, CA, USA.

11. Narayan JR, Sekar B (1991) Computation of Turbulent High Speed Mixing Layers Using a Two-Equation Turbulence Model. Computational Fluid Dynamics Symposium on Aeropropulsion 409-428.

12. Djavareshkian MH (2004) Pressure-based compressible calculation method utilizing normalized variable diagram scheme. Iran J Sci Technol 28: 495-500.

13. Issa RI (1986) Solution of the implicitly discretised fluid flow equations by operator-splitting. J Comput Phys 62: 40-65.

14. Landon RH (1982) NACA0012 oscillation and transient pitching," in Compendium of Unsteady Aerodynamic Measurements, Advisory Report 702.

15. Uzun, A (1999) Parallel Computations of Unsteady Euler Equations on Dynamically Deforming Unstructured Grid. M.Sc. Thesis Purdue University USA.

5

Carrier-phase GNSS Attitude Determination and Control for Small Unmanned Aerial Vehicle Applications

Roberto Sabatini[1]*, Anish Kaharkar[1], Celia Bartel[1] and Tesheen Shaid[2]

[1]*Department of Aerospace Engineering, Cranfield University, Cranfield, Bedford, MK43 0AL, UK*
[2]*Leopoldo Rodríguez Salazar, Aerotech Systems Ltd, Milton Keynes, Buckinghamshire, MK11 1BY, UK*

Abstract

As part of our recent research to assess the potential of low-cost navigation sensors for Unmanned Aerial Vehicle (UAV) applications, we investigated the potential of carrier-phase Global Navigation Satellite System (GNSS) for attitude determination and control of small size UAVs. Recursive optimal estimation algorithms were developed for combining multiple attitude measurements obtained from different observation points (i.e., antenna locations), and their efficiencies were tested in various dynamic conditions. The proposed algorithms converged rapidly and produced the required output even during high dynamics manoeuvres. Results of theoretical performance analysis and simulation activities are presented in this paper, with emphasis on the advantages of the GNSS interferometric approach in UAV applications (i.e., low cost, high data-rate, low volume/weight, low signal processing requirements, etc.). The simulation activities focussed on the AEROSONDE UAV platform and considered the possible augmentation provided by interferometric GNSS techniques to a low-cost and low-weight/volume integrated navigation system (presented in the first part of this series) which employed a Vision-Based Navigation (VBN) system, a Micro-Electro-Mechanical Sensor (MEMS) based Inertial Measurement Unit (IMU) and code-range GNSS (i.e., GPS and GALILEO) for position and velocity computations. The integrated VBN-IMU-GNSS (VIG) system was augmented using the inteferometric GNSS Attitude Determination (GAD)sensor data and a comparison of the performance achieved with the VIG and VIG/GAD integrated Navigation and Guidance Systems (NGS) is presented in this paper. Finally, the data provided by these NGS are used to optimise the design of a hybrid controller employing Fuzzy Logic and Proportional-Integral-Derivative (PID) techniques for the AEROSONDE UAV.

Keywords: GNSS attitude determination; Attitude determination and control; Unmanned aerial vehicle; Low-cost navigation sensors; Fuzzy logic controller; PID controller

Introduction

Technological developments in the realm of satellite navigation have led to innovative concepts in the mission management of current and next generation air, land and sea vehicles. Navigation systems including GNSS or integrated GNSS/INS are being used extensively today in most aerospace platforms around the world and new promising technologies are being explored. The great majority of current manned and unmanned aerial vehicles perform attitude determination tasks by using inertial sensors (ring laser gyros, fibre optics gyros, accelerometers, etc.), packaged into Attitude and Heading Reference Systems (AHRS) or into Inertial Navigation Systems (INS). Although AHRS/INS technologies are well established [1], they have some disadvantages. High accuracy class products are costly when compared with emerging alternative technologies (e.g. MEMS based Inertial Measurement Units), AHRS/INS position data accuracy degrades with time and their attitude accuracy is strongly dependent on platform dynamics. Furthermore, a significant amount of data processing is required to "smooth-out" sensor errors and extensive simulation, laboratory and ground/flight test activities are often required in order to properly design and calibrate the Kalman Filter parameters. The use of inexpensive GNSS technology for aiding AHRS/INS has been extensively investigated over the past decades, and integrated GNSS/INS systems are the state-of-the-art for aerospace platform navigation applications [2-4]. The concept of replacing traditional attitude sensors with GNSS interferometric processing (carrier-phase) has been also considered in recent years, mostly for spacecraft applications (replacing or aiding traditional sun-sensors, horizon-trackers, star-trackers, magnetometers, etc.), and for manned aircraft [5-8] and ship applications [9]. Due to the low volume/weight of current carrier-phase GNSS receivers, and the extremely high

accuracy attainable notwithstanding their lower cost, interferometric GNSS technology is becoming an excellent candidate for future UAV applications [10]. The accuracy of the GNSS Attitude Determination (GAD) systems is affected by several factors including the selected equipment/algorithms and the specific platform installation geometry, with the baseline length and multipath errors being the key elements dominating GAD systems performance [10-12] developed an extension of the known Least-squares Ambiguity Decorrelation Adjustment (LAMBDA) method [13] for solving nonlinearly constrained ambiguity resolution problems associated to GNSS attitude determination.

One of the main challenges of implementing GAD systems for attitude determination in UAV and other aerospace platforms is the need of resolving integer ambiguity in real-time in order to obtain reliable attitude estimations [10]. In recent years several techniques have been developed for integer ambiguity resolution. Giorgi et al. [10] In terms of data rate, Pinchin [8] suggests that a typical AHRS/INS system provides attitude measurements upwards of 100Hz where as a GAD system output is in the order of 1-5Hz which is too low for high dynamics platform applications. In small UAV platforms a simple solution that integrates a low cost GNSS/MEMS-IMU system for attitude determination may be also affected by vibrations and aerodynamic

---

*Corresponding author: Roberto Sabatini, Department of Aerospace Engineering, Cranfield University, Cranfield, Bedford, MK43 0AL, UK
E-mail: r.sabatini@cranfield.ac.uk

effects acting on the platform itself (e.g., aeroelasticity). Therefore, a very accurate initial heading estimate or integration with other sensors is often required for stable filter performancein such applications [13]. As a consequence, the integration of additional augmenting sensors such as Vision-based Navigation (VBN) sensors [14,15] can provide significant improvements in the accuracy and continuity of the measurements. Several methods have been developed in the past decades for GAD systems. The classical method, developed by Cohen [14], involves two main steps. The first step is to find a matrix that transforms the baseline configurations to an equivalent orthonormal basis and the second step is the use of fast algorithms (e.g., QUEST and FOAM) for attitude determination. An alternative method is to adopt recursive algorithms to minimize a cost function that links all available carrier phase measurements. Independently from the method selected, since GAD errors are dominated by lengths of the baselines used, some efficient geometric algorithmsare proposed for baseline selection in the presence of redundant satellite measurements. Various controller schemes have also been applied in the past to the design of autonomous control/servoing systems for UAVs. Some of these techniques include Adaptive Control [16-19], Fuzzy Control [17,20], Neural Networks, Genetic Algorithms and Lyapunov Theory [21]. Beyond studying the possible synergies attainable from integration of GAD systems with other low-cost and low-weight/volume navigation sensors (e.g. VBN and MEMS-INS), and additional objective of our research is to develop an hybrid Fuzzy/PID controller using INS, GNSS and GAD input data and also capable of VBN guidance (visual servoing) during the final approach and landing phases of the flight. This is allowing the development of an integrated Navigation and Guidance System (NGS) capable of providing the required level of performance in all flight phases of a small UAV.

GNSS Attitude Determination

In the fundamental concept of interferometric GNSS Attitude Determination (GAD), the measurement of the phase of the GNSS signal carrier allows to determine the relative displacement of the antennae in the body reference frame. This information is directly related to the attitude of the vehicle. The displacement of the antenna baseline (b) with respect to the LOS of the GNSS signal is given by:

$$\theta = \cos^{-1}\left[\frac{k + \frac{\ddot{A}\phi}{360}}{b} \cdot \lambda\right] \quad (1)$$

Where the phase difference $\Delta\phi/360$ is proportional to the projection of the baseline (b) on the Line-of-Sight (LOS).Since the antennae are placed at different locations, the phase measurements of the incoming GNSS signal carrier are different for each antenna. By knowing the integer number of cycles travelled by the carrier (N), it is possible to determine the vehicle attitude. When using GNSS for attitude determination it is sufficient that only two satellites are in view due to the following considerations:

- **Common time reference:** measurements are independent from the error at the receiver clock as it is the same for the measurements performed by each antenna.

- **Baseline setting:** the relative position of the antennae on the vehicle is known a priori; this eliminates another unknown factor which reduces the number of satellites required.

GAD algorithms

Knowing the coordinates,both in the body reference frame and in the North-East-Down (NED) frame, of the unit vectors of the LOS to the S_n satellites, and the unit vector perpendicular to the plane containing three antennae $\hat{A}$, it is possible to determine the attitude of the vehicle. In the body axis reference frame (x, y, z) any combination of 3 not aligned antennae located at the points P_1, P_2, P_3 originates a plane π. This plane is the locus of points P with coordinates that satisfy the equation:

$$\begin{vmatrix} x & y & z & 1 \\ x_1 & y_1 & z_1 & 1 \\ x_2 & y_2 & z_2 & 1 \\ x_3 & y_3 & z_3 & 1 \end{vmatrix} = 0 \begin{vmatrix} x-x_1 & y-y_1 & z-z_1 \\ x_2-x_1 & y_2-y_1 & z_2-z_1 \\ x_3-x_1 & y_3-y_1 & z_3-z_1 \end{vmatrix} = 0 \rightarrow ax+by+cz+d=0 \quad (2)$$

Since the plane π is represented by equation $ax + by + cz + d=0$, the vector of components (a, b, c) is orthogonal to the plane. Therefore, the coordinates of the unit vector $\hat{A}$ orthogonal to the plane are:

$$A_x = \frac{a}{\sqrt{a^2+b^2+c^2}}; A_y = \frac{b}{\sqrt{a^2+b^2+c^2}}; A_z = \frac{c}{\sqrt{a^2+b^2+c^2}} \quad (3)$$

From the three antennae located on the plane π, a master antenna M and two "slaves" B with components (B_1, B_2, B_3) and C with components (C_1, C_2, C_3) are defined (Figure 1).

Using the relations to determine the angle between two vectors and between a vector and a plane, the unit vectors from the LOS to satellites (S_n) are those for which the following conditions apply:

$$\cos\theta_1 = \frac{B_1 S_1 + B_2 S_2 + B_3 S_3}{\sqrt{B_1^2+B_2^2+B_3^2}\sqrt{S_1^2+S_2^2+S_3^2}} \quad (4)$$

$$\cos\beta = \sin S\pi = \frac{aS_1 + bS_2 + cS_3}{\sqrt{a^2+b^2+c^2}\sqrt{S_1^2+S_2^2+S_3^2}} \quad (5)$$

$$\cos\theta_2 = \frac{C_1 S_1 + C_2 S_2 + C_3 S_3}{\sqrt{C_1^2+C_2^2+C_3^2}\sqrt{S_1^2+S_2^2+S_3^2}} \quad (6)$$

$$\cos\beta = \frac{A_1 S_1 + A_2 S_2 + A_3 S_3}{\sqrt{A_1^2+A_2^2+A_3^2}\sqrt{S_1^2+S_2^2+S_3^2}} \quad (7)$$

From Eq. (5) and Eq. (7), a system of 3 equations with 3 unknowns (S_1, S_2, S_3) is obtained only if the magnitude of the LOS vector is known. The unknowns are the coordinates of vector LOS in the body frame. Then, the angle β, which is the angle between the LOS vector to the

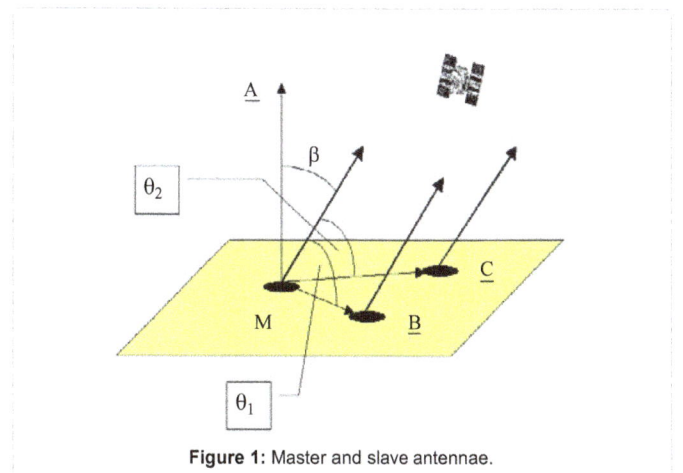

Figure 1: Master and slave antennae.

satellite $\hat{S}_n$ and the perpendicular $\hat{A}$ to the plane π, can be obtained directly from equations Eq. (5) and Eq. (7). The unit vectors $\hat{S}_n \rightarrow$ LOS, known in the body frame, are fully defined in the NED frame (CG, x_N, y_N, z_N). In fact, the receiver extracts the coordinates of the satellite from the navigation message. From these parameters, it computes the unit vector of the LOS in ECI frame. Since the NED frame is always defined with respect to the ECI-frame the unit vectors $\hat{S}_N$ are then properly defined in the NED frame. In particular if C_E^N is the transformation matrix from ECI-frame to NED, the unit vector $\hat{S}_{nN}$ in the NED frame is given by the following transformation:

$$\hat{S}_{nN} = \begin{bmatrix} S_{x_N} \\ S_{y_N} \\ S_{z_N} \end{bmatrix} = C_E^N \begin{bmatrix} S_{X\,ECI} \\ S_{Y\,ECI} \\ S_{Z\,ECI} \end{bmatrix} \tag{8}$$

The next step is to determine the coordinates of $\hat{A}$ in the NED frame in order to have a full set of vectors that will be used for attitude determination. Analytically this geometric problem can be represented by a system of 3 equations with 3 unknowns A_1, A_2, A_3. These are the components of vector A in the auxiliary reference frame (x_1, x_2, x_3):

$$\begin{cases} \hat{A} \cdot \hat{S}_1 = \cos\cos\beta_1 \\ \hat{A} \cdot \hat{S}_2 = \cos\cos\beta_2 \\ \hat{A}.\hat{A} = 1 \end{cases} \tag{9}$$

By getting

$$p = \frac{\cos\beta_1 - \hat{S}_2.\hat{S}_1\cos\beta_2}{1 - \left(\hat{S}_2 \cdot \hat{S}_1\right)^2}; \quad q = \frac{\cos\beta_2 - \hat{S}_2.\hat{S}_1\cos\beta_1}{1 - \left(\hat{S}_2 \cdot \hat{S}_1\right)^2}; \quad r = \pm\sqrt{\frac{1 - p\cos\beta_1 - q\cos\beta_2}{1 - \left(\hat{S}_2 \cdot \hat{S}_1\right)^2}} \tag{10}$$

The solution of $\hat{A}$ becomes

$$\hat{A} = p\,\hat{S}_1 + q\,\hat{S}_2 + r\,\hat{S}_x \tag{11}$$

Eq. (11) generates 2 possible ambiguous solutions. In order to solve this ambiguity the following steps can be performed:

• Compare the possible solution with an estimation made in advance.

• Compare more attitude solutions that can be accumulated in a certain observation time discarding those which are dispersed.

• Use a third satellite.

The analytical solution of the system with three satellites is given by:

$$\begin{cases} \hat{A} \cdot \hat{S}_1 = \cos\beta_1 \\ \hat{A} \cdot \hat{S}_2 = \cos\beta_2 \\ \hat{A} \cdot \hat{S}_3 = \cos\beta_3 \\ \hat{A}.\hat{A} = 1 \end{cases} \rightarrow \begin{cases} \hat{A} = p\,\hat{S}_1 + q\,\hat{S}_2 + r\,\hat{S}_x \\ \hat{A} \cdot \hat{S}_3 = \cos\beta_3 \end{cases} \tag{12}$$

Although the system Eq. (12) has a unique solution for $\hat{A}$ in a real system it is necessary to take into account the possible errors in the

determination of the values of $\hat{S}_n$ and β_n. The geometry with three satellites and the error values is illustrated in Figure 2.

With the methodology described above, the input data required to determine the attitude states of the vehicle is defined (i.e., the coordinates of the vectors $\hat{S}_n$ and the coordinates of the vectors $\hat{A}$ in the body frame and in the NED frame). Then two approaches can be used for attitude determination, one is a variant of the classical method [16] that allows the determination of the attitude states by considering one single pair of vectors (e.g., $\hat{A}$ and $\hat{S}_1$, $\hat{S}_1$ and $\hat{S}_2$).In order to select the optimal pair of vectors, the errors associated to such combination are considered (the pair with the minimum RMS/RSS error is selected. The recursive algorithm method, uses all available information from 3 nonaligned antennae and 3 satellites ($\hat{A}$, $\hat{S}_1$, $\hat{S}_2$, $\hat{S}_3$), to obtain an estimation of the attitude of the vehicle by minimizing the following cost function:

$$CJ[C] = \frac{1}{2}a_1\left|\hat{A}_b - C\hat{A}_i\right|^2 + \frac{1}{2}a_2\left|\hat{S}_{1b} - C\hat{S}_{1i}\right|^2 - \frac{1}{2}a_3\left|\hat{S}_{2b} - C\hat{S}_{2i}\right|^2 + \frac{1}{2}a_4\left|\hat{S}_{3b} - C\hat{S}_{3i}\right|^2 \tag{13}$$

Where a_1, a_2 and a_3 and a_4 are 4 non-negative weights. Therefore, for a number of N measurements, such a cost function can be generalized as follows:

$$J[C] = \frac{1}{2}\sum_{k=1}^{N} a_k \left|\hat{W}_k - C\hat{V}_k\right|^2 \tag{14}$$

Where $\hat{W}_k$ is a vector determined in the body axis frame and $\hat{V}_k$ is the corresponding vector in the inertial frame. In the ideal case of absence of errors, each term of Eq. (13) would be cancelled in correspondence to a certain proper orthogonal matrix C. As this does not occur in reality, it is necessary to assign appropriate weights in order to minimize the cost function by considering the accuracy of the measurements. Since only 3 of 9 elements are independent, it is acceptable to minimize the cost function for a minimum number of parameters (e.g., Euler angles), in order to reduce the complexity on the calculation.

GAD accuracy

Similarly to Geometric Dilution of Precision (GDOP, the Attitude Dilution of Precision (ADOP) is a parameter that indicates how accurate the attitude solution is. The ADOP is related to the error in attitude calculation σ_θ, the error in range σ_r and the baseline length b by the following equation

$$\sigma_\theta = ADOP \cdot \frac{\sigma_r}{b} \tag{15}$$

Where

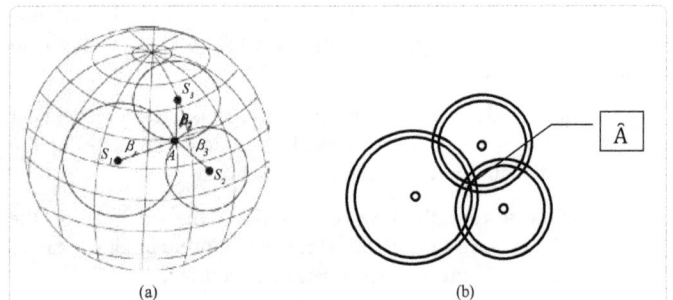

Figure 2: Solution with three satellites (a) and errors in the determination (b) of $\hat{S}_n$ for the computation of $\hat{A}$.

$$ADOP = \sqrt{trace\left[\left(nI - SS^T\right)^{-1}\right]} \qquad (16)$$

And $S = [s_1, s_2, \ldots s_n]$ is the matrix $1 \times n$ of the LOS to the satellite, n is the number of satellites in view and l the identity matrix. The value of ADOP is generally equal to 1 or less. This indicates that the GNSS constellation guarantees favorable geometry for the attitude determination. Therefore, it is possible to make an approximation of the attitude error by assuming that ADOP=1. With this assumption the relationship is simplified to

$$\sigma_\theta = \frac{\sigma_r}{b} \qquad (17)$$

By knowing the error σ_θ associated with each measure, assuming that the measures are statistically independent, it is possible to calculate the total RSS error σ_{TOT} by the relation

$$\sigma_{TOT} = \sqrt{\sigma_{\theta 1}^2 + \sigma_{\theta 2}^2 + \sigma_{\theta 3}^2} \qquad (18)$$

The error in attitude determination is a function of the instantaneous orientation of the aircraft, the satellite geometry and the selected baselines. The range error σ_r in the Eqs. 15 and 17 includes the following contribution:

• **Multipath:** This is the main source of error. Even though the error is highly deterministic, previous research [22] shows that even with the most careful study on the location of the antennae the error cannot be reduced below the 5 mm threshold. This error is directly dependant on different non-controlled variables such as the environment itself; other variables also influence this source of error, such as materials, antennae gain, geometry, etc. The control of these variables to reduce the error is often complex and expensive.

• **Structural distortion:** In high temperature applications the vehicle surface may experience thermal deformation. This will cause a relative displacement between antennae with consequent errors in the attitude solution. Aeroelastic effects also introduce structural distortions.

• **Tropospheric error:** The troposphere is often considered a source of error for the transmission of electromagnetic signals [23,24]. The error becomes more significant with the increase of the refraction index. This increase becomes significant at altitudes . The refraction index causes a deflection of the GNSS signal [25]. The refraction index can be modelled according to Snell's law. Therefore, an error is introduced when the phase measurements are converted to attitude angles.

• **Signal-to-Noise Ratio (SNR):** In high dynamics applications the tracking loop bandwidth needs to be extended. By extending it, the bandwidth of the associated error is also increased [26]. Many stochastic models have been proposed based on the SNR reported by the receiver [27,28].

• **Specific errors in the receiver:** This source of error can become significant if it is not considered at an early design stage. Nowadays, technology allows to have precise models of it [22,29]. There are several examples of those errors such as crosstalk, which is common in antennae with high gain, line bias, which is the phase offset between one antenna and another and inter-channel bias, which results of the phase measurements from different satellites that use a different channel.

• **Total error:** From the analysis on the different source of errors in range, considering that multipath is the dominant error, a rough approximation to this error is given by:

$$\sigma_\theta(rad) \cong 0.5 / L \ \ (cm) \qquad (19)$$

Where L is the longitude of a given baseline. In Eq. (19), it is shown that the error appears inversely proportional to the length of the baseline used for attitude determination. Hence it is always preferred to use longer baselines which allow a more accurate attitude solution. A detailed discussion of the sources of errors can be found in the literature [30,31].

Geometric algorithm for antennae selection

As a first step the antennae with less than 2 satellites in view are discarded by using a masking algorithm. It is then when the baselines are measured between the remaining antennae.

$$\bar{b} = \bar{a}_0 - \bar{a}_1 \qquad (20)$$

By ordering the baselines in descending order there is a selection of the first two that are associated with the greater area of the triangle formed by the baselines and their links. The common antenna with respect to these baselines is identified as possible Master M antenna while the other two are possible slaves: Sl_1 and Sl_2. Once the process is repeated for all antennae with at least 2 satellites in view the optimal combination of three antennae is selected for those, whose the following function is maximum

$$R = \frac{\rho_1 + \rho_2}{\rho_1 - \rho_2} \qquad (21)$$

Where ρ_1 and ρ_2 are the lengths of baselines $M-Sl_1$ and $M-Sl_2$

Multisensor Data Fusion

Employing the geometric algorithm for optimal selection of the antenna baselines and the recursive algorithm (Eq. 20,21) for over-determined attitude computations, the resulting error analysis is presented in Table 1.

Then the GNSS attitude determination is integrated to the VIG Navigation System as Illustrated in Figure 3.

It can be observed that the output of the GNSS Attitude Determination System (GAD) is integrated to the navigation system extended Kalman Filter for data fusion. The details of the EKF implementation can be consulted in [14,15].

Controller Design

The AEROSONDE model from Unmanned Dynamics LLC was used in the simulation. The AEROSONDE UAV is a small autonomous aircraft used in weather-reconnaissance and remote-sensing missions [32]. This model is part of the Aero Sim Blockset implemented in MATLAB/Simulink® [30]. In addition to the basic dynamic blocks, complete aircraft models are present which can be configured as required. The library also includes Earth models (geoid references, gravity and magnetic fields) and atmospheric models. The inputs to the AEROSONDE model include control surface deflections in radians,

Configuration	1-σ Pitch Error (°)	1-σ Roll Error (°)	1-σ Yaw Error (°)
3 Antennae	1.37	0.93	1.77
4 Antennae	0.47	0.32	0.76
5 Antennae	0.38	0.52	0.54
6 Antennae	0.32	0.45	0.36
7 Antennae	0.29	0.34	0.31
8 Antennae	0.27	0.23	0.22

Table 1: GNSS attitude determination errors.

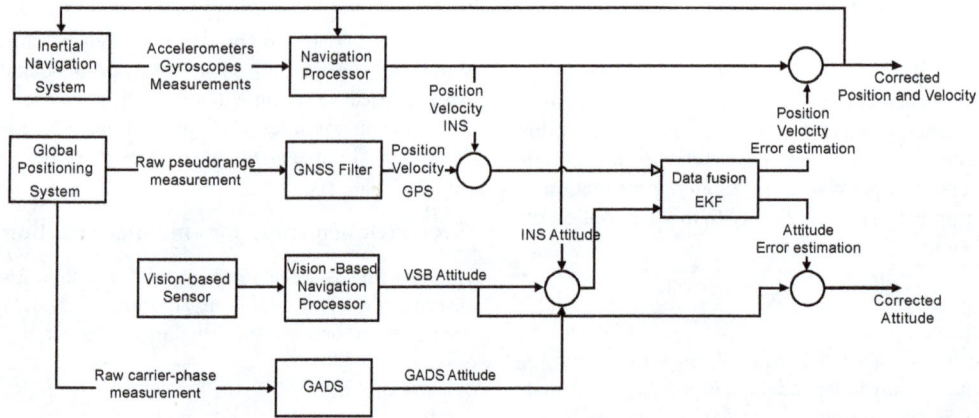

Figure 3: GNSS Attitude Algorithm Integrated to VIG Navigation System.

Antennae	1	2	3	4	5
1		100	180	120	200
2	100		100	100	140
3	180	100		100	100
4	120	100	100		130
5	200	140	100	130	

Table 2: Baseline Length (cm) of Antennae in AEROSONDE UAV.

throttle input, mixture and ignition. Wind disturbances can be added to the model to simulate variable atmospheric conditions. The model outputs the various aircraft states such as the position in the Earth-fixed frame, attitude and attitude rates. In order to perform the GNSS attitude determination for the AEROSONDE, 5 GNSS antennae were selected to optimize the length of the baselines. The baseline lengths are defined in Table 2.

The position of the antennae on the AEROSONDE is shown in Figure 4. For the design of the control system, an hybrid approach was adopted allowing the controller to take advantage of the VIG/VIG/GAD integrated navigation sensors during the other phases of flight. To achieve this, fuzzy logic and PID control strategies were adopted for controlling the UAV. PID is the simplest type of linear controller and is used in most UAV control systems. It is easy to implement and is effective for simple systems. On the other hand, fuzzy logic is a form of multi-value logic based on a representation of knowledge and reasoning of a human operator. In contrast to conventional PID controllers, Fuzzy Logic Controllers (FLC) do not require a model of the system. Therefore, it can be applied to non-linear systems or various ill-defined processes for which it is difficult to model the dynamics. The process consists of four components: fuzzification, fuzzy rule base, inference engine and defuzzification. Fuzzification refers to transforming a crisp set into a fuzzy set using linguistic terms. A fuzzy set is a set without crisp, clearly defined boundary. It can contain elements with only a partial degree of membership. A Membership Function (MF) is defined as a curve that classifies how each point in the input space is mapped to a membership value (or a degree of membership) between 0 and 1. Different types of fuzzy logic membership function exist which include s-function, π-function, z-function, triangular function, trapezoidal function, flat π function rectangle and singleton. An example of this is given in Figure 5a. Let 'input1' be a crisp set for the input to the system with fuzzy sets 'short', 'medium' and 'long'. Triangular membership functions are used in this case. It is observed that for 'medium', the

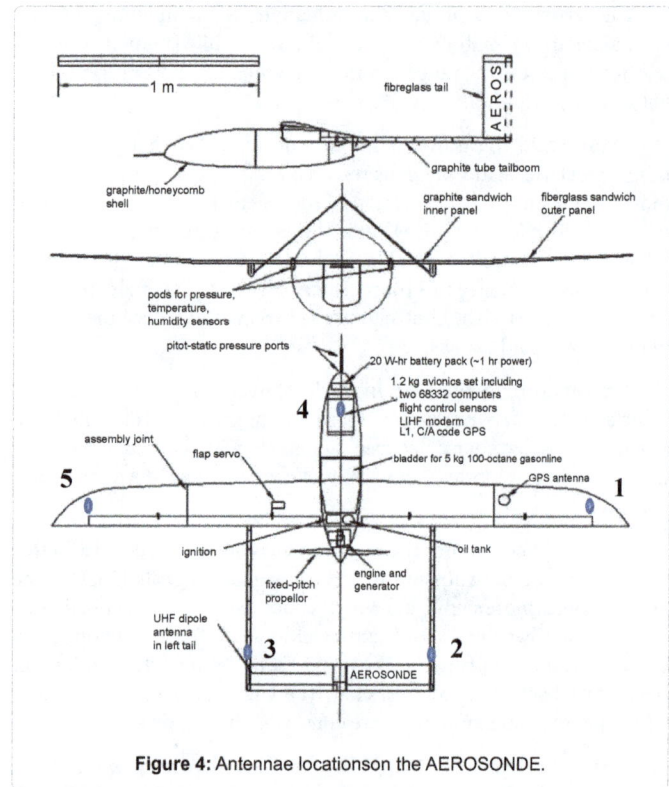

Figure 4: Antennae locationson the AEROSONDE.

value 5 has a membership function of 1. The value 3 has a membership function 0.3. Therefore it can be inferred that 3 has a lesser belonging to the fuzzy set 'medium' than 5. Similarly an output function 'output1' is defined with fuzzy sets 'left', 'centre' and 'right' as shown in Figure 5b.

The second component, that is the Fuzzy Rule base, forms the main part of fuzzy logic. It is based on if-then rules that tell the controller how to react to the inputs. The inference engine applies the fuzzy rule base to the inputs and output. It calculates the output required from the rules and passes this to defuzzification. Defuzzification is the method to obtain the output from the controller. It converts the output fuzzy set value to a crisp set using its membership functions. The UAV controller design was approached by decoupled the dynamic models of the aircraft. This resulted in two complimentary controllers, one for

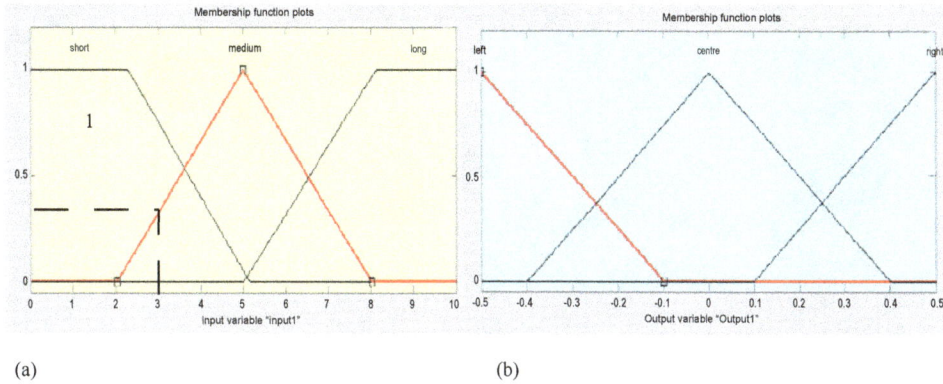

(a)   (b)

Figure 5: Input and output fuzzy sets and their membership functions.

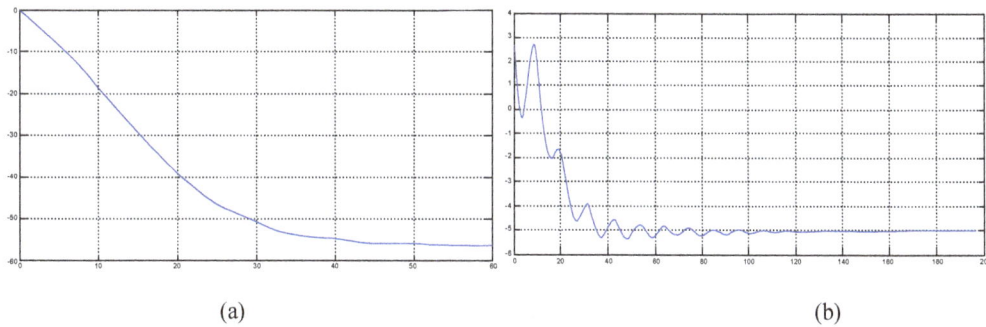

(a)   (b)

Figure 6: Roll and Pitch angle open-loop response (spiral mode).

lateral motion and one for longitudinal motion. Before initiating the controller design, the open-loop response of the system was first tested. In open-loop flight, the control inputs were set to a fixed value without any feedback from the aircraft states. It is observed that the UAV is unstable in this condition and settles in a constant bank turn and pitch angle as shown in Figure 6a and Figure 6b. This is due to the propulsion system which causes an unbalanced roll moment and excites the spiral mode.

The lateral controller was first designed to stabilize the lateral dynamics of the UAV. This was followed by the longitudinal controller to control the pitch angle. The overall design was then adapted to perform servoing using the information from the VBN sensors and integrated VIG/VIG/GAD navigation systems. The lateral and longitudinal controllers were implemented on MATLAB using the Fuzzy Logic Toolbox. The Mamdani fuzzy inference system (FIS) from the toolbox was used to create the membership functions. Based on the input and output membership functions, the fuzzy rules were developed that relate the inputs and the output. The membership functions and the rules were modified by trial and error to obtain better responses. Triangular and trapezoidal membership functions were used for the membership functions due to their simplicity and ease of implementation. A rough estimate of the membership functions was used for all the variables which were then modified as required. The membership functions which gave the best results for the roll and pitch responses were selected. Linguistic variables were used to define the fuzzy sets of inputs and the outputs of the controller. The fuzzy sets and the range of the inputs and outputs are shown in Table 3, where

Input Variable	Fuzzy Set	Range
Roll Error (in)	VN, SN, Z, SP, VP	-180° to 180°
Roll Rate (in)	VN, SN, Z, SP, VP	-40°/s to 40°/s
Pitch Error (in)	VL, SL, Z, SH, VH	-90° to 90°
Deviation (in)	VN, SN, Z, SP, VP	-512 pixels to 512 pixels
Deviation Rate (in)	VN, SN, Z, SP, VP	-600 pixels/s to 600 pixels/s
Aileron Deflection (out)	VN, SN, Z, SP, VP	-60° to 60°
Elevator Deflection (out)	VN, SN, Z, SP, VP	-60° to 60°
Required Roll to correct Deviation (out)	VN, SN, Z, SP, VP	-60° to 60°

Table 3: Fuzzy sets and range of inputs and outputs.

VN=Very Negative, VP=Very Positive, VH=Very High, VL=Very Low, SN=Slightly Negative, SP=Slightly Positive, SH=Slightly High, SL=Slightly Low, Z=Zero.

The lateral controller design was designed with the aim of stabilising the roll of the aircraft during the landing phase. This was required to maintain zero roll during touchdown at the centre of the runway so as to avoid wing-strike on the runway. It also controlled the position of the aircraft with respect to the centreline of the runway. Inputs to the controller were the Roll Error, Roll Rate, Deviation and the Deviation Rate and the output was the Aileron Deflection in degrees. The difference between the current roll angles given by the AEROSONDE model with the required value was used to represent the Roll Error. A gain of $(\pi/180)$ was applied to the Aileron Deflection to convert it into radians. The flap and elevator deflection were set to zero while the throttle was set to full (one). The mixture, ignition and wind were kept

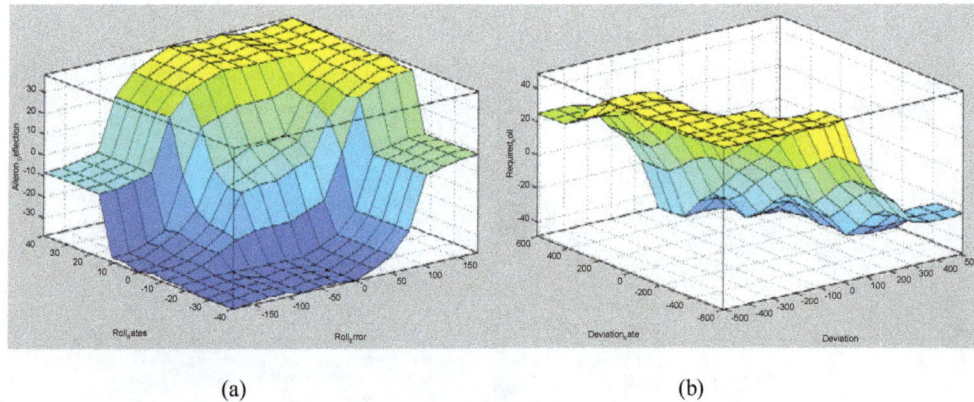

(a)   (b)

Figure 7: Fuzzy rules for roll (a) and deviation control (b).

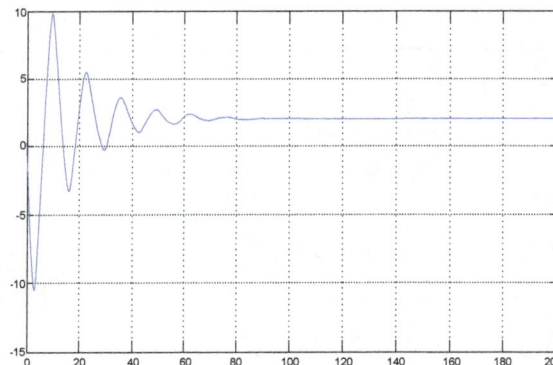

Figure 8: Partially stabilised pitch response due to lateral controller.

at their default settings. The system was simulated for 200 iterations on Simulink with a required roll of 0°. Various membership functions of the Roll Error and Aileron Deflections were considered in order to identify the most optimal FLC for stabilization. The simulation was then repeated with a required roll of 15°. The fuzzy rules used are as follows:

- If (Roll is Z) then (Aileron_Deflection is Z)

- If (Roll is SP) then (Aileron_Deflection is SP)

- If (Roll is SN) then (Aileron_Deflection is SN)

- If (Roll is VN) then (Aileron_Deflection is VN)

- If (Roll is VP) then (Aileron_Deflection is VP)

The Roll Rate was added to the controller so as to give it a higher degree of control. The membership functions for the Roll Rates were developed using the same methodology used for Roll Error and Aileron Deflection. 25 fuzzy rules were developed for the FLC and their surface representation is given in Figure 7a. A steady-state error and overshoot were observed from the roll response of the aircraft. Therefore, a PID controller was desgined to eliminate these errors. PID tuning was carried out to find the values for the gains which gave the optimal roll response. The deviation from the centerline of the runway was controlled using the roll of the aircraft. The value of the Deviation and Deviation Rate was used by the controller to calculate the Required Roll. A surface representation of the fuzzy rules is given in Figure 7b.

The longitudinal controller was used to stabilize and control the Pitch of the aircraft using Elevator Deflections. Prior to design, it was observed that the pitch angle was stabilized to some extent due to the lateral controller as shown in Figure 8

The design process of the longitudinal controller followed the same methodology as that of the lateral controller. The FLC was first designed using trial-and-error for the membership functions of Pitch Error and Elevator Deflections followed by the PID controller. A derivative gain was used instead of pitch rates. The fuzzy rules used for the longitudinal controller are given below:

- If (Pitch is Z) then (Elevator_Deflection is Z)

- If (Pitch is SH) then (Elevator_Deflection is SP)

- If (Pitch is SL) then (Elevator_Deflection is SN)

- If (Pitch is VH) then (Elevator_Deflection is VN)

- If (Pitch is VL) then (Elevator_Deflection is VP)

The pitch and roll responses of the controller are shown in Figure 9.

The results show that the pitch and roll converge rapidly towards the required value of zero after a short initial instability. Comparing these results with the uncontrolled response in Figure 6, we can confirm that the controller gives satisfactory results. The simulation showed that the controller is able to correct the attitude disturbances caused by moderate to high wind speeds. However, it was observed that

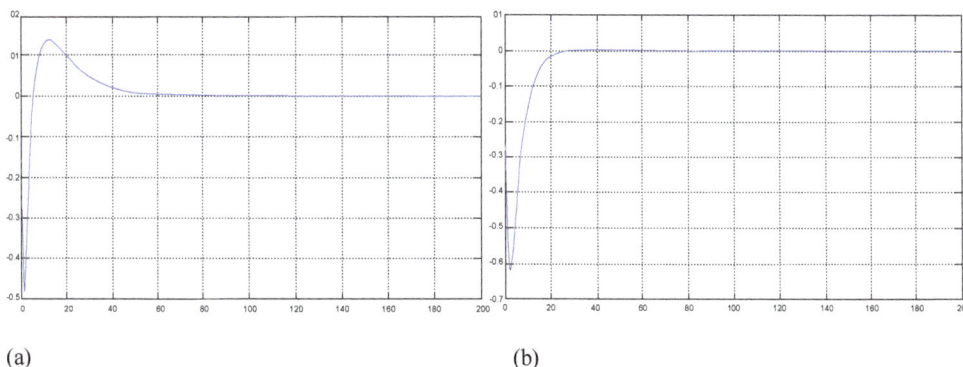

(a)   (b)

Figure 9: Pitch (a) and Roll (b) response with controller.

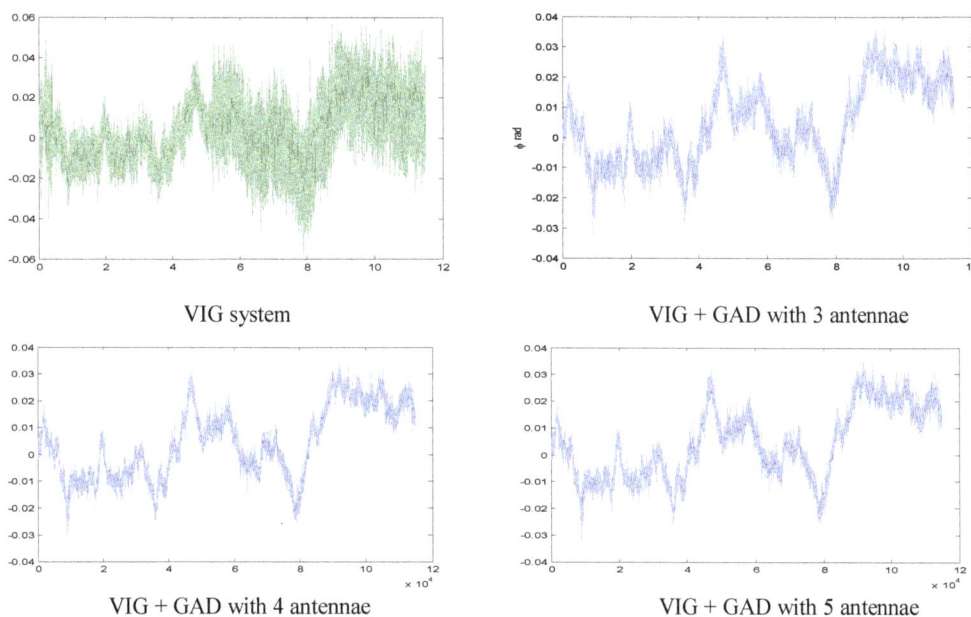

VIG system   VIG + GAD with 3 antennae

VIG + GAD with 4 antennae   VIG + GAD with 5 antennae

Figure 10: Roll (φ) error time histories.

Phases of Flight	VIG		VIG/GAD 3 Antennae		VIG/GAD 4 Antennae		VIG/GAD 5 Antennae	
	Mean	σ	Mean	σ	Mean	σ	Mean	σ
Straight Climb	2.13E-01	3.04E-01	2.23E-01	1.66E-01	2.19E-01	1.24E-01	2.20E-01	1.35E-01
Right Turn Climb	5.47E-01	3.41E-01	5.55E-01	1.88E-01	5.56E-01	1.85E-01	5.55E-01	1.85E-01
Straight and Level	2.32E-01	3.73E-01	2.53E-01	2.01E-01	2.52E-01	1.49E-01	2.52E-01	1.63E-01
Level Left Turn	1.12E-01	2.04E-01	1.27E-01	1.61E-01	1.19E-01	1.34E-01	1.21E-01	1.39E-01
Straight Descent	1.07E-01	2.57E-01	9.03E-02	2.05E-01	9.57E-02	1.78E-01	9.32E-02	1.83E-01
Level Right Turn	-8.86E-01	2.81E-01	-9.18E-01	2.69E-01	-9.23E-01	2.42E-01	-9.21E-01	2.48E-01
Left Turn Descent	-5.71E-01	1.98E-01	-6.12E-01	1.48E-01	-6.11E-01	1.33E-01	-6.11E-01	1.34E-01

Table 4: Roll (φ) error statistics (degrees).

the aircraft became unstable with lateral wind speeds exceeding 20 m/s.

VIG and VIG/GAD Simulation

In order to evaluate the performance of the integrated VIG/GAD system in conjunction with the Fuzzy/PID controller, a simulation was carried out using the AEROSONDE UAV platform. A suitable flight profile was defined including a number of representative flight manoeuvres [15]. The duration of the simulation is 1150 seconds.

Figure 10 shows a graphical comparison of the φ (roll) error obtained with the VIG and the VIG/GAD systems. It is observed that the VIG/GAD system, with 3, 4 and 5 antennae provides a significant improvement over the VIG system. Table 4 provides the roll error mean and standard deviation values. The performance achieved with 4 and 5 antennae is similar

Figure 11 presents a similar comparison for the θ (pitch) angle. There is a significant improvement with the GAD integration. In this

VIG system

VIG + GAD with 3 antennae

VIG + GAD with 4 antennae

VIG + GAD with 5 antennae

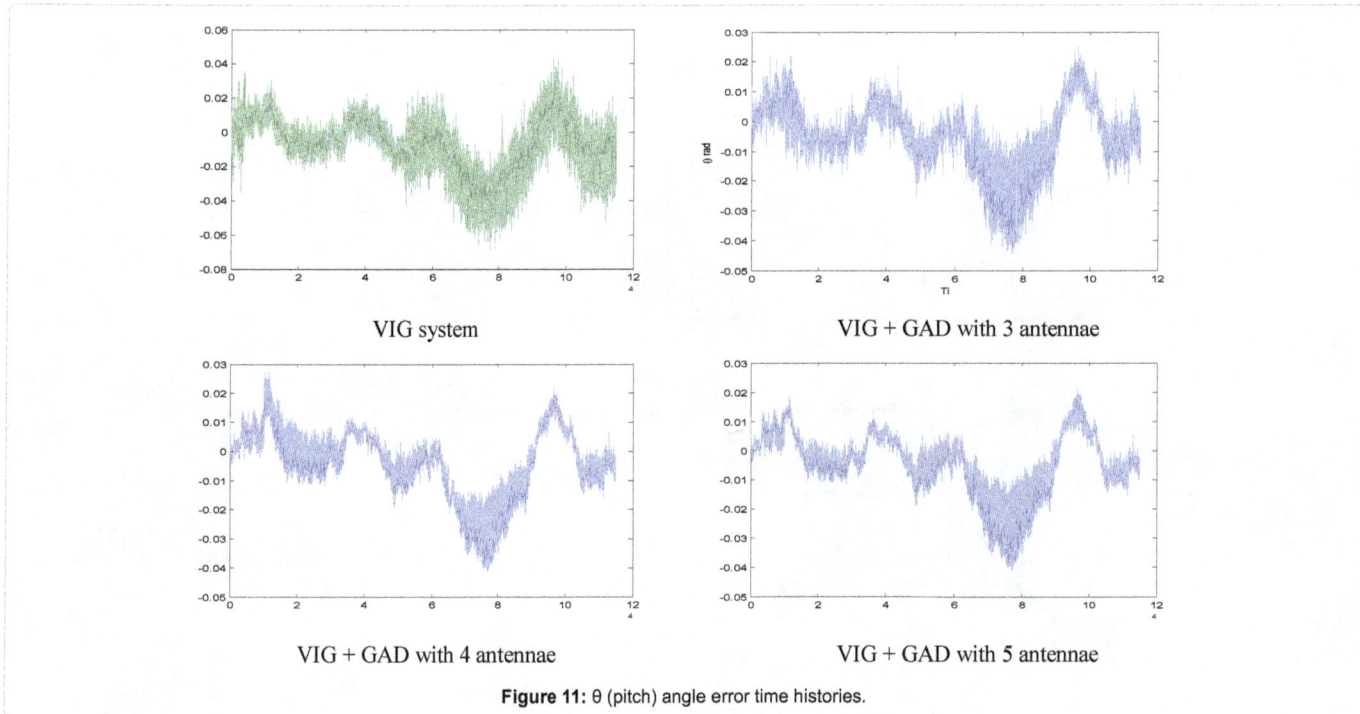

Figure 11: θ (pitch) angle error time histories.

Phases of Flight	VIG		VIG/GAD 3 Antennae		VIG/GAD 4 Antennae		VIG/GAD 5 Antennae	
	Mean	**σ**	**Mean**	**σ**	**Mean**	**σ**	**Mean**	**σ**
Straight Climb	-6.17E-02	2.25E-01	-1.27E-02	1.81E-01	-2.81E-02	1.09E-01	-2.32E-02	1.07E-01
Right Turn Climb	1.45E-01	2.23E-01	1.28E-01	1.24E-01	1.22E-01	7.84E-02	1.27E-01	7.34E-02
Straight and Level	3.15E-01	3.67E-01	2.89E-01	2.78E-01	2.85E-01	2.41E-01	2.89E-01	2.40E-01
Level Left Turn	4.74E-01	1.27E-01	4.21E-01	1.86E-01	4.06E-01	1.02E-01	4.06E-01	9.67E-02
Straight Descent	4.17E-01	1.55E-01	3.44E-01	2.21E-01	3.47E-01	1.21E-01	3.50E-01	1.12E-01
Level Right Turn	4.26E-01	1.43E-01	3.73E-01	2.16E-01	3.60E-01	1.19E-01	3.63E-01	1.09E-01
Left Turn Descent	6.48E-01	1.40E-01	5.03E-01	2.57E-01	6.62E-01	1.89E-01	5.96E-01	1.15E-01

Table 5: Pitch (θ) error statistics (degrees).

Phases of Flight	VIG		VIG/GAD 3 Antennae		VIG/GAD 4 Antennae		VIG/GAD 5 Antennae	
	Mean	**Σ**	**Mean**	**σ**	**Mean**	**σ**	**Mean**	**σ**
Straight Climb	-7.63E-01	2.21E-01	-1.01	2.17E-01	-8.35E-01	2.16E-01	-8.52E-01	2.05E-01
Right Turn Climb	1.08	4.24E-01	1.15	3.79E-01	1.14	3.71E-01	1.14	3.71E-01
Straight and Level	4.74E-01	3.67E-01	5.40E-01	3.93E-01	5.40E-01	3.07E-01	5.40E-01	2.93E-01
Level Left Turn	2.35E-01	2.87E-01	2.94E-01	3.06E-01	2.79E-01	2.60E-01	2.76E-01	2.58E-01
Straight Descent	2.26E-01	3.79E-01	2.09E-01	3.94E-01	2.18E-01	3.46E-01	2.20E-01	3.42E-01
Level Right Turn	-1.74	5.74E-01	-1.84	5.40E-01	-1.85	4.95E-01	-8.18E-01	4.90E-01
Left Turn Descent	-1.07	3.95E-01	-1.22	3.32E-01	-1.21	3.15E-01	-1.21	3.18E-01

Table 6: Yaw (ψ)error statistics (degrees).

case it is also observed that the error decreases significantly when the number of antennae is increased. Table 5 confirms such improvement by showing the values of means and standard deviation for different phases of flight.

Finally in Figure 12 a similar behaviour is observed for the yaw error. The tendency to improvement versus the VIG system is observed for all phases of flight. Table 6 provides the mean and standard deviation values.

Conclusions

In this paper we have investigated the potential of GAD systems for integration in small size UAVs. Processing algorithms have been proposed, which allow a fast and reliable computation of the vehicle attitude data. A recursive algorithm has been proposed for combining multiple attitude measurements obtained from different antenna locations, and its efficiency has been analysed in various dynamic conditions using the AEROSONDE UAV platform as a representative test case. Modelling and simulation activities also considered the possible augmentation provided by GAD to a low-cost and low-weight/ volume VIG integrated navigation system employing a VBN, MEMS-IMU and code-range GNSS (i.e., GPS and GALILEO) for position and velocity computations. Integration of the GAD with the VIG system

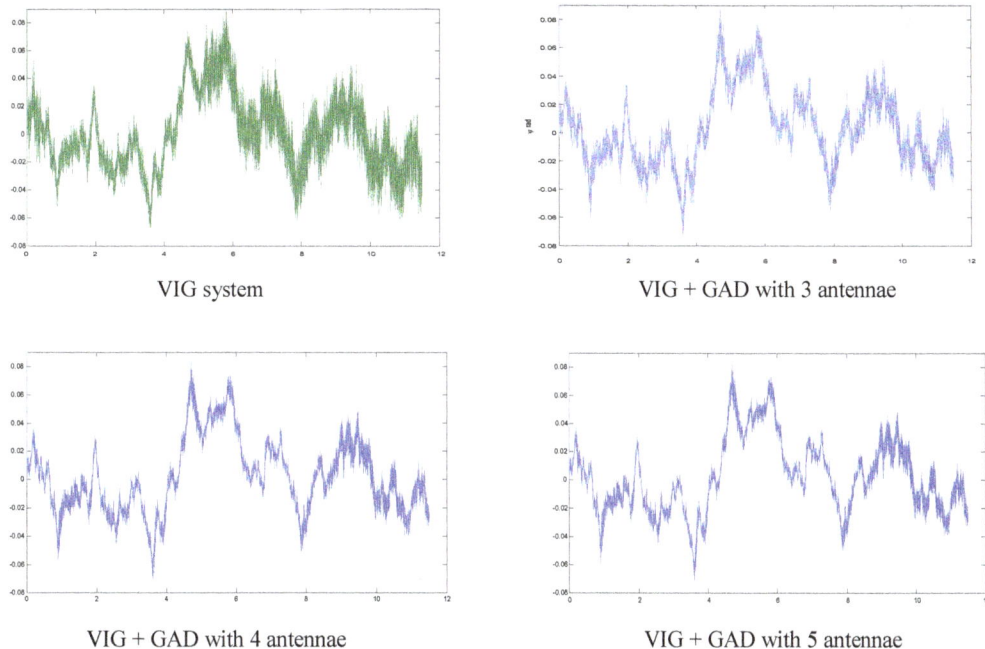

VIG system

VIG + GAD with 3 antennae

VIG + GAD with 4 antennae

VIG + GAD with 5 antennae

Figure 12: Yaw (ψ)error time histories.

using an EKF was accomplished. Considering the AEROSONDEUAV and a number of possible GNSS antenna network configurations, it was demonstrated that, in a variety of dynamics conditions, the accuracy of the VIG/GAD attitude solution was comparable to the accuracy obtainable with traditional inertial sensors. However, the accuracy could be significantly influenced by the chosen antenna network geometry and the number of antennae available. Compared to the VIG system, the VIG/GAD shows an improvement of the accuracy in all three attitude angles. The magnitude of this improvement varies for each angle and for different flight phases. As expected, as the number of antennae increases, also the accuracy improves. The design of the Fuzzy/PID controller was successfully accomplished. However, during the test activities, it was observed that the Fuzzy/PID controller becomes unstable at wind speeds greater than 20 m/s. In case of pure visual servoing during the approach and landing phase, this would lead to the impossibility of tracking the desired features from the surrounding. Current research activities at Cranfield University are investigating the potential of low-cost GNSS attitude sensors (two or more antennae) in various classes of UAVs and Unmanned Space Vehicles (USVs). Additionally, multipath and shielding problems are being carefully modelled and adequate algorithms are being developed in order to cope with these effects during high dynamics manoeuvres.

References

1. Gebre-Egziabher D, Hayward R, Powell J (1998) A Low-Cost GPS/Applications. UCGE Reports Number 20183, University of Calgary, Alberta, Canada.

2. Godha S (2006) Performance Evaluation of Low Cost MEMS-Based IMU Integrated With GPS for Land Vehicle Navigation Application. UCGE Report Number 20239, University of Calgary, Alberta, Canada.

3. Molina P, Wis M, Pares ME, Blazquez M, Tatjer JC, et al. (2008) New Approaches to IMU Modeling and INS/GPS Integration for UAV-Based Earth-Observation. Proceedings of the 21st International Technical Meeting of the Satellite Division of The Institute of Navigation Savannah, GA USA 1335-1344.

4. Clark CE, Bradford PW, David MB (1994) Flight Tests of Attitude Determination Using GPS Compared Against an Inertial Navigation Unit. Navigation 41: 83-98.

5. Frank VG, Michael B (1991) GPS Interferometric Attitude and Heading Determination: Initial Flight Test Results. Navigation 38: 297-316.

6. Brown AK, Thorvaldsen TP, Bowles WM (1983) Interferometric attitude determination using the global positioning system-A new gyrotheodolite. Proceedings of 3rd International Geodetic Symposium on Satellite Doppler Positioning Las Cruces, NM, USA.

7. Giorgi G, Teunissen PJG, Gourlay TP (2012) Instantaneous Global Navigation Satellite System (GNSS)-Based Attitude Determination for Maritime Applications. IEEE J Oceanic Eng 37: 348-362.

8. Pinchin JT (2011) GNSS Based Attitude Determination for Small Unmanned Aerial Vehicles. University of Canterbury, New Zealand.

9. Han KJ, Gerard L (1999) Determining heading and pitch using a single difference GPS/GLONASS approach. University of Calgary, Canada.

10. Giorgi G, Teunissen PJG (2010) Carrier phase GNSS attitude determination with the Multivariate Constrained LAMBDA method. IEEE Aerospace Conference 6-13 March Big Sky, MT USA.

11. Teunissen PJG (1995) The least-squares ambiguity decorrelation adjustment: a method for fast GPS integer ambiguity estimation. J Geodesy 70: 65-82.

12. Sabatini R, Kaharkar A, Shaid T, Bartel C, Jia H, et al. (2012) Vision-based Sensors and Integrated Systems for Unmanned Aerial Vehicles Navigation and Guidance. Proceedings of the SPIE Conference Photonics Europe 2012 Brussels Belgium.

13. Sabatini R, Kaharkar A, Shaid T, Bartel C, Jia H, et al. (2012) Low-cost Vision Sensors and Multisensor Systems for Small to Medium Size UAV Navigation and Guidance. Proceedings of the European Navigation Conference 2012 Gdansk Poland.

14. Cohen CE (1992) Attitude Determination Using GPS. Stanford University, USA.

15. Saripalli S, Montgomery JF, Sukhatme G (2002) Vision-based autonomous landing of an unmanned aerial vehicle. Proceedings of IEEE International Conference on Robotics and Automation 11-15 May Washington, DC USA.

16. Gong X, Fleming G (2006) A Survey of Techniques for Detection and Tracking of Airport Runways. 44th AIAA Aerospace Sciences Meeting and Exhibit.

17. Chen Z, Birchfield ST (2009) Qualitative Vision-Based Path Following. IEEE Transactions on Robotics 25: 749-754.

18. Nixon M, Aguado AS (2008) Feature Extraction & Image Processing, Feature Extraction and Image Processing Series. Academic Press, USA.

19. Sangyam T, Laohapiengsak P, Chongcharoen W, Nilkhamhang I (2010) Path tracking of UAV using self-tuning PID controller based on fuzzy logic. Proceedings of SICE Annual Conference 18-21 August 1265-1269.

20. Parkinson B (1996) GPS error analysis," Global Positioning System: Theory and applications.

21. Hopfield HS (1971) Tropospheric Effect on Electromagnetically Measured Range: Prediction from Surface Weather Data. Radio Sci 6: 357-367.

22. McGraw GA (2012) Tropospheric error modeling for high integrity airborne GNSS navigation. IEEE/ION Position Location and Navigation Symposium (PLANS) 23-26 April Myrtle Beach, SC, USA.

23. Kleijer F (2004) Troposphere Modeling and Filtering for Precise GPS Leveling. NCG, Nederlandse Commissie voor Geodesie, Netherlands Geodetic Commission, Delft, Netherlands.

24. Lau L, Cross P (2006) A New Signal-to-Noise-Ratio Based Stochastic Model for GNSS High-Precision Carrier Phase Data Processing Algorithms in the Presence of Multipath Errors. Proceedings of the 19th International Technical Meeting of the Satellite Division of The Institute of Navigation, Fort Worth, TX, USA.

25. Brunner FK, Hartinger H, Troyer L (1999) GPS signal diffraction modelling: the stochastic SIGMA-δ model. J Geodesy 73: 259-267.

26. Wieser A, Gaggl M, Hartinger H (2005) Improved Positioning Accuracy with High-Sensitivity GNSS Receivers and SNR Aided Integrity Monitoring of Pseudo-Range Observations. Proceedings of the 18th International Technical Meeting of the Satellite Division of The Institute of Navigation September 13-16 Long Beach, CA, USA.

27. Misra P, Enge P (2006) Global positioning system: signals, measurements, and performance. Ganga-Jamuna Press, India.

28. Park C, Kim I (2000) An Error Analysis of 2-Dimensional Attitude Determination System Using Global Positioning System. IEICE T Commun E83-B: 1370-1373.

29. Park C, Cho DJ, Cha EJ, Hwang DH, Lee SJ (2006) Error Analysis of 3-Dimensional GPS Attitude Determination System. International Journal of Control, Automation, and Systems 4: 480-485.

30. AeroSim aeronautical simulation blockset, Version 1.01, User's Guide.

31. Maurer J (2002) Polar remote sensing using an unpiloted aerial vehicle (UAV).

32. Sabatini R, Rodríguez L, Kaharkar A, Bartel C, Shaid T (2012) GNSS Data Processing for Attitude Determination and Control of Unmanned Aerial Vehicles. Proceedings of the European Navigation Conference 2012 Gdansk Poland.

6

The Buckling Response of Lattice Fuselage Structures: Validation of Finite Element Models by Using Smeared Unit Cell Analytical Methodology

Kostopoulos V*, Kotzakolios T and Vlachos DE

Applied Mechanics Laboratory, Department of Mechanical Engineering and Aeronautics, University of Patras, Patras University Campus, GR-26500 Patras, Greece

Abstract

Composite lattice structures are shells that are reinforced by unidirectional helical and hoop ribs. Their main advantage over contemporary composite structures is their superior stiffness to mass ratio. However, their application in industry is still limited. In this paper, the lattice structure concept was applied for the case of a small business aircraft. Emphasis here is given at the initial stages of the design. More specifically, the buckling modes for bending loads were calculated by utilizing a continum unit cell model which was correlated with finite element models for a cylindrical small fuselage structure and two scaled down versions.

Introduction

Composite lattice strictures are composed of helical and hoop ribs that reinforce a shell. They are characterized by high specific strength and stiffness, making them promising for lightweight applications. The history and development of this kind of structures can be found in literature, mostly published by Vasiliev et al. [1-5].

Lattice structures are using reinforcing ribs in a regular pattern, which allows to be analyzed by smearing the ribs over the skin surface. Thus, the lattice structure can be analyzed as a continuous layer with calculated effective stiffness. Stress and strain equations that are based on theory of orthotropic shells can be used. Such continuum models are published by Vasiliev [6] and Vasiliev and Morozov [7]. Strength and buckling analysis of cylindrical lattice shells based on different continuum models are described in articles from Slinchenko and Verijenko [8], Totaro and Gurdal [9], Buragohain and Velmurugan [10], Paschero and Hyer [11], Totaro [12,13], and Zheng et al. [14].

Additionally, composite lattice structures have been analyzed by using finite element analysis. Results from FE models can be found in articles published by Hou et al., Zhang et al., Frulloni et al., Fan et al., Morozov et al. and Azarov et al. [15-20].

However, in most applications, both methodologies are used, where the basic specifications of the structure are calculated by employing the continuum models. The parameters are then further refined by using the finite element models. Correlation between these methodologies has been done successfully by Azarov [20].

Despite the aforementioned developments in designing composite lattice structures, their application is still limited in some spacecraft applications [3,21].

In this work, the case of a fuselage section of a small business aircraft is examined. The fuselage section has an outer diameter of 1.8 m and a length of 4 m. The fuselage section was assumed to be subjected to buckling loads and its buckling limit was calculated by a smeared unit cell method and finite element analysis. Furthermore, a scaling down approach was followed for estimating the buckling loads for scaled down prototypes, namely 0.5 and 1 m in diameter. The scaling down approach is useful since many times full scale testing is prohibited for this kind of structures since they are integrated with no assembly required, making testing of single subcomponents not feasible. The correlation between the smeared unit cell theory and finite element analysis was very good.

Unit Cell Model Development

In developing the analytical model, a unit cell of the stiffener structure has to be defined first. The unit cell is chosen such that the whole grid structure can be reproduced by repetition of this unit cell.

The equivalent stiffness parameters of this unit are determined and then applied to the whole cylinder panel. Validation comes from the generation of the panel by repetition of the cell.

Assumptions

1. The transverse modulus of the unidirectional stiffeners is much lower than the longitudinal modulus, and cross sectional dimensions are also very small compared to the length dimension, therefore the stiffeners are assumed to support axial loads only.

2. The strain is uniform across the cross sectional area of the stiffeners. Hence a uniform stress distribution is assumed.

3. Load is transferred through shear forces between the stiffeners and the shell.

Strains used as the matching condition of the stiffener and the shell (inner surface of the shell as interface) is given by:

$$\varepsilon_x = \varepsilon_x^o + \kappa_x \frac{t}{2} \tag{1}$$

$$\varepsilon_\theta = \varepsilon_\theta^o + \kappa_\theta \frac{t}{2} \tag{2}$$

$$\varepsilon_{x\theta} = \varepsilon_{x\theta}^o + \kappa_{x\theta} \frac{t}{2} \tag{3}$$

Strain relationship along the stiffeners' directions is given by:

*Corresponding author: Kostopoulos V, Applied Mechanics Laboratory, Department of Mechanical Engineering and Aeronautics, University of Patras, Patras University Campus, GR-26500 Patras, Greece
E-mail: kostopoulos@mech.upatras.gr

$$\begin{bmatrix} \varepsilon_l \\ \varepsilon_t \\ \varepsilon_{lt} \end{bmatrix} = \begin{bmatrix} c^2 & s^2 & sc \\ s^2 & c^2 & -sc \\ -2sc & 2sc & c^2-s^2 \end{bmatrix} \begin{bmatrix} \varepsilon_x \\ \varepsilon_\theta \\ \varepsilon_{x\theta} \end{bmatrix} \qquad (4)$$

$$\overset{Assumption\ I}{\Rightarrow} \varepsilon_l$$

$$= c^2\varepsilon_x + s^2\varepsilon_\theta + sc\varepsilon_{x\theta}$$

Where c=cos(ϕ), s=sin(ϕ) and ϕ is the stiffener orientation angle from vertical direction.

In this specific case of study, $\phi\pm26°$ was selected for the helical ribs and $\phi=90°$ was selected for the hoop ribs (Figure 1).

Resultant Forces of the force diagram is as shown below (Figure 2):

$$F_1 = AE_{l1}\varepsilon_{l1} = AE_l\left(c^2\varepsilon_x + s^2\varepsilon_\theta - sc\varepsilon_{x\theta}\right) \qquad (5)$$

$$F_2 = AE_{l2}\varepsilon_{l2} = AE_l\left(c^2\varepsilon_x + s^2\varepsilon_\theta - sc\varepsilon_{x\theta}\right) \qquad (6)$$

$$F_3 = AE_{l3}\varepsilon_{l3} = AE_l\left(\varepsilon_\theta\right) \qquad (7)$$

Summing up the forces of each opposite sides in both directions, F$_x$ and F$_\theta$ are obtained:

$$F_x = F1\cos(\varphi) + F2\cos(\varphi) \qquad (8)$$

$$F_\theta = F1\sin(\varphi) + F2\sin(\varphi) + 2F3 \qquad (9)$$

$$F_{x\theta} = F_{\theta x} = F2\cos(\varphi) - F1\cos(\varphi) \qquad (10)$$

Resultant Forces on the unit cell is given by:

$$N_x = \frac{AE_l}{a}\left[2c^3\varepsilon_x^o + 2c^3\kappa_x\left(\frac{t}{2}\right) + 2s^2c\varepsilon_\theta^o + 2s^2c\kappa_\theta\left(\frac{t}{2}\right)\right] \qquad (11)$$

$$N_\theta = \frac{AE_l}{a}\left[2sc^2\varepsilon_x^o + 2sc^2\kappa_x\left(\frac{t}{2}\right) + \left(2s^3+2\right)\varepsilon_\theta^o + \left(2s^3+2\right)\kappa_\theta\left(\frac{t}{2}\right)\right] \qquad (12)$$

$$N_{\theta x} = \frac{AE_l}{a}\left[2sc^2\varepsilon_x^o + 2sc^2\kappa_\theta\left(\frac{t}{2}\right)\right] \qquad (13)$$

Following the same procedure as the force analysis on the unit cell,

resultant moments are computed with respect to the moment diagram (Figure 3):

$$M_x = M1\cos(\varphi) + M2\cos(\varphi) \qquad (14)$$

$$M_\theta = M1\sin(\varphi) + M2\sin(\varphi) + 2M3 \qquad (15)$$

$$M_{x\theta} = M_{\theta x} = M2\cos(\varphi) - M1\sin(\varphi) \qquad (16)$$

Resultant Moments on the unit cell:

$$M_x = \frac{AE_l t}{2a}\left[2c^3\varepsilon_x^o + 2c^3\kappa_x\left(\frac{t}{2}\right) + 2s^2c\varepsilon_\theta^o + 2s^2c\kappa_\theta\left(\frac{t}{2}\right)\right] \qquad (17)$$

$$M_\theta = \frac{AE_l t}{2a}\left[2sc^2\varepsilon_x^o + 2sc^2\kappa_x\left(\frac{t}{2}\right) + \left(2s^3+2\right)\varepsilon_\theta^o + \left(2s^3+2\right)\kappa_\theta\left(\frac{t}{2}\right)\right] \qquad (18)$$

$$M_{\theta x} = \frac{AE_l t}{2a}\left[2sc^2\varepsilon_x^o + 2sc^2\kappa_\theta\left(\frac{t}{2}\right)\right] \qquad (19)$$

The functions of the mid plane strains of the shell then become:

$$\begin{bmatrix} N_x^S \\ N_\theta^S \\ N_{x\theta}^S \\ M_x^S \\ M_\theta^S \\ M_{x\theta}^S \end{bmatrix} = AE_l \begin{bmatrix} \dfrac{2c^3}{a} & \dfrac{2s^2c}{a} & 0 & \dfrac{c^3t}{a} & \dfrac{s^2ct}{a} & 0 \\[2mm] \dfrac{2sc^2}{b} & \dfrac{(2s^3+2)}{b} & 0 & \dfrac{sc^2t}{b} & \dfrac{(2s^3+2)t}{b} & 0 \\[2mm] 0 & 0 & \dfrac{2sc^2}{b} & 0 & 0 & \dfrac{sc^2t}{b} \\[2mm] \dfrac{c^3t}{a} & \dfrac{s^2ct}{a} & 0 & \dfrac{c^3t^2}{2a} & \dfrac{s^2ct^2}{2a} & 0 \\[2mm] \dfrac{sc^2t}{b} & \dfrac{(2s^3+2)t}{b} & 0 & \dfrac{sc^2t^2}{2b} & \dfrac{(2s^3+2)t^2}{4b} & 0 \\[2mm] 0 & 0 & \dfrac{sc^2t}{b} & 0 & 0 & \dfrac{sc^2t^2}{2b} \end{bmatrix} \begin{bmatrix} \varepsilon_x^o \\ \varepsilon_\theta^o \\ \varepsilon_{x\theta}^o \\ \kappa_x \\ \kappa_\theta \\ \kappa_{x\theta} \end{bmatrix} \qquad (20)$$

The superscript 's' denotes the force and moment contributions of the stiffener.

By applying the Rule of Mixtures, the forces and the moments are superimposed according to the volume fractions of the stiffeners (V_s) and the shell (V_{Sh}).

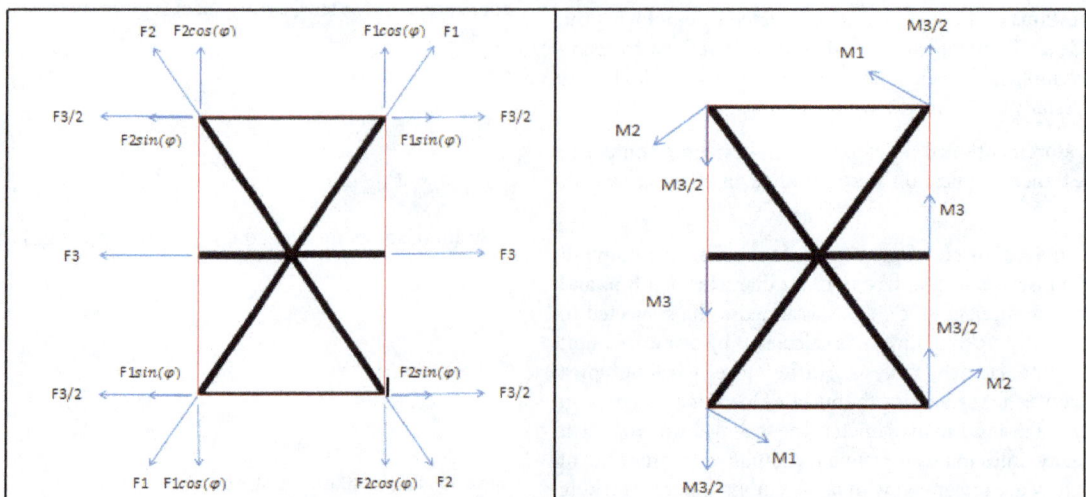

Figure 1: Forces and moments diagram for the lattice unit cell.

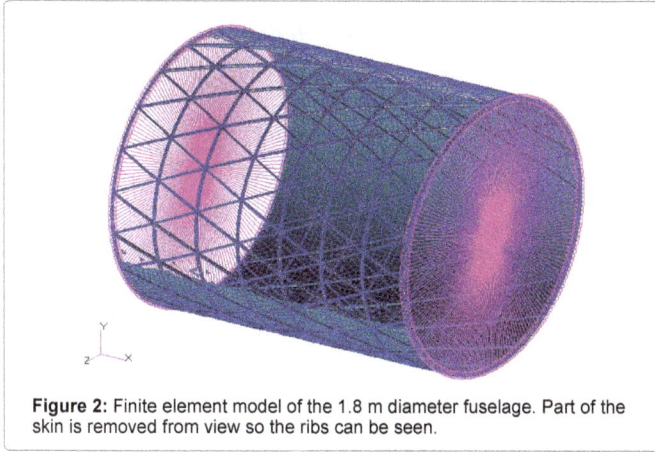

Figure 2: Finite element model of the 1.8 m diameter fuselage. Part of the skin is removed from view so the ribs can be seen.

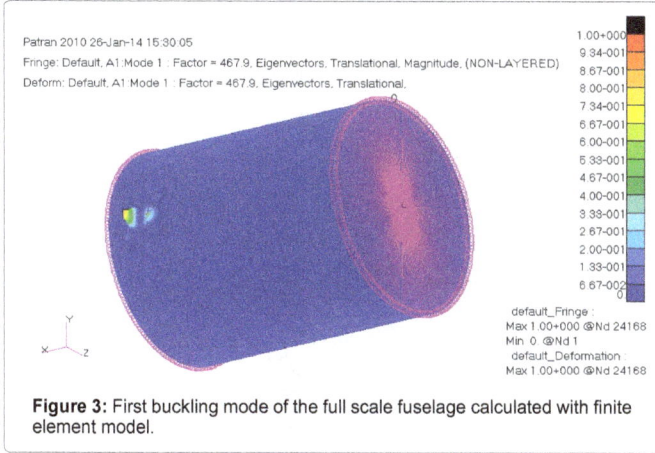

Figure 3: First buckling mode of the full scale fuselage calculated with finite element model.

$$\begin{bmatrix} N \\ M \end{bmatrix} = \begin{bmatrix} V_s A^s + V_{sh} A^{sh} & | & V_s B^s + V_{sh} B^{sh} \\ ----- & - & ----- \\ V_s B^s + V_{sh} B^{sh} & | & V_s D^s + V_{sh} D^{sh} \end{bmatrix} \begin{bmatrix} \varepsilon^0 \\ \kappa \end{bmatrix} \qquad (21)$$

The resultant stiffness parameters obtained from the above equation are the equivalent stiffness parameters of the whole panel.

For the calculation of the critical buckling load, the Ritz method (Przemieniecki, 1968) was used which is based on the minimization of the total energy Π of the cylinder. The total energy of the cylindrical structure is the sum of the strain energy U and the external force work V.

$$\Pi = U + V \qquad (22)$$

The strain energy of an orthotropic cylinder with length L is given by the following equation:

$$U = \frac{1}{2} \int_0^{2\pi r} \int_0^L \left\{ A_{11} \left(\frac{\partial u}{\partial x} \right)^2 + 2A_{12} \frac{\partial u}{\partial x} \left(\frac{\partial v}{\partial \theta} + \frac{w}{r} \right) + A_{22} \left[\frac{\partial v}{\partial \theta} \left(\frac{\partial v}{\partial \theta} + \frac{w}{r} \right) + \left(\frac{w}{r} \right)^2 \right] \right.$$

$$+ 2\left[A_{16} \frac{\partial u}{\partial \theta} + A_{26} \left(\frac{\partial v}{\partial \theta} + \frac{w}{r} \right) \right] \left(\frac{\partial u}{\partial \theta} + \frac{\partial v}{\partial x} \right) + A_{66} \left(\frac{\partial u}{\partial \theta} + \partial \frac{v}{\partial x} \right)^2 - B_{11} \frac{\partial u}{\partial x} \frac{\partial^2 w}{\partial x^2}$$

$$- 2B_{12} \left[\left(\frac{\partial v}{\partial \theta} + \frac{w}{r} \right) \frac{\partial^2 w}{\partial x^2} + \frac{\partial u}{\partial x} \frac{\partial^2 w}{\partial u^2} \right] - B_{22} \left(\frac{\partial v}{\partial \theta} + \frac{w}{r} \right) \frac{\partial^2 w}{\partial \theta^2}$$

$$- 2B_{16} \left[\frac{\partial^2 w}{\partial x^2} \left(\frac{\partial u}{\partial \theta} + \frac{\partial v}{\partial x} \right) + 2 \frac{\partial u}{\partial x} \frac{\partial^2 w}{\partial x \partial \theta} \right] - 2B_{26} \left[\frac{\partial^2 w}{\partial \theta^2} \left(\frac{\partial u}{\partial \theta} + \frac{\partial v}{\partial x} \right) + 2 \left(\frac{\partial v}{\partial \theta} + \frac{w}{r} \right) \frac{\partial^2 w}{\partial x \partial \theta} \right] \qquad (23)$$

$$- 4B_{66} \frac{\partial^2 w}{\partial x \partial \theta} \left(\frac{\partial u}{\partial x} + \frac{\partial v}{\partial \theta} \right) + D_{11} \left(\frac{\partial^2 w}{\partial x^2} \right)^2 + 2D_{12} \frac{\partial^2 w}{\partial x^2} \frac{\partial^2 w}{\partial \theta^2} + D_{22} \left(\frac{\partial^2 w}{\partial \theta^2} \right)^2$$

$$\left. + 4 \left(D_{16} \frac{\partial^2 w}{\partial x^2} + D_{26} \frac{\partial^2 w}{\partial \theta^2} \right) \frac{\partial^2 w}{\partial x \partial \theta} + 4D_{66} \left(\frac{\partial^2 w}{\partial x \partial \theta} \right)^2 \right\} dxd\theta$$

U is dependent from the total stiffness matrix of the unit cell, radius r of the cylinder and axial, circumferential and radial displacements u, v, w respectively.

The dynamic energy term V, due to the external work for length L is given by:

$$V = \frac{1}{2} \int_0^{2\pi r} \int_0^L N_\theta \left(\frac{\partial w}{\partial x} \right)^2 dxd\theta \qquad (24)$$

The displacement field u, v and w are defined by kinematically admissible functions and they are approximated by a double Fourier series with clamped boundary condition:

$$u = \sum_{m=1}^{\infty} \sum_{n=1}^{\infty} A_{mn} \cos(\bar{m}x) \sin(\bar{n}s) \qquad (25)$$

$$v = \sum_{m=1}^{\infty} \sum_{n=1}^{\infty} B_{mn} \sin(\bar{m}x) \cos(\bar{n}s) \qquad (26)$$

$$u = \sum_{m=1}^{\infty} \sum_{n=1}^{\infty} C_{mn} (1 - \cos(\bar{m}x)) \sin(\bar{n}s) \qquad (27)$$

$$\bar{m} = \frac{m\pi}{L}, \bar{n} = \frac{n}{R}, s = R\theta,$$

$$m, n = 1, 2, 3 \dots$$

The total energy expression is a function of the stiffness matrix elements of the equivalent laminate and the unknown displacement field coefficients A_{mn}, B_{mn} and D_{mn}. For the equilibrium to be stable, the total potential energy of the system must be minimum. This can be satisfied by finding the first derivative of the total potential energy with respect to the unknown constants A_{mn}, B_{mn} and D_{mn} and equating to zero. This results in an eigenvalue problem. The resulting Equation is then solved for the unknown in-plane load Nx.

Applications for the Selected Cases of Interest

The specifications for the lattice with 1.8 m diameter are summarized in Table 1, whereas the material properties in Table 2 (Tables 1 and 2).

By using the above requirements and theory, a MATLAB code was written where the buckling load was calculated and verified by finite element analysis using MSC NASTRAN. The Finite element model consisted of 4400 beam elements for the hoop and helical ribs and

Parameter	[90/±26°] ribs orientation
Material	HTS5631
Skin Thickness, mm	1.4
No of Reels (each side)	20
No. of Helical ribs	40
No of Circum. Hoops	6
Area of Rib Section, mm²	134
Mass, Kg	48.8

Table 1: Specifications for lattice fuselage structures.

Property	Value
Young's modulus E1	128.9 GPa
Young's modulus E2	10.4 GPa
Shear modulus G12	4.11 GPa
v12	0.34
Thickness t	0.35 mm

Table 2: HTS-5631 material properties.

12000 shell elements for the composite skin. The one side of the model was clamped whereas the other one was subjected to a compressive load which was distributed to the nodes via multi-point constraints (MPCs). The first buckling mode from the finite element model of the full scale fuselage is shown in Figure 3, whereas the comparison with the unit cell analysis is shown in Table 3 (Figures 2 and 3).

In order to scale down the wafer to the desired main dimensions, the analysis was based on the of buckling load/mass ratio (P_{cr}/M) of the scaled down structure versus the cross-sectional area (A_{sc}) of the ribs. Moreover, the number of reels was decreased with respect to manufacturing restrictions, such as keeping the spacing between the reels the same. Finally, the skin thickness remained constant, since further scaling down its thickness below the four layers with total thickness of 1.4 mm was not possible because it could lead to an unsymmetrical fuselage skin layup. Given the above descriptions, a parametric analysis was done with the unit cell theory, in order to extract the required ribs cross section for two scaled down fuselage geometries with 1 m and 0.5 m in diameter. The results for both scaled down structures are summarized in Figures 4 and 5, where the ratio of critical buckling load to mass (Pcr/M) against ribs cross section is shown. The results were used with final goal to achieve the same Pcr/M ratio with the full scale fuselage (Figures 4 and 5).

The final step was the correlation of the unit cell models of the scaled down structures with finite element models. Figure 6 depicts the first buckling mode for the 1 m diameter prototype whereas Figure 7 shows the first buckling mode of the 0.5 m diameter scaled down structure. Table 4 shows the comparison of the critical buckling loads by finite element models and unit cell method (Figures 6 and 7).

Calculation method	Full Scale prototype 1.8 m in diameter, 2 m in length
Buckling load calculated by FEM	467.9 kN
Buckling load calculated analytically	463.9 kN

Table 3: Computation of full scale fuselage buckling load by unit cell method and finite element analysis.

	Scaled down prototype 1m in diameter, 1m in length	Scaled down prototype 0.5m in diameter, 1m in length
Buckling load calculated by FEM	263.9 kN	136.5 kN
Buckling load calculated analytically	260.6 kN	132.1 kN

Table 4: The comparison of the critical buckling loads by finite element models and unit cell method.

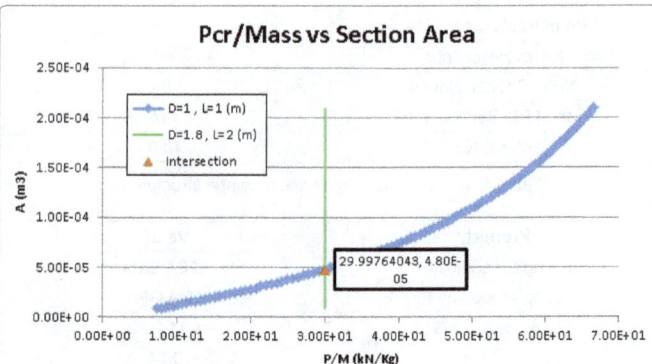

Figure 4: Buckling load/mass ratio against cross-sectional area for the scaled down prototype with 1 m diameter.

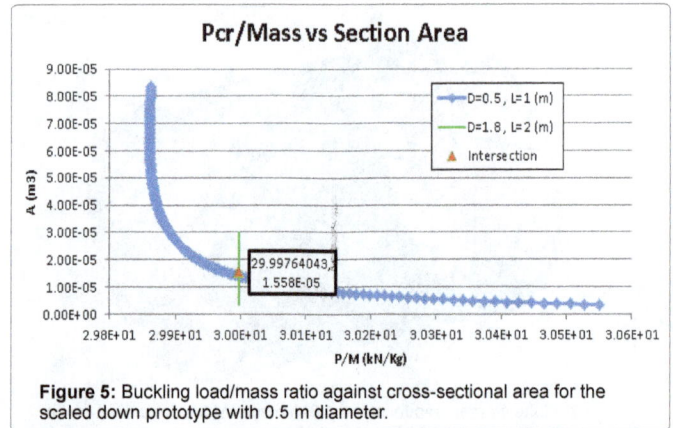

Figure 5: Buckling load/mass ratio against cross-sectional area for the scaled down prototype with 0.5 m diameter.

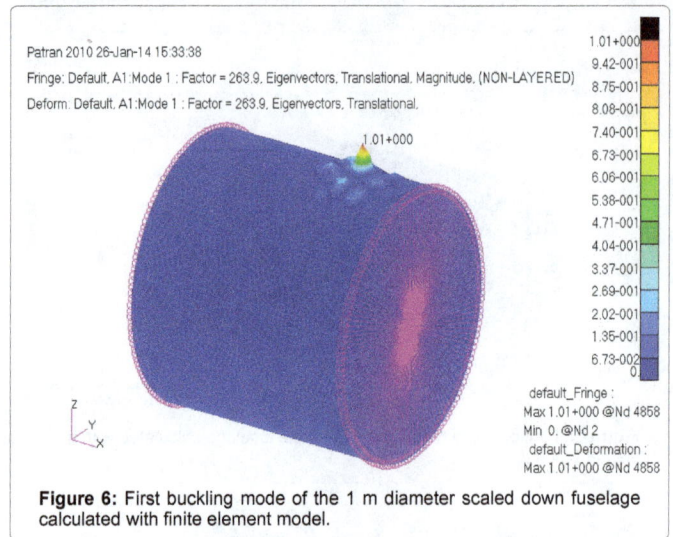

Figure 6: First buckling mode of the 1 m diameter scaled down fuselage calculated with finite element model.

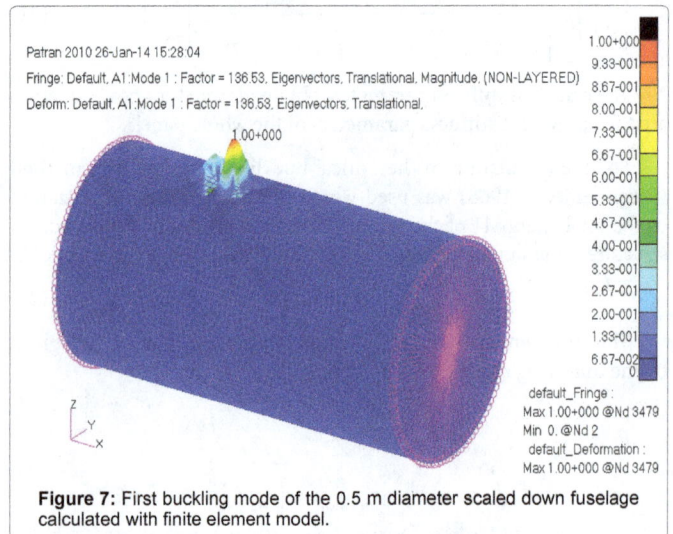

Figure 7: First buckling mode of the 0.5 m diameter scaled down fuselage calculated with finite element model.

Conclusion

A smeared unit cell analytical solution has been developed for the case of a small business aircraft fuselage section, which was also correlated with finite element analysis approach. Afterwards, a scaling down approach was followed for two scaled down fuselage structures.

By keeping the skin thickness the same and reducing the full scale fuselage diameter, the ribs cross section and distance were calculated in order to achieve the same buckling load to mass ratio. All models have shown an excellent correlation between the unit cell approach and the finite element analysis.

Moreover, a close to linear relationship was found between the buckling load and the diameter of the structures, enabling testing and validation of computational tools in scaled down prototypes.

References

1. Vasiliev VV, Barynin VA, Rasin AF (2001) Anisogrid lattice structures – Survey of development and application. Compos Struct 54: 361–370.

2. Vasiliev VV, Razin AF (2006) Anisogrid composite lattice structures for spacecraft and aircraft applications. Compos Struct 76: 182–189.

3. Vasiliev VV, Barynin VA, Rasin AF, Petrokovskii SA, Khalimanovich VI (2009) Anisogrid composite lattice structures – development and space applications. Compos Nanostruct 3: 38–50.

4. Vasiliev VV, Barynin VA, Razin AF (2012) Anisogrid composite lattice structures – development and aerospace applications. Compos Struct 94: 1117–1127.

5. Vasiliev VV, Razin AF, Nikityuk VA (2014) Development of geodesic composite fuselage structure. Int Rev Aerosp Eng 7(1): 48–54.

6. Vasiliev VV (1993) Mechanics of composite structures. Washington: Taylor & Francis.

7. Vasiliev VV, Morozov EV (2013) Advanced mechanics of composite materials and structural elements. 3rd ed. Elsevier, Amsterdam.

8. Slinchenko D, Verijenko VE (2001) Structural analysis of composite lattice shells of revolution on the basis of smearing stiffness. Compos Struct 54: 341–348.

9. Totaro G, Gurdal Z (2009) Optimal design of composite lattice shell structures for aerospace applications. Aerosp Sci Technol 13: 157–164.

10. Buragohain M, Velmurugan R (2009) Buckling analysis of composite hexagonal lattice cylindrical shell using smeared stiffener model. Defence Sci J 50(3): 230–238.

11. Paschero M, Hyer MW (2009) Axial buckling of an orthotropic circular cylinder: Applications to orthogrid concept. Int J Solids Struct 46: 2151–2171.

12. Totaro G (2013) Local buckling modelling of isogrid and anisogrid lattice cylindrical shells with triangular cells. Compos Struct 94: 446–452.

13. Totaro G (2013) Local buckling modelling of isogrid and anisogrid lattice cylindrical shells with hexagonal cells. Compos Struct 95: 403–410.

14. Zheng Q, Ju S, Jiang D (2014) Anisotropic mechanical properties of diamond lattice composites structures. Compos Struct 109: 23–30.

15. Hou A, Gramoll K (1998) Compressive strength of composite latticed structures. J Reinforc Plastics Compos 17: 462–483.

16. Zhang Y, Xue Z, Chen L, Fang D (2009) Deformation and failure mechanisms of lattice cylindrical shells under axial loading. Int J Mech Sci 51: 213–221.

17. Frulloni E, Kenny JM, Conti P, Torre L (2007) Experimental study and finite element analysis of the elastic instability of composite lattice structures for aeronautic applications. Compos Struct 78: 519–528.

18. Fan H, Jin F, Fang D (2009) Uniaxial local buckling strength of periodic lattice composites. Mater Des 30: 4136–4145.

19. Morozov EV, Lopatin AV, Nesterov VA (2011) Finite-element modelling and buckling analysis of anisogrid composite lattice cylindrical shells. Compos Struct 93: 308–323.

20. Azarov AV (2012) Continuum and discrete models of lattice composite cylindrical shells. Mech Compos Mater Struct 18: 121–130.

21. Lopatin AV, Morozov EV, Shatov AV (2015) Deformation of a cantilever composite anisogrid lattice cylindrical shell loaded by transverse inertia forces, Composite Structures 129: 27–35.

7

Flight Dynamics and Control of a Vertical Tailless Aircraft

Bras M[1], Vale J[1], Lau F[1] and Suleman A[2]*

[1]Instituto Superior Técnico, Lisbon, Portugal
[2]University of Victoria, Victoria BC, Canada

Abstract

The present work aims at studying a new concept of a vertical tailless aircraft provided with a morphing tail solution with the purpose of eliminating the drag and weight created by the vertical tail structure. The solution consists on a rotary horizontal tail with independent left and right halves to serve as control surfaces. Different static scenarios are studied for different tail configurations. The proposed morphing configurations are analyzed in terms of static and dynamic stability and compared with a conventional configuration. The stability derivatives defining the limits of static stability are calculated for the whole range of tail rotation angles. The aircraft's dynamic model is developed and feedback control systems are implemented. A sideslip suppression system, a heading control system and a speed and altitude hold system are studied for three different configurations, MC1, MC2 and MC3 configurations. Static results show that the aircraft is longitudinally stable for a wide range of tail rotation angles. Variation of tail dihedral and rotation angles are two mechanisms able to maintain directional and lateral stability but only the last is able to produce lateral force and yawing moment. Dynamic stability results demonstrate no spiral nor Dutch-roll modes due to the absence of the vertical stabilizer. The increase in tail rotation produces an appearance of the spiral mode and an unstable Dutch-roll mode that quickly degenerates into two unstable real roots with the increase in tail rotation. The addition of dihedral to the tail increases the stability of the overall modes while decreasing their variation amplitude with the tail rotation. The morphing tail configuration proved to be a feasible control solution to implement in an aircraft such as a small UAV, with the MC1 configuration being the most simple of the three morphing configurations and also the most reliable one.

Keywords: Morphing aircraft; Vertical tailless aircraft; Tail rotation mechanism; Tail dihedral change mechanism; Automatic flight control system

Introduction

In the field of aeronautics, shape morphing has been used to identify those aircraft that undergo substantial geometrical changes in their external shape to enhance or adapt to their mission profiles during flight. This creates superior system capabilities not possible without morphing shape changes. The objective of morphing concepts is to develop high performance aircraft with lifting surfaces designed to change shape and performance substantially during flight to create a multiple-regime, aerodynamically efficient, and shape-changing aircraft. Compared to conventional aircraft, morphing aircraft become more competitive as the demand for improved cost efficient aircraft increases.

The concept of implementing shape morphing in aircraft isn't new. In fact, the use of retractable flaps or slats for increased lift during take-off and landing, retractable landing gear for reduced drag during flight, variable sweep wings in fighters to reduce shock waves in transition from subsonic to supersonic speeds and variable incidence noses, as used in the Concorde for better pilot visibility during take-off and landing, are just a few examples of morphing solutions that aircraft have been using in the past [1]. However, recent research in smart materials and adaptive structures led to the development of new flexible skins and improved structural mechanization allowing substantial shape changes, particularly in wing area and twist and in airfoil camber, directly benefiting the airplane efficiency in each mission while expanding its overall flight envelope [2,3].

There are many challenges in the design of morphing aircraft: the integrity of structures needs to be ensured, the system should be designed so the required actuation force is realizable, the skin has to be designed to give a smooth aerodynamic surface while supporting the aerodynamic loads, the design process should be extended to encompass multiple flight regimes, engines need to be designed for efficient low and high speed operation, and control systems have to be designed with highly coupled control effects in mind. While many questions remain unanswered regarding the utility of morphing air vehicles, enough evidence of improved performance and new abilities has been established to warrant further consideration of the prospects of morphing aircraft, both for multiple flight regimes and for flight control [4].

The most common configuration of an aircraft used nowadays is built upon an empennage, or tail assembly, structure. Most aircraft feature empennage incorporating vertical and horizontal stabilizing surfaces which stabilize pitch and yaw, as well as housing control surfaces. Aircraft empennage designs vary with the number of tailplanes (stabilizers) and fins used and their location. Tailplanes can either have movable elevator surfaces or be single combined (stabilator or flying tail). There can also be alternative approaches as V and X tails and the case of tailless aircraft (flying wing) having all its horizontal and vertical control surfaces on its main wing surface.

Despite these different tail configurations, they all serve the purpose of providing an aircraft with pitch and yaw stability and control. The purpose of such a component in an aircraft mimics the one of a tail

---

*Corresponding author: Suleman A, University of Victoria, Victoria BC, Canada
E-mail: suleman@uvic.ca

in birds. In fact, birds seem to adjust their tail to optimize their flight rather than just using them uniquely as a stabilizing and control surface [5]. Thomas studied the influence of bird tails on profile and induced drag. He concluded that by using the tail to generate lift, birds can have the small wings needed for fast flight (with the tail closed) and still have good performance in slow flight (with the tail spread), during turns, or when accelerating [5,6]. Evans et al. conducted wind tunnel tests on barn swallows and compared the results with deltawing theory (slender-wing theory). He observed that at low speeds, the tail was spred and held at a high angle of attack, and wingspan was maximized. At high airspeeds, the tail was furled; held parallel to the airflow and wingspan was reduced [7]. However, their empirical observations failed to provide robust support for the variable-geometry application of delta-wing theory.

Birds don't have a vertical tail stabilizer and yet they are capable of controlling yaw motion. A study carried by Sachs revealed that, on one hand, bodies of birds are aerodynamically well integrated in the wing. The integration of the body is supported by its smaller size relative to the wing. As a consequence, the effect of the integrated body on the tendency to sideslip when yawing may be reduced when compared with a case where the body is considered alone without a wing. On the other hand, birds have a fast restoring capability in the yaw axis in terms of dynamic stiffness. This is due to the fact that the yawing moment of inertia is more reduced with a size decrease than the restoring aerodynamic moment, leading to a reduction in the required aerodynamic yawing moment in birds. This suggests that in such a case birds do not need a vertical tail as the wing alone can provide the required aerodynamic yawing moment [8].

A later study carried out by the same author regarding the specific tail effects on yaw stability in birds with different tail shapes revealed that elongated delta shaped tails can produce yawing moment in case of sideslip. This is due to the asymmetry in the airflow at the tail, because of the delta shape. This asymmetry leads to an asymmetrical lift distribution which also causes a correspondingly asymmetrical induced drag distribution forming a couple that yields a yawing moment [9]. The case of birds with forked tails was also studied and such tails showed drag forces at the elongated elements. By controlling the spread angle of each half tail, birds with such tails are able to control yaw due to the drag forces with different lever arms, forming a couple and hence a yawing moment. A further ability for producing stabilizing yawing moments is due to the legs and feet, according to Sachs. Depending on their length, they can stretch out in rearward direction to a considerably larger extent than the tail to control the couple produced by the asymmetry in drag produced by both feet. Sachs also suggests that as what happens with an aircraft flying at low speeds (take-off and landing situations), where flaps are used to increase drag, birds also lower their feet so that they are exposed to the airflow and generate drag for low speed flight conditions, while keeping them in a streamlined position for high speed flight, producing little drag.

The present work focuses on a new concept of morphing applied to the aircraft horizontal stabilizer with the purpose of eliminating the vertical component of the aircraft tail hence reducing weight and parasitic drag caused by the vertical stabilizer, while still maintaining the aircraft's ability to control yaw. This is achieved by means of a rotary horizontal stabilizer that can generate horizontal forces when needed.

Starting from simple flight mechanics theory, the static aircraft body equations are assembled considering the aerodynamic loading of the wing and tail. The equations are kept non-linear for the tail angles as the tail rotation amplitude needed to provide trim and control is unknown. Different static scenarios are studied and the static stability of the aircraft is determined for different tail rotation and dihedral angles.

The chosen aircraft model is a Subsonic Business Jet Model (SBJ) from reference [10] that contains the majority of the aerodynamic derivatives needed, with the remaining ones being estimated. As the static model is built considering each one of the aircraft's components, the aerodynamic derivatives that refer to the whole aircraft are not needed as they can easily be obtained from the assembled equations. Introducing dynamic stability and control theory, the aircraft's dynamic model is developed followed by the implementation of the feedback control system that provides control for the aircraft's tail. The dynamic model was also kept non-linearized but linear control techniques were used. Three different configurations are studied: the MC1 configuration, whith both halves of the tail rotating together collinearly; the MC2 configuration with only one half rotating at a time, with the other one steady and the MC3 configuration with both halfs free to move independently.

Static Analysis

The aerodynamic model of forces and moments acting on the aircraft, disregarding dynamic effects of the aircraft, was built by taking advantage of flight mechanics theory [11]. Four coordinate reference frames were defined: the

Earth, wind, stability and body coordinate frames. Both Earth-Centered Inertial (ECI) and Aircraft-Body Coordinate

(ABC) frames were of particular interest as they were used to write the aircraft equations of motion. These reference lines are depicted in Figure 1. To study the particular problem of a rotary horizontal tail, another coordinate system was introduced that has its x axis aligned with the aircraft body x axis and its y axis aligned with the horizontal tail span direction so it can rotate along with the tail, the Aircraft-Horizontal tail Coordinate (AHC) frame (Figure 1). The aerodynamic forces applied to the aircraft by each lifting surface, specified in the wind axes frame, were calculated and transferred to the aircraft-body reference frame.

Assembly of equations

The aircraft wind, stability and body axes are related with the wind velocity components by the aircraft's angle of attack α and sideslip β. Following the rules for finding rotation matrices, the transformation from body to stability axes of the velocity vector V_{body} specified in the body axes is

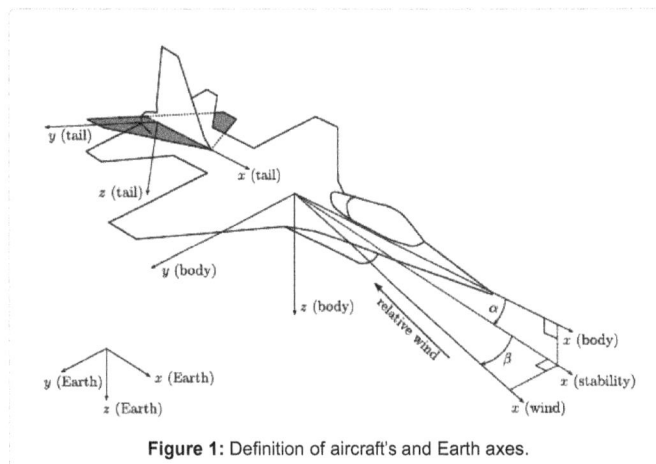

Figure 1: Definition of aircraft's and Earth axes.

$$V_{stability} = \begin{bmatrix} \cos\alpha & 0 & \sin\alpha \\ 0 & 1 & 0 \\ -\sin\alpha & 0 & \cos\alpha \end{bmatrix} V_{body} \tag{1}$$

and the rotation from stability axes to wind axes is

$$V_{wind} = \begin{bmatrix} \cos\beta & \sin\beta & 0 \\ -\sin\beta & \cos\beta & 0 \\ 0 & 0 & 1 \end{bmatrix} V_{stability} \tag{2}$$

With these rotations defined by S_α and S_β, respectively.

The rotation from wind to body axes can be calculated by inverting the rotation matrix as follows

$$V_{body} = (S_\beta S_\alpha)^T V_{wind} = \begin{bmatrix} V_T \cos\alpha \cos\beta \\ V_T \sin\beta \\ V_T \sin\alpha \cos\beta \end{bmatrix} \tag{3}$$

Considering the angle of incidence of the aircraft's wing, the wing angle of attack is

$$\alpha_w = \alpha + i_w \tag{4}$$

where i_w is the aircraft's wing angle of incidence.

The wing velocity vector can be determined by applying a rotation about the aircraft's y axis to account for its incidence

$$V_{wing} = \begin{bmatrix} \cos i_w & 0 & -\sin i_w \\ 0 & 1 & 0 \\ \sin i_w & 0 & \cos i_w \end{bmatrix} V_{body} \tag{5}$$

where the rotation matrix is defined as S_{ih}.

The relation II.3 is now rewritten for the conventional horizontal tail as

$$V_{horiz\,tail} = \begin{bmatrix} V_{Th} \cos\alpha_{\bar{h}} \cos\beta \\ V_{Th} \sin\beta \\ V_{Th} \sin\alpha_{\bar{h}} \cos\beta \end{bmatrix} \tag{6}$$

where $\alpha_{\bar{h}}$ is the angle of attack of a conventional horizontal tail with no incidence angle.

It's now necessary to represent the velocity vector at the horizontal tail in the rotated tail surface by applying the same rotation rules as applied before. As the tail is usually built into the aircraft with an incidence angle we must also apply a second rotation to take this into consideration.

Figure 2 shows the rotations applied to transform from horizontal tail axes to a rotated tail axes as well as the rotation needed to account for the incidence angles of both wing and tail.

The transformation from the horizontal tail axes to rotated tail axes is

$$Vrot\,tail = \begin{bmatrix} 1 & 0 & 0 \\ 0 & \cos\delta_h & \sin\delta_h \\ 0 & -\sin\delta_h & \cos\delta_h \end{bmatrix} V_{horiz\,tail} \tag{7}$$

And the rotation to account for the tail incidence is

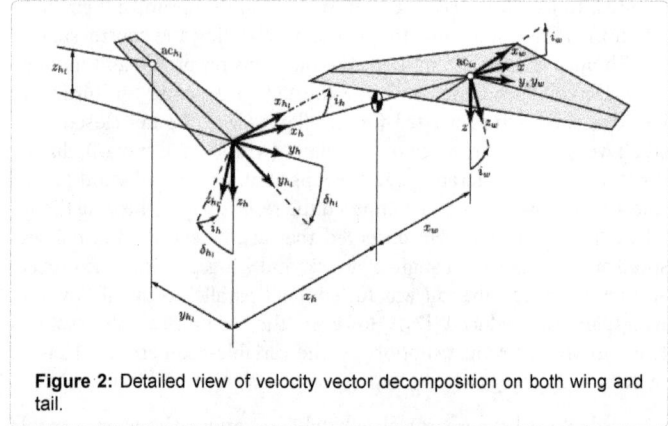

Figure 2: Detailed view of velocity vector decomposition on both wing and tail.

$$V_{rot\,tail\,with\,inc} = \begin{bmatrix} \cos i_h & 0 & -\sin i_h \\ 0 & 1 & 0 \\ \sin i_h & 0 & \cos i_h \end{bmatrix} V_{rot\,tail} \tag{8}$$

with these rotations defined by $S_{\delta h}$ and S_{ih}, respectively.

The complete rotation from horizontal tail to rotated tail axes including the tail incidence is

$$V_{rot\,tail\,with\,inc} = S_{ih} S_{\delta h} V_{horiz\,tail} \tag{9}$$

where δ_h is replaced by δ_{hl} for the left half horizontal tail and by δ_{hr} for the right half.

Considering small angles of attack and sidelip, the angles of attack for both left and right half tails are given by

$$\alpha_{hl} \approx \frac{i_h - \beta \sin\delta_{hl} + (\alpha - \varepsilon)\cos\delta_{hl}}{1 + i_h\beta \sin\delta_{hl} - i_h(\alpha - \varepsilon)\cos\delta_{hl}}$$
$$\alpha_{hr} \approx \frac{i_h - \beta \sin\delta_{hr} + (\alpha - \varepsilon)\cos\delta_{hr}}{1 + i_h\beta \sin\delta_{hr} - i_h(\alpha - \varepsilon)\cos\delta_{hr}} \tag{10}$$

Having defined the wing and tail angles of attack, the aircraft body equations can be assembled. Trigonometric functions are written abbreviated.

$$\begin{bmatrix} X \\ Y \\ Z \end{bmatrix}_{wing} = \begin{bmatrix} c\alpha_w & 0 & -s\alpha_w \\ 0 & 1 & 0 \\ s\alpha_w & 0 & c\alpha_w \end{bmatrix} \begin{bmatrix} -D_w \\ 0 \\ -L_w \end{bmatrix} \tag{11}$$

$$\begin{bmatrix} L \\ M \\ N \end{bmatrix}_{wing} = \begin{bmatrix} x_w \\ 0 \\ z_w \end{bmatrix} \times \begin{bmatrix} X \\ Y \\ Z \end{bmatrix}_{wing} \tag{12}$$

$$\begin{bmatrix} X \\ Y \\ Z \end{bmatrix}_{l,r\,tail} = \begin{bmatrix} c\alpha_h & 0 & -s\alpha_h \\ -s\delta_h s\alpha_h & c\delta_h & -c\alpha_h s\delta_h \\ c\delta_h s\alpha_h & s\delta_h & c\delta_h c\alpha_h \end{bmatrix} \begin{bmatrix} -D_h \\ 0 \\ -L_h \end{bmatrix} \tag{13}$$

$$\begin{bmatrix} L \\ M \\ N \end{bmatrix}_{l,r\,tail} = \begin{bmatrix} -x_h \\ y_h \\ z_h \end{bmatrix} \times \begin{bmatrix} X \\ Y \\ Z \end{bmatrix}_{l,r\,tail} \tag{14}$$

where the subscript h is replaced by hl for the left half tail and by hr for the right half, accordingly.

Angles of attack and elevators deflection angles to trim

For the case of zero tail dihedral and rotation angles, a trimmed longitudinal flight is obtained for $\alpha_{trim}=2.42$ deg and $\delta_{etrim}=1.99$ deg. As the tail rotates, the angle of attack to trim keeps constant and the elevators angle varies (Figure 3). The elevator behavior reverses with the increase in tail dihedral. As the dihedral angle increases the elevator angle to trim shows a larger variation throughout the entire range of tail rotation. The variation in dihedral angle also doesn't change the angle of attack to trim. For a dihedral angle of about 30 deg the variation in elevators angle with the tail rotation angle is minimum for the range -45 deg $\lesssim \delta h \lesssim 45$ deg.

Variation of static stability with tail rotation and dihedral

The variation of the $C_m{}^\alpha$ stability derivative was studied with both tail rotation and dihedral variation for a longitudinally trimmed flight. For each angle of tail rotation, the necessary elevator deflection was calculated.

It's possible to observe in Figure 3b that although a longitudinally trimmed flight was always present for each tail rotation angle, the aircraft could only be static longitudinal stable for tail rotation angles in the range -67 deg $\lesssim \delta h \lesssim 67$ deg for the case of a tail with no dihedral. The addition of an initial dihedral angle greater than 30 deg maintains static longitudinal stability for the whole range of tail rotation angles.

It is possible to observe from Figure 3c two mechanisms of increasing the value of C^β_n: variation of tail rotation and variation of tail dihedral. Both mechanisms have the same purpose of increasing the aircraft wet and hence increasing its directional stability. The addition of an initial

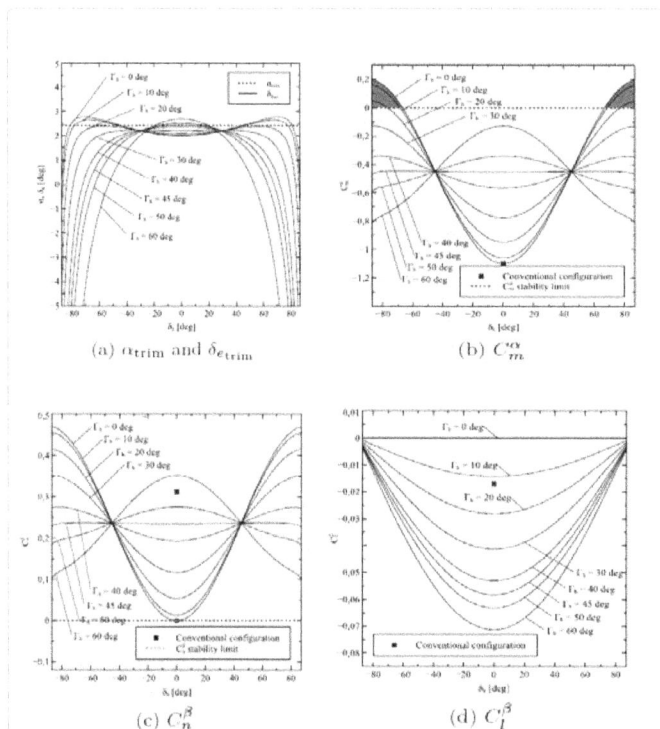

Figure 4: Variation of Cn with tail rotation and dihedral, with longitudinal trim maintained, for the SBJ aircraft.

dihedral to the tail reduces the amplitude variation of C^β_n; when using the tail rotation as a directional stability mechanism. The increase in tail dihedral angle produced an increase, in magnitude, in the value of C^β_l (Figure 3d). Also, the variation of the tail rotation angle for a tail with no initial tail dihedral maintained a zero value of C^β_l. The addition of an initial dihedral to the tail contributed to the variation of C^β_l with the tail rotation and hence the lateral stability was increased. However, the tail rotation mechanism was much less effective in producing lateral stability than the tail dihedral variation. When the tail was considered to be above the c.g. the value of C^β_l proved to vary using the tail rotation mechanism with no initial dihedral and its effectiveness became similar to the dihedral variation mechanism.

Yawing moment produced by the tail rotation

Although the tail dihedral change mechanism was able to create the necessary directional and lateral stability, it was not able to to produce lateral force and yawing moment because when dihedral is changed the aircraft maintained its symmetry in the xz plane. However, the tail rotation mechanism, besides being able to produce directional and lateral stability, was also able to produce lateral force and yawing moment because when the tail rotated, asymmetries arise in aircraft's xz plane.

Figure 4 depicts the variation of C_n with the tail rotation, for different tail dihedral angles, and also a comparison with a conventional configuration of horizontal and vertical tail with rudder. The yawing moment generated by a deflection of rudder was much more effective than a tail rotation. This happened because a longitudinal trim condition was being maintained by the elevators as the tail rotated and their deflection decreased the tail effectiveness.

Dynamic Stability Analysis

The linear state-space model was built with the tail rotation non-linearites included in the aircraft stability derivatives, as follows

$$\dot{x}(t) = A(\delta_{hl},\delta_{hr})x(t) + B(\delta_{hl},\delta_{hr})u(t)$$
$$\dot{y}(t) = C(\delta_{hl},\delta_{hr})x(t) + D(\delta_{hl},\delta_{hr})u(t) \quad (15)$$

and the system eigenvalues were calculated as a function of the tail rotation angles.

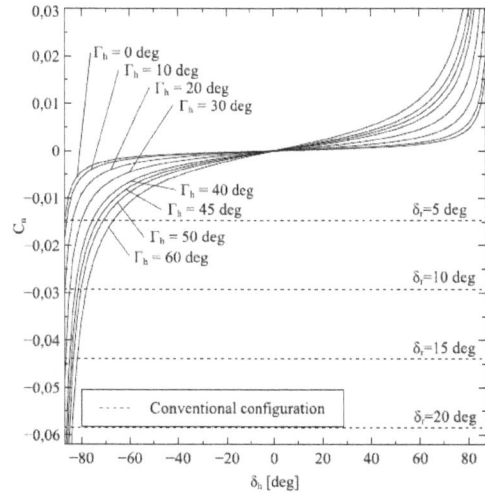

Figure 3: Variation of angle of attack and elevators deflection angle to trim as well as the stability derivatives, with tail rotation and dihedral, for the SBJ aircraft. The shaded areas represent the static instability zone.

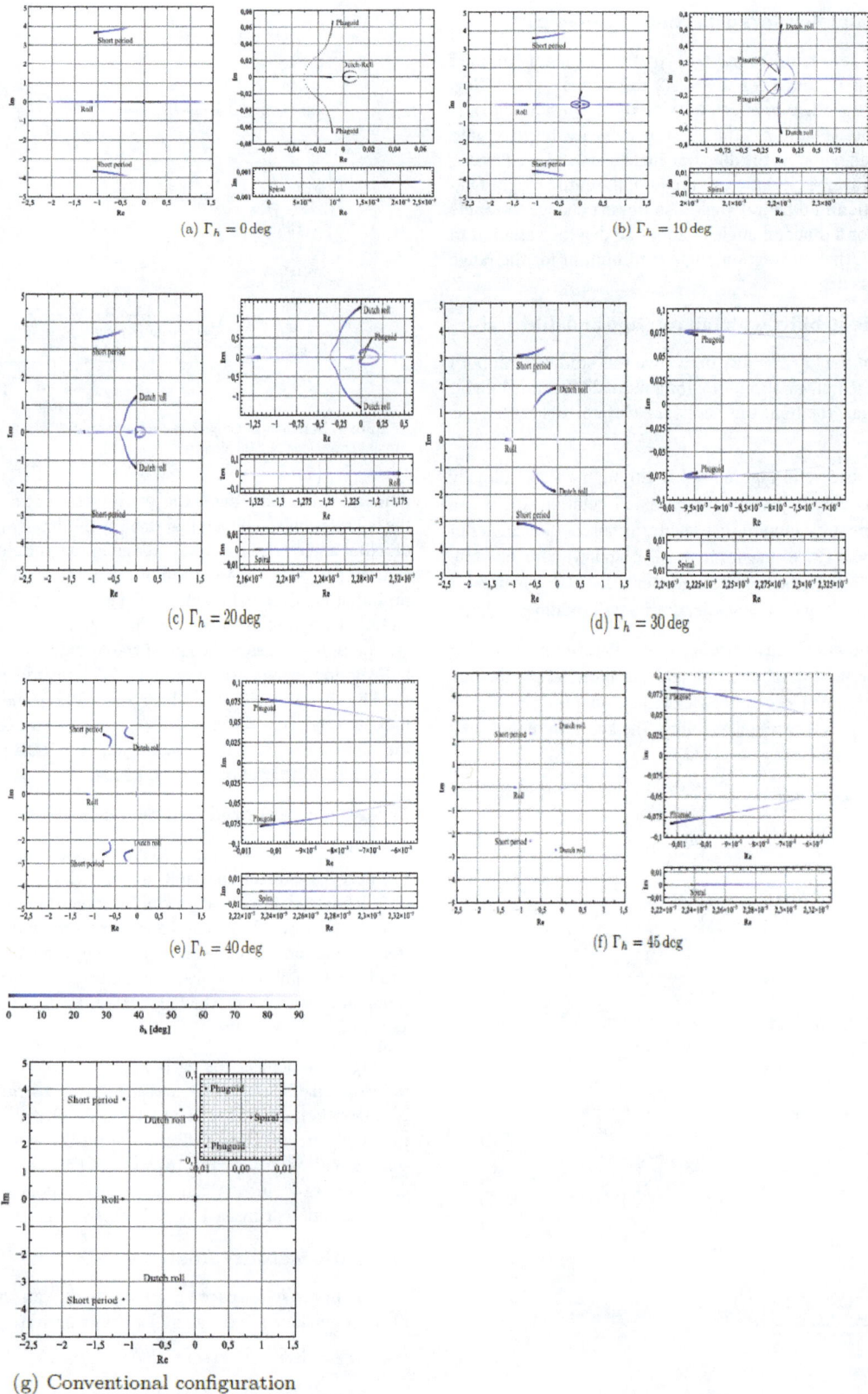

(a) $\Gamma_h = 0\,\mathrm{deg}$

(b) $\Gamma_h = 10\,\mathrm{deg}$

(c) $\Gamma_h = 20\,\mathrm{deg}$

(d) $\Gamma_h = 30\,\mathrm{deg}$

(e) $\Gamma_h = 40\,\mathrm{deg}$

(f) $\Gamma_h = 45\,\mathrm{deg}$

(g) Conventional configuration

Figure 5: State-space root travel with tail rotation for different initial dihedral angles (a)-(f) and for a conventional configuration (g) for the SBJ aircraft.

Complete state-space roots

For the case of the rotary tail configuration, the system roots were calculated as a function of the tail rotation and dihedral angles. Figure 5 represents the roots calculated for the whole range of tail rotation for different tail dihedral configurations. Looking at the case of a tail with no dihedral (Figure 5a) the short-period roots became less stable as the tail increased its rotation angle. The phugoid poles kept stable for the whole range of rotation angles and eventually degenerated into two stable real roots with the increase in the tail rotation. As for the roll pole, it appeared to become more stable with the increase in tail rotation angle. No spiral nor Dutch-roll poles were present for zero tail rotation due to the absence of a vertical tail surface. The addition of tail rotation produced an appearance of an unstable spiral pole and two unstable complex poles, the Dutch-roll poles that quickly degenerated into two unstable real roots with the increase in tail rotation.

For the case of a tail with 10 deg of initial dihedral angle (Figure 5b), it is possible to observe that the two complex Dutch-roll roots exist, due to the presence of a vertical surface, but are unstable. These roots tended to stabilize for small rotation angles and destabilize for higher rotation angles and eventually degenerated into two unstable real poles. Again, the phugoid kept stable for the whole range of tail rotation and eventually degenerated into two stable real roots. As for the short-period and roll roots, their behavior was also the same as for the tail with no dihedral. The spiral pole was present and tended to destabilize with the increase of the tail rotation.

When a dihedral of 20 deg was initially added to the tail, the Dutch-roll roots tended to stabilize whereas the phugoid ones became unstable very quickly, and were only stable for small angles of tail rotation, degeneration into two unstable real roots. Also, an interference with the Dutch-roll roots and the roll ones seemed to happen but both kept stable.

The addition of 30 deg of dihedral to the tail increased the overall stability of the modes while decreasing their variation amplitude with the tail rotation (Figures 5d to 5f). All the roots but the spiral kept stable for the whole range of tail rotation angles. The same happened when 45 deg of dihedral was initially added to the tail. The stability of the roll mode increased with the tail rotation for initial dihedral angles up to 30 deg. For higher initial dihedral values it decreased with the tail rotation, but its variation amplitude was also reduced. Nevertheless, it always kept stable.

Control Analysis

Having determined the system stability characteristics, the possible control mechanisms using the tail were established and the feedback control laws for the closed-loop system were built.

The aircraft was considered to be in a steady cruise flight with no sideslip angle. The initial flight conditions used in the simulations are given in Table 1.

Parameter	Variable	Value
Altitude	h_0	9744 m
Aircraft Speed Components	$[u_0 \, v_0 \, w_0]$	[182 0 7.6] m/s
Angle of attack	α_0	2.42 deg
Pitch angle	θ_0	2.42 deg
Elevators angle	δ_0	1.99 deg
Thrust coefficient	C_{T_0}	0.0295 deg

Table 1: Initial flight conditions used for the simulations.

Mode	Eigenvalue	ω_n[rad/s]	ζ	Level
Short-period	-2.16 ± 3.67i	4.260	0.507	1
Phugoid	-0.057 ± 0.037i	0.068	0.843	1

Table 2: Dynamic characteristics and flying qualities of the closed-loop linear longitudinal modes of both conventional and morphing configuration, for the SBJ aircraft.

Mode	Eigenvalue	ω_n[rad/s]	T[s]	ζ	Level
Spiral	0.003	0.003	336	-1	1
Dutch-roll	-3.7 ± 3.6i	5.17	-	0.717	1
Roll	-1.090	1.092	0.92	1	1

(a) Conventional Configuration

Mode	Eigenvalue	ω_n[rad/s]	T[s]	ζ	Level
Spiral	0.002	0.002	412	-1	1
Dutch-roll	-3.0 ± 2.7i	4.010	-	0.749	1
Roll	-1.090	1.092	0.92	1	1

(b) Morphing tail configuration

Table 3: Dynamic characteristics and flying qualities of the closed-loop linear longitudinal modes of both conventional and morphing configuration, for the SBJ aircraft.

Stability augmentation

To understand the effect of the tail actuation on the augmentation of the system's stability, the root-locus method was used to calculate the new system poles with the augmentation system added [12,13]. To be able to do that the system was linearized.

Furthermore, the contribution of the fuselage to the aircraft stability derivatives was added at this point. This was not done previously because the aircraft poles obtained from reference [10] did not include the fuselage contribution on the aircraft stability derivatives and so, for comparison purposes, this contribution was not taken into account. However, the absence of vertical wet area on the morphed aircraft configuration led to the disappearance of the Dutch- Roll and Spiral modes. This proved to be problematic when using the root-locus method. Also, the fuselage contribution is of great importance to account for the effects of wind on the aircraft and to implement the necessary control, as described in section IV-E.

For the case of the longitudinal motion, the root-locus was evaluated for different feedback relations. The best solution proved to be the simultaneous feedback of the pitch rate q and the pitch angle θ to the elevators δ_e with feedback gains of $kq=k_\theta=0.147$. These values guaranteed level 1 flying qualities to both the short-period and phugoid modes (Table 2).

For the case of the lateral motion, the best feedback solution proved to be the feedback of the yaw rate r to the rudder δ_r, for the case of the conventional configuration, and to the tail rotation δ_{h}, for the case of the morphing configuration, with feedback gains of $k_r=1.2$ for the conventional configuration and $k_r=42.4$ for the morphing configuration. These values guaranteed level 1 flying qualities to the Dutch-roll and roll modes. Despite the unstable spiral mode, its large time constant still guaranteed level 1 flying qualities (Table 3).

The foregoing results demonstrated that the rotary tail configuration was effective as a lateral augmentation system. The results are only valid for small rotation angles and don't take into account the effect of the tail rotation in the elevators effectiveness. Also, the actuators dynamics are not taken into account and, for the case of the rotary tail configuration, the tail size is of great importance on the tail actuation time since rapid rotations of the tail may compromise its structural integrity.

The foregoing results demonstrated that the rotary tail configuration was effective as a lateral augmentation system. The results are only valid for small rotation angles and don't take into account the effect of the tail rotation in the elevators effectiveness. Also, the actuators dynamics are not taken into account and, for the case of the rotary tail configuration, the tail size is of great importance on the tail actuation time since rapid rotations of the tail may compromise its structural integrity.

Control configurations

The attitude control systems tested consisted on a pitch and roll control and sideslip suppression systems while the flight path control systems tested consisted on a height and speed hold systems and also a heading and direction control systems [12].

The first morphing configuration considered (MC1) consisted on an aircraft with a rotary tail that can have a fixed dihedral angle. In this way, the control systems affecting pitch, altitude and speed were controlled using the elevators that rotate with the tail and the roll is controlled using the ailerons. The sideslip suppression and heading control systems were controlled using the tail rotation.

The second morphing configuration (MC2) consisted on an aircraft with independently rotary half tails that could also have a fixed initial dihedral angle between them. In this way, the roll was also controlled using the ailerons while the control systems affecting pitch, altitude and speed were controlled using one elevator from one half tail that doesn't rotate. The other half tail rotated to provide the control needed for sideslip suppression and heading control systems. The half tail that rotated was chosen by the signal of the rate of yaw variable.

The third morphing configuration (MC3) also consisted on an aircraft with independently rotary half tails that could also have a fixed initial dihedral angle between them. The roll was again controlled using the ailerons but the control systems affecting pitch, altitude and speed were controlled using tail dihedral change. The sideslip suppression and heading control systems were controlled using the tail rotation.

These configurations were compared with the Conventional Configuration (CC) where pitch, altitude and speed were controlled using both elevators, roll was controlled using the ailerons and the sideslip suppression and heading control systems were controlled using the rudder deflection.

Figure 6 illustrates the different control configurations tested.

Sideslip suppression during a banked turn with pitch controller system

The direction control system used consisted on a roll control using the ailerons and a simultaneously sideslip suppression using the rudder, for the conventional configuration, and tail rotation, for the morphing configurations.

All configurations revealed similar times to complete the turn but the CC configuration was slightly faster than the remaining two and the

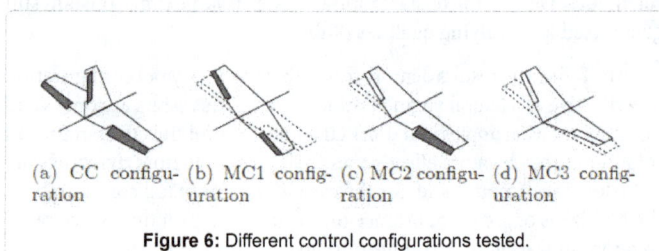

Configuration		Maximum pitch θ variation		Maximum sideslip β variation		Time t to complete turn	
	Γ_h	Φ_{sat}=30 deg	Φ_{sat}=60 deg	Φ_{sat}=30 deg	Φ_{sat}=60 deg	Φ_{sat}=30 deg	Φ_{sat}=60 deg
MC1	0 deg	0.22 deg	0.87 deg	2.13 deg	4.26 deg	273.7 s	169.6 s
	15 deg	0.12 deg	0.46 deg	1.51 deg	2.93 deg	272.1 s	166.9 s
	30 deg	0.05 deg	0.17 deg	1.10 deg	2.08 deg	271.4 s	167.0 s
	45 deg	0.01 deg	0.03 deg	0.84 deg	1.55 deg	271.4 s	167.0 s
MC2	0 deg	0.25 deg	0.74 deg	2.56 deg	4.17 deg	276.2 s	169.5 s
	15 deg	0.09 deg	0.39 deg	1.38 deg	3.07 deg	272.8 s	168.1 s
	30 deg	0.09 deg	0.85 deg	1.20 deg	1.65 deg	271.6 s	167.1 s
	45 deg	0.10 deg	0.19 deg	0.92 deg	1.71 deg	271.4 s	167.0 s
CC		0.00 deg	0.00 deg	0.36 deg	0.67 deg	270.9 s	166.8 s

Table 4: Results of maximum pitch and sideslip variation values during a banked turn as well as the time to complete the turn, using sideslip suppression and pitch controller systems for MC1, MC2 and CC configurations.

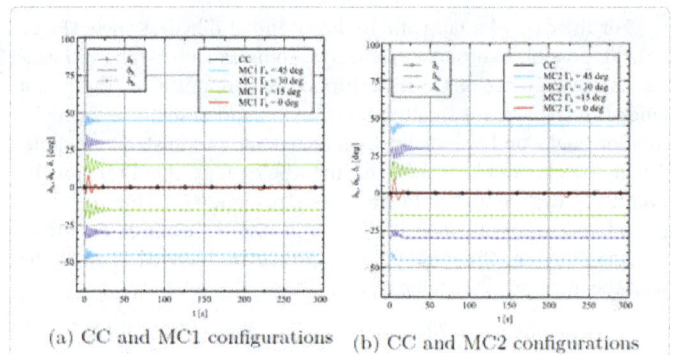

(a) CC and MC1 configurations   (b) CC and MC2 configurations

Figure 7: Control input responses during a 360 deg banked turn with a maximum bank angle of ϕ_{sat}=30 deg for the MC1, MC2 and CC configurations.

MC2 slightly slower than the MC1 configuration. The pitch variation during the turn was minimal but again the CC configuration performed slightly better in controlling pitch than the remaining ones. The sideslip was effectively controlled by the rotary tail for either the MC1 or MC2 configurations but its maximum variation was higher than the CC configuration. This variation was only observable when the aircraft was entering and leaving the turn. Table 4 presents the maximum amplitude variation values of pitch and sideslip as well as the time to complete the turn for the two bank angles and different initial tail dihedral angles.

The increase in the initial tail dihedral demonstrated an overall benefit in the aircraft during the turn: lower pitch and sideslip variation, shorter time to complete the turn and fewer requests to tail rotation (Figure 7).

The trajectory plot of Figure 8 shows no significant differences in the trajectory of the three different configurations with Γ_h=0 deg. It is evident the increase of approximately 200 m in altitude during the turn with ϕ_{sat}=30 deg and approximately 400 m for the turn with ϕ_{sat}=60 deg. The trajectory plots for the configurations with different initial dihedral angles are not illustrated but the results show no significant variation. The next control system discussed attempts to eliminate this altitude variation during the turn.

Sideslip suppression during a banked turn with altitude hold system

The second set of control systems tested was similar to the previous one but in this case the pitch control system was replaced by an altitude hold system. The altitude control using the MC2 configuration was difficult to implement due to the fact that the system was more

(a) CC configuration   (b) MC1 configuration   (c) MC2 configuration   (d) MC3 configuration

Figure 6: Different control configurations tested.

(a) $\phi_{sat} = 30\deg$

(b) $\phi_{sat} = 60\deg$

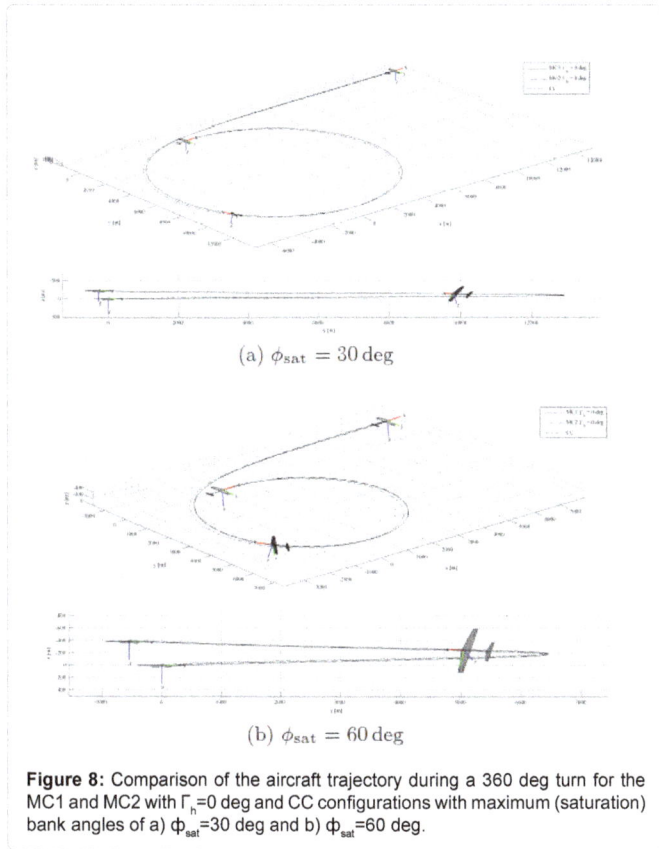

Figure 8: Comparison of the aircraft trajectory during a 360 deg turn for the MC1 and MC2 with Γ_h=0 deg and CC configurations with maximum (saturation) bank angles of a) ϕ_{sat}=30 deg and b) ϕ_{sat}=60 deg.

oscillatory and so the yaw rate signal. As this signal served as the trigger to the half tail rotation, as described previously, the tail became prone to oscillations and eventually the system became uncontrollable. To overcome this problem, the MC2 configuration was modified so that the right half tail was left fixed and the left half provided the necessary rotation to control the sideslip, independently on the yaw rate signal. This modified configuration, described as MC2* configuration, was also used for heading control. All configurations revealed similar times to complete the turn. The altitude variation during the turn was minimal for the banked turn with ϕ_{sat}=30 deg and the CC configuration performed better than the remaining configurations. The altitude variation for the case of the banked turn with ϕ_{sat}=60 deg was slightly higher but also minimal, except for the MC3 configuration with initial tail dihedral angles of 30 deg and 45 deg. This happened because of the high initial tail dihedral angles which eventually caused the tail to reach its saturation limits and hence unable to keep the altitude within a reasonable range. Despite this, the MC3 configuration performed well with the remaining initial tail dihedral angles.

As what happened with the the previous set of control systems, the increase in the initial tail dihedral demonstrated an overall benefit in the aircraft during the turn, mainly in pitch and sideslip variations, diminishing their oscillation amplitudes.

Pitch and sideslip variations were also small for the MC3 configuration, performing better than the MC1 and MC2* configurations.

Despite the small variations in pitch and sideslip, the controller for the MC3 configurations was harder to tune and eventually the request to the tail was high and very oscillatory for the initial request. The MC1 and MC2* configurations were easier to control and demonstrated

lower oscillation amplitudes. These results are depicted in Figure 9.

Table 5 presents the maximum amplitude variation values of altitude and sideslip as well as the time to complete the turn for the two bank angles and different initial tail dihedral angles.

The trajectory plot of Figure 10 shows no significant differences in the trajectory of the four different configurations with Γ_h=0 deg. It is possible to observe that the altitude is maintained within reasonable limits for all four configurations. It is also evident that the aircraft does not maintain a constant radius of curvature because of the slight variation of sideslip during the turn.

Heading control system

A different kind of flight path controller tested was a heading controller together with a pitch and roll controllers. This system was tested under two different situations: first, the aircraft was set to follow a desired heading of 20 deg, controlling it with either the ruder (CC

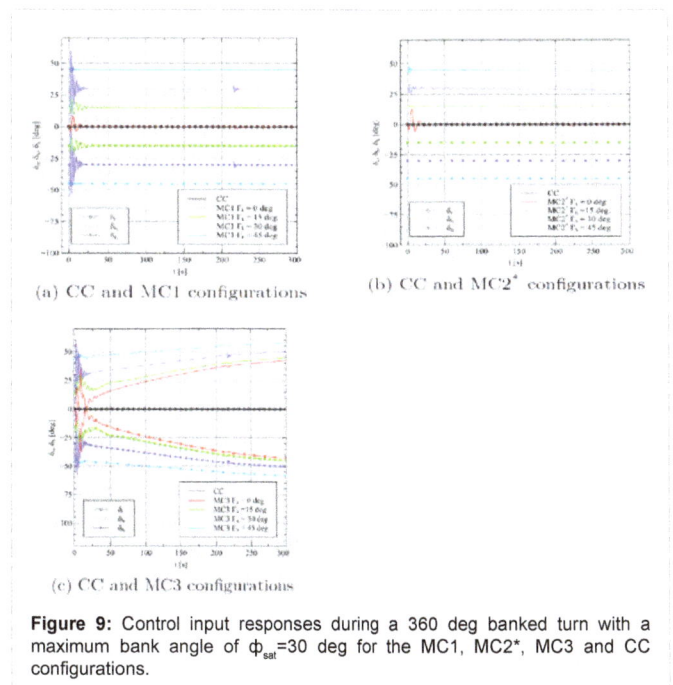

(a) CC and MC1 configurations

(b) CC and MC2* configurations

(c) CC and MC3 configurations

Figure 9: Control input responses during a 360 deg banked turn with a maximum bank angle of ϕ_{sat}=30 deg for the MC1, MC2*, MC3 and CC configurations.

Configuration		Maximum pitch h variation		Maximum sideslip β variation		Time t to complete turn	
	Γ_h	Φ_{sat}=30 deg	Φ_{sat}=60 deg	Φ_{sat}=30 deg	Φ_{sat}=60 deg	Φ_{sat}=30 deg	Φ_{sat}=60 deg
MC1	0 deg	3.78 m	13.10 m	2.28 deg	5.91 deg	273.3 s	169.1 s
	15 deg	2.38 m	5.67 m	1.46 deg	2.97 deg	271.6 s	167.2 s
	30 deg	1.48 m	5.21 m	0.74 deg	1.75 deg	271.4 s	167.0 s
	45 deg	0.63 m	6.00 m	0.58 deg	1.53 deg	271.0 s	167.0 s
MC2*	0 deg	4.95 m	11.47 m	3.86 deg	7.30 deg	280.1 s	162.1 s
	15 deg	2.20 m	7.42 m	0.69 deg	1.10 deg	268.7 s	162.0 s
	30 deg	2.14 m	5.46 m	0.90 deg	1.83 deg	267.5 s	161.8 s
	45 deg	1.47 m	5.92 m	0.72 deg	1.21 deg	269.6 s	164.7 s
MC3	0 deg	7.28 m	28.25 m	1.57 deg	2.28 deg	271.5 s	167.0 s
	15 deg	8.70 m	16.39 m	1.34 deg	2.06 deg	271.4 s	167.0 s
	30 deg	5.12 m	98.45 m	0.80 deg	1.66 deg	271.4 s	167.0 s
	45 deg	4.87 m	246.7 m	0.83 deg	1.46 deg	271.4 s	167.0 s
CC		0.44 m	1.32 m	0.39 deg	0.66 deg	271.5 s	167.1 s

Table 5: Results of maximum altitude and sideslip variation values during a banked turn as well as the time to complete the turn, using sideslip suppression and pitch controller systems for the MC1, MC3 and CC configurations.

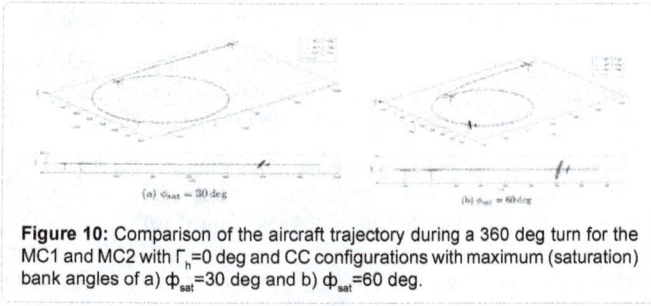

Figure 10: Comparison of the aircraft trajectory during a 360 deg turn for the MC1 and MC2 with Γ_h=0 deg and CC configurations with maximum (saturation) bank angles of a) ϕ_{sat}=30 deg and b) ϕ_{sat}=60 deg.

Configuration		Maximum sideslip β variation		Time t to reach constant heading	
	Γ_h	λ=20 deg	λ=0 deg hold with wind	λ=20 deg	λ=0 deg hold with wind
MC1	0 deg	1.83 deg	6.18 deg	279.5 s	108.9 s
	15 deg	0.78 deg	6.11 deg	556.5 s	180.3 s
	30 deg	0.21 deg	6.03 deg	>1000 s	783.8 s
	45 deg	0.01 deg	5.85 deg	>1000 s	903.7 s
MC2*	0 deg	0.18 deg	6.13 deg	869.2 s	498.2 s
	15 deg	0.32 deg	6.04 deg	848.3 s	>1000 s
	30 deg	0.40 deg	-	580.4 s	-
	45 deg	0.46 deg	-	407.5 s	-
CC		7.14 deg	10.29 deg	172.3 s	103.3 s

Table 6: Results of maximum sideslip variation values for a heading request of 20 deg and a heading hold with lateral wind as well as the time to reach a constant heading value, using heading and pitch controllers for the MC1, MC2* and CC configurations.

configuration) and tail rotation (MC1, MC2* and MC3 configurations); second, a lateral wind component of 10 m/s (19.4 knots) was introduced and the aircraft was forced to maintain the initial heading. The rudder and tail rotation were used, for the conventional and morphing configurations, respectively, to give the aircraft the required sideslip to overcome the lateral wind.

The results of both situations tested are displayed in Table 6. Regarding the results of the MC2* configuration some remarks must be pointed out. For the first situation of heading request, the addition of initial dihedral to the tail caused an increase in the sideslip value, as opposed to what happened so far. This was mainly due to the high saturation limits of the tail which created some negative dihedral values during some periods. These negative values of dihedral contributed to the increase of the sideslip variations. Also, an increase in the static error of the heading angle was observed for initial tail dihedral angles of 30 deg and 45 deg that could not be eliminated by tuning the controller gains. This is reflected in the time required to reach a constant heading which diminishes with the increase of initial tail dihedral.

For the second situation of heading hold with lateral wind, the MC2* tests were only possible to obtain with lower tail saturation limits of ± 30 deg so that only the first two MC2* configurations with lower initial tail dihedral values were possible to test. It was observed that the maximum sideslip variation values were similar to the ones of the MC1 configuration but the time to reach a constant heading value was much higher, approximately 500 s for Γ_h=0 deg and greater than 1000 s for Γ_h=15 deg.

The control input responses for the MC1, MC2*and CC configurations for both situations tested are plotted in Figures 11 and 12. A trajectory plot of both situations tested is illustrated in Figure 13 for the MC1 with Γ_h=0 deg, 15 deg and 30 deg and CC configurations.

Speed hold system

The last flight path control system tested was a speed hold system for the MC3 configuration, i.e. using the tail dihedral as a control of the longitudinal speed, for different initial tail dihedral angles. Two situations were tested, one with a tailwind component of 10 m/s and another with a headwind component of -10 m/s.

It is possible to observe from Table 7 that for the case of positive wind component (tailwind), the tail was only able to control the longitudinal speed when an initial dihedral angle of 45 deg was present. This happened because in order to be able to produce positive and negative variations in longitudinal velocity using dihedral change, one has to provide some initial dihedral to the tail.

Regarding the case of negative wind component (headwind), the tail was able to control the longitudinal speed regardless of the initial tail dihedral angle. The addition of dihedral proved to minimize the maximum speed variations and also reduced the time needed to reach a constant speed value. Nevertheless, these times were much higher than the one obtained with the conventional configuration.

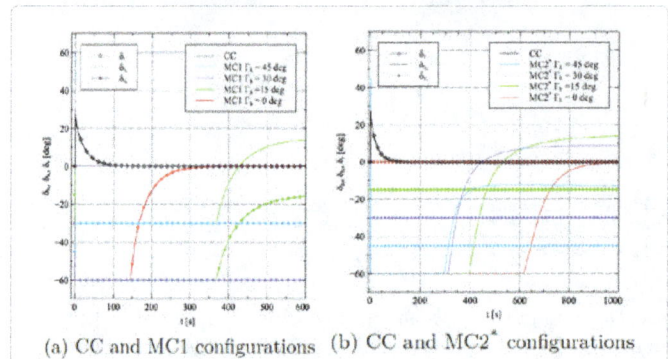

(a) CC and MC1 configurations (b) CC and MC2* configurations

Figure 11: Control input responses during a heading request of 20 deg for the MC1, MC2* and CC configurations.

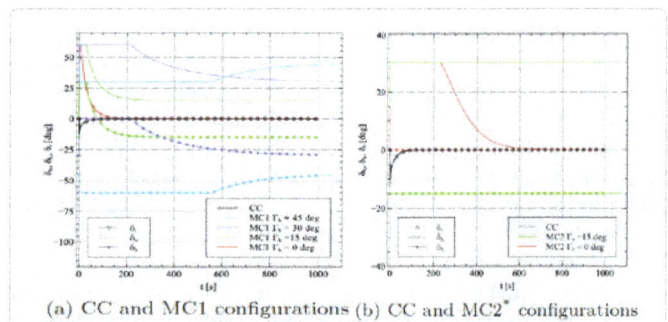

(a) CC and MC1 configurations (b) CC and MC2* configurations

Figure 12: Control input responses during a heading hold with lateral wind for the MC1, MC2* and CC configurations.

Configuration		Maximum speed u variation		Time t to reach constant speed	
	Γ_h	Tailwind of 10 m/s	Headwind of -10 m/s	Tailwind of 10 m/s	Headwind of -10 m/s
MC1	0 deg		5.75 m/s	-	>200 s
	15 deg		3.49 m/s	-	193.0 s
	30 deg		2.43 m/s	-	>200 s
	45 deg	2.17 m/s	1.39 m/s	191.1 s	121.8 s
CC		0.38 m/s	0.42 m/s	25.4 s	25.5 s

Table 7: Results of maximum longitudinal speed variation values as well as the time to reach a constant speed for the MC3 configuration with a speed hold system.

able to maintain directional and lateral stability but only the last was able to produce lateral force and yawing moment and its effectiveness was affected by the coupling with the elevators deflection.

The dynamic stability results demonstrated no spiral nor Dutch-roll modes due to the absence of the vertical stabilizer. The increase in tail rotation produced an appearance of the spiral mode and an unstable Dutch-roll mode that quickly degenerated into two unstable real roots with the increase in tail rotation. The addition of dihedral to the tail increased the stability of the overall modes while decreasing their variation amplitude with the tail rotation.

A Stability Augmentation System (SAS) was implemented in a linearized model of the morphing aircraft so that the root-locus method could be used. Both longitudinal modes guaranteed level 1 flying qualities. The Dutch-roll and roll modes of the both configurations were stabilized and were level 1 in terms of flying qualities. Although the spiral kept unstable, its large time constant also guaranteed level 1 flying qualities.

From the control analysis results, the morphing tail configuration proved to be a feasible solution to implement in an aircraft such as a small UAV. The MC2 configuration proved to be problematic when used for heading control. The MC3 configuration demonstrated good results in all situations but the MC1 configuration was the most simple of the three morphing configurations and also the most reliable one. The addition of a dihedral component helped controlling sideslip and heading but increased the time needed to reach a steady state so that the choice of a tail dihedral angle must be a trade-off solution between the required maximum amplitude variation of the controlled variables and the desired time to reach a steady state in a given situation.

Future work will focus on implementing a non-linear control system, improve the model and redefine it to account for a bendable and torsional tail instead of a rigid one.

(a) $\lambda = 20\,\mathrm{deg}$

(b) $\lambda = 0\,\mathrm{deg}$ hold with lateral wind of $10\,\mathrm{m/s}$ (19.4 knots)

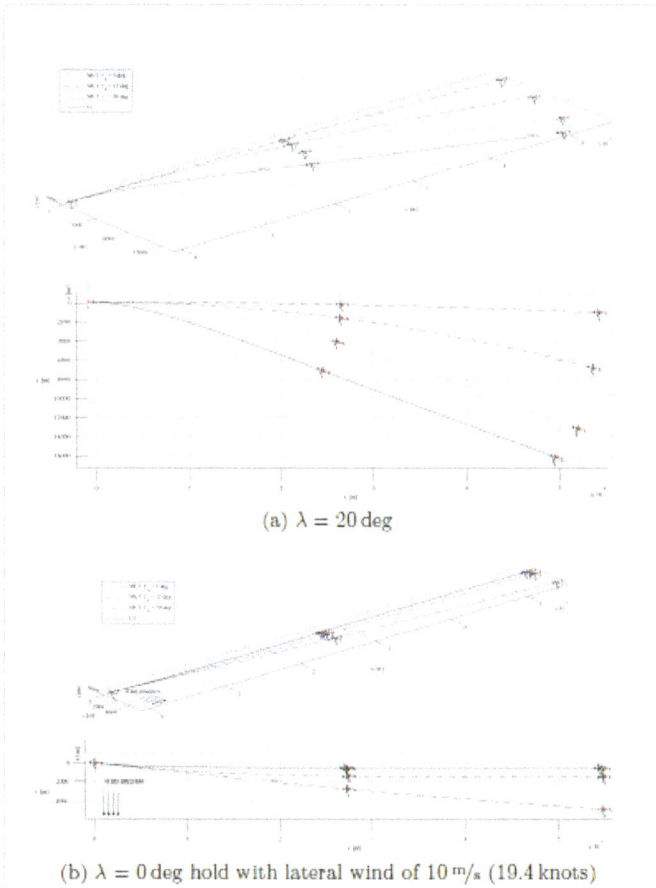

Figure 13: Comparison of the aircraft trajectory during a) a heading solicitation of 20 deg and b) 0 deg heading hold with lateral wind of 10 m/s (19.4 knots) for the MC1 with Γ_h=0 deg, 15 deg and 30 deg and CC configurations.

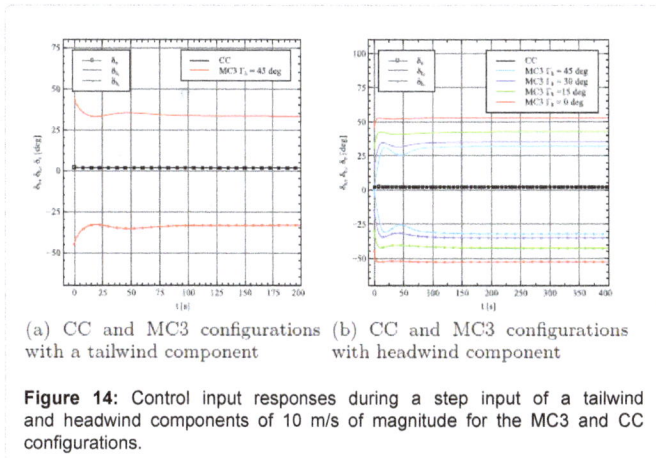

(a) CC and MC3 configurations with a tailwind component

(b) CC and MC3 configurations with headwind component

Figure 14: Control input responses during a step input of a tailwind and headwind components of 10 m/s of magnitude for the MC3 and CC configurations.

Figure 14 depicts the tail input response for both cases of headwind and tailwind components. It is possible to observe that the dihedral request amplitude decreases with the increase in the initial tail dihedral but is much higher when compared with the request to the elevators.

Conclusions

The static results showed that the aircraft was longitudinal stable for tail rotation angles in the range $-67\,\mathrm{deg} \lesssim \delta h \lesssim 67\,\mathrm{deg}$. Variation of tail dihedral angle and tail rotation angle are two mechanisms that were

References

1. Weisshaar TA (2006) Morphing Aircraft Technology - New Shapes for Aircraft Design.

2. Barbarino S, Bilgen O, Ajaj RM, Friswell MI, Inman DJ (2011) A Review of Morphing Aircraft. J Intel Mat Syst Str 22: 823-877.

3. Anna-Maria MR, Dan VD, Ronald BC, Andrew HS (2009) Perspectives on Highly Adaptive or Morphing Aircraft. NASA Technical Reports Server.

4. Friswell MI, Inman DJ (2006) Morphing Concepts for UAVs. 21st Bristol UAV Systems Conference.

5. Thomas ALR (1996) Why do Birds have Tails? The Tail as a Drag Reducing Flap, and Trim Control. J Theor Biol 183: 247-253.

6. Thomas ALR (1996) The Flight of Birds that have Wings and a Tail: Variable Geometry Expands the Envelope of Flight Performance. J Theor Biol 183: 237-245.

7. Evans MR, Rosén M, Park KJ, Hedenström A (2002) How do birds' tails work? Delta-wing theory fails to predict tail shape during flight. Proc Biol Sci 269: 1053-1057.

8. Sachs G (2005) Yaw stability in gliding birds. Journal of Ornithology 146: 191-199.

9. Sachs G (2007) Tail effects on yaw stability in birds. J Theor Biol 249: 464-472.

10. Hull DG (2007) Fundamentals of Airplane Flight Mechanics. Springer Berlin Heidelberg Germany.

11. Etkin B, Reid LD (1995) Dynamics of Flight: Stability and Control. 3 Edn Wiley, John & Sons, Incorporated USA.

12. McLean D (1990) Automatic flight control systems. Prentice Hall USA.

13. Stevens BL, Lewis FL (1992) Aircraft control and simulation. Wiley USA.

8

Development and Flight Test of an Avionics Lidar for Helicopter and UAV Low-Level Flight

Roberto Sabatini[1]*, Mark A Richardson[2] and Ermanno Roviaro[3]

[1]*Department of Aerospace Engineering, Cranfield University, Cranfield, Bedford MK43 0AL, UK*
[2]*Defence Academy of the UK, Cranfield University, Shrivenham, Swindon SN6 8LA, UK*
[3]*Electro-Optics R&D Laboratories, SELEX-ELSAG, Cogoleto (Genova) 16016, Italy*

Abstract

In recent years, laser radar (LIDAR) has become a promising technology for navigation and obstacle avoidance in helicopters and UAV, mainly because of its good wire detection performance on a wide range of incidence angles, and also due to its outstanding range and accuracy. In this paper we describe the activities carried out for the design, integration and test of the Laser Obstacle Avoidance System "Marconi" (LOAM) on helicopter and UAV platforms. After a brief description of the system architecture and sensor characteristics, emphasis is given to the performance models and processing algorithms required for obstacle detection/classification and calculation of alternative flight paths, as well as to the ground and flight test activities performed on various platforms.

Keywords: LIDAR; Laser obstacle detection; Laser warning system; UAV low-level flight; Helicopter low-level flight

Introduction

In order to achieve mission effectiveness in the present threat environment, military helicopters and small Unmanned Aerial Vehicles (UAVs) operations are focusing on low-level or nap-of-the-earth flying. This is the tactic of employing the aircraft in such a manner as to utilize the terrain profile to enhance survivability by degrading the enemy's ability to visually, optically or electronically detect or locate the aircraft. In these scenarios, radar is normally required to maintain the aircraft flight at a present altitude above the terrain. Since the adoption of this philosophy, the incidence of obstacle strike accidents has grown. The main restrictions for low-level navigation and terrain following operations with helicopters and UAV are due to adverse weather conditions. Low visibility is the main reason that prevents flight/ground crews from safely controlling the aircraft and from identifying possible obstacle collision hazards. The first laser experiment directed towards a laser obstacle detection and avoidance system started in 1965 with a Nd:YAG laser [1]. This system demonstrated the feasibility of using lasers to detect obstacles such as wires. Semiconductor lasers, such as GaAs and GaAlAs have been experimented with since 1966 [2]. These lasers radiate in the wavelength region of 0.84 to 0.9 µm. The experience gained with these experimental systems pointed out many features that are now being incorporated into present day research. Due to eye-safety and adverse weather (fog) propagation concerns, further development with Nd:YAG and the various semiconductor lasers has been substantially reduced, in favour of CO_2 lasers. One of the first heterodyne detection CO_2 system was the LOWTAS, developed by UTRC. More recent developments include CLARA, the Anglo-French compact laser radar demonstrator program [3]; HIWA, a German system built and tested by Eltro and Dornier [4]; and OASYS, developed in the U.S. by Northrop [5]. Current research efforts are concentrating on 1.54 µm (Raman-shifted Nd:YAG and Er:glass) solid state lasers. One 1.54 µm system is being developed for the Italian Military Forces by Marconi S.p.A., in cooperation with the Air Force Flight Test Centre. The equipment, here named LOAM (Laser Obstacle Avoidance System "Marconi"), is a low-weight/volume navigation aid system for rotary-wing/UAV platform specifically designed to detect potentially dangerous obstacle placed in or nearby the flight trajectory and to warn the crew in suitable time to implement effective avoiding manoeuvres.

The first airborne prototype of the LOAM has been assembled and extensive laboratory and field tests have been performed on the various sub-units, in order to refine the system design (both hardware and software components). Furthermore, the overall system is now being tested in flight.

Operational Requirements for an OWS

For an Obstacle Warning System (OWS) to be effective it must meet certain requirements. The first and most important requirement is reliable detection of all obstacles at almost all angles of incidence of radiation with a very high probability of detection and very low false alarm rate. By all obstacles, it is meant terrain masses, buildings, poles, towers, power cables and indeed any structure which may pose a hazard to low/fast flying aircraft.

The need for a high probability of detection is obvious since no obstacle must go undetected. A low false alarm rate is required to prevent spurious warnings that would cause the pilot to increase his altitude without real need, thus making him a better target.

Another operational requirement is the minimum detection range. This will depend upon the aircraft speed, climb angle capability (different for helicopter, UAV and airplane platforms), and pilot reaction time. As an example, for an airplane flying straight and level at 300 m/sec and allowing a reasonable pilot reaction time and aircraft response time of between five to ten seconds, detection ranges of about two to three kilometres are adequate. For helicopter and small UAV applications, this range is generally reduced by an order of magnitude or more.

---

*Corresponding author: Roberto Sabatini, Department of Aerospace Engineering, Cranfield University, Cranfield, Bedford MK43 0AL, UK
E-mail: r.sabatini@cranfield.ac.uk

The system should, ideally, perform all of its required functions in all weather, day and night. In practice however, laser radiation is not capable of all-weather operation and the best trade-off of system characteristics must be looked at.

LOAM General Description

The LOAM system is capable of detecting obstacles placed in or nearby the helicopter/UAV trajectory, classifying/prioritising the detected obstacles, and providing obstacle warnings and information to the crew (both aural and visual). The system laser beam scans periodically the area around the flight trajectory inside a FOV of 40° in azimuth and 30° in elevation with field of regard capability of ± 20° both on azimuth and elevation, centred on the optical axis of the system (Figure 1).

Furthermore, LOAM allows the operator to select the azimuth orientation of the FOV among three possible different directions (Figure 2).

This allows for the optical axis to be oriented either in the same direction of the platform "heading" (normal flight envelope), or 20°

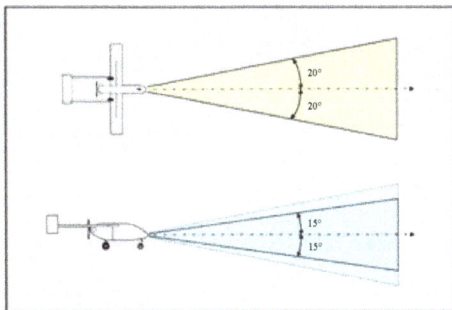

Figure 1: LOAM Horizontal and Vertical FOV.

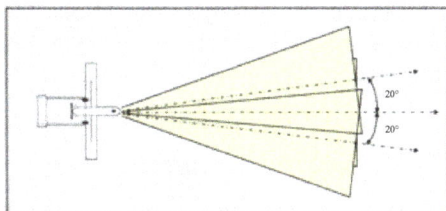

Figure 2: LOAM FOV Orientation.

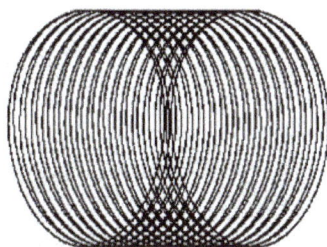

Figure 3: LOAM Scan Pattern.

left/right with respect to the platform "heading" (to optimise coverage during turning manoeuvres at high angular speed). During every scan (4 Hz repetition frequency), the laser beam changes its orientation producing an elliptical pattern across the FOV with the characteristics shown in Figure 3.

After various experiments performed with different patterns, the scanned elliptical pattern was selected. The main advantages of this scanning pattern are:

• It is well adapted to the detection of most dangerous obstacles, like wires, due to the several and evenly qually spaced vertical lines

• It can be realised with a very reliable scan mechanics with reduced weight.

Using dedicated signal processing algorithms optimised for low-level obstacle detection, the system holds an inherently high capacity to detected various types of obstacle independently from the platform motion during the frame acquisition period, providing the possibility of reconstructing the obstacle shape without using navigation data (stand-alone integration) in slow-moving platforms with a benign attitude envelope. Additionally, LOAM can be integrated with the aircraft navigation sensors if required, especially in platforms with high dynamics envelopes [6,7].

LOAM performs echo detection through an analogue process comprising an optical-electrical conversion, a signal pre-amplification and a threshold comparison. The signal pre-amplification is achieved by an automatic controlled gain amplifier to increase the system sensitivity as the elapsed time from the laser emission increases in order to adjust the sensitivity on the basis of the expected return signal power in connection with the obstacle range. Furthermore, an adjustable threshold level is provided to take into account the background conditions. These features reduce the probability of false echo detection due to the atmospheric back-scatter near the laser beam output and optimise the system sensitivity in various operational weather conditions.

LOAM performs echo analysis in order to determine the presence of possible obstacles and to determine their geometrical characteristics and position. For this purpose, LOAM operates through two sequential analysis process: local analysis and global analysis.

The "local analysis" process is performed on the single echoes in order to determine range, angular coordinates and characteristics of the obstacle portion generating them. The "global analysis" process manages groups of echoes, detected during a scan period, with the related information provided by the "local analysis" process, in order to perform the obstacle detection as a whole and determine the related shape and type. LOAM is capable to automatically classify obstacles according to the following classes [6]:

• Wire: This class groups all thin obstacles like wires and cables (e.g., telephone cables, electrical cables and cableway);

• Tree: This class groups vertical obstacles of reduced dimensions (e.g., trees, poles and pylons);

• Structure: This class groups extended obstacles (e.g., bridges, buildings and hills).

Furthermore, LOAM performs automatic prioritisation of the detected obstacles in function of the risk represented according to the relevant range, and provides the crew with timely warnings and information of the detected obstacles in order to allow the

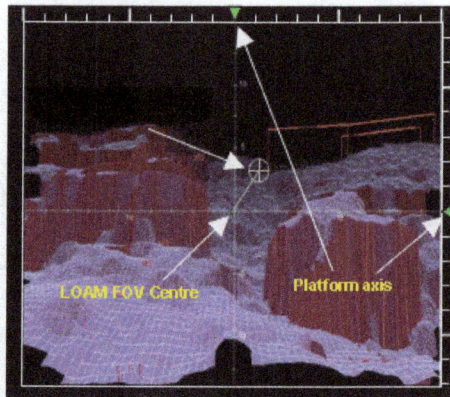

Figure 4: LOAM 3-D Display Format.

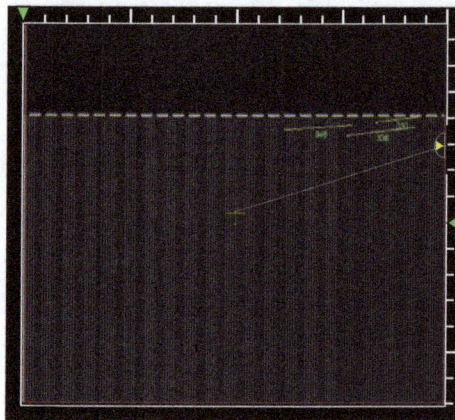

(a)

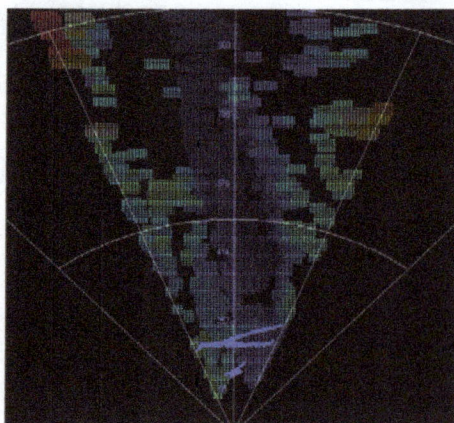

(b)

Figure 5: LOAM 2-D and Altimetric Display Format.

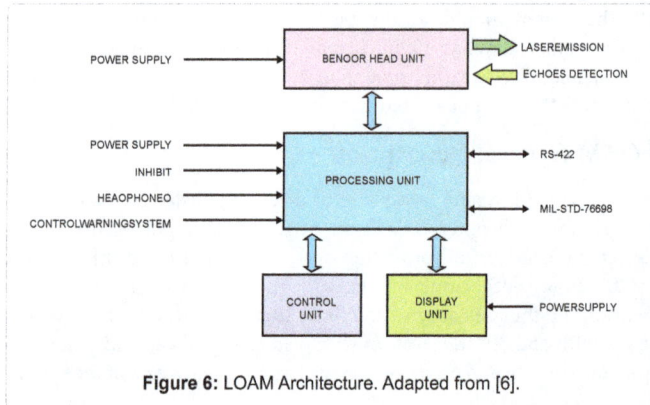

Figure 6: LOAM Architecture. Adapted from [6].

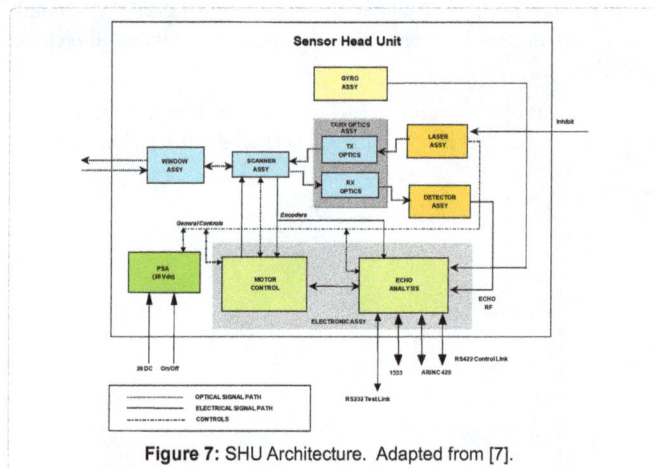

Figure 7: SHU Architecture. Adapted from [7].

together with an altimetric profile format. An example of a 3-D LOAM display format is shown in Figure 4.

The LOAM 2-D and altimetric display formats are shown in Figure 5a and 5b respectively The general architecture of LOAM system is shown in Figure 6. LOAM main components are the Sensor Head Unit (SHU), the Control Panel (CP) and the Display Unit (DU). In the following paragraph a brief description of LOAM SHU is given, together with an outline of the main EPU functions.

LOAM Sensor Head Unit Architecture

The SHU performs the following main functions [7]:

• To generate a laser beam and scan the area around the flight trajectory;

• To detect return echoes;

• To analyse detected echoes in order to compute range, coordinates and local geometrical characteristics (attributes) of the obstacles they come from;

• To communicate echo range, coordinates and attributes to LOAM Processing Unit, or to other on board systems, via a RS-422 high speed serial data link.

The architecture of the SHU is shown in Figure 7.

As illustrated above, the SHU scans a laser beam in the area around the flight trajectory, performs echo detection through an analogue process comprising an optical-electrical conversion (by means of an avalanche photodiode-APD), a signal pre-amplification and a threshold

implementation of effective avoidance manoeuvres. For this purpose, LOAM system is able to deliver both visual and audio warnings.

LOAM information relative to the detected obstacles can be provided on a dedicated display (NVG compatible), whose screen represents the FOV of the system. Both 3-D and a 2-D are possible,

comparison (adjustable threshold).

The SHU performs echo analysis in order to compute range, coordinates (azimuth, elevation with respect to LOAM reference frame) and local geometrical characteristics (attributes) of the obstacles they come from. Particularly, the following functions are performed:

• The echo angular coordinates are determined on the grounds of the scanner orientation;

• The echo range is calculated computing the "two-way" travelling time of the scan laser pulse;

• The geometrical characteristics of the echo are determined with a local "geometrical" analysis of nearby echoes along the scanner pattern and on the ground of the "absolute" power returned.

The SHU interfaces include

• One RS422 serial link to the Control Panel for controls and BIT activation

• One RS232 serial link for off-line test purpose

• One ARINC 429 to acquire H/C navigation data

• One MIL-1553-BUS to deliver obstacle properties and coordinates

• One discrete input signal to inhibit laser emission

• Two discrete output signal for audio warning.

The Window Assembly allows the transmission and the reception of the laser beam across the SHU chassis. The Window Assembly is realised with a slice of synthetic fused silica.

The Scanner Assembly integrates the H/W resources necessary to scan the laser beam, and the virtual input pupil of the detector, throughout the overall FOV. It also allows Line of Sight (LOS) orientation. For this purpose, the Scanner Assembly comprises:

• A swash mirror mounted on an azimuth turret

• One brushless motor to allow the swash mirror motion

• One brushless motor to allow the azimuth turret motion

• One brushless motor to allow the tilt turret motion.

The swash mirror rotates at a constant speed around its axis reflecting the laser beam thus to draw a quasi-elliptical pattern in space. The turret periodically sweeps in azimuth the FOV. The composition of these two movements allows to produce the required quasi elliptical scan pattern previously described. Change in LOS orientation is achieved offsetting the central position of the periodical sweep of the

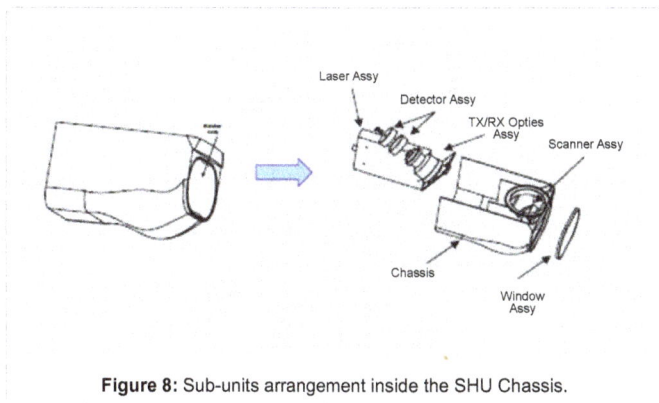

Figure 8: Sub-units arrangement inside the SHU Chassis.

turret by an angular value equal to the required change.

According to the SHU architecture shown in Figure 8, the TX/RX Optics Assembly integrates the optical components necessary:

In transmission

• To collect via fiber optics the laser output power from the Laser Assembly

• To generate the scan laser beam with the required optical divergence and dimensions

• To projecting the scan laser beam on the swashing mirror of the Scanner Assembly

In reception

• To collect the echo return power reflected by the swashing mirror of the Scanner Assembly

• To focalise the collected power on the photodiode of the Detector Assembly.

To this purpose, the TX/RX Optics Assembly comprises:

For transmission

• One beam expander that provides to collect the laser output power via fiber optics and to expand and parallelise it

• One prism that allows to reflect the generated beam onto the swashing mirror with the due alignment

For reception

• One telescope that collects the echo returns power reflected by the swashing mirror of the Scanner Assembly and to focalise it on the photodiode of the Detector Assembly.

The Detector Assembly provides to detect laser echoes on the grounds of the laser return power received through the TX/RX Assembly. To this purpose, the Detector Assembly comprises an Avalanche Photodiode (APD) with related bias circuitry, a controlled gain amplifier and the threshold circuitry necessary for the echoes detection, all integrated in a single mechanical module straight connected to the telescope of the TX/RX Assembly.

The Electronic Assembly performs the following functions:

• Analyses detected echoes, received as an RF signal from the Detector Assembly

• Controls the scanner assembly motors

• Handles SHU general controls and BIT operations

• Processes the detected echoes, in order to analyse and recognise obstacle class, in order to properly alert H/C crew

• Processes the acquired information in order to detect, isolate and calculate position and characteristics of potential obstacles

• Computes display information and symbols data

• Makes available the warning information to the Display Unit.

All the relevant electronics to accomplish the above mentioned functions is integrated in two printed circuit board.

The location of the Laser Assembly, the Detector Assembly, the TX/RX Optics Assembly, the Scanner Assembly and the Window Assembly inside the Chassis is shown in Figure 9. The Laser Assembly provides

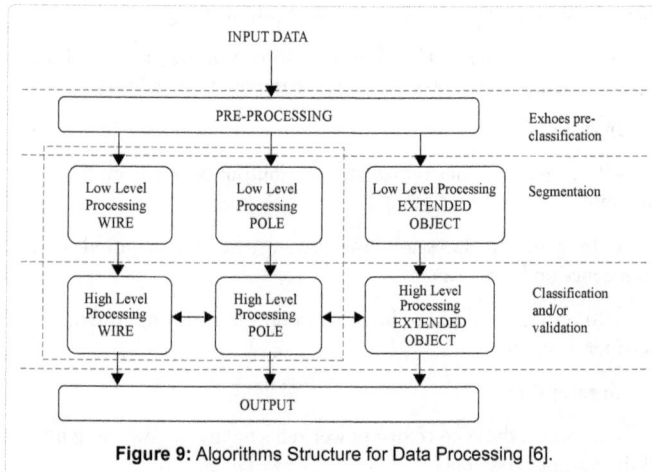

Figure 9: Algorithms Structure for Data Processing [6].

the required laser power necessary to scan operations. It comprises an Erbium doped fiber (Er:fiber) laser, the related control circuitry and power supply, all integrated in a single mechanical module. The laser power delivery to the "TX/RX Optics assembly" is provided via a fiber optics connected to the beam expander.

The Power Supply Assembly provides the power required by all SHU sub-units, except for the Laser Assembly which is directly connected to the platform mains. To this purpose, the Power Supply Assembly comprises in a single mechanical module all the circuitry necessary to interface with the platform mains and to generate output voltages as required by the SHU sub-units. The Gyro Assembly provides reference signals to the Electronic Assembly to uncouple echoes coordinates with respect to the helicopter/UAV vibration to compensate echoes angular co-ordinates, to allow a correct obstacles geometry reconstruction. The Gyro Assembly is composed by 3 gyros integrated in a single mechanical module. The Chassis is realised by a casting aluminium mechanical envelope that encloses and protect all the SHU sub-units.

In an obstacles detection and warning system, there is the need to provide only essential information to the pilot. The scanner system, in fact, detects the position of every potential obstacle in the environment where the helicopter/UAV is moving. In a generic scenario, with many obstacles in the field of view of the warning system, the pilot may have difficulties to control them. For this reason, a system capable of discriminating the most dangerous obstacles and supplying the relative information is required. To solve this problem, three algorithms have been developed for future incorporation in LOAM:

- Calculation of future trajectory

- Calculation of intersections with the obstacles

- Determination of alternative (optimal) trajectory.

Another area of development consists in the definition of efficient processing algorithms for performing Obstacle Detection and Classification Processing (ODC) at a very high speed and with high precision. In LOAM, a pre-processing is performed that, according to the range contrast between consecutive laser returns, allows a pre-classification of detected obstacles. This pre-classification defines the attributes of the echoes. The subsequent processing, taking into account these attributes, performs final classification and also straight-line recognition algorithms, which extract from the echoes list those related to the same structure. Particularly, two different algorithms have been developed: the first is optimised to process echoes generated by

thin objects, like wires and poles, the second is optimised to process all the echoes generated by extended obstacles, like houses, trees, woods and other solid objects. These algorithms identify the boundaries of the obstacles; additional neighbourhood criteria allow distinguishing "wire-class" from "extended object" obstacle classes.

Dedicated simulation activities and actual flight tests were required to verify and refine the performance of the various processing algorithms described, as well as to determine the sensor performance in favourable and adverse weather conditions.

Some electro-optical parameters relative to the laser sub-unit are listed in Table 1.

Obstacle Detection and Classification Algorithms

As described before, LOAM anti-collision system performs obstacle detection on the basis of laser radar technique. Once the echoes energy has been optically collected, the obstacle detection is performed on the basis of an echoes analog detection and of two successive analysis processes of the detected echoes.

The first process, referred as preprocessing, is performed at a very high rate straight during the echo acquisition in order to achieve single echo specific data and to characterize it on the basis of local range contrast analysis with respect to nearby echoes. The preprocessing analyzes only echoes that are generated by obstacles that are within the fixed range not to compel the processing assembly to process unnecessary echoes.

The second process, referred as processing, is performed at a lower rate and manages groups of preprocessed echoes with the related information in order to achieve, by a two-step analysis, the final obstacle detection and the related shape and type.

Considering the current helicopter/UAV operational scenarios, wires represent by far, the most dangerous obstacle due to their low visibility to the naked eye even in good weather conditions. No present on board navigation/vision system is able to effectively aid the pilot in being warned of their presence.

In solving the problem of obstacle detection and classification for a laser radar system, the main difficulty was to find an algorithm capable to perform the processing at a very high speed and at the same time able to provide result responding to the precision requirements.

The pre-processing algorithms elaborate the range contrast between consecutive laser returns and perform a pre-classification of detected obstacles. This pre-classification defines the attributes of the echoes. The processing algorithms, taking into account these attributes performs final classification and also straight-line recognition algorithms, which extract from the echo's list, those related to the same structure.

There are two different types of processing algorithms: the first is optimized to process echoes generated by thin objects, like wires and poles, the second is optimized to process all the echoes generated by extended obstacles, like houses, trees, woods and other solid objects. These algorithms identify the boundaries of the obstacles; additional

Parameter	Description	Value
Wavelength	Laser emission wavelength	1.55 µm
Peak Power	Laser pulse power at the "Laser Assembly" output	10 kW
Pulse Duration	Laser pulse duration	3 to 5 ns
Frequency	Laser pulse repetition frequency	60 kHz

Table 1: Laser Parameters.

neighborhood criteria allow distinguishing "wire-class" obstacles from "extended object-class" obstacles.

In order to perform its tasks the processing algorithms make use of image and data segmentation and data validation. Figure 10 shows the three levels of the processing algorithms. The processing algorithms are conceived and optimized for the quasi elliptical scanning pattern described before.

The thin object-processing algorithm works on a subset of echoes of the current frame. This algorithm processes only the echoes whose attributes, defined by pre-processing assembly, are weak echo and thin object. To process the incoming data it is not necessary to acquire the entire frame; the single echo, in fact, is processed as soon as it is acquired.

Image segmentation is the process of dividing the image into areas where the echoes are relatively uniform in range value and possible thin obstacles are created from this subset of data. Then all possible generated thin obstacles are processed so that some redundant cluster can be merged.

After the image segmentation the different clusters must be validated. This means that the detected echoes are processed by a statistical algorithm to determine if the obstacles are generated by real aligned echoes or by noising data. Also the algorithm dedicated at the detection of the extended object is divided in two different phases: echoes classification and segmentation. The echoes already classified as extended object need to be processed by a dedicated selection algorithm because many of these are not generated by a real extended obstacle (like, for example, the ground). A well-defined number of echoes, acquired in a short time range, have some geometrical characteristics. These additional attributes permit to decide the validation of data, which are passed to the segmentation algorithm. With the segmentation the different founded clusters are rearranged and validated.

The results of the developed processing algorithms are tested with experimental data, acquired with a sensor prototype, and then displayed and analyzed with a debugging interface. The user may change the key parameters, which define the algorithms, with commands available in the debugging interface, so their values can be determined observing the experimental results. In Figure 11 an example of the debugging

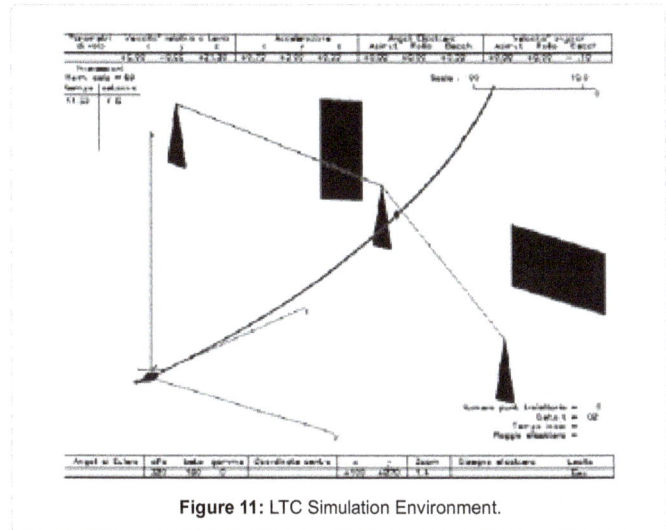

Figure 11: LTC Simulation Environment.

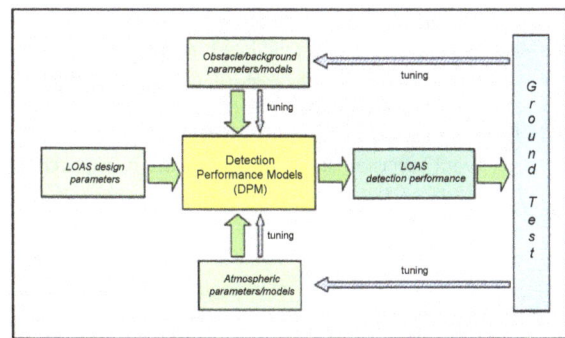

Figure 12: Detection performance models and ground test.

interface is shown. With the processing of currently available data, we see that the algorithms are capable to detect and classify the different obstacles, thanks to the flight testing currently performed, the parameters are being definitively set and optimized.

Processing of currently available data has shown that the algorithms now implemented are capable to detect and classify the different obstacles of interest, and flight-testing being carried out on helicopters and UAVs is giving the opportunity to definitively set and optimize the processing parameters.

Calculation of Alternative Flight Paths

In a laser obstacle detection and warning system, there is also the problem of providing only essential information to the pilot. LOAM scanner system, in fact, detects the position of every potential obstacle in the environment where the platform is moving. In a generic scenario, with many obstacles in the field of view of the warning system, it may be difficult for the pilot to monitor all of them. For this reason, a system able to discriminate the more dangerous obstacles and to supply the relative information is required.

To solve this problem, three algorithms have been implemented: calculation of next trajectory segment, calculation of possible intersections with obstacles and determination of an alternative (optimal) trajectory. A simulation environment was required to test and refine the algorithms performance. Therefore, to simulate the platform flight in a generic scenario, a three-dimensional environment

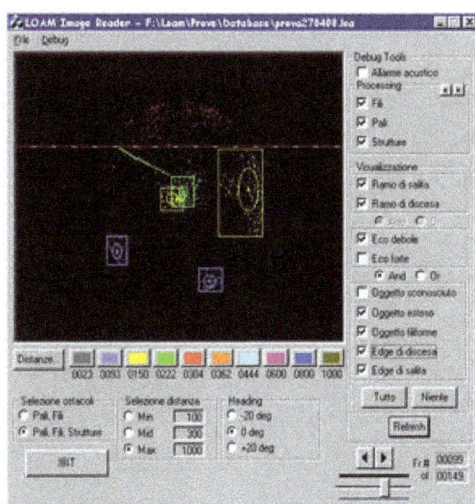

Figure 10: Debugging Interface with displayed results.

Figure 12 has been implemented that allows seeing, from different point of view, the scenario scanned from laser sensor. In addition, this environment permits to observe the trajectory's trace followed by the platformr (trajectory from proper algorithm calculated) and verify the intersection with the obstacles. The simulation allows to modify smoothly the flight parameters and to observe the platform motion and the forces acting on it.

The main functions performed by the algorithms are essentially two: calculation of the next trajectory and determination of possible intersections between the platform and the obstacles located in the scenario and detected by laser scanner.

In the case of manned and unmanned rotorcraft, the calculation problem for the trajectory is a more difficult task due to the complexity of this type of aircraft and implies a careful study of his flight dynamic. The trajectory followed by the platform and the supposed flight path in the subsequent twenty seconds, is extrapolated from motion's data supplied by navigation's system: the velocity relative to earth, the velocity relative to air, the acceleration and the angles describing the platform orientation.

The second algorithm is developed to search a possible intersection between the trajectory previously calculated and the obstacles acquired by laser scanner. Through the analysis of every single intersection, the algorithm provides a priority list of the most dangerous obstacles located in front of the aircraft. The algorithm is optimized to satisfy the performance, about precision and speed of execution, required.

LOAM History Function

Due to the restricted system field of view, some information, acquired in the previous frame, may be lost successively. To keep obstacles information when they are outside the present frame, it is necessary to store the position of every object detected and then update the coordinates with respect to the platform body-fixed reference system. LOAM history function stores the data of the obstacles for a defined time interval and deletes them when they are outside the platform possible trajectories (outside its flight envelope).

Since the motion data supplied from navigation system are, like every measure, affected by an error, we evaluate how these errors affect the positions calculated for every obstacle. To do so, a Gaussian error is added to every data and a statistic of the position error is calculated for obstacles near and far from the aircraft. When the impact warning processing establishes that the trajectory currently flown by the aircraft has a collision risk, the algorithm searches the corrections necessary to avoid the obstacles, and provides the pilot an indication about the alternative (optimal) direction to fly. The optimal trajectory is the one having the smaller possible correction (necessary to avoid the obstacles) and which is compatible with a safe flight path.

LOAM Performance Prediction and Ground Test

Ground trials of the LOAM system were performed in order to verify the system detection performance in various weather conditions, and to test the validity of the mathematical models used for performance calculations. This was particularly important for preparing the LOAM flight test activity. It was in fact necessary to define a criteria for determining the system detection range performances in the worst environmental conditions, and with the worst obstacle scenarios (i.e., small wires with low reflectivity), even without performing real tests in these conditions (i.e., using experimental data collected in fear weather and with average obstacles). Mathematical modeling and ground

testing of the LOAM detection performance were therefore required in order to give proper weights to the parameters playing a role in realistic operational scenarios, and to determine the target LOAM detection performances to be demonstrated in flight. Figure 13 illustrates the process involved.

As the ground test activities permitted to validate the models developed, it was then possible to identify reference sets of obstacle, background and atmospheric parameters giving the absolute minimum performance of the LOAM system. This is illustrated in Figure 14. Obviously, the successive flight test activities were performed only in a small portion of the LOAM operational envelopes, but the results obtained could be extended to the entire envelopes by using the validated mathematical models.

For initial design calculations, the wire obstacle detection capability of the LOAM is modelled by the following simplified Signal to Noise Ratio (SNR) equation:

$$SNR = \frac{4E_p A_r L_T L_r e^{-2\gamma R} d_W \rho}{\pi P_D R^2 (\alpha R + D) NEP} \tag{1}$$

where:

EP=output laser pulse energy

Ar=receiver aperture

LT=transmission losses (including beam shaping)

Lr=reception losses (including optical filter)

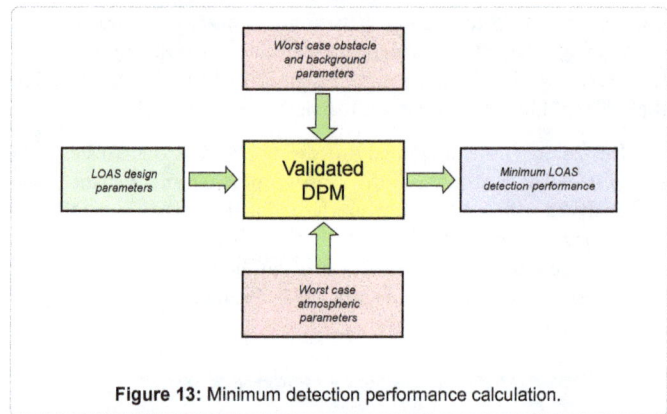

Figure 13: Minimum detection performance calculation.

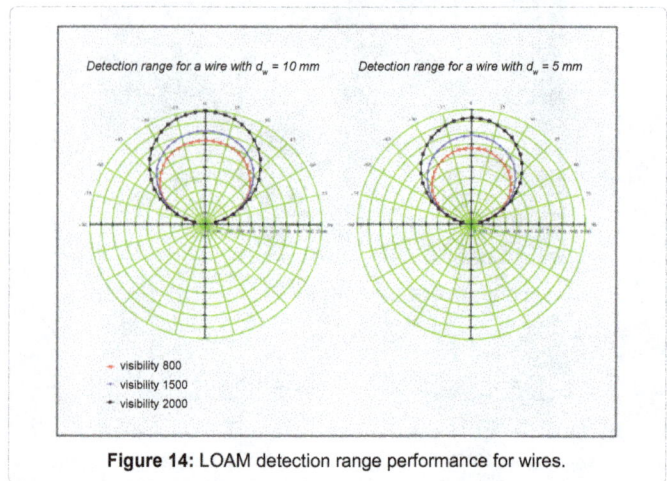

Figure 14: LOAM detection range performance for wires.

γ=atmospheric extinction coefficient dW=wire diameter

ρ=wire reflectivity

PD=pulse duration

R=obstacle range

α=beam divergence (l/e2)

D=initial beam diameter

NEP=noise equivalent power

The extinction coefficient (γ) is calculated using the empirical model suggested by Elder and Strong [8] and modified by Langer [9]. Additionally, for propagation in rainy conditions, the equations developed by Middleton were adopted [10]. This approach (ESLM model) is particularly useful because it provides a means of relating the atmospheric transmission of the ith window to the atmospheric visibility, relative humidity and rainfall-rate (i.e., readily measurable parameters).

The first assumption made [8,9] is that variations in the transmission are caused by changes in the water content of the air. Specifically, changes in the concentration of H_2O cause changes in the absorption, and changes in the size and number of water droplets with humidity cause changes in the scattered component. This is a valid assumption since the other atmospheric constituents have a reasonably constant effect on the transmittance of a given atmospheric window.

It is customary to express the number of H_2O molecules encountered by the beam of light in terms of the number of precipitable millimetres of water in the path. Specifically, the depth of the layer of water that would be formed if all the water molecules along the propagation path were condensed in a container having the same cross-sectional area as the beam is the amount of precipitable water. A cubic meter of air having an absolute humidity of ρ grams per m³ would yield condensed water that cover an l m² area and have a depth of:

$$w' = 10^{-3}\rho \tag{2}$$

where w' is the precipitable water having units of mm per meter of path length. For a path length of z meters eq. (2) becomes:

$$w = 10^{-3}\rho \cdot z \tag{3}$$

where w is now the total precipitable water in millimetres. The value of ρ, the density of water vapour, can be obtained using the following equation [6], which is convenient for computer code implementation:

$$\rho = 1322.8 \cdot \frac{RH}{T} e^{\left[\frac{25.22\cdot(T-273.16)}{T}-5.31\cdot\ln\left(\frac{T}{273.16}\right)\right]} \tag{4}$$

Where RH is the relative humidity (as a fraction), and T is the absolute temperature (°K). Two empirical expressions, developed by Langer [9], can be used to calculate the absorptive transmittance τ_{ai} for the ith window for any given value of the precipitable water content. These expressions are:

$$\tau_{ai} = e^{-A_i\sqrt{w}} , \text{ for } w < w_i \tag{5}$$

$$\tau_{ai} = k_i\left(\frac{w_i}{w}\right)^{\beta_i}, \text{ for } w > w_i \tag{6}$$

Where A_i, k_i, β_i and w_i are constants whose values for each atmospheric window are listed in [8,9]. For the LOAM laser wavelength

(λ=1550 nm-4th atmospheric window), A_i=0.211, k_i=0.802 , β_i=0.111 and w_i=1.1.

In summary, eq. (5) and eq. (6), together with eq. (3) and eq. (4), provide information that can be used to obtain an estimate of the absorptive transmittance (τ_{ai}) of laser beams having wavelengths that fall within the various atmospheric windows. The results apply to horizontal paths in the atmosphere near sea-level and for varying relative humidity. To obtain the total atmospheric transmittance we must multiply τ_{ai} by τ_{si} (i.e., the transmittance due to scattering only).

Based on rigorous mathematical approaches, the scattering properties of the atmosphere due to the aerosol particles are difficult to quantify, and it is difficult to obtain an analytic expression for the scattering coefficient that will yield accurate values over a wide variety of conditions. However, an empirical relationship that is often used to model the scattering coefficient [5] has the form:

$$\beta(\lambda) = C_1\lambda^{-\delta} + C_2\lambda^{-4} \tag{7}$$

Where C_1, C_2, and δ are constants determined by the aerosol concentration and size distribution, and λ is the wavelength of the radiation. The second term accounts for *Rayleigh* scattering. Since for all wavelengths longer than about 0.3 μm the second term is considerably less than the first, it may be neglected. It has been found that produces reasonable results when applied to aerosols with a range of particle sizes. An attempt has also been made to relate δ and C_1 to the meteorological range. The apparent contrast C_z, of a source when viewed at λ=0.55 μm from a distance z is by definition:

$$C_z = \frac{R_{sz} - R_{bz}}{R_{bz}} \tag{8}$$

where R_{sz} and R_{bz} are the apparent radiances of the source and its background as seen from a distance z. For λ=0.55 μm, the distance at which the ratio:

$$V = \frac{C_z}{C_0} = \frac{\dfrac{R_{sz} - R_{bz}}{R_{bz}}}{\dfrac{R_{s0} - R_{b0}}{R_{b0}}} = 0.02 \tag{9}$$

is defined as the meteorological range V (or visual range). It must be observed that this quantity is different from the standard observer visibility (V_{obs}). Observer visibility is the greatest distance at which it is just possible to see and identify a target with the unaided eye. In daytime, the object used for V_{obs} measurements is dark against the horizon sky (e.g., high contrast target), while during night time the target is a moderately intense light source. The International Visibility Code (IVC) is given in Table 2.

Designation	Visibility
Dense Fog	0–50 m
Thick Fog	50–200 m
Moderate Fog	200–500 m
Light Fog	500–1 km
Thin Fog	1–2 km
Haze	2–4 km
Light Haze	4–10 km
Clear	10–20 km
Very Clear	20–50 km
Exceptionally Clear	>50 km

Table 2: International Visibility Code (IVC).

It is evident that, while the range of values for each category is appropriate for general purposes, it is too broad for scientific applications. Visibility is a subjective measurement estimated by a trained observer and as such can have large variability associated with the reported value. Variations are created by observers having different threshold contrasts looking at nonideal targets. Obviously, visibility depends on the aerosol distribution and it is very sensitive to the local meteorological conditions. It is also dependent upon the view angle with respect to the sun. As the sun angle approaches the view angle, forward scattering into the line-of-sight increases and the visibility decreases. Therefore, reports from local weather stations may or may not represent the actual conditions at which the experiment is taking place. Since meteorogical range is defined quantitatively using the apparent contrast of a source (or the apparent radiances of the source and its background) as seen from a certain distance, it eliminates the subjective nature of the observer and the distinction between day and night. Unfortunately, carelessness has often resulted in using the term "visibility" when meteorological range is meant. To insure that there is no confusion, "observer-visibility" (V_{obs}) will be used in this thesis to indicate that it is an estimate.

If only V_{obs} is available, the meteorological range (V) can be estimated [11] from:

$$V \approx \left(1.3 \pm 0.3\right) \cdot V_{obs} \tag{10}$$

From eq. (10), if we assume that the source radiance is much greater than the background radiance (i.e., $R_s \gg R_b$) and that the background radiance is constant (i.e., $R_{bo} = R_{bz}$), then the transmittance at $\lambda = 0.55$ μm (where absorption is negligible) is given by:

$$\frac{R_{sv}}{R_{s0}} = e^{-\beta V} = 0.02 \tag{11}$$

Hence, we have:

$$\ln\left(\frac{R_{sv}}{R_{s0}}\right) = -\beta V = -3.91 \tag{12}$$

and also:

$$\beta = \frac{3.91}{V} = C_1 \lambda^{-\delta} \tag{13}$$

It follows from eq. (13) that the constant C_1 is given by:

$$C_1 = \frac{3.91}{V} \cdot 0.55^{\delta} \tag{14}$$

With this result the transmittance at the centre of the ith window is:

$$\tau_{si} = e^{-\frac{3.91}{V}\left(\frac{\lambda_i}{0.55}\right)^{-\delta} \cdot z} \tag{15}$$

where λ_i must be expressed in microns. If, because of haze, the meteorological range is less than 6 km, the exponent δ is related to the meteorological range by the following empirical formula:

$$\delta = 0.585 \sqrt[3]{V} \tag{16}$$

where V is in kilometres. When $V \geq 6$ km, the exponent δ can be calculated by:

$$\delta = 0.0057 \cdot V + 1.025 \tag{17}$$

For exceptionally good visibility $\delta = 1.6$, and for average visibility $\delta \approx 1.3$. In summary, eq. (17), together with the appropriate value for δ, allows to compute the scattering transmittance at the centre of the ith window for any propagation path, if the meteorological range V is known. It is important to note here that in general the transmittance will, of course, also be affected by atmospheric absorption, which depending on the relative humidity and temperature may be larger than τ_{si}.

For LOAM, it is also very important to model propagation through haze and precipitation. Haze refers to the small particles suspended in the air. These particles consist of microscopic salt crystals, very fine dust, and combustion products. Their radii are less than 0.5 μm. During periods of high humidity, water molecules condense onto these particles, which then increase in size. It is essential that these condensation nuclei be available before condensation can take place. Since salt is quite hygroscopic, it is by far the most important condensation nucleus. Fog occurs when the condensation nuclei grow into water droplets or ice crystals with radii exceeding 0.5 μm. Clouds are formed in the same way; the only distinction between fog and clouds is that one touches the ground while the other does not. By convention fog limits the visibility to less than 1 km, whereas in a mist the visibility is greater than 1 km. We know that in the early stages of droplet growth the Mie attenuation factor K depends strongly on the wavelength. When the drop has reached a radius a $\approx 10\,\lambda$ the value of K approaches 2, and the scattering is now independent of wavelength, i.e., it is non-selective. Since most of the fog droplets have radii ranging from 5 to 15 μm they are comparable in size to the wavelength of infrared radiation. Consequently the value of the scattering cross section is near its maximum. It follows that the transmission of fogs in either the visible or *IR* spectral region is poor for any reasonable path length. This of course also applies to clouds. Since haze particles are usually less than 0.5 μm, we note that for laser beams in the *IR* spectral region $a/\lambda \ll 1$ and the scattering is not an important attenuation mechanism. This explains why photographs of distant objects are sometimes made with infrared-sensitive film that responds to wavelengths out to about 0.85 μm. At this wavelength the transmittance of a light haze is about twice that at 0.5 μm. Raindrops are of course many times larger than the wavelengths of laser beams. As a result there is no wavelength-dependent scattering. The scattering coefficient does, however, depend strongly on the size of the drop.

Middleton [11] has shown that the scattering coefficient with rain is given by:

$$\beta_{rain} = 1.25 \cdot 10^{-6} \frac{\Delta x / \Delta t}{a^3} \tag{18}$$

where $\Delta x / \Delta t$ is the rainfall rate in centimetres of depth per second and a is the radius of the drops in centimetres. Rainfall rates for four different rain conditions and the corresponding transmittance (due to scattering only) of a 1.8-km path are shown in Table 3.

These data are useful for order of magnitude estimates. In order to obtain accurate estimates, the concentrations of the different types of rain drops (radius) and the associated rainfall rates should be known. In this case, the scattering coefficient can be calculated as the

Rainfall (cm/h)	Transmittance (1.8 km path)
0.25	0.88
1.25	0.74
2.5	0.65
10.0	0.38

Table 3: Transmittance of a 1.8-km path through rain.

Rain Intensity	Rainfall (mm/hour)
Mist	0.025
Drizzle	0.25
Light	1.0
Moderate	4.0
Heavy	16
Thundershower	40
Cloud-burst	100

Table 4: Representative rainfall rates.

sum of the partial coefficients associated to the various rain drops. A simpler approach, used in LOWTRAN, gives good approximations of the results obtained with eq. (18) for most concentrations of different rain particles. Particularly, in LOWTRAN, the scattering coefficient with rain has been empirically related only to the rainfall rate $\Delta x/\Delta t$ (expressed in mm/hour), as follows [12]:

$$\beta_{rain} \approx 0.365 \cdot \left(\frac{\Delta x}{\Delta t}\right)^{0.63} \qquad (19)$$

Table 4 provides representative rainfall rates which can be used in eq. (18) and (19), when no direct measurements are available, to obtain order of magnitude estimations of β_{rain}.

In the presence of rain, in addition to the scattering losses calculated with eq. (18) or (19), there are, of course, losses by absorption along the path, and these must be included in the calculation of the total atmospheric transmittance with rain.

In order to estimate the SNR from experimental LOAM detector current measurements (iSIG), obtained with certain obstacle ranges (R) and incidence angles (θ), SNR was expressed as follows:

$$SNR = 20\log\left(\frac{i_{SIG}(R,\theta)}{i_{NOISE}}\right) \qquad (20)$$

The noise current terms in eq. (20) was modeled as:

$$i_{NOISE} = \sqrt{i_{TH}^2 + i_{BK}^2 + i_{DK}^2 + i_{RA}^2} \qquad (21)$$

where:

i_{TH}=thermal noise current

i_{BK}=background noise current

i_{DK}=dark noise current

i_{RA}=receiver amplifier noise

According to the LOAM design characteristics, we have:

$$i_{BK} = \sqrt{2qP_S P_h M_A (2 + kM_A) B} \qquad (22)$$

$$i_{TH} = \sqrt{4K_B \frac{T_k Bk}{R_L}} \qquad (23)$$

$$i_{DK} = 0.5 \cdot 10^{-12} \qquad (24)$$

$$i_{RA} = 1.5 \cdot 10^{-12} \qquad (25)$$

where:

P_S=received solar power

P_h=amplifier gain

M_A=avalance multiplier

k=noise factor of the avalance photodiode

B=electronic bandwidth

K_B=Boltzmann constant (1.39×10-23 J/°K)

T_k=absolute temperature (°K)

R_L=amplifier load resistance

The following characteristics were defined for a wire type obstacle according to LOAM operational requirements:

• Diameter: 5 mm ≤ DW ≤ 70 mm

• Shape: twisted or round

• Reflection: Purely diffuse (Lambertian)

• Reflectivity: ≥ 20% (θ=0)

The reference environmental parameters were set as follows:

• Visibility: V ≥ 800 m

• Humidity: RH ≤ 100%

• Temperature: T ≤ 50°C

• Rain: Light/Medium/Heavy

• Background: P_B=50 W/m² sr μm

For calculation purposes, the $i_{SIG}(R,\theta)$ term in eq. (26), was modelled as:

$$i_{SIG} = \sqrt{\frac{P_T d_W \rho D_a^3 \eta e^{-2\gamma R}}{4R^3 \lambda} \cdot \frac{P_h}{K_a} \cdot \frac{1}{R_L}} \qquad (26)$$

where:

P_T=transmitted power

P_h=amplifier gain

D_a=aperture diameter

K_a=aperture illumination constant=$sen(\theta)^{5.4}$

Results of range performance calculations with various visibilities and with all other parameters set to the worst case are shown in Figure 15.

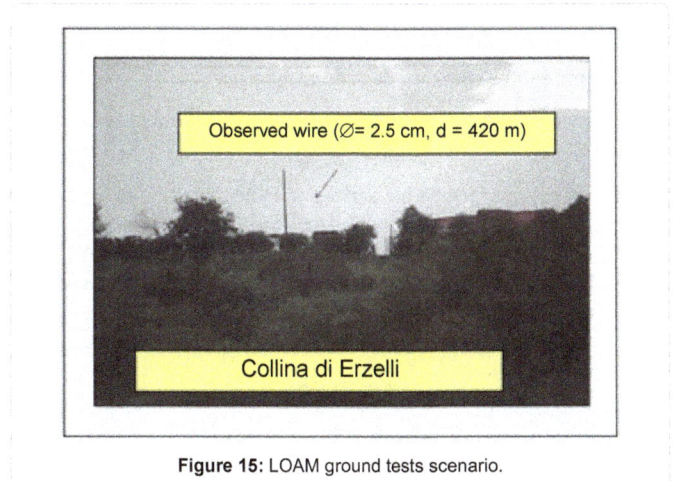

Observed wire (∅= 2.5 cm, d = 420 m)

Collina di Erzelli

Figure 15: LOAM ground tests scenario.

The false alarm probability is modelled as:

$$P_{fa} = \frac{1}{B \cdot T_{fa} \cdot \eta} \tag{27}$$

where:

B=receiver bandwidth

T_{fa}=mean time between false alarms

η=maximum useful/non-ambiguous range

The mean time between false alarms corresponds to elementary electrical false alarms at the receiver level. The probability to have several false alarms on a straight line pattern is much lower. Statistically, these phenomena are described by the False Alarm Rate (FAR) and Detection Probability (P_d). If the noise and signal distributions are known, the SNR can be estimated and the corresponding D_p and FAR can be determined. According to the Rice calculation [11], the average FAR for the LOAM system is given by:

$$\overline{FAR} = \frac{1}{2\tau\sqrt{3}}\exp\left(-\frac{I_t^2}{2I_n^2}\right) \tag{28}$$

where:

τ=Electrical pulse length

I_t=Threshold current

I_n=Average noise current The LOAM P_d is determined using pure Gaussian statistics [13,14]:

$$P_d = \frac{1}{\sqrt{\pi}} \int\limits_{\frac{I_t - I_n}{\sqrt{2}I_n}}^{\infty} \exp\left(-\frac{i_n^2}{2I_n^2}\right) d\left(\frac{i_n}{\sqrt{2}I_n}\right) \tag{29}$$

I_n=average signal current

i_n=instantaneous noise current

The false alarm probability (P_{fa}) is given by:

$$P_{fa} = \tau \cdot \overline{FAR} \tag{30}$$

and the cumulative detection probability (P_D) is given by:

$$P_D = 1 - \sum_{i=0}^{m} C_M^i P_d^i \left(1 - P_d\right)^{M-i} \tag{31}$$

where:

M=number of possible detections

m=minimum number of detections required

The scenario in which ground tests were performed is shown in Figure 16.

Tests were performed in various weather conditions (i.e., clear weather with $10 \leq V \leq 15$ km, and light/medium/heavy rain), using a wire of known section and reflectivity (D_w=2.5 cm and ρ=40%). The sets of data collected in clear and rainy weather conditions are shown in Figure 17. Form these data, it was evidenced that the returned signal power fluctuates independently from pulse to pulse according to a Gaussian distribution.

A comparison between the SNR predicted (SNR$_p$) with γ calculated using the ESLM model (0.19 km^{-1} $\leq \gamma \leq$ 0.22 km-1 for clear weather and 1.23 km^{-1} $\leq \gamma \leq$ 2.94 km^{-1} for rainy conditions), assuming a

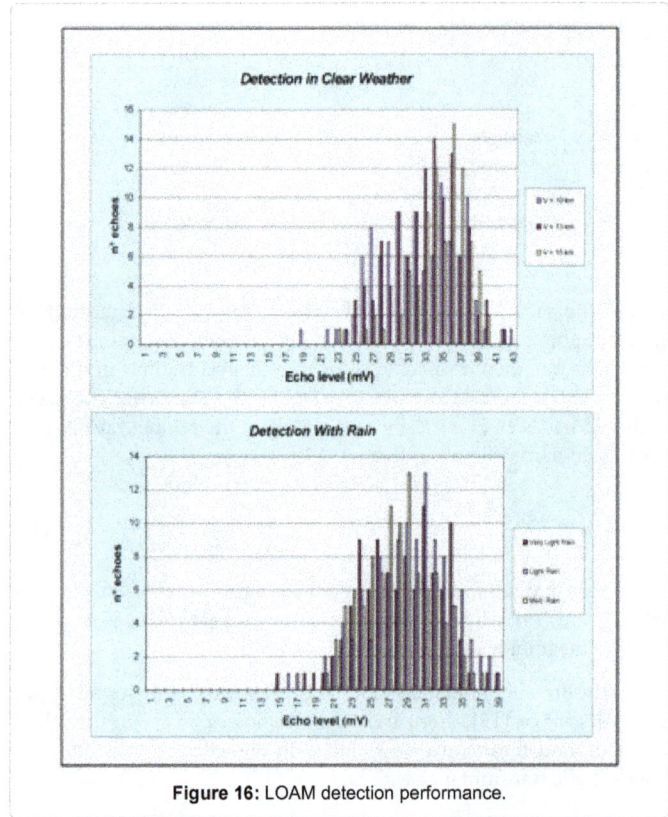

Figure 16: LOAM detection performance.

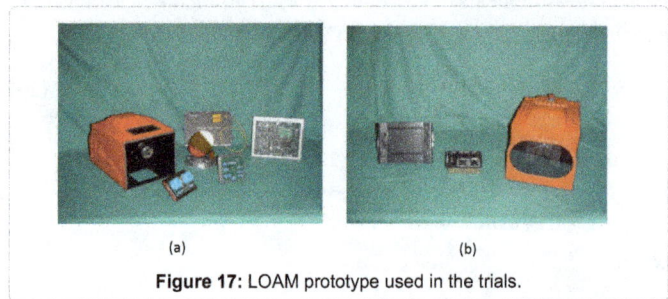

Figure 17: LOAM prototype used in the trials.

	Clear Weather			Rain		
	V=10 km	V=12.5 km	V=15 km	Light	Medium	Heavy
SNR$_P$	4.90×10⁴	4.95×10⁴	5.02×10⁴	3.14×10⁴	1.83×10⁴	1.45×10⁴
SNR$_E$	3.35×10⁴	3.80×10⁴	4.27×10⁴	2.87×10⁴	2.47×10⁴	2.13×10⁴

Table 5: LOAM predicted and measured SNR's.

background power of 10 Watt/m²/sr/μm and ρ=0.5, and estimated from experimental data (SNR$_E$), is shown in Table 5.

LOAM Flight Test Activities

Figure 18 shows the LOAM prototype used for the flight trials. Particularly, the LOAM sub-units are shown in Figure 18a, while the pilot interface units are shown in Figure 18b.

Two different platforms were used for the tests. Figure 19 shows the LOAM installed on the NH-300 helicopter.

Figure 14 LOAM prototype units installed on NH-500.

Figure 20 shows the LOAM Sensor Head mounted on the second test platform (Agusta AB-212 helicopter).

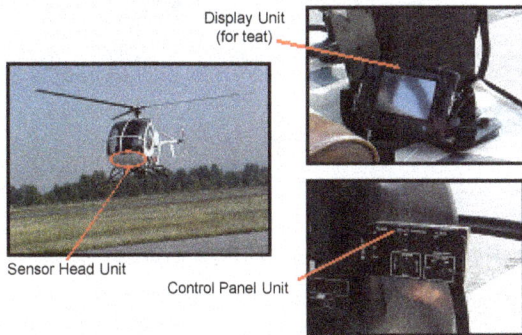

Figure 18: LOAM prototype units installed on NH-500.

Figure 19: LOAM installed on AB-212.

Figure 20: LOAM Display Unit on AB-212.

Figure 21: LOAM Main Control Unit on AB-212.

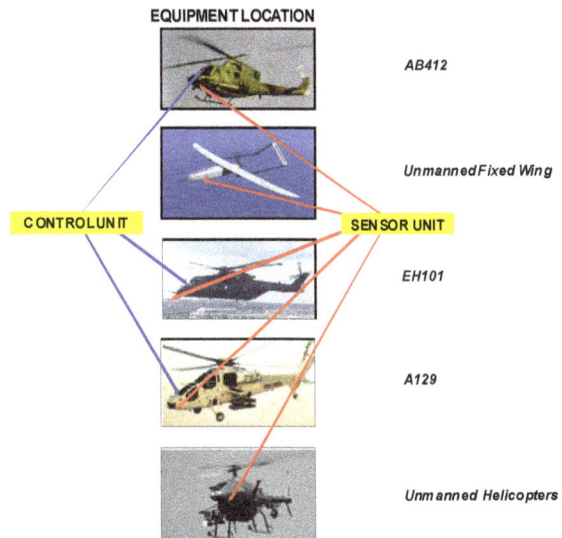

Figure 22: Candidate platforms for future LOAM flight tests.

The Cockpit Display Unit (CDU) used for the trials is shown in Figure 21a. As shown in Figure 21b, the LOAM CDU was installed at the centre of the AB-212 glareshield, in order to be accessible to both pilot and co-pilot.

For the AB-212 test campaign, the LOAM Main Control Unit (MCU) was installed in the centre of the middle-console, as shown in Figure 22 (in a position accessible to both pilot and co-pilot). During the test flights, a Flight Test Engineer also operated a computer, linked to the LOAM system and displaying in real-time a 3-dimensional image reconstructed using the LOAM data. All images were recorded for the successive data analysis.

The results of this test campaign were very satisfactory. Particularly, the LOAM range performance were in accordance with the predictions and the LOAM detection/classification data processing algorithms were validated (detection and classification of all obstacles encountered was performed successfully). Furthermore, it was verified that the LOAM "History Function" was correctly implemented.

Future tests will be performed in order to finally assess the LOAM system performance (sensor and processing algorithms) in day/night with various weather/environmental conditions and to optimize the system Human Machine Interface both in helicopters and UAV (ground operator HMI). The candidate platforms for future LOAM integration and flight test activities are shown in Figure 18.

For future UAV platform flights, a dedicated control unit is being designed. Its characteristics are similar to the MCU developed for the initial helicopter flights. However, as this MCU has to be operated by the ground UAV pilot, in this case the LOAM operating modes are activated using two different communication data links for Line-of-Sight (LOS) and Beyond LOS (BLOS) operations. Additionally, the LOAM display functions will be fully integrated in the UAV remote control position and the required LOAM display formats displayed to the UAV pilot in real-time [15-17].

Acknowledgements

This activity was funded by the Italian Ministry of Defence (MoD) under R&D contract No. 2097-22-12-2000. The Italian Air Force Flight Test Centre developed, during a PhD research conducted at Cranfield University, the mathematical models, simulation tools and test facilities required for system ground and flight test. The authors wish to thank the personnel of SELEX-ELSAG, LOT-ORIEL and the Italian MoD Laser Test Range for helping in the preparation and execution of the ground and flight test activities.

References

1. Kellington CM (1975) An Optical Radar System for Obstacle Avoidance and Terrain Following.

2. Goldstein BS, Dalrymple GF (1967) Gallium arsenide injection laser radar. Proceedings of the IEEE 55: 181-188.

3. The Anglo-French Compact Laser Radar demonstrator programme.

4. Buechtemann W, Eibert M (1995) Laser Based Obstacle Warning Sensors for Helicopters. AGARD.

5. Holder S, Branigan R (1994) Development and Flight Testing of an Obstacle Avoidance System for US Army Helicopters.

6. Selex SPA (2001) Technical Document TS021. LOAM System Characteristics.

7. Selex SPA (2001) Technical Document TS022. LOAM Sensor Head Unit Description.

8. Elder T, Strong J (1953) The infrared transmission of atmospheric windows. J Franklin I 255: 189-208.

9. Langer RM (1957) Signal Corps Report n° DA-36-039-SC-72351.

10. Kneizys FX, Shuttle EP, Abreau LW, Chetwynd JH, Anderson GP, et al. (1988) Users Guide to LOWTRAN 7. Air Force Geophysical Laboratory Report AFGL-TR-88-0177 Hansom AFB.

11. Middleton WEK (1952) Vision through the atmosphere. University of Toronto Press Canada.

12. Selex SPA (2001) Report V81, Programma di prove del sistema LOAM su elicottero AB212.

13. Skolnik M (2008) Radar Handbook. 3rd edn. McGraw-Hill Education, India.

14. Jelalian AV (1992) Laser Radar Systems. Artech House, United Kingdom.

15. Sabatini R (1998) Tactical Laser Systems Performance Analysis in Various Weather Conditions. RTO SET Symposium on "E-O Propagation, Signature and Syatem Performance Under Adverse Meterological Conditions Considering Out-of-Area Operations.

16. Thomas ME, Donald DD (1996) Atmospheric Transmission. ERIM-SPIE IR&EO Systems.

17. Sabatini R, Richardson MA (2010) Airborne Laser Systems Testing and Analysis. NATO Research and Technology Organization Flight Test Techniques Series 26.

9

CFD Analysis on MAV NACA 2412 Wing in High Lift Take-Off Configuration for Enhanced Lift Generation

Arvind Prabhakar* and Ayush Ohri

Department of Mechanical and Manufacturing Engineering, Manipal Institute of Technology, Karnataka, India

Abstract

In a high lift take off configuration an MAV wing utilizes partially extended flaps and slats. Slats and Flaps are high lift devices installed on a wing for the purpose of augmenting Coefficient of Lift (C_L). While Slats are installed on the leading edge of a wing the Flaps may be installed on the trailing edge or the leading edge of a wing. In this paper the effect of Slot size created by the Slats in percentage of wing chord 'c' and Double Slotted Flaps on C_L for a MAV NACA 2412 has been studied using CFD analysis at $2*10^5$ Reynolds Number. It is found that the maximum value of C_L achieved is 67.134% higher than the plain NACA 2412 wing at 4 degrees angle of attack when slats are extended at 1.7 percent of wing chord 'c'. The Stall angle of the MAV NACA 2412 wing in high lift take-off configuration was found to be 54 degrees whereas the plain NACA 2412 wing stalled at 20 degrees angle of attack.

Keywords: Angle of attack; Augmentation; Computational fluid dynamics

Abbrevations: NACA: National Advisory Committee on Aeronautics; MAV: Micro Air Vehicle; CFD: Computational Fluid Dynamics; C_L: Coefficient of Lift; (c): Wing Chord Slats; DSP: Double Slotted Flaps; Slot; Stall Angle

Nomenclature

C_L-Coefficient of Lift

'c'-wing chord

d-Perpendicular slat distance from a point on the airfoil nose

L/D-Lift to drag ratio

Introduction

Flaps are high lift devices attached to the leading or trailing edge of a wing. They help to increase the value of C_L and the Stall Angle during the take-off phase of an aircraft. Stall Angle is the angle between the chord line of an airfoil and the undisturbed relative airflow at which stalling occurs where stalling refers to the condition when there is a sudden reduction in the lift generated by the wing [1,2]. If the stalling angle is higher compared to plain airfoils it allows the aircraft to take off at lower speeds and hence it can even take off from shorter runways [3].

Flaps when fully extended during the landing phase of an aircraft tend to increase the drag so that the aircraft can land on the runway with a safe speed depending upon the shape, size and weight of the aircraft [4]. In this paper application of flaps during take-off phase has been highlighted.

Slats are referred to as high lift devices which are attached to the leading edge of a wing. Their function is similar to that of the flaps. They help in increasing the coefficient of lift and stalling angle by re-energizing the airflow over the wings surface so that the airflow remains streamline upto high angles of attack. However, it may start becoming turbulent at much higher angles. Once the aircraft takes-off and begins to cruise at a particular altitude the slats and flaps are made to retract back into the leading and trailing edges of the wing. If the slats and flaps remain extended from the wing during the cruise period of the aircraft they tend to increase the drag produced and thereby increase the fuel consumption of the aircraft [5].

Flaps are of various types such as plain flaps, split flaps, fowler flaps, slotted flaps, double slotted flaps, zap flaps, Krueger flaps etc. The type of flap that needs to be used for a particular aircraft depends upon size, speed and shape of the aircraft. Generally in modern day jetliners Krueger flaps are used which are attached to the leading edge of the wing.

It has been investigated that the lift of a MAV can be enhanced by using moveable passive flaps at the extrado of the airfoil at low Reynolds number of about $3*10^4$ and it was found that the lift of an SD8020 airfoil wing increased by 50% [6]. In this paper double slotted flaps and slats were attached to the wing of a micro air vehicle and their effect was studied. The plain NACA 2412 wing was 6 inches in chord and it had a span of 12 inches which is ideally suitable for a MAV. The chord of flap was 25 percent of the plain NACA 2412 wing. The analysis was carried out at $2*10^5$ Reynolds Number. Standard K-epsilon turbulence model was used to carry out CFD simulation. The results obtained were compared with the results of plain wing at different angles of attack. In armin ghoddoussi's, conceptual study of airfoil performance enhancements it has been proved that at slot size equal to 1.7% of wing chord 'c' maximum amount of lift is generated [7]. In this paper the comparison of results in terms of CL and stall angle proved that the wing which had slat extended by distance'd' equal to 1.7 percent of plain NACA 2412 wing chord and double slotted flap extended at an angle of 40 degrees generated maximum value of CL. hence, this configuration was assumed as the MAV NACA 2412 high lift take off configuration (Figure 1).

*Corresponding author: Arvind Prabhakar, Department of Mechanical and Manufacturing Engineering, Manipal Institute of Technology, Karnataka, India
E-mail: arvind.prabhakar1988@gmail.com

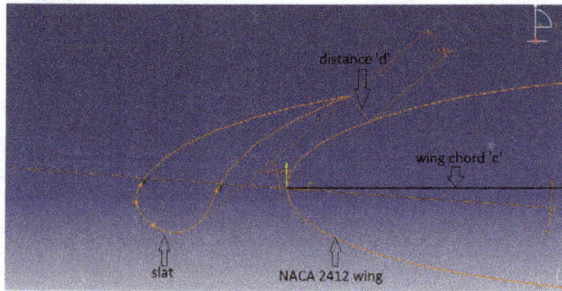

Figure 1: MAV NACA 2412 slot nomenclature.

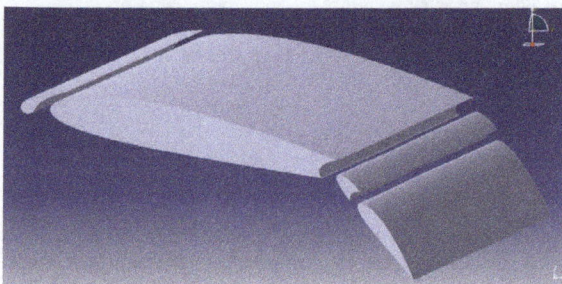

Figure 2: 3d modeling of the MAV NACA 2412 wing on CATIA v5.

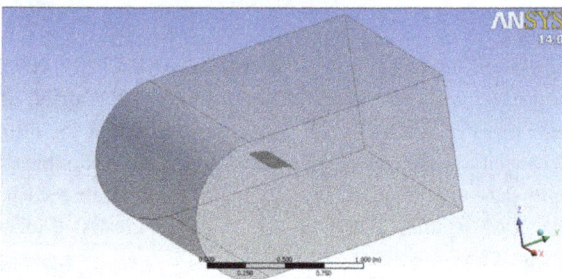

Figure 3: Design of fluid domain and Boolean subtraction on ANSYS workbench.

Theoretical Formulation and Method of Solution

The objective of the analysis is to study the effect of Slot size in percentage of wing chord 'c' as a measure of high lift take-off configuration. The following assumptions are made during the analysis.

• The computational domain is assumed to be made of solid-air interface.

• Turbulent flow is assumed under fully developed conditions.

• Solution is reached at steady state which is assumed to be obtained after all the residuals are bought to prescribed constant values.

Methodology

The methodology for computation involved the designing of plain NACA2412 airfoils, slats, double slotted flaps and a bullet shaped fluid domain on commercial cad software CATIA V5. The meshing of the designs for grid generation was done on ICEM CFD ANSYS 14.0 workbench. The CFD analysis was carried out using commercial code ANSYS FLUENT 14.0. The post processing of the results was done using ANSYS CFD POST.

Design of MAV NACA 2412 airfoil

The 2D airfoil co-ordinates were exported to part module of CATIA V5 which is a commercial bench marked software for modeling. The 2d airfoil coordinates sketch was extruded using the pad command. The slat is of NACA 2415 profile and the airfoil and flaps are of NACA 2412 profile (Figure 2).

Design of fluid domain

A bullet shaped fluid domain was created in part design module of CATIA V5 to replicate the experimental wind tunnel whose results have been used to validate the CFD analysis (Figure 3).

Grid generation

Tetrahedral meshing was done for generating the grid. Refinements were done on the leading and trailing edges of the airfoil, slats and double slotted flaps. Coarse type sizing was used to generate the mesh. Since tetrahedral meshing takes lesser computational time and keeps the accuracy of the results obtained after analysis within acceptable limits, it was used for generating the grid. The Minimum Size was set to 0.0018 whereas the Maximum Size, Maximum Face size and Growth Rate were set to 3 meters, 3 meters, 1.072 respectively. The Minimum Edge Length was set to 7.8474*10-5.

The mesh quality depends upon the value of smoothness which defines the rate of change of cell size. For a good quality mesh the rate of change of cell size should be smooth. The value of smoothness has been set to medium to ensure a good quality mesh (Figure 4).

Grid Generation Outline

After the wing had been accurately modeled in CATIA V5 as per the dimensions, a 3D grid was generated on the modeled wing which divides the volume surrounding the wing into several smaller volumes. The type of grid layout that was generated on different areas of the wing depended upon the type of test the grid was required to solve. Since the air flow whose behavior had to be studied in order to study the phenomenon of turbulence was majorly stuck to the upper portion of the wing, grid with high cell density was generated on the upper surface of the wing. Since substantial amount of air causing re-energization of turbulent flow over the upper surface of the wing passes through slot and the gap between the main wing and the double slotted flaps, the leading and trailing edges of the slats, flaps and wing were refined with a Minimum size of 0.0018 m to generate a grid of very high cell density so that the phenomenon of re-energisation could be studied (Figure 5). Shows grid generation outline.

The Standard K-epsilon model is one of the most common Turbulence model used in CFD analysis. This is a 2 equation model

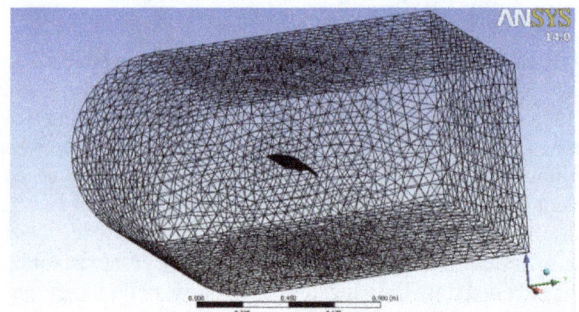

Figure 4: Generation of grid on ICEM CFD.

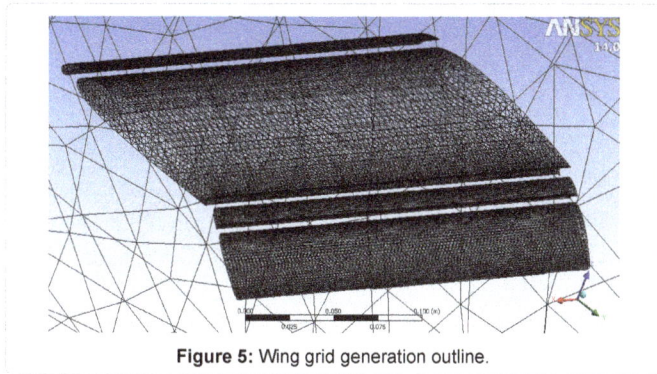

Figure 5: Wing grid generation outline.

which represents the turbulent properties of flow. This model takes into account the effect of convection and diffusion of turbulent flow. The first equation gives the turbulent kinetic energy of the flow which is represented by symbol K whereas the second equation gives the turbulent dissipation in the flow which is represented by symbol $\in$. This model does not give accurate results in case the pressure gradients are large and is mainly used in case of free-shear layer flows with relatively small pressure gradients.

First order upwind scheme is selected for spatial discretization of the Reynolds Average Navier Stoke (RANS) equations as well as energy and turbulence equations [8,9]. Converged results are obtained after the residuals were found to be less than the specified values. A converged result renders an energy residual of 10^{-6} and momentum and turbulence kinetic energy residuals being 10^{-5} [9].

Turbulent Kinetic Energy Equation

Turbulence dissipation equation

$$\frac{\delta(\rho \in)}{\delta t} + \frac{\delta(\rho \in u_i)}{\delta x_i} = \frac{\delta}{\delta x_j}[\frac{\mu_t}{\sigma_k}\frac{\delta k}{\delta x_j}] + C_{1\in}\frac{\in}{k}2\mu_t E_{ij} - C_{2\in}\rho\frac{\in^2}{k}$$

u_i represents velocity component in corrosponding direction

E_{ij} component of rate of deformation

μ_t represents eddy viscosity

$$\mu t = \rho C_\mu \frac{k^2}{\in}$$

Grid independence test

A grid independence test was carried out to ensure that there is no effect on the solution due to the size of the grid and to study the effect of quality of mesh on accuracy of solution. This was achieved by considering three different grid configurations and studying their convergence behavior for the values of coefficient of lift at Reynolds Number $2*10^5$.

Grid configuration 1 was a coarse grid that had 513211 numbers of cells and a Minimum Orthogonal Quality=0.0485750 it showed inaccurate results due to a deviation of about 15.36% from the experimental value.

Grid configuration 2 was finer grid that had 2853877 number of cells with Minimum Orthogonal Quality=0.18607 it consisted of refinement at the leading and trailing edges of the wing, this showed realistic behavior and solution obtained was closer to the experimental data. A deviation of about 4.41% was observed.

Grid configuration 3 was the finest mesh and had 4217840 number of cells with a Minimum Orthogonal Quality=0.188692 and showed a behavior not much different than configuration 2 and also, produced a solution similar to configuration 2 hence, grid configuration 2 was selected for all further analysis, considering computation time and size constraints in mind. The plot of solutions obtained with these three different grid configurations and experimental results has been shown in Figure 6.

Validation of the predicted values from CFD Analysis

for the purpose of CFD validation The values obtained from the CFD Analysis of the plain NACA 2412 wing at Reynolds Number $2*10^5$ for angles of attack 0, 4, 8, 12, 16, and 18 degrees were compared with the experimental results from the wind tunnel data on the airfoil of similar dimensions and profile at the same Reynolds Number [9].

It can be observed from figure 4 that the values of C_L obtained are in good agreement with the experimental values [9] hence, this has enabled the analysis to be extended for the analysis on MAV NACA 2412 wing in high take off configuration for predicting the values of C_L (Figure 7).

Once the CFD analysis results were obtained for the 3D wing with flaps and slats attached to it, a comparison on the basis of the values of C_L and Lift to Drag ratio (L/D) obtained for various angles of attack was made. Moreover the value of Stall Angle obtained in case of plain MAV NACA2412 wing was lesser compared to that of MAV NACA2412 wing having slats and flaps attached to it. The simulation was carried out at a Reynolds Number of $2*10^5$ which corresponded to a velocity of 20.5826 m/s. The values of C_L and L/D ratio were calculated at 4, 8, 12,

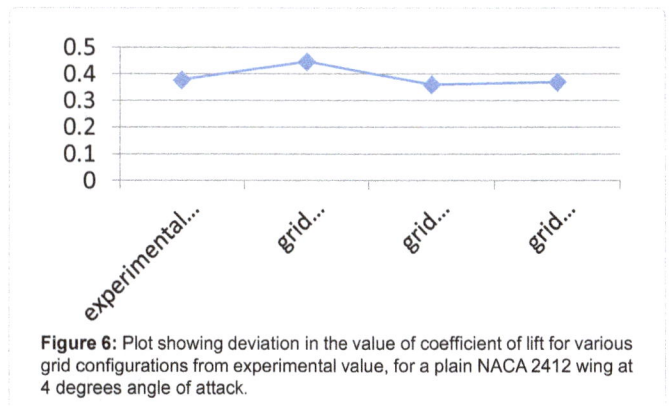

Figure 6: Plot showing deviation in the value of coefficient of lift for various grid configurations from experimental value, for a plain NACA 2412 wing at 4 degrees angle of attack.

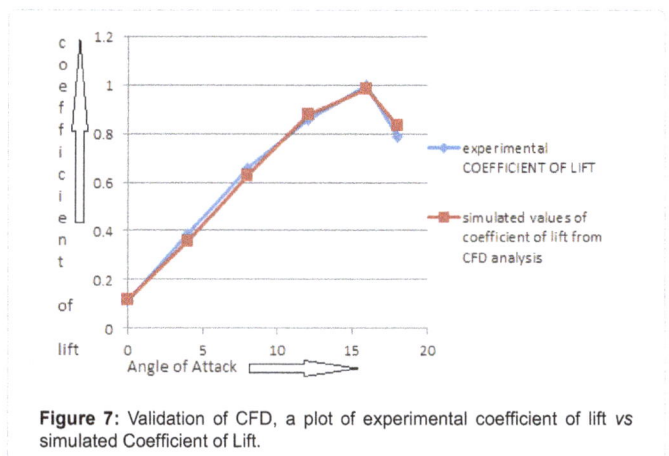

Figure 7: Validation of CFD, a plot of experimental coefficient of lift vs simulated Coefficient of Lift.

18 and 20 degrees angle of attack for the plain MAV wing. Whereas the C_L and L/D ratio values for the MAV wing with flaps and slats attached to it was calculated at 4, 812, 16, 18, 20, 22, 24, 26, 30, 34, 38, 42, 46, 50 and 54 degrees angle of attack. It was observed from the values of C_L obtained after CFD simulation that the plain NACA 2412 wing began to stall at 20 degrees whereas the MAV wing having flaps and slats attached to it began to stall at 54 degrees angle of attack. All the analysis were carried out with double slotted flaps extended at 40 degrees since it corresponded to high take off configuration. Armin ghoddoussi in his conceptual study of airfoil performance enhancements studied and proved that at slot size equal to 1.7% of wing chord 'c' maximum amount of lift is generated [8]. In consideration of this aspect the present paper extends the work to study the additional effect of double slotted flaps at 40 degrees attached to the MAV NACA 2412 wing. A study on the slat gap size when extended at 0.5%, 1.2%, 1.7%, 3.2%, 3.7%, 4.2% and 4.7% of plain MAV NACA 2412 wing chord was done to obtain maximum value of C_L. It was observed that when slats were extended at 1.7% of plain MAV NACA 2412 wing chord with double slotted flaps attached to the wing extended at 40 degrees, maximum value of C_L was obtained.

Lift-coefficient analysis

Angle of attack 4^0: At $2*10^5$ Reynolds Number the value of C_L for MAV NACA 2412 wing with a configuration of slot size equal to 1.7% wing chord 'c' and double slotted flaps extended at 40 degrees was found to be 0.600765 which was the maximum value of C_L obtained in comparison to the C_L values obtained with 0.5%, 1.2%, 3.2%, 3.7%, 4.2%, 4.7% slot size configuration, this value of C_L obtained was found to be higher by 67.134% compared to plain NACA 2412.

Angle of attack 8^0: At $2*10^5$ Reynolds Number the value of C_L for MAV NACA 2412 wing with a configuration of slot size equal to 1.7% wing chord 'c' and double slotted flaps extended at 40 degrees was found to be 0.80005 which was the maximum value of C_L obtained in comparison to the C_L values obtained with 0.5%, 1.2%, 3.2%, 3.7%, 4.2%, 4.7% slot size configuration, this maximum value of C_L was found to be higher by 27.349% compared to plain NACA 2412.

Angle of attack 12^0: At $2*10^5$ Reynolds Number the value of C_L for MAV NACA 2412 wing with a configuration of slot size equal to 1.7% of wing chord 'c' and double slotted flaps extended at 40 degrees was found to be 1.052494 which was the maximum value of C_L obtained in comparison to the C_L values obtained with 0.5%, 1.2%, 3.2%, 3.7%, 4.2%, 4.7% slot size configuration, this maximum value of C_L obtained was found to be higher by 19.889% compared to plain NACA 2412.

Angle of attack 16^0: At $2*10^5$ Reynolds Number the value of C_L for MAV NACA 2412 wing with a configuration of slot size equal to 1.7% of wing chord 'c' and double slotted flaps extended at 40 degrees was found to be 1.254877 which was the maximum value of C_L obtained in comparison to the C_L values obtained with 0.5%, 1.2%, 3.2%, 3.7%, 4.2%, 4.7% slot size configuration.

Angle of attack 18^0: At $2*10^5$ Reynolds Number the value of C_L for MAV NACA 2412 wing with a configuration of slot size equal to 1.7% of wing chord 'c' and double slotted flaps extended at 40 degrees was found to be 1.355955 which was the maximum value of C_L obtained in comparison to the C_L values obtained with 0.5%, 1.2%, 3.2%, 3.7%, 4.2%, 4.7% slot size configuration (Figures 8-13) (Table 1).

Effect of MAV NACA 2412 wing on stall angle

At $2*10^5$ Reynolds Number it was found that the MAV NACA

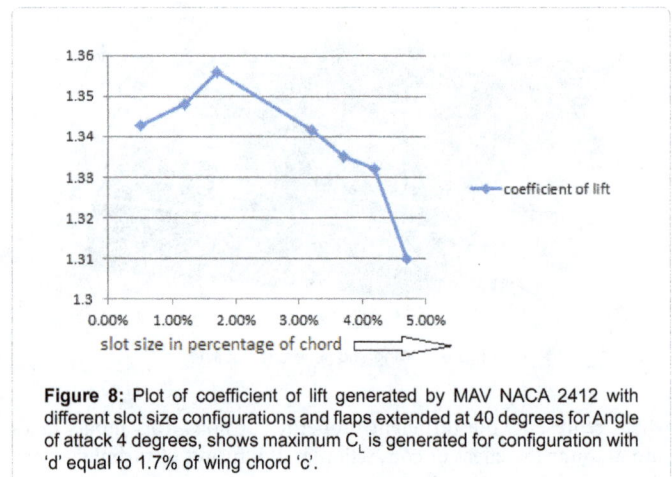

Figure 8: Plot of coefficient of lift generated by MAV NACA 2412 with different slot size configurations and flaps extended at 40 degrees for Angle of attack 4 degrees, shows maximum C_L is generated for configuration with 'd' equal to 1.7% of wing chord 'c'.

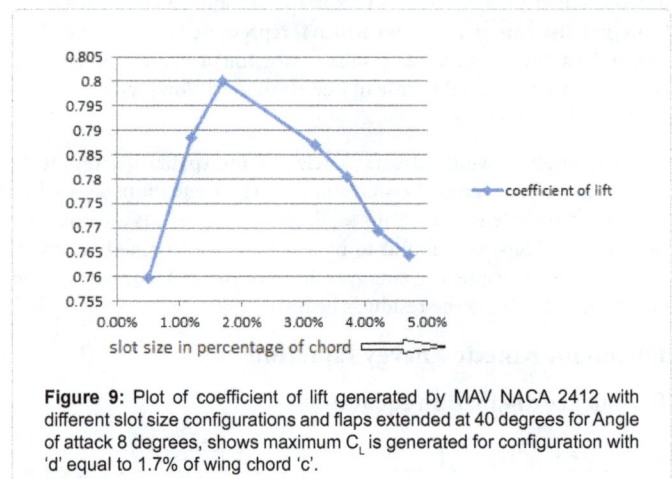

Figure 9: Plot of coefficient of lift generated by MAV NACA 2412 with different slot size configurations and flaps extended at 40 degrees for Angle of attack 8 degrees, shows maximum C_L is generated for configuration with 'd' equal to 1.7% of wing chord 'c'.

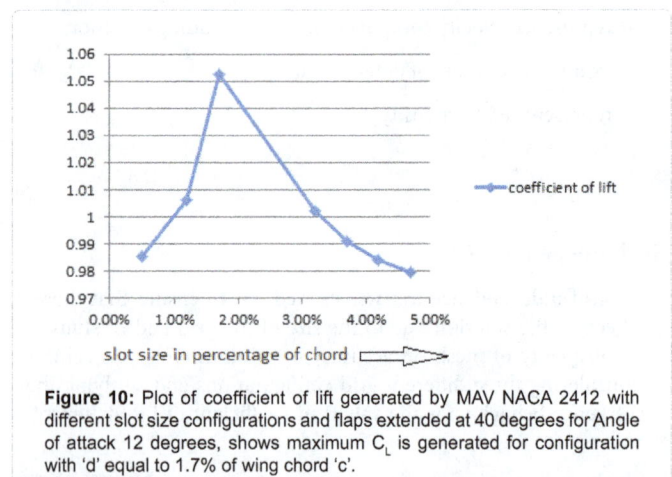

Figure 10: Plot of coefficient of lift generated by MAV NACA 2412 with different slot size configurations and flaps extended at 40 degrees for Angle of attack 12 degrees, shows maximum C_L is generated for configuration with 'd' equal to 1.7% of wing chord 'c'.

2412 wing with 'd' equal to 1.7% of wing chord 'c' and double slotted flaps extended at an angle of 40 degrees increased stalling angle upto 54 degrees compared to a stalling angle of 18 degrees achieved by plain NACA 2412 airfoil of same aspect ratio (Table 2).

Lift to Drag (L/D) ratio analysis

Angle of attack 4^0: at $2*10^5$ Reynolds Number the value of L/D ratio for high lift configuration with 'd' equal to 1.7% of wing chord 'c'

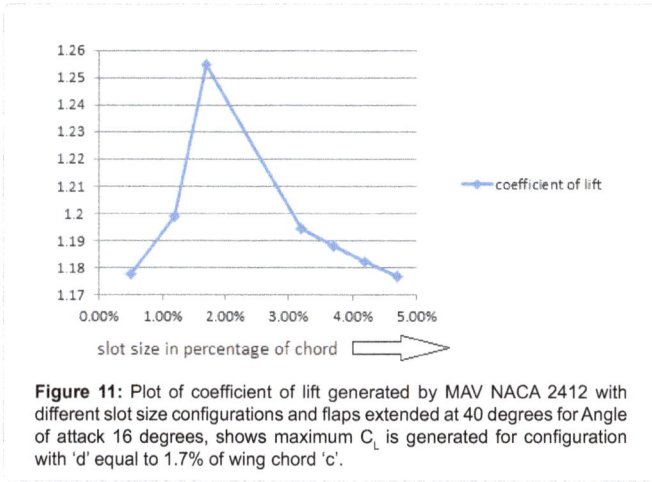

Figure 11: Plot of coefficient of lift generated by MAV NACA 2412 with different slot size configurations and flaps extended at 40 degrees for Angle of attack 16 degrees, shows maximum C_L is generated for configuration with 'd' equal to 1.7% of wing chord 'c'.

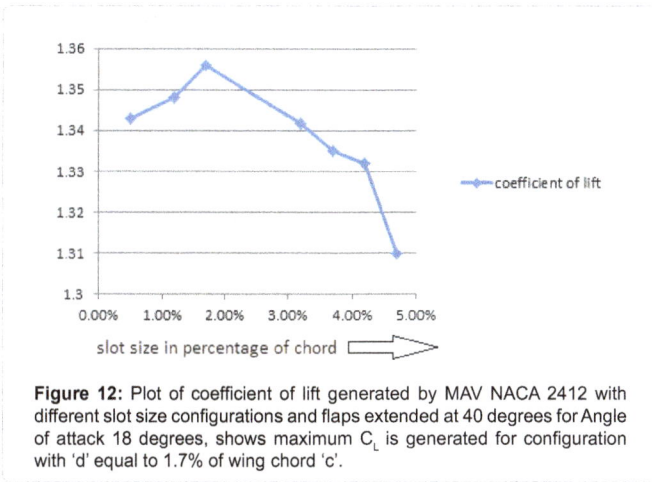

Figure 12: Plot of coefficient of lift generated by MAV NACA 2412 with different slot size configurations and flaps extended at 40 degrees for Angle of attack 18 degrees, shows maximum C_L is generated for configuration with 'd' equal to 1.7% of wing chord 'c'.

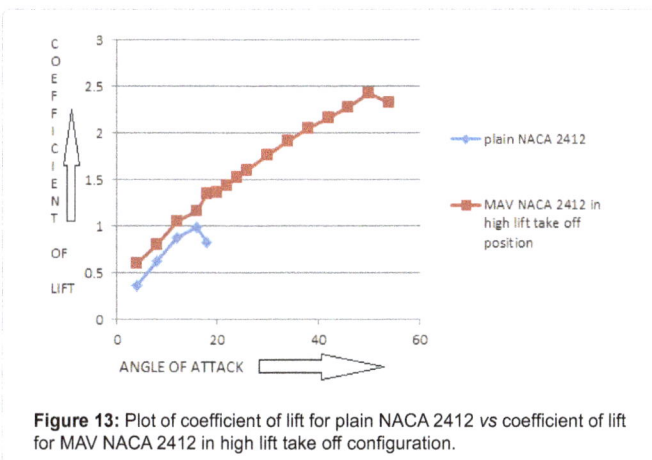

Figure 13: Plot of coefficient of lift for plain NACA 2412 *vs* coefficient of lift for MAV NACA 2412 in high lift take off configuration.

Angle of Attack (In Degrees)	Coefficient of Lift (C_L)						
	MAV NACA 2412 With d=.5% of CHORD 'C'	MAV NACA 2412 With d=.2% of CHORD 'C'	MAV NACA 2412 With d=1.7% of CHORD 'C'	MAV NACA 2412 With d=3.2% of CHORD 'C'	MAV NACA 2412 With d=3.7% of CHORD 'C'	MAV NACA 2412 d=4.2% of CHORD 'C'	MAV NACA 2412 With d=4.7% of CHORD 'C'
4	0.525731	0.578133	0.600765	0.576324	0.58268	0.566233	0.564747
8	0.759913	0.788417	0.80005	0.78697	0.780398	0.769281	0.764446
12	0.985432	1.006522	1.052494	1.002013	0.990952	0.984119	0.979466
16	1.177815	1.198905	1.254877	1.194396	1.188335	1.182502	1.176849
18	1.343061	1.348144	1.355955	1.341745	1.335	1.332122	1.310071

Table 1: Lift Coefficient comparison for various wing configurations, Reynolds Number=$2*10^5$.

Wing Configuration	Coefficient of Lift				
	4^0	8^0	12^0	16^0	18^0
Plain NACA 2412	0.35945018	0.62823054	0.87788515	0.983365	0.83125631
MAV NACA 2412 In High Lift Take Off Configuration	0.600765	0.80005	1.052494	1.194396	1.341745

Table 2: Lift Coefficient comparison between plain NACA 2412 and MAV NACA 2412 with 'd' equal to 1.7% of chord 'c' and flaps extended at 40 degrees (i.e. MAV NACA 2412 in high lift take off configuration) at Reynolds Number=$2*10^5$.

Angle of Attack (In Degrees)	Plain NACA 2412	MAV NACA 2412 In High Lift Take Off Configuration
4	0.35945018	0.60076541
8	0.62823054	0.800504
12	0.87788515	1.0524937
16	0.983365	1.1648765
18	0.83125631	1.3559547
20	STALL.	1.3644363
22		1.4416201
24		1.5277626
26		1.599189
30		1.7714788
34		1.9146608
38		2.0519581
42		2.1697813
46		2.276407
50		2.4338166
54		2.329211
55		STALL.

Table 3: Stall angle determination by Lift Coefficient comparison between plain NACA 2412 and MAV NACA 2412 with 'd' equal to 1.7% of wing chord 'c' and flaps extended at 40 degrees (i.e. MAV NACA 2412 in high lift take off configuration) at Reynolds Number=$2*10^5$.

and double slotted flaps extended at 40 degrees obtained was 2.0998 which is 76.109% lower than the plain NACA 2412 wing.

Angle of attack 16⁰: at $2*10^5$ Reynolds Number the value of L/D ratio for high lift configuration with 'd' equal to 1.7% of wing chord 'c' and double slotted flaps extended at 40 degrees obtained was 2.1083 which is 78.1281% lower than the plain NACA 2412 wing.

Angle of attack 18⁰: at $2*10^5$ Reynolds Number the value of L/D ratio for high lift configuration with 'd' equal to 1.7% of wing chord 'c' and double slotted flaps extended at 40 degrees obtained was 2.12810 which is 79.3441% lower than the plain NACA 2412 wing (Table 4).

Contours of coefficient of pressure

The plot of Coefficient of Pressure for a MAV NACA 2412 wing

and double slotted flaps extended at 40 degrees obtained was 2.2981 which is 77.3429% lower than the plain NACA 2412 wing (Table 3).

Angle of attack 8⁰: at $2*10^5$ Reynolds Number the value of L/D ratio for high lift configuration with 'd' equal to 1.7% of wing chord 'c' and double slotted flaps extended at 40 degrees obtained was 2.3686 which is 77.1193% lower than the plain NACA 2412 wing.

Angle of attack 12⁰: at $2*10^5$ Reynolds Number the value of L/D ratio for high lift configuration with 'd' equal to 1.7% of wing chord 'c'

Wing Configuration	LIFT/DRAG RATIO				
	4⁰	8⁰	12⁰	16⁰	18⁰
Plain Rectangular NACA 2412	10.143	10.352	8.864	9.639	10.30
MAV NACA 2412 In High Lift Take Off Position	2.2981	2.3686	2.099	2.1083	2.128

Table 4: LIFT TO DRAG ratio comparison between plain NACA 2412 and MAV NACA 2412 in high lift take off configuration.

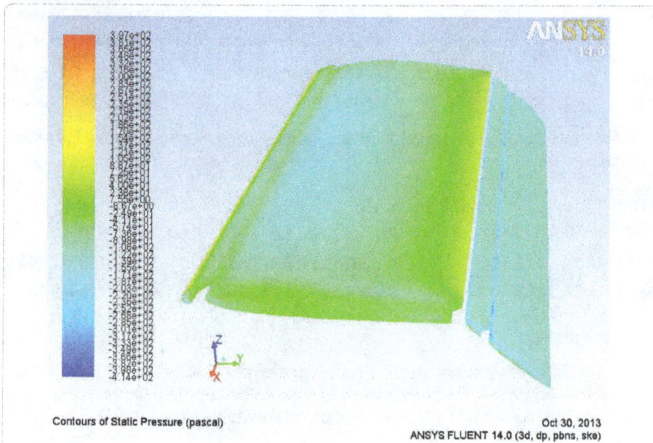

Figure 14: Top surface Coefficient of Pressure contours on MAV NACA 2412 wing at Angle of Attack 4 degrees with 'd' equal to 1.7% of wing chord (c) and double slotted flaps extended at 40 degrees at Reynolds Number=$2*10^5$.

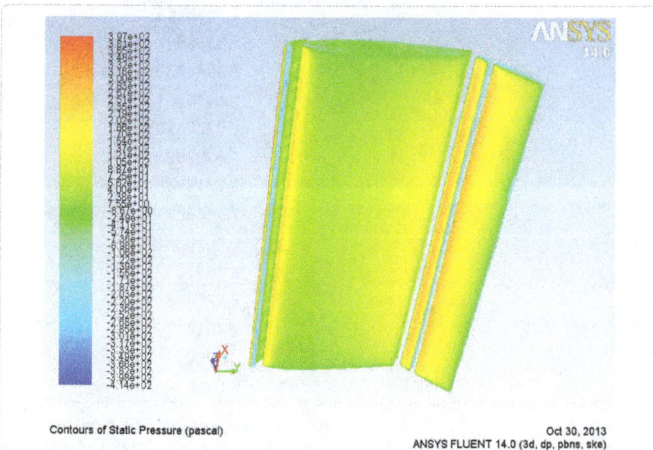

Figure 15: Bottom surface Coefficient of Pressure contours on MAV NACA 2412 wing at Angle of Attack 4 degrees with 'd' equal to 1.7% of wing chord (c) and double slotted flaps extended at 40 degrees at Reynolds Number=$2*10^5$.

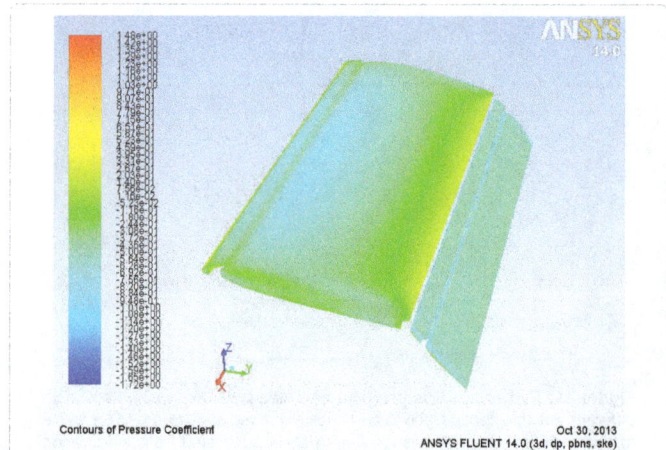

Figure 16: Top surface Coefficient of Pressure contours on MAV NACA 2412 wing at Angle of Attack 8 degrees with 'd' equal to 1.7% of wing chord (c) and double slotted flaps extended at 40 degrees at Reynolds Number=$2*10^5$.

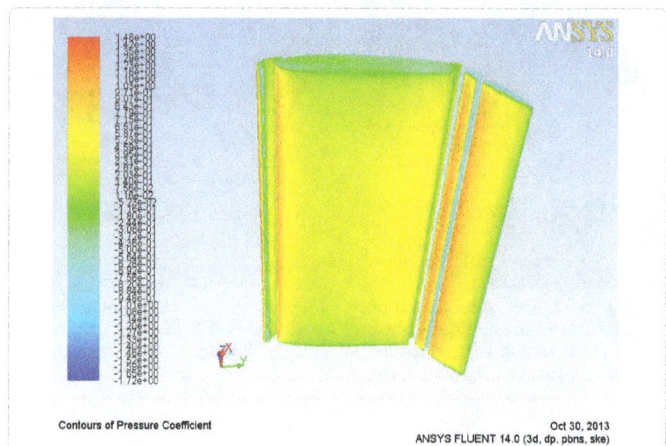

Figure 17: Bottom surface Coefficient of Pressure contours on MAV NACA 2412 wing at Angle of Attack 8 degrees with 'd' equal to 1.7% of wing chord (c) and double slotted flaps extended at 40 degrees at Reynolds Number=$2*10^5$.

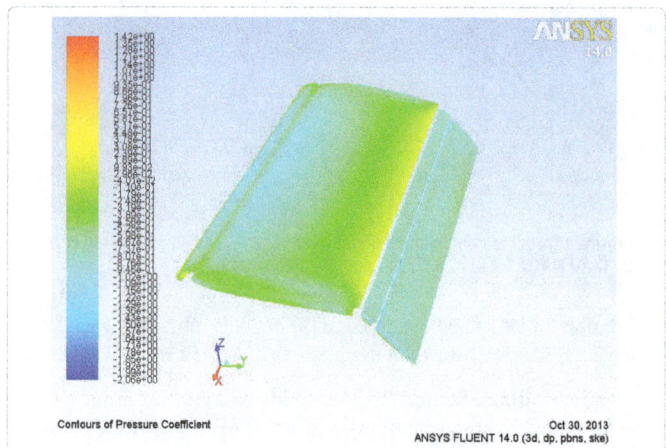

Figure 18: Top surface Coefficient of Pressure contours on MAV NACA 2412 wing at Angle of Attack 12 degrees with 'd' equal to 1.7% of wing chord (c) and double slotted flaps extended at 40 degrees at Reynolds Number=$2*10^5$.

with double slotted flaps and slats attached to it is shown from figures 14 to 21. The slats are extended to maximum lift generation position that is 'd' is equal to 1.7% of wing chord (c). The figures suggest that with increasing angle of attack from 4 to 18 degrees the low pressure coefficient area tends to shift towards the leading edge on the upper surface. The magnitude of this Pressure Coefficient is lesser than the Pressure Coefficient on the plain NACA 2412 wing thus, causing higher lift generation. The intensity of this Pressure Coefficient decreases with increasing angle of attack suggesting lower values of Pressure Coefficient and indicating a high lift generation at such positions. The bottom surface contours for coefficient of pressure suggest increasing higher pressure coefficient at the

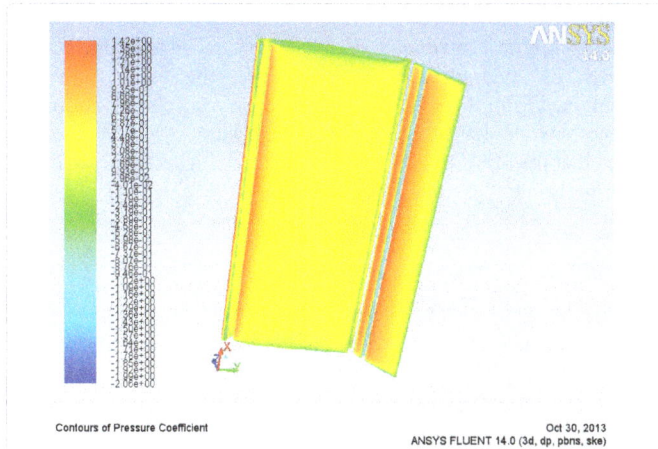

Figure 19: Bottom surface Coefficient of Pressure contours on MAV NACA 2412 wing at Angle of Attack 12 degrees with 'd' equal to 1.7% of wing chord (c) and double slotted flaps extended at 40 degrees at Reynolds Number=$2*10^5$.

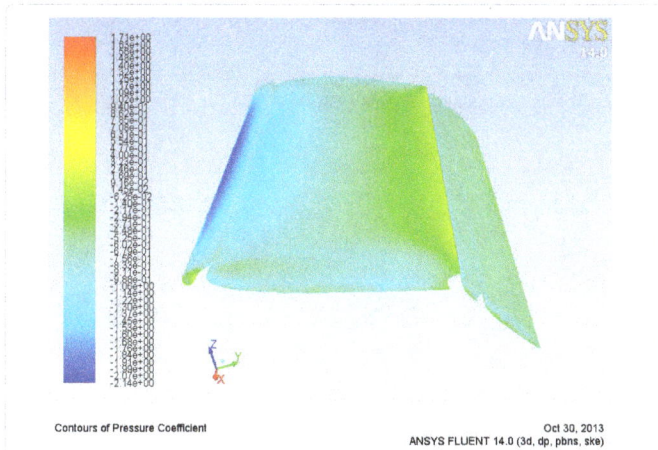

Figure 20: Top surface Coefficient of Pressure contours on MAV NACA 2412 wing at Angle of Attack 18 degrees with 'd' equal to 1.7% of wing chord (c) and double slotted flaps extended at 40 degrees at Reynolds Number=$2*10^5$.

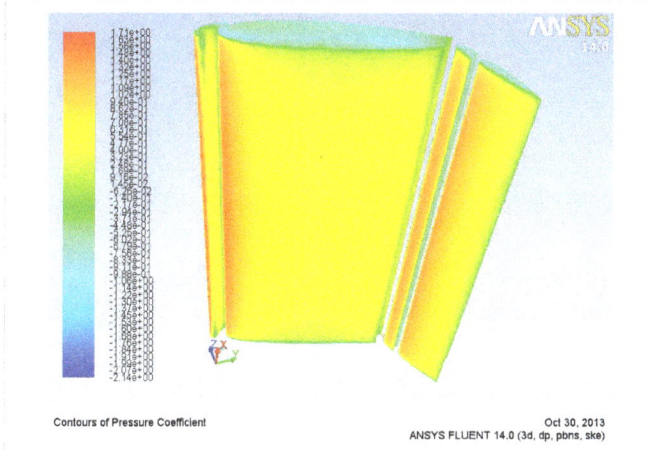

Figure 21: Bottom surface Coefficient of Pressure contours on MAV NACA 2412 wing at Angle of Attack 18 degrees with 'd' equal to 1.7% of wing chord (c) and double slotted flaps extended at 40 degrees at Reynolds Number=$2*10^5$.

Pathlines

Pathlines are used to study the path of a fluid as it moves over a structure or interacts with it. Lines generated during post processing of CFD results individually define the flow path of a fluid on the design. These lines are graphically represented by various colour. Each colour represents a particular magnitude of velocity, pressure, energy etc. The figures below show Pathlines of velocity magnitude. It is observed from the figures 19 to 23 that the turbulent flow in case of plain NACA 2412 wing as shown in figure 19 gets shifted towards the trailing edge of the airfoil and on the top surface of double slotted flaps of MAV NACA 2412 wing in high lift configuration as shown in figure 20. Therefore the flow over the wing portion of the MAV NACA 2412 wing becomes laminar and gives a higher value of CL and increases Stall angle in comparison to the plain NACA 2412 wing. The shift of turbulent flow takes place in case of MAV wing due to re-energization phenomenon which occurs through gap size created by the slats and double slotted flaps shown in Figure 21 to Figure 25 respectively [8]. Generally in this phenomenon bleed air from engine exhaust is made to pass over the rear upper surface of the wing and flaps through the gaps created by the flaps and slats. This air allows the air flow to remain attached to the wings surface at even higher angles of attack causing the Stall angle to rise considerably [8].

Conclusion

The results obtained from the CFD analysis of various wing

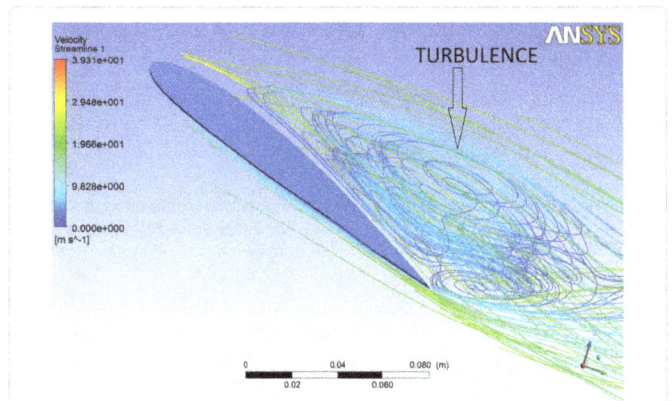

Figure 22: Pathlines of velocity magnitude on plain NACA 2412 airfoil at Angle of Attack 18 degrees at Reynolds number=$2*10^5$ showing high turbulence.

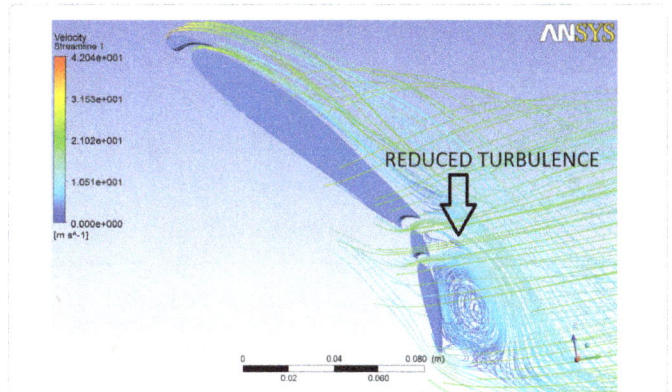

Figure 23: Pathlines of velocity magnitude on MAV NACA 2412 airfoil at Angle of Attack 50 degrees with 'd' equal to 1.7% of wing chord and flaps extended at 40 degrees at Reynolds number=$2*10^5$.

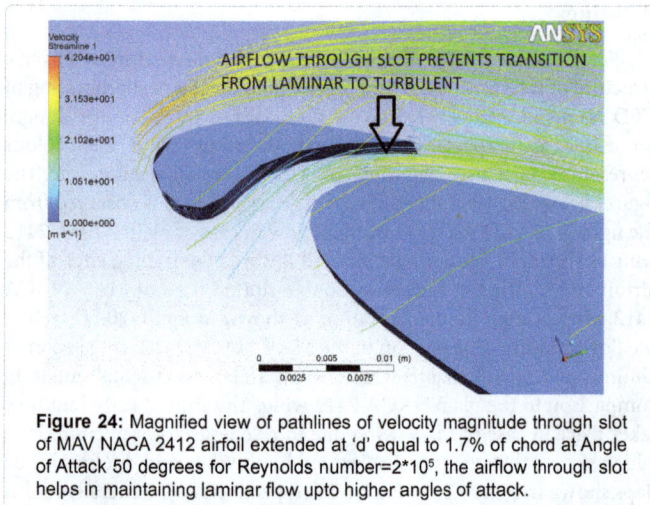

Figure 24: Magnified view of pathlines of velocity magnitude through slot of MAV NACA 2412 airfoil extended at 'd' equal to 1.7% of chord at Angle of Attack 50 degrees for Reynolds number=$2*10^5$, the airflow through slot helps in maintaining laminar flow upto higher angles of attack.

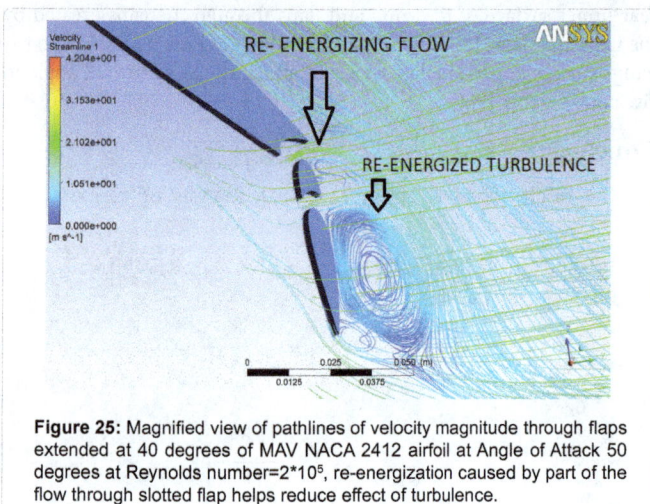

Figure 25: Magnified view of pathlines of velocity magnitude through flaps extended at 40 degrees of MAV NACA 2412 airfoil at Angle of Attack 50 degrees at Reynolds number=$2*10^5$, re-energization caused by part of the flow through slotted flap helps reduce effect of turbulence.

configurations proved that the wing configuration having 'd' equal to 1.7% of wing chord and double slotted flaps extended to 40 degrees was the ideal MAV high lift take-off configuration. The Stalling Angle that was obtained from this MAV high lift take-off wing configuration was 54 degrees while that of plain NACA 2412 wing was 20 degrees this proved that the MAV could sustain lift at much higher angles of attack in comparison to the plain NACA 2412 wing.

References

1. Burke R (2005) Principles of Flight.

2. http://en.wikipedia.org/wiki/Stall_%28flight%29

3. http://www.faa.gov/regulations_policies/handbooks_manuals/aviation/pilot_handbook/media/PHAK%20-%20Chapter%2010.pdf

4. Ali J (2012) Wing Flaps for Lift Augmentation in Aircraft. Decoded Science.

5. http://en.wikipedia.org/wiki/Flap_%28aircraft%29

6. Schlüter JU (2009) Lift Enhancement at Low Reynolds Numbers using Pop-Up Feathers. 39th AIAA Fluid Dynamics Conference 22-25 June San Antonio, Texas, USA.

7. Ghoddoussi A (1998) A Conceptual Study of Airfoil Performance Enhancements Using CFD. Sojo University, Japan.

8. Ali J (2012) Slats, Slots and Spoilers–Lift Modifying Devices on Airplane Wings. Decoded Science.

9. Saha N (1999) Gap Size Effect on Low Reynolds Number Wind Tunnel Experiments. Blacksburg, Virginia USA.

10

Design Considerations and Requirements for In-flight Refueling of Unmanned Vehicles

Ian R McAndrew[1]* and Kenneth Witcher[2]

[1]*Embry Riddle Aeronautical University, Brandon England, UK*
[2]*Embry Riddle Aeronautical University, Las Vegas Nevada USA*

Abstract

The need to refuel in-flight has become a significant part of military strategy for air forces to work at further distances from safe shores. The use of Unmanned Vehicles is increasing and expected to be the principal part of military deployment. This paper will address the concepts and requirements for applying refueling unmanned vehicles in a military context for supporting fixed and rotor aircraft. Design aspects of human factors in the process are considered, reviewed and solutions proposed to allow for the first generation of designs to be developed. Furthermore, the practical and operational limitations will be addressed as part of the human factors implications. Finally, the design parameters are proposed for the first stage developments to achieve Unmanned Vehicle refueling.

Keywords: Inflight refueling; Design considerations; Unmanned vehicles; Human factors

Introduction to the Problem

Inflight refueling has been a main aspect of supporting superiority in the skies and has evolved greatly since its conception. It relies on pilot skill and systems to deliver fuel to military aircraft in extreme conditions in hostile environments. As the role of manned aircraft decreases and unmanned increases the need to refuel inflight is becoming more important for unmanned vehicles. There are instances where very long operations have been required and in 1982 the initial bombing of the Falklands airport involved 11 refueling tankers to support the lead aircraft. Such requirements are extreme and current unmanned aircraft will operate in different environments. Their significantly lower costs to design, develop and operate make them viable for higher risk sorties. To maximize their roles they need to be operational for extended periods. Against this backdrop there is the problem that when no local airfield is available these increased distances will take significantly longer times to arrive and deploy.

There have been instances where unmanned aircraft have flown in excess of 24 hours, albeit with a limited or zero cargo [1]. As their roles expand with more advanced sensors, systems and weapons the take-off weight will increase and the endurance levels will reduce. Each new generation of unmanned vehicles will possibly have greater payloads and advanced weapons systems. Those that are now being designed to operate at altitudes above 60,000 ft will offer on ground commander's flexibility and instant information of terrain and activities without risking unnecessary loss of piloted aircraft. Currently, a 24 hour cycle is the norm, allowing for changes in ground operating personnel on a 3 shift basis. This may seem a long flight, but with limited top speed and the need for continuous monitoring of the ground their range is limited and need to be used locally. Given an advanced flight speed of 90–120 mph a 24 hour deployment will only have travelled under 3,000 miles. Or practically, it could take a whole day to travel to its needed deployment region. Thus, re-fueling is needed to just arrive and added to that more refueling to operate in-situ. As usage depends on proximity the only alternative to local bases and risks involved are sea launches, which prevents immediate and surprise usage. It may be necessary for these unmanned aircraft to fly extended periods before arriving at their destination.

To deploy unmanned aircraft at further distances and for extended periods of time they need more fuel, which will limit payload, or to be refueled in-flight as other military aircraft [2]. This paper will address the current problems, potential changes and aerodynamics of solutions for the next generation of usage. In addition, the practical human factors of operation are included to offer a practical description of the first generation of unmanned refuelers. These aspects are needed to set design parameters to develop prototypes for development.

Refueling Concerns

Cruising speeds of the majority of unmanned vehicles are within the range of 80–100 mph with top speeds around 120 mph [3]. New unmanned aircraft designs are not focusing on increasing these speeds as they will affect the effective usage and operation, there are some that are designed to have increased altitude ceilings of 60,000 ft to give a wider view of the ground with improved visual sensors and cameras. Slow speeds are good for flight stability and the sensors used onboard will produce results with higher resolutions at these speeds; it can be argued that there are justifiable reasons to increase these speeds [4]. If these unmanned cruising speeds are doubled to the region of 200 mph they are significantly below the stall speeds of current military refueling aircraft (KC-135 and A 330). Indeed, C-130 aircraft are refueled flying near maximum speeds and their refuelers flight paths are in descent to assist in maintain aerodynamics stability.

The possibility of using smaller designated manned aircraft to refuel inflight may be possible. An existing aircraft could be modified for carrying fuel, a delivery system and procedures to refuel unmanned aircraft [5]. There are human factors that need addressing, not least the need for potentially very long flights to remote regions and returning, which will exceed flight time capabilities and endurance limits unless numerous pilots. A manned aircraft being docked with an unmanned aircraft raises the concerns of safety incidents that cannot be ignored or

*Corresponding author: Ian R McAndrew, Embry Riddle Aeronautical University, Brandon England, UK, E-mail: McAnd4f1@erau.edu

re-engineered [6]. Docking could be done automatically or remotely; either way an unmanned aircraft approaching a fuel source is a safety risk. Likewise, if they are intending to operate in hostile areas for as long as practical then refueling with manned aircraft is a high risk factor. Two lost unmanned aircraft is both financially and ethically more justifiable.

Currently, refueling is carried out using one of two principal types of systems. The *Boom* operated version that requires the aircraft to be refueled effectively parking at the rear of the refueler and a boom operators positions the refueling arm in place to transfer fuel. A *Drogue* system is one that will trail a funnel cone behind the aircraft from a feeder pipe for the incoming vehicle to dock and received fuel. This latter system has the advantage of being suitable for both fixed wing and rotary aircraft. It could be argued that this has many advantages with rotor unmanned vehicles will play significant roles and to remove its capability before use would be counter-productive.

All the human factors for each of the above mentioned systems vary considerably. Boom arms are controlled by within the refueler by an independent operator at the rear of the aircraft for visible recognition, or by electronic screen, who positions the boom into the receiver aircraft. Skill and dexterity are required to accurately position in various types of weather and atmospheric conditions. The pilots of the refueler and receiver have a separate task to keep flying level and identical speeds with little room for error. The boom arm is designed for lateral and vertical displacements to overcome slight movements and aerodynamic effects, for example *dutch roll*. To work effectively both pilots and a boom operator are required to work in unison and need instant vocal links.

All drogue refueling systems work with the refueler flying straight and level at a pre-arranged altitude and flight pattern. The drogue is deployed into position by the refueler and docking would not be attempted until the feed system is in-situ and not being adversely affected by air conditions. The human factors are directed to the receiver pilot positioning their aircraft to dock, whether remotely of a pilot if manned. One person is responsible to control the system, no second party [7]. As with all refueling the speeds need to be maintained within tolerances of altitude, speed and cross winds. Fast reduction in gross weight and gain as fuel is transferred will require power adjustment to maintain joint constant speed between each aircraft. The drogue system has more flexibility both laterally and vertically and is considered more forgiving by pilots. This system has been automated and can easily be added to an unmanned aircraft for auto docking without major weight additions.

Drogue refueling

Applying *Drogue* refueling for unmanned aircraft has several advantages. First, the refueler can fly straight and level whist the unmanned aircraft positions itself accordingly. Secondly, the drogue can be positioned at a distance to minimize flying in the wake of the air turbulence. Thirdly, the docking can be achieved for control of one aircraft with the possibility of this being achieved automatically. Positional movement of either aircraft is accommodated by the free movement of the fuel line between the refueler and receiver. The disadvantages are that this method supplies fuel at a lower flow rate; nevertheless unmanned aircraft do not require the large volumes for piloted aircraft. To transfer 500 kg of fuel could be achieved within less than 3 minutes. Figure 1 shown the height drop below the aircraft that the drogue operates, which a parameter is to be determined according to aerodynamic requirements of the aircraft. The drogue is collapsible and the storage pod under the wing clearly visible.

Figure 1: Drogue (below) and Boom (above) refueling.

Boom refueling

A *Boom* system works in an opposite process to drogue ones. The aircraft to be refueled positions itself behind the refueler and a boom arm is positioned manually to attach the nozzle and supply fuel. Its fast flow rate reduces the refueling time and positional movements are compensated by lateral movement allowances in the boom arm. When large volumes of fuel are transferred the supply aircraft has to incrementally reduce the power to the engines whilst the receiver aircraft has to increase the power to allow for both to maintain the same airspeed. The principal disadvantage is that accurate positional reliability is needed by both aircraft, possible with unmanned aircraft but advanced control systems needed. For the purpose of unmanned refueling this system has to be superseded by the drogue one for practical and design reasons.

Unmanned Requirements and Human Factors

For unmanned aircraft to refuel in-flight the same practical concerns exist with safety, docking and undocking control. The human factors influence current processes and will if unmanned refueling is used. For example, remote control will require totally on sensors and relayed video links. No true situational awareness can be achieved. In bad weather conditions visibility may be poor and as many unmanned aircraft have lower ceiling limits this can remove flying over bad weather that are available for conventional aircraft. Alternatively, refueling in cloud cover may be desirable.

Ground based control of unmanned aircraft may be flown by trained pilots, but increasingly they are specifically trained unmanned aircraft operators. These people are not fully aware or experienced to the same situational awareness of pilots; although this can be addressed long-term it does leave a shortcoming within the usage. An alternative would be to have the receiver aircraft hand over control to the refueler for this particular operation if the refueler is manned. It cannot be emphasized that even without manning any accident in flight will result is potentially dangerous situations that may have serious consequences for areas on the ground under any safety concern incident [8].

Ideally, remote computer control from the refueler is probably the safest procedure and sensors will overcome visibility and weather constraints. Here, when in close proximity to each other, the flight control will be released from the ground operators and resumed by them when refueling complete and undocked. The control could even be operated with infrared so as not to be detectable from afar.

Aerodynamic Considerations

An unmanned aircraft flying in the wake of a larger refueler will have stability control issues that may significantly inhibit docking or maintaining level flight for safe fuel transfer regardless of method used. Boom methods will increase the likelihood of stability in level flight and lower speed further compound the concerns. The concerns of human comfort for the receiver aircraft are not relevant and can be ignored in this scenario and remove some design considerations. Likewise, unmanned aircraft are not generally designed to operate at high or supersonic speeds; thus, aerodynamic changes to produce stability, extra lift or reducing the effect of turbulence can be incorporated that will not have adverse influences on use or range. Low flight requires its own unique design parameters especially at low altitudes where inclement weather is likely to be influential [9]. Consequently, low speed flight requires high lift and this is easily achieved by large wing area without additional lift devices, *e.g.*, flaps or slot.

Dutch roll

A problem that occurs in all aircraft, regardless of design, operational use or weather and a result of a weaker positive stability compared to positive lateral stability. It is more obvious when flying behind another aircraft with a focus point. In effect, the aircraft oscillates in an elliptical manner–a major concern when marrying the flight path of two aircraft. This is a major concern for the aerodynamics of aircraft at low speeds and increasing altitude. Traditionally this has been engineered out as a concern by using yaw damping sensors–albeit that this would add extra complexity to the aircraft system and weight. The influence of *Dutch Roll* from the wake of a smaller aircraft could significantly reduce the effect and possibly removed the need for an added yaw system to be used. Even if the *Dutch Roll* cannot be removed the damping systems could be designed to eliminate its need as without human occupants the certification process would be different. Alternatively a wing structure with a very high wing aspect ratio would reduce the effects; which is a typical design parameter currently used.

Design consideration for unmanned refuelers

Currently most unmanned aircraft cannot operate at speeds approaching those needed by the principal refuelers currently employed. Thus any refueling tanker that could be used to work at lower speeds would need to be aerodynamically capable for low speeds, high lift and a significant fuel carrying capacity. Basing a full fuel system for a typical unmanned aircraft at 500 kg, and the capability to re-fuel 4 craft in one sortie the fuel requirements would be at least 3,000 kg if you include the fuel needed for the tanker. An initial design specification would use the assumption of 4,000 kg as a probable total weight at take-off. This could be increased if more than 4 refuels were required; however, this paper's focus is on the principles of unmanned refuelers are a concept. Thus, the fundamental lift needed is based upon:

$$L = \frac{1}{2}\rho v^2 A C_L \qquad (1)$$

Where L is the Lift of 4,000 kg, ρ=1.225 kg m^{-2}, take off speed of 35 m s^{-1}, C_L=0.3 (estimate) at take-off speed with a low angle of attack. A≈ 15m^2 wing area. As mentioned above, this would be ideally used with a high wing aspect ratio to assist in stable flight control. The maximum width would be limited to the runway width, and a balance needed to ensure it would not be limited to use depending on the available runways [10]. However, if the wing cord length was only 0. 5m the wing tip length would still only be 30 m, still smaller than a B 737 and suitable for most commercial and military airfields. This would not necessarily limit its use from small remove fields or clearings if needed.

Propeller layout

Unmanned aircraft are usually of a push design, rear mounted propeller (Figure 2). A refueler would require a front mounted propeller, tractor, in order not to interfere with the refueling system and reduce any wash created from the propeller and in the wake if only one used. Alternatively, a twin rear mounted engine combination with a centrally deployed drogue would allow for an operational layout with advantages. For example, if *canards* were incorporated this would assist in reducing *dutch roll* and allow potentially for more stability in straight and level flight (Figure 3).

Fuel delivery type

Boom systems may deliver fuel faster (less than 1 minute is possible after docking) but require accurate flight control of both aircraft. There are no recorded examples of automatic boom control on any aircraft. To automate a joint remote design on small unmanned aircraft would likely take the design to a level without any guarantee of success. A boom arm would add significantly more weight, extra control and complexity. A drogue system would be less complex and could be deployed independently, leaving the receiver to be adjusted in-flight for docking. A drogue will need to be centre mounted, unlike typical refuelers that have them mounted on both wings. There are applications

Figure 2: Unmanned Vehicle with a push propeller system.

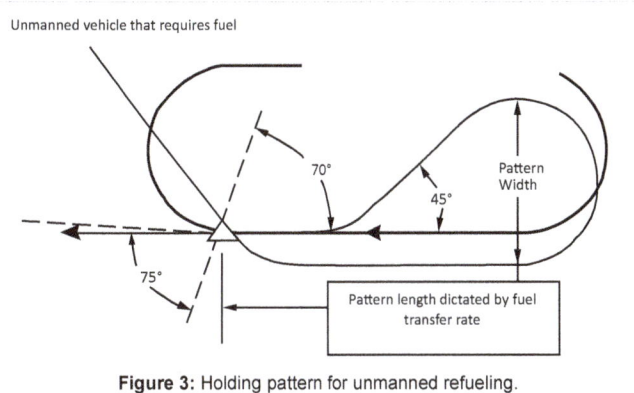

Figure 3: Holding pattern for unmanned refueling.

of centrally mounted drogues on large scale refuelers; however, it does not allow for twin refueling that is required for certain applications. For a small unmanned aircraft the weight and aerodynamic loads would require wing structures heavier than needed otherwise if twin refueling was an aim.

Operational use

Refueling would have to occur at straight and level flight; all air-to-air requires this procedure for optimum easy and safety. A holding pattern similar to those waiting to land at airports would be needed and expected to be a parallel pattern with minimum turning radius between turns [11]. Taking a drogue design as the slowest refueling time a transfer status of 4 minute would need to be linked to a docking pattern. If this allows for 4 minutes then an 8 minute combined time, at 100 mph, needs a minimum of 14 miles. The approaching aircraft needs to accelerate to a position behind the refueler then slow to align.

Unmanned Refueler Specifications

This paper has addressed the principal aspects and operational requirements of unmanned refuelers and how conventional design aspects can be conflated to produce a specification for prototype design.

The key assumptions are that the unmanned vehicle does not need to be redesigned from its current layout. A tube mounted centrally at the front to receive fuel is the only specification recommended. It may be necessary to add canard wings to assist in flight stability when transferring fuel; and this could dictate the fuel transfer rates.

The refueling unmanned vehicle will for the reasons stated above be of a central drogue fuel delivery system. A high wing stricture with probably twin wing mounted engines and a lower central refueling tank to allow stability. The fundament parameters of designing an unmanned vehicle are not complex [12]. The additional supplementary procedures are where detail, agreement and application will dictate. For example, the protocol of holding patterns to enable receiving aircraft to dock efficiently and safely must be finalized [13]. More so, if you think NATO might need to share operational use of deployed refuelers. Likewise, dis-engagement procedures for emergencies, i.e., if attacked, when transferring fuel. Nevertheless these are not leading edge considerations but aspects to verify.

Finally, the significant aspect to develop and refine now is the extension of the drogue and the drop from the refueler to the receiver. Space will be limited for storage of the drogue. Figure 2 does highlight that its collapsible design does allow for safe storage within a small device. Currently, this aspect is being researched to allow for more data, subsequently to allow for further detail design.

Summary

Requirements to allow unmanned refueling have been addressed in this paper. All the principal practical difficulties of using current aircraft has be reviewed and shown that is it not feasible or possible to reconfigure current refueling aircraft. Thus, smaller unmanned aircraft are possible but not a practical solution unless key aspects are considered and accommodated. An unmanned refueler is both practical and possible given these constraints are addressed. A front mounted engine or twin high wing engine mounting is required for the refueling envelope. Drogue systems are the most suitable method for fuel transfer and designs can accommodate the low speed and conditions. Whether the docking is achieved remotely or automatically is unimportant, either way the human factor risks remove risk to life in an operational environment.

References

1. Lerner JC, Boldes U (2012) Applied Aerodynamics. InTech, USA.

2. Raymer DP (2006) A Conceptual Approach. Aircraft Design 4th Edn. Aiaa Education Series.

3. Bertin JJ, Cummings RM (2008) Aerodynamics for Engineers. 5th Edn. Prentice Hall, Inc. USA.

4. Anderson JD (2005) Introduction to Flight. McGraw-Hill Higher Education, USA.

5. Dole CE, Lewis JE (2000) Flight theory and aerodynamics: a practical guide for operational safety. John Wiley & Sons, USA.

6. Hagan P, Krieger GR, Montgomery JF, O'Reilly JT (2013) Accident Prevention Manual: Engineering & Technology. NSC, USA.

7. Ferguson M, Nelson S (2013) Aviation Safety: A Balanced Industry Approach. Blackwells, UK.

8. Reason J (1990) Human Error. Cambridge University Press, UK.

9. Simons M (1999) Model aircraft aerodynamics. Nexus Special Interests, UK.

10. Moir I, Seabridge A (2012) Design and Development of Aircraft Systems. Wiley, USA.

11. http://www.wikihow.com/Fly-a-Holding-Pattern

12. http://www.as.northropgrumman.com/products/globalhawk/index.html

13. Fahlgren G (2011) Human Factors. AuthorHouse, USA.

The Study by Visualization of Vortex Structures on a Warhead Wing, Apex Angle (ß)=68.6°

Abene Abderrahmane*

Laboratory of Aerodynamics, University of Valenciennes, France

Abstract

This study has been given prominence and confirmed by extensive documentation while recent visualizations observed in a wind tunnel have revealed that rectilinear vortex flows from different angles interacting with one another above warheads. My experimental research has indicated the many and very specific series, by visualizations of wisps of smoke trails, of vortex structures of a preferential nature above a warhead where β=68.6°. The positioning of those structures has thus been determined with regard to the application of very general criteria governing the stability of vortex flows. What have also been exposed are the specific aspects of vortex flows together with some comments concerning the bursting of vortices above the warhead. The purpose of this study is to optimize the connection between marine, air and land forms such as sails, the wings of aircraft, cars and trains – and the stability of fluid flows around such forms.

Keywords: Visualization; Aircraft; Hydro-dynamics; Aerodynamic

Nomenclature

$\theta_{\ell,m}$: preferential angle associated with the whole numbers l and m;

ℓ and m: whole numbers such as $m>\theta$ and $\ell \geq m$;

β: apex angle;

Co: length of wing;

i: incidence;

Vo: speed of the flow at upstream infinite;

Re: Reynolds number;

α_1: the main or interior inter vortex angle for warheads;

α_2: the secondary or exterior inter vortex angle for warheads;

ω_1: the main or inter vortex angle for warheads or cones;

$\omega2$: the secondary or exterior inter vortex angle for warheads or cones

Introduction

Quite a considerable number of studies have been carried out to date into delta wings, gothic wings, cones and also into more or less simple slender bodies formed from combinations of such components; the findings have dealt as much with the development of approximate theories [1] as with the definition of models specifying vortex lift by unit area.

Visualisations of hyper lifting vortex structures, mainly those carried out by Werle [2-8], the analysis of pressure and speed fields created by these vortices, with or without their bursting–notably the analysis by Solignac [9-11] provide quite outstanding studies that are the standard works in their fields. Already described fully in such papers as, for example, those by Werle, Solignac and Stahl [2-17], these findings offer today in their entirety a thorough knowledge of the properties of various types of slender bodies. However, given that the character of most of the aspects referred to remains empirical and limited to this or that degree of incidence [2-17] or to a numeric range [18], the way lies open, starting out from experimental data and various factors of analysis [19], for new attempts to be undertaken to examine the fundamental problems related to the position of vortices created by such slender bodies.

A large number of photographic visualisations, concerning vortex flows developed on the upper surface of delta or gothic wings and cones, have been carried out in the laboratory at Valenciennes University (France) [1,20] in such a way as to provide a better understanding of the development and positioning of vortex structures at not only low and mean incidence but also at high incidence. These visualisations have enabled priority to be accorded to the study of examples of the most elementary shaped bodies, i.e. delta and gothic wings. The results obtained in these two cases, and already fully described in previous papers [19,21], may be acknowledged to have remarkable simplicity and consequently convey the fundamental nature of these studies.

The angles between the vortices have been found, under experimental conditions, to have a preferential nature thereby underlining a simple angular characterisation of their positioning relative to single or double vortex torques.

Visualisation Techniques

The visualisations in their entirety were carried out within the dimensions of the air tunnel (45 cm×45 cm section) at the Valenciennes University Aerodynamics and Hydro-dynamics laboratory. The vortex systems developed on the upper surface were rendered visible by providing for emissions of wisps of white smoke at the tip of the warhead: these emissions were obtained by injecting oil under pressure through a tube of small diameter and then, by means of an integrated electrical system, vaporising the oil immediately as it left the probe.

*Corresponding author: Abene Abderrahmane, University of Valenciennes, Aulnoy les valenciennes, 59300, France, E-mail: a.abene@yahoo.fr

The warhead was fixed onto an axis relayed to a cursor graduated from 0° to 360° which enabled the incidence of cone rotation to be varied. The visualisations were captured on photographs and video film which are today stored in the department's data base.

Vortex Structures of a Warhead Wing

Geometrical description

The profile under investigation in the wind tunnel is a warhead having an apex angle $\beta=68.6°$ and a chord Co=240 mm. It is 1 mm thick. 11 and 12=vortex rotation in opposite directions. 1, 2 and 3, V_O=speed (Figures 1a and 1b).

1) Tablecloth horn wrapping around the main vortex from the apex

2) Low secondary winding around vortices in the same direction but opposite to the main vortex

3) Primary separation line and secondary line of separation

4) Areas swept by previous 1-2-3 vortices

5) Partitioning lines within the fluid

6) Allure lines parietal current.

Analysis of the Results

The evolution of the vortex phenomena was traced in terms of that parameter which exerts the greatest influence on them, namely the angle of incidence of the configuration in relation to the flow. The visualisations were carried out at an upstream speed of flow of 3 m/s.

Reminder of the main phenomena taking place at low and mean incidences:

incidence=0°: the flow is uniform on the upper side of the warhead. The boundary layer is observed but there is no flow separation as yet.

incidence=5°: the upstream flow skirts around the leading edges of the profile. Three zones become organised into a central zone and two external ones. The hyper lifting vortices, resulting from the separation of the boundary layer, begin to appear as increasingly organised structures.

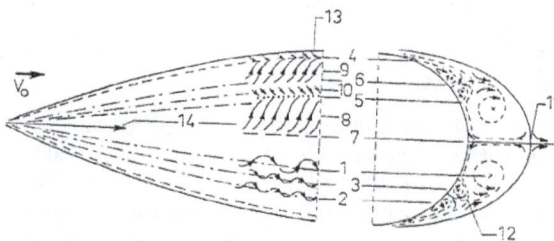

V_0: Blower speed
i: Angle of inclination of the wing

Figure 1b: Inclination of the warhead wing.

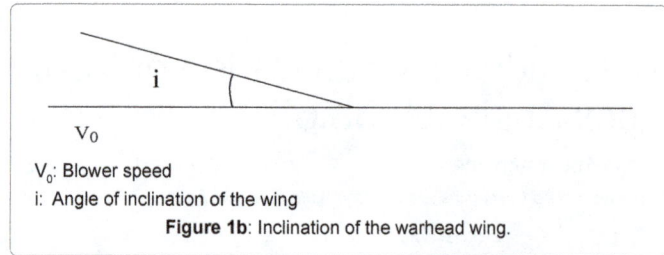

Figure 2: View of the upper surface: $\beta=68.6°$, i=3°, 41000<Re<88000.

Figure 3: View of the upper surface: $\beta=68.6$, i=22°, $\alpha=45°$, 41000<Re<88000.

incidence=8°: the main and secondary vortices are clearly detected and have now become individualised, concentrated and separated from the boundary layer. Visualization of the central zone has shown it to be gradually fading out.

incidence=10°, i=15°, i=20°: the vortices increase in strength. Both the vortex flow and the direction of the rotation of the vortices are clearly seen. The central zone has disappeared. The presence of tertiary vortices is to be noted although they are extremely difficult to visualise. There is no bursting as yet (Figure 2).

incidence=3°, i=5°: the upstream flow skirts around the leading edges of the profile. Three zones become organised: a central zone and two external ones. The hyperlifting vortices resulting from the separation of the boundary layer begin to appear as increasingly organised structures (Figure 3).

i=25°: the bursting phenomenon makes its appearance. The main vortices commence bursting a long way downstream from the profile: in fact, a more diffused mass of smoke is observed in this zone. The secondary vortices have burst upstream from the trailing edge; as for the tertiary ones, which are difficult to observe because of their positioning at the edge of the boundary layer, it seems that they burst in the area close to the apex and coil around the main and secondary vortices. Once the secondary vortices have burst, they also coil around

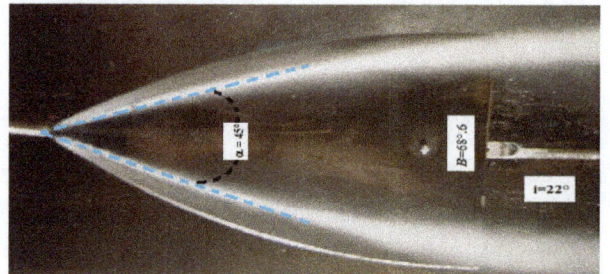

1 –The vortex tablecloth.

2 – 3 The vortex tablecloth rotates in the same direction.

4 – 5 Row from primary and secondary detachment.

6 – 7 Row from lateral and medial detachment.

8 – Partitioning Line within the fluid.

9 – 10 the main vortex tablecloth.

11 – 12 The rotation of the vortex reverse.

13 – Beginning of bursting of secondary vortex.

14 – Vortex streamline.

Figure 1a: Influence of the impact on the flow around the wing warhead legend.

the main ones. The asymmetry of the bursting point of the main and secondary vortices is to be noted (Figure 4).

incidence=40°: the secondary vortices burst near the apex whereas the point of bursting of the main ones has advanced to a third of the way along the chord. A sudden expansion at the core of the main vortices is still discernible, followed by an unstable zone showing quite considerable turbulence (Figure 5).

incidence=30°: the tertiary vortices have now completely disappeared. The secondary ones commence bursting in the area of the trailing edge of the profile (Figure 6).

NB: the position of the bursting point is estimated on the basis of a mean reading of the respective bursting points of the right-hand and left-hand main vortices.

incidence=45°: the main vortices break down at the fore quarter of the chord; the secondary ones are absorbed by the main vortices at the apex and are no longer discernible.

incidence=50°: a total bursting of the vortices takes place at the

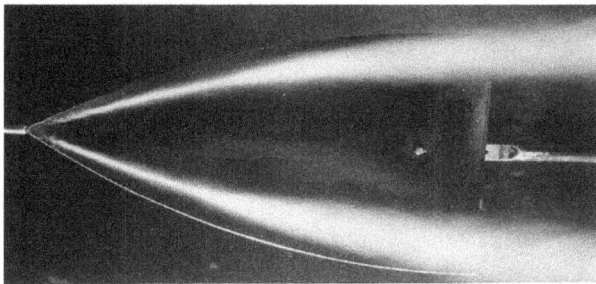

Figure 4: View of the upper surface: β=68.6°, incidence=25°, α=45°, 41000<Re<88000.

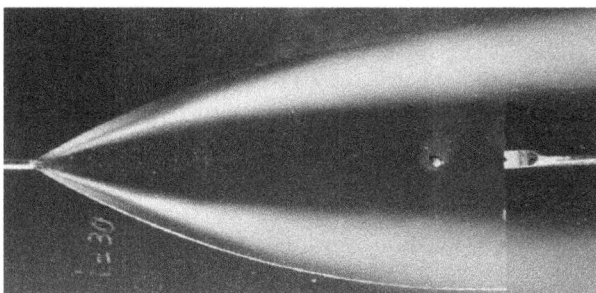

Figure 5: View of the upper surface: β=68.6°, incidence=30°, α=45°, 41000<Re<88000.

Figure 6: View of the upper surface: β=68.6°, i=35°, 41000<Re<88000.

apex. Intense turbulence is observed at about the trailing edge, entailing diffusion of the smoke trail.

Interpretation

Comments

The bursting of the secondary vortices is not as impressive as that of the main ones. In fact, the vortex cores of the latter, after having maintained a cone-shaped form increasing in diameter towards the downstream side, undergo a sudden expansion into a brush-shaped form (the rate of expansion is approximately three to four times the diameter of the vortex core upstream from the bursting point). The point of bursting is immediately followed by a zone in which the flow circulates again and by an area of intense turbulence. The thin wisps of smoke, indicating the main vortices, seem to change direction at the point of bursting and to follow a spiral trajectory downstream.

1. The pulsation phenomenon of the bursting of the vortex has become highly conspicuous from this point in time; the bursting points undergo quite considerable positional fluctuations. The burst vortices seem "to dance".

2. On placing an object in the axis of the burst vortices, it can be seen that the bursting points "rise again" in an upstream direction but it is also noticed that the bursting point of the secondary vortices is still located upstream from that of the main vortices.

The presence of preferential intervortex angles is observed. The intervortex angles on the warhead are: $10° \leq \iota \leq 45° \Rightarrow \alpha_\iota = 45°$. The experiment has shown the following:

1. A flow on the upper surface of the warhead as regards low incidences;

2. The start of the formation of vortex structures with steadily flowing thin streams becoming separated from each other and heading towards the leading edges;

3. The formation of two vortex systems because the warhead under investigation has a non-preferential apex angle;

4. The deterioration of the systems occasioned by the bursting phenomenon;

5. The unsettling of the profile under investigation on the appearance of the torch phenomenon (i.e. the "cornet-like" spiral coiling of the flow) constituting the disappearance of the intervortex zone.

The exterior secondary system is less significant than, and not so dense, as the core system and is the first one to deteriorate.

The law of filiation is examined between the two systems: the angle of the secondary system is produced from that of the core system (akin to a "father and son" relationship). The bursting evolves in relation to the incidence. It is a function of the apex angle: the greater the apex angle, the sooner the bursting takes place. When the angle of attack increases, the vortices coming from the canard plan evolve closer to the surface of the rear section of the aerofoil wing and towards its exterior. As regard high incidence, the merging process of the two vortex structures begins at about the aerofoil wing: this leads to the fact that no more than one sole vortex structure exists on the surface of the aerofoil wing being the result of the joining of the other two structures (Figure 7).

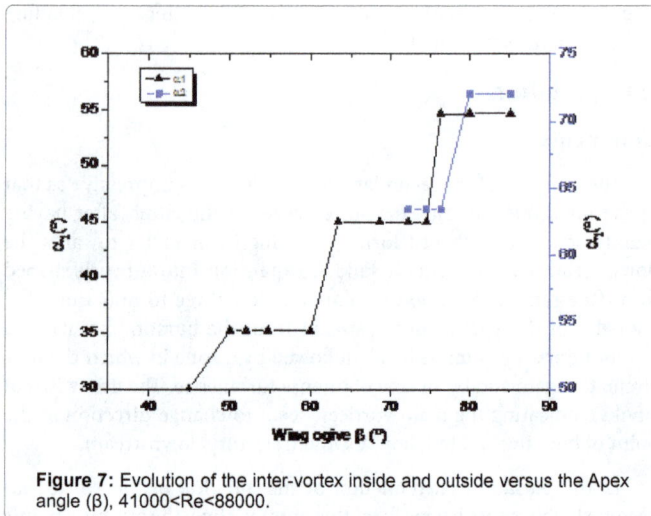

Figure 7: Evolution of the inter-vortex inside and outside versus the Apex angle (β), 41000<Re<88000.

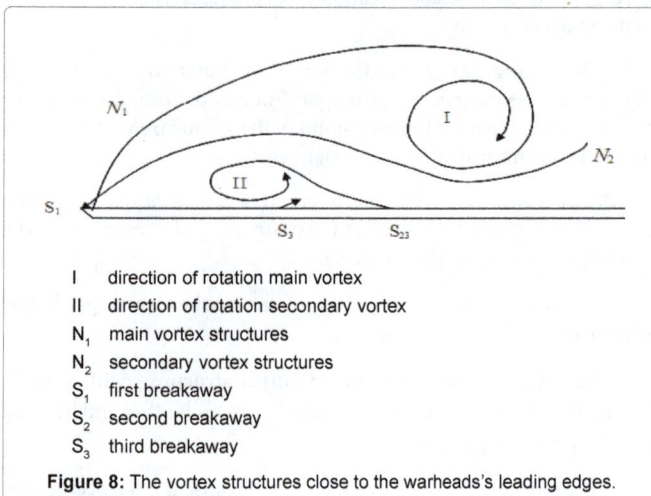

I	direction of rotation main vortex
II	direction of rotation secondary vortex
N_1	main vortex structures
N_2	secondary vortex structures
S_1	first breakaway
S_2	second breakaway
S_3	third breakaway

Figure 8: The vortex structures close to the warheads's leading edges.

The vortex structures close to the warheads's leading edges

The secondary vortex (II) flows in an opposite direction to that of the main one (I). As the structures are located at the top of the warhead, between S_1 and S_2, it is very difficult to visualize them by wisps of smoke trails, because their rotational forces, according to Solignac et al. are from 3 to 5 times lower than one of the main vortex (Figure 8).

Comparison with vortex structures formed on the upper surface of thin gothic wings

The results of the study of the behaviour of vortex structures on the upper surface of gothic wings [13] can be compared with the findings which can be obtained through another examination of the photographs published by Werle [5]. The figures numbered 11a, 11c and 11e in that publication reveal that, for the apex angle β, those structures can easily be calculated from the constituents E (namely, the relationship between the median chord at the edges of the flow and the parabolic shape of the leading edges), each of which being individually and in the order mentioned 37°, 56.4° and 72.1° as indicated by Werle [8].

Conclusion

The progressive evolution from elementary vortices of the sheared flow before bursting towards a particularly stable vortex system,

wherein spatial positioning reveals an original organisation, still remains today an enigma. As regards those warhead wings studied, which have included apex angles over a very wide range, the major results from visualisations–the latter made possible by producing smoke streams at the apex of those warheads may be resumed thus :

1. If the angle between the leading edge at their point of intersection (apex angle) β is a preferential one of the first grouping (i.e. 20.7°; 26.6°; 30°; 35.3°; 45°), only two curvilinear vortices are formed and these are to be found above the warhead starting from the apex and thereby constituting a preferential angle between them at the apex point.

2. However, where the apex angle β is either a non-preferential one, or a preferential angle of the second group (i.e. 54.7°; 63.4°), four vortices are observed, the two interior of which form between them a preferential angle $α_1$, while the two exterior vortices form between them another preferential angle $α_2$.

For example, above the warhead having an apex angle β of 52°, the interior vortices create an angle $α_1$ of 35.3° (=$θ_{lm}$: this corresponds with $l=m=1$) whilst the exterior vortices also create a preferential angle, immediately inferior to the angle β, being an angle $α_2$ of 45° (corresponding with $l=m=1$). The angle (or angles) between vortices is (or are) constant, either throughout the range of incidences – i.e. from the first appearance of the vortices up to their point of bursting–or at a maximum of two or three levels within the range [13,15].

However, it is important to remember that several of these preferential angles, either separately or in groups of two or three, are equally to be found in the widely accepted standard theories of hydrodynamics and aerodynamics such as those pertaining to the wake of ships and to aerodynamic drag.

One of the central ideas of this present paper is the following: the leading edge of the warhead is at the same time a line along which the borderline layer of flow is especially not very thick and this is because the leading edge is more often than not very close to the bursting point, or to the partitioning line along which the flow is divided between currents on the lower and upper surfaces. On the other hand, this same leading edge is equally a line along which the speed is very high at the boundary of the borderline layer. This is a result of the narrowing of thin fluid streams associated with the sharp curvature of the wall.

A leading edge is therefore a line around which the transverse variation rate of speed is especially high and where, consequently, is to be found a concentration of very high values of the module of the vortex vector whose direction should, in a stationary flow before bursting, coincide with that of the leading edge.

References

1. Jones RT (1990) Wing Theory. Princeton University Press, New York.

2. Werle H (1986) Flows of visualizations in stationary in the water tunnels ONERA using quantitative methods facilitating their exploitation. Technical Note 3, ONERA.

3. Werle H (1986) Detachment structures on cylindrical wings. The Aerospace Research 3: 13-19.

4. Werle H (1987) Transition and turbulence (hydrodynamic visualizations). ONERA, Technical Note 7.

5. Werle H (1965) Vortices thin wings very slender. The Aerospace Research 109:3-12.

6. Werle H (1980) Vortex interactions on delta wings fixed or oscillating (hydrodynamic visualizations). The Aerospace Research 2: 43-48.

7. Werle H, Gallon M (1976) Study by hydrodynamic visualization separated flows of various control processes. The Aerospace Research 2: 75-94

8. Werle H (1990) Peeling on the body of low speed revolution. Aeronautics Research 6: 49-72.

9. Solignac JL, Pagan D, Molton P (1988) Review of certain properties of the flow on the upper surface of a delta wing.

10. Solignac JL, Pagan D, Molton P (1982) Fundamental study on the formation and flow of vortex structures, basic experience and modeling. Aerodynamic Department, ONERA.

11. Solignac JL, Pagan D, Molton P (1989) Experimental study of the flow in the extraction of a delta wing in incompressible regime. The Aerospace Research 6: 47-65.

12. Delery J, Pagan D, Solignac JL (1987) The bursting of the vortex generated by the delta wing. Baden: 6.

13. Stahl W (1993) Experimental investigations of asymmetric vortex flow behind elliptic cones at incidences. AIAA Journal 31(5): 966-968.

14. Stahl W, Asghar A (1990) Suppression of vortex asymmetry behind circular cones. AAIA Journal 28(6): 1138-1140.

15. Stahl W, Mahmood M, Asghar A (1990) Experimental investigations of the vortex flow on very slender, sharp-edged delta wings at high incidence. German Aerospace Research Establishment, Cologne, Germany.

16. Stahl W, Mahmood M, Asghar A (1992) Experimental investigations of the vortex flow on delta wings at high incidence. AIAA Journal 30(4): 1027-1032.

17. Stahl W, Hartmann K (1990) Entwicklung und erprobung einer Nasengeometric fuer flugkoerper bei grosser anstelling. German Aerospace Research Establishment, Cologne, Germany.

18. Morteveille A, Tournier L (1985) Behavior of slender cones at high incidence. University of Valenciennes.

19. Pagan D, Benay R (1988) Numerical study of the vortex breakdown subjected to pressure gradients. The Aerospace Research.

20. Leray M, Deroyon MJ, Deroyon JP, Minair C (1985) Angular stability criteria of a helical vortex or a couple of rectilinear vortices role of vantage points in the optimization of wing sails, hulls of aircraft and ships. Bulletin of the ATMA, Paris 85: 11-529.

21. Abene A (1988) Graduate videographic and photographic materials vortices formed on the upper surface of various slender bodies. University of Valenciennes.

12

Effects of Air Vitiation on Scramjet Performance Based on Thermodynamic Cycle Analysis

Juntao Chang[1]*, Shibin Cao[1], Junlong Zhang[1], Jianfeng Z[1], Wen Bao[1], Wenqing S[2] and Zhixin Li[2]

[1]*Academy of Fundamental and Interdisciplinary Sciences, Harbin Institute of Technology, Harbin 150001, China*
[2]*Beijing Research Institute of Mechanical and Electrical Engineering, 100074, Beijing, China*

Abstract

The influences of air vitiation on scramjet performance were studied in this paper based on thermodynamic cycle analysis. By the aid of NASA Glenn's computer program Chemical Equilibrium with Applications (CEA), parameters of vitiated air at combustion heater exit could be calculated. Based on thermodynamic cycle analysis, scramjet performance, including internal thrust, specific thrust, and specific impulse, were analyzed with 9 simulation criterions under Mach 4.5 and 6.5 conditions, respectively. The results show that, (1) specific heat and specific heat ratio of vitiated air are mainly affected by combustion heating medium, and also have minor connection with temperature criterions. Moreover, choosing different simulation criterions influences the temperature or pressure between clean air and vitiated air significantly; (2) the internal thrust and specific thrust of scramjet with hydrogen heated airstream are greater than that of scramjet with alcohol and kerosene heated airstream; (3) matching static pressure and dynamic pressure will obtain higher internal thrust compared to total pressure, and the effect of pressure criterions on internal thrust is more remarkable than temperature criterions. On the other hand, these findings proved that the method based on thermodynamic cycle analysis is a simple, dependable, and effective approach to evaluate the scramjet performance affected by air vitiation.

Keywords: Air vitiation; Scramjet; Performance; Thermodynamic cycle

Introduction

A large number of researches have been made to develop the hypersonic airbreathing propulsion systems (especially scramjet). One of effective research methods is conducting meaningful ground tests. With the information gained from experimental tests, Computational Fluid Dynamics (CFD) simulation can not only describe the details of complex flow field accurately, but also expand flight envelop which is not achievable at ground tests. In addition, theoretical analysis provides a quick and reliable approach to assess scramjet. Based on above three methods, lots of critical techniques and problems, such as combustion oscillations in supersonic combustor, combustion mode analysis, re-cooled cycle and thermal management system of hydrogen-fueled scramjet, were investigated in recent years [1-4]. Inevitably, the research of scramjet will be more and more intensive and extensive in the future.

In order to simulate the real flight conditions at ground tests, generating high-enthalpy flow is necessary. For example, the stagnation temperature of engine airstream is about 1100 K at Mach 4.5 and the corresponding stagnation enthalpy reach over 1 MJ/kg. The methods of air heating are currently as follows [5]: combustion heating, shock tube heating, arc heating, storage heating, and electric heating. Because of the ease of operation, the low operating cost and the supply of long test times, combustion heating has become the most widely used method of producing high-enthalpy air.

The main disadvantage of combustion heating is the introduction of contaminating species, which primarily include H_2O and CO_2, and little other radical species, such as OH, O, H and NO. As a result, the vitiation contaminates lead to differences of physical property and flow parameter between clean air and combustion-heated (vitiated) air. For instance, when hydrogen is used to heating air, the mole fraction of H_2O in vitiated air could reach 12% at 1200 K total temperature simulation, which changes the specific heat of vitiated air to 1280 $Jkg^{-1}K^{-1}$ [6].

When selecting hydrocarbon fuel to combust air, the contaminating species will contain not only water but also carbon dioxide. Changing the thermodynamic properties of engine inflow through vitiation has been shown to lower both the combustion-induced temperature rise and the internal thrust in a scramjet engine [7]. Extrapolating vitiated air measurements directly to flight could result in over fueling of the combustor and possible inlet unstart [8,9].

Previous studies related to vitiation effects focus mainly on pressure and temperature variation in supersonic combustor, mode transition of dual-mode scramjet, and ignition and flame holding of scramjet, etc. It was discovered that as the increase of the level of H_2O, CO_2 or H_2O and CO_2 in vitiated air, the pressure rise due to combustion and isolator shock train length decreased and the combustion mode might be changed [10]. The fuel equivalence ratio at which mode transition takes place raised when compared with clean air [11,12], and the influence of water and carbon dioxide on combustor performance was nonlinear as the concentration of vitiation species increased [13]. Li Jianping et al. [14] made two-dimensional calculations of kerosene-fueled supersonic combustion, finding that the presence of vitiation components lead to combustor performance deterioration characterized by the decrease of temperature rise, combustion efficiency and stream wise impulse relative to those in clean air. The effects of vitiated air on ignition and flame holding were also studied broadly [15,16]. Pellet et al. [15] reviewed vitiation effects on scramjet ignition and flame holding, and

*Corresponding author: Juntao Chang, Academy of Fundamental and Interdisciplinary Sciences, Harbin Institute of Technology, Harbin, China
E-mail: changjuntao@hit.edu.cn

revealed that the existence of H_2O may change the reaction kinetics. Generally, all of the above studies matched total pressure and total temperature between clean air and vitiated air, so matching other parameters, such as static pressure, total enthalpy, become another research focus [11,13,17]. It was revealed that matching total enthalpy rather than total temperature get closer mode transition critical equivalence ratio related to clean air [11].

However, ground test and numerical simulation cost too much and need a lot of time. It is necessary to evaluate the effects of different heating mediums and matching schemes on scramjet performance quickly and easily. To achieve this, a calculation method based on thermodynamic cycle analysis introduced was introduced. Via this method, the qualitative tendencies and rules of vitiation effects on scramjet performance are available, which supply effective guidance on the selection and validation of heating mediums and simulation criterions. More important, this method will save plenty of costs and time.

This paper assesses the impact of simulation criterions on vitiated air properties entirely and evaluates three combustion heaters by the help of well-developed chemical equilibrium applications and thermodynamic cycle analysis method. The results will be in favor of understanding the characteristics of vitiated air and its effects on scramjet performance. The article begins with calculation and analysis of parameters of vitiated air at combustion heater exit. Then, method based on thermodynamic cycle analysis is introduced, and the results will be represented in detail.

Parameters of Vitiated Air

This section gives an example of how to calculate the components of vitiated air at combustion heater exit, firstly. Then, parameters of vitiated air, including specific heat, specific heat ratio, static temperature, and static pressure are discussed amply. Meantime, comparison of Mach 4.5 and 6.5 conditions is also presented.

Method of calculating vitiated air parameters

Before estimating scramjet performance, the parameters of vitiated air at combustion-air preheater exit must be calculated. By mixing, burning fuel and air with ambient temperature, combustion heater supply engine with simulated total temperature inflow. Due to the introduction of heating fuel, the components of vitiated air will differ from clean air, resulting in the differences in specific heat, specific heat ratio, and so on. One method of acquiring the components of vitiated air is to make use of the NASA Glenn's computer program *Chemical Equilibrium with Applications* (CEA) [18].

Because of the differences of thermodynamic properties between clean air and vitiated air, matching part of their parameters will lead to the distinction of other parameters. For instance, it is impossible to match static temperature and total temperature or total enthalpy at the same time. Therefore, it is significant to estimate the effect of different simulation criterions on parameters of vitiated air and scramjet performance. Table 1 lists 9 simulation criterions, where T_t is total temperature, T_s is static temperature, H_t is total enthalpy, p_t is total pressure, p_s is static pressure, p_d is dynamic pressure, Ma is flight Mach number. Table 2 presents parameters of clean air as the standard conditions. In clean air, the mole fractions of N_2 and O_2 are 78% and 21%, respectively. Little CO_2, almost 0.03%, and other species also exist in clean air, but they are negligible. As for an example, the first simulation criterion, T_t - p_t - Ma was chose to show how the parameters of vitiated air are obtained. To ensure the consistence between clean

air and vitiated air, the mole fraction of O_2 was fixed at 21%. Then, the Mach number, total temperature and total pressure of vitiated air were set with the same value of clean air. There are three fuels for heating air, that is, alcohol, hydrogen, and kerosene. After choosing one of the heating fuels, the compositions of vitiated air were gained by using CEA. This paper focuses only on two vitiation species, H_2O and CO_2.

Table 3 gives the mole fractions of H_2O and CO_2 in vitiated air at Mach 4.5 and 6.5, respectively, corresponding to T_t - p_t - Ma criterion. As shown in Table 3, hydrogen heater can only generate H_2O, but alcohol heater and kerosene heater will produce H_2O and CO_2, whose concentration are determined by the level of simulated total temperature. Using other programs, the specific heat C_p, specific heat ratio γ and other parameters might be obtained. Table 4 shows some calculated parameters of vitiated air at Mach 4.5 and 6.5.

Calculated results

Specific heat of clean air and vitiated air at Mach 4.5 and Mach 6.5 are shown in Figures 1 and 2, respectively. Because of containing

Criterions	Parameters
1	T_t-p_t-Ma
2	H_t-p_t-Ma
3	T_s-p_t-Ma
4	T_t-p_d-Ma
5	H_t-p_d-Ma
6	T_s-p_d-Ma
7	T_t-p_s-Ma
8	H_t-p_s-Ma
9	T_s-p_s-Ma

Table 1: Simulation criterions.

Ma	T_s, K	p_s, kPa	T_t, K	p_t, kPa
4.5	216.7	6.936	1049	2164
6.5	223.6	1.879	1901	6340

Table 2: Standard conditions of clean air at Mach 4.5 and Mach 6.5.

Heating fuel	O_2, %		H_2O, %		CO_2, %	
	Mach 4.5	Mach 6.5	Mach 4.5	Mach 6.5	Mach 4.5	Mach 6.5
Alcohol	21.00	21.00	6.25	16.72	4.20	11.22
Hydrogen	21.00	21.00	10.50	27.60	0.02	0.02
Kerosene	21.00	21.00	4.08	10.76	4.27	11.30

Table 3: Mole fractions of O_2 and vitiation species in vitiated air with criterion 1.

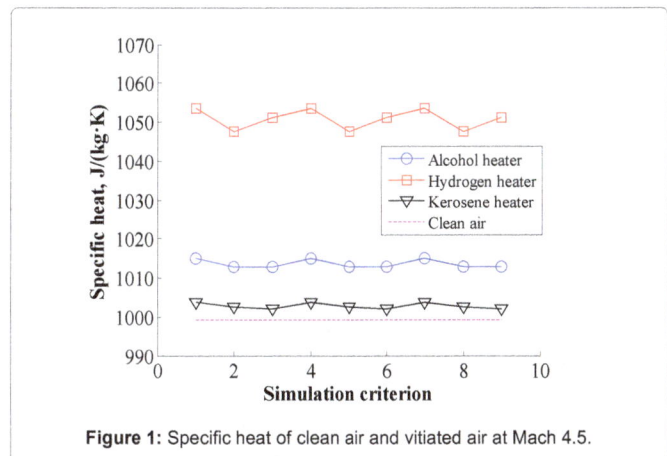

Figure 1: Specific heat of clean air and vitiated air at Mach 4.5.

Heating fuel	C_p, Jkg⁻¹K⁻¹		γ		T_s, K		p_s, Pa	
	Mach 4.5	Mach 6.5	Mach 4.5	Mach 6.5	Mach 4.5	Mach 6.5	Mach 4.5	Mach 6.5
Alcohol	1015.0	1048.8	1.388	1.370	224.54	251.03	6489.44	1250.92
Hydrogen	1053.7	1156.4	1.391	1.377	222.17	242.95	6659.03	1463.74
Kerosene	1003.7	1016.9	1.390	1.374	223.61	247.11	6535.46	1315.19

Table 4: Calculated parameters of vitiated air with criterion 1.

Conditions	Clean airstream	Alcohol heated airstream	Hydrogen heated airstream	Kerosene heated airstream
Ma 4.5	0.65	0.58	0.60	0.58
Ma 6.5	0.65	0.50	0.55	0.51

Table 5: Combustion efficiency at different conditions.

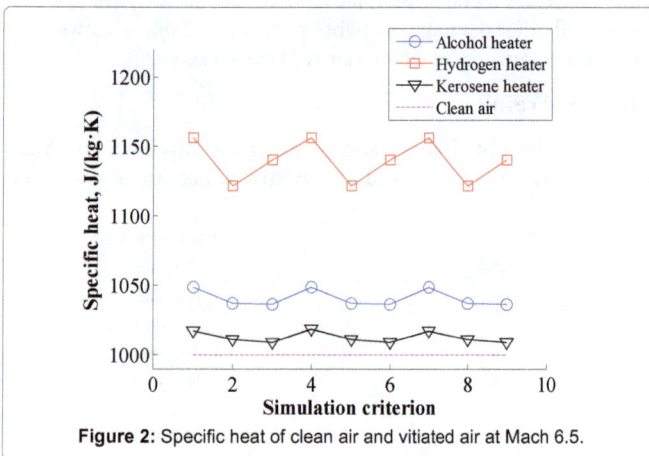

Figure 2: Specific heat of clean air and vitiated air at Mach 6.5.

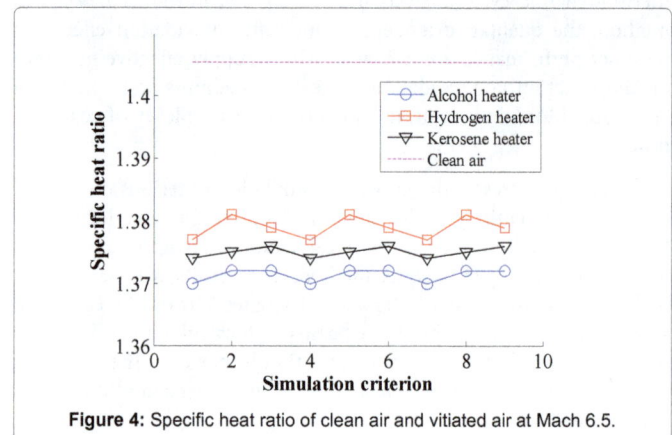

Figure 4: Specific heat ratio of clean air and vitiated air at Mach 6.5.

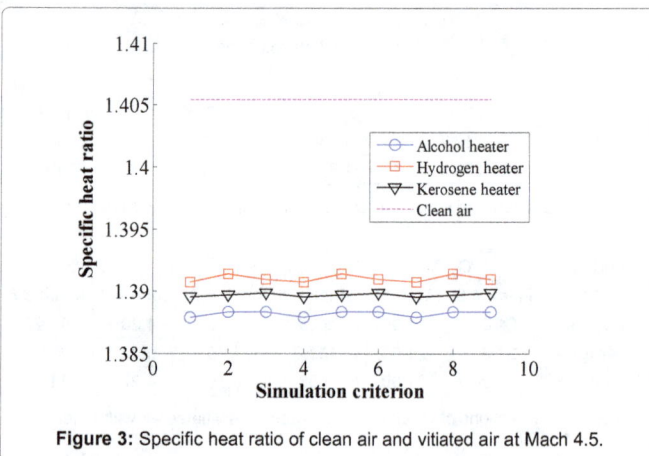

Figure 3: Specific heat ratio of clean air and vitiated air at Mach 4.5.

a certain amount of H_2O, whose specific heat is the largest among all components of air, the specific heat of vitiated air is greater than that of clean air, which is set at 1006 Jkg⁻¹K⁻¹, at all criterions. Furthermore, as the volume of water generated by hydrogen heater is bigger than others, the specific heat of vitiated air produced by hydrogen heater is the highest, as shown in (Figures 1 and 2), followed by alcohol heater, then kerosene heater. Evidently, because more water was brought into vitiated air at Mach 6.5, the corresponding specific heat of vitiated air is bigger than those at Mach 4.5 condition.

Figures 3 and 4 show specific heat ratio of clean air and vitiated air at Mach 4.5 and Mach 6.5, respectively. As we known, the specific heat ratio is determined by both mean molecular weight and specific heat of gas. The presence of H_2O will decrease mean molecular weight of vitiated air and increase specific heat of vitiated air. Inversely, CO_2 can enhance mean molecular weight and reduce specific heat of

contaminated air. As a result, the specific heat ratio of vitiated air has minor distinction (± 0.01) between 3 combustion heaters. However, the specific heat ratios of vitiated air are smaller than that of clean air, as the specific heat ratios of H_2O and CO_2 are fewer than that of N_2.

When different temperature criterions are matched for the same combustion heater, the volume of heating fuel varies, causing the disparities of concentration of H_2O and/or CO_2 in vitiated air. Generally, the difference in hydrogen heater is moderately obvious and selecting total temperature will produce more water. However, for alcohol or kerosene heater, volumes of H_2O and CO_2 have few distinctions at different criterions, leading to the approximate value of specific heat and specific heat ratio, as shown in Figures 1-4. What's more, no matter which pressure criterions are matched, the specific heat and specific heat ratio are just identical.

Static temperatures of vitiated air and clean air at Mach 4.5 and Mach 6.5 are shown in Figures 5 and 6, respectively. When simulating total temperature (criterion 1, 4 and 7), the static temperature is inversely proportional to specific heat ratio, causing that static temperatures of vitiated air are greater than that of clean air. While matching total enthalpy (criterion 2, 5 and 8), the static temperature is inversely proportional to both specific heat and specific heat ratio. The results with Mach 6.5 show the same trends.

Figures 7 and 8 show static pressure of vitiated air and clean air at Mach 4.5 and Mach 6.5, respectively. Static pressure is in proportion to γ when matching total pressure and in inverse proportion to γ when simulating dynamic pressure. As a result, the static pressure of vitiated air is smaller than that of clean air when matching total pressure and greater than that of clean air when simulating dynamic pressure.

It can be concluded from this section that the combustion products introduced to vitiated air have considerable effects on its properties. The specific heat of vitiated air is greater, but the specific heat ratio is

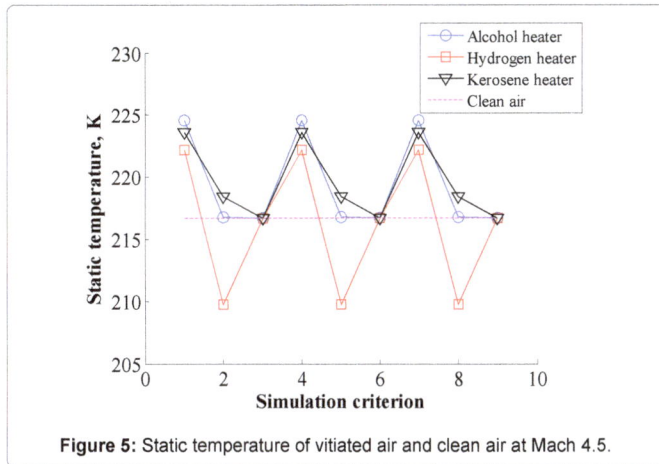

Figure 5: Static temperature of vitiated air and clean air at Mach 4.5.

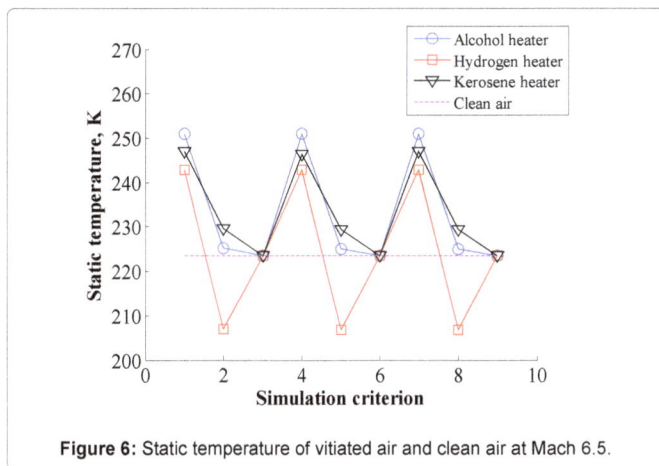

Figure 6: Static temperature of vitiated air and clean air at Mach 6.5.

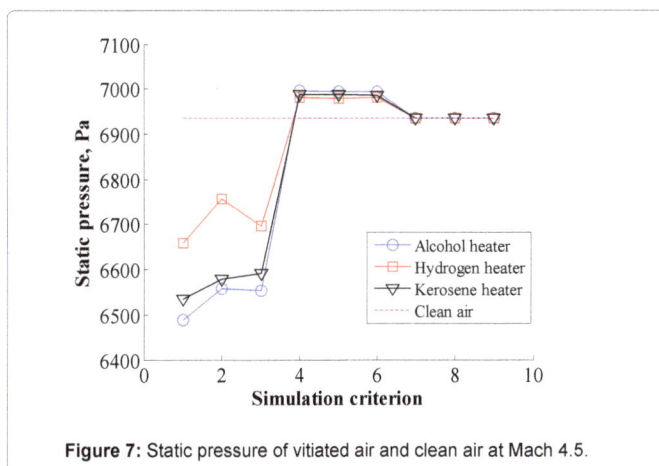

Figure 7: Static pressure of vitiated air and clean air at Mach 4.5.

Assessment of Scramjet Performance

Method of calculating scramjet performance

The method of estimating scramjet performance is based on the 4th chapter of [19], which deals with hypersonic airbreathing engine performance analysis. As described in [19], the real thermodynamic cycle of scramjet is composed of four process, 1) adiabatic compression, 2) constant pressure combustion, 3) adiabatic expansion, 4) constant pressure heat rejection. By means of thermodynamic closed cycle analysis and first law analysis, it is capable to find closed solutions for performance of real ramjets and scramjets and use them to expose important trends and sensitivities. Furthermore, stream thrust analysis can easily account for several phenomena that can have a significant influence on airbreathing engine performance, such as the mass, momentum, and kinetic energy fluxes contributed by the fuel, the geometry of the burner, and exhaust flows that are not matched to the ambient pressure [19]. The above methods provide sufficient detail and accuracy to assess the performance of scramjet. The formulas for calculating scramjet internal thrust, specific thrust, and specific impulse are as follows:

$$Fs = (1+f)S_{a9} - S_{a3} - \frac{R_3 T_3}{c_3}\left(\frac{A_9}{A_3} - 1\right) \text{ \textbackslash* MERGEFORMAT} \qquad (1)$$

$$F = Fs \cdot \dot{m}_a \text{ \textbackslash* MERGEFORMAT} \qquad (2)$$

$$Is = F / \dot{m}_f \text{ \textbackslash* MERGEFORMAT} \qquad (3)$$

where F_s is specific thrust of scramjet, F is internal thrust of scramjet, I_s means specific impulse of scramjet, f means fuel/air ratio, Sa is stream thrust function, reference station 3 corresponds to combustor entrance, reference station 9 corresponds to external expansion ends, R is the gas constant, T is static temperature, c is velocity, m_a means airstream mass flow rate, m_f means fuel mass flow rate. It should be noted that in the adiabatic compression process, the calculated airs parameters at inlet exit is inconsistent of the real shock wave compress process. However, if experiments were conducted at direct-connect facility, the process of compression can be omitted. On the other hand, with the help of numerical simulation, if necessary, the real air parameters at inlet exit (that is, combustor entrance) are obtainable.

As presented before, the main difference between vitiated air and clean air lies in their thermodynamic properties and flow parameters. In addition, the parameters of airstream vary along with the process of

smaller, than that of clean air. Specific heat and specific heat ratio of vitiated air are mainly determined by combustion heating medium, and there also are few distinctions for the same combustion heater when select different temperature criterions. Besides, simulation criterion affects the temperature and pressure between clean air and vitiated air significantly. What's more, the vitiation effects are more obvious at Mach 6.5 condition. With the parameters of vitiated airstream, the scramjet performance can be calculated at next.

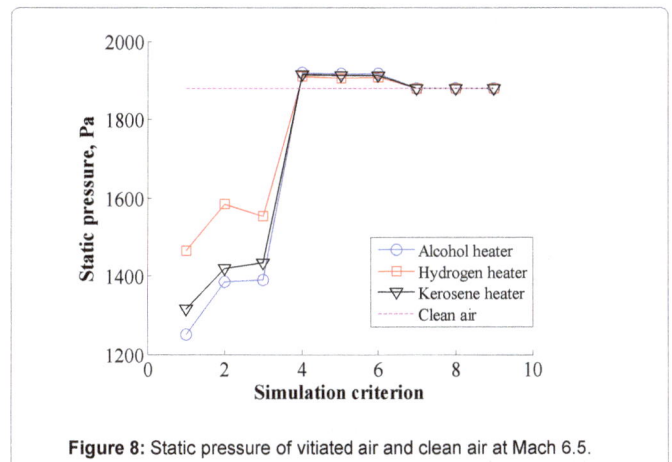

Figure 8: Static pressure of vitiated air and clean air at Mach 6.5.

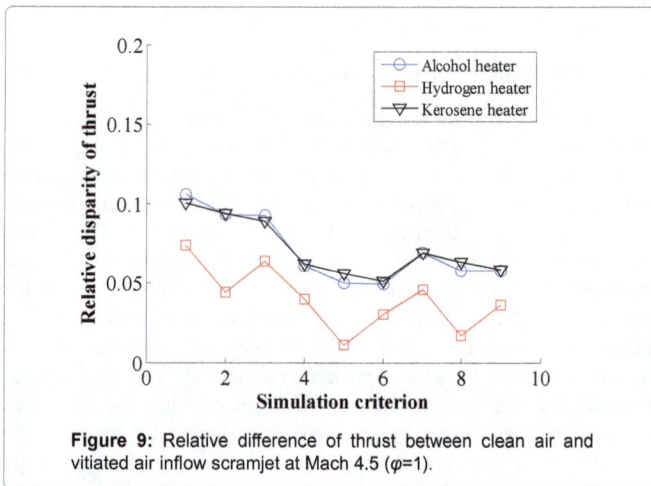

Figure 9: Relative difference of thrust between clean air and vitiated air inflow scramjet at Mach 4.5 ($\varphi=1$).

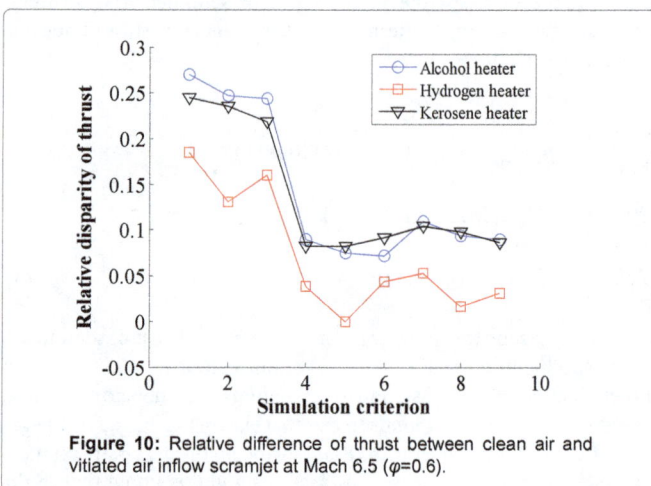

Figure 10: Relative difference of thrust between clean air and vitiated air inflow scramjet at Mach 6.5 ($\varphi=0.6$).

heat addition and expansion. Hence, the effects of parameter variation on engine performance must be taken into account. In this article, the heat addition process and expansion process were divided into a number of sub processes. At each sub process, the properties of flow were regarded as constant. In view of this, the approach can be used to evaluate scramjet performance related to air vitiation. Considered that H_2O and CO_2 have significant chemical kinetic effects on combustion process, the combustion efficiency of combustor must be taken into consideration. Referring to [9] and [20], the combustion efficiency of combustor with clean airstream and vitiated airstream are presented in Table 5. The fuel of scramjet is $C_{12}H_{23}$.

Calculated results

Figures 9-14 show relative disparities of thrust (all thrust means internal thrust at next), specific thrust and specific impulse between clean air inflow scramjet and vitiated air inflow scramjet at Mach 4.5 and 6.5 conditions, which have different fuel equivalence ratios φ. As shown in Figures 9 and 10, thrust corresponding to hydrogen heater is larger than others at the same criterion. Also, the thrust of vitiated airstream scramjet is smaller than that of clean airstream scramjet. As listed in Table 3, there are about 10% H_2O (mole fraction) in vitiated air produced by hydrogen heater and 6% $H_2O+4\%$ CO_2 in alcohol heater generated air at Mach 4.5. Hence, it proved that vitiation species decrease the thrust of scramjet and the effect of CO_2 on thrust is greater than H_2O. What's more, the effects become intenser at Mach 6.5

condition for more introduced vitiation species.

For the same heater, selecting static pressure or dynamic pressure as criterion can obtain larger thrust compared to total pressure, as shown in Figures 9 and 10. For alcohol heater or kerosene heater, matching static temperature get greater thrust. And for hydrogen heater, choosing total enthalpy will gain bigger thrust. As calculated above, when matching static temperature for alcohol and kerosene heater or total enthalpy for hydrogen heater, the total temperature of vitiated air is the lowest. In this case, therefore, the quantity of vitiation species is lower than other schemes. In summary, the influence of pressure criterions on thrust is stronger than temperature criterions and the effects will be enhanced with the increasing Mach number.

Because the specific thrust is mainly affected by mass fraction of oxygen in airstream, which is in proportion to fuel/air ratio f, the specific thrust of scramjet vitiated by hydrogen heater is greater than others, as shown in Figures 12 and 13. Clearly, the specific impulse with vitiated air inflow is less than that of clean air. It should be considered that the combustion efficiency of scramjet could be inconsistent at different simulation criterions. Therefore, the trends of specific impulse will also be affected by combustion efficiency, which is our main future work.

Compared with experimental and numerical results referred to [6-14], the essential tendency of vitiation effects on scramjet performance is identical, which evidenced that the method based on thermodynamic

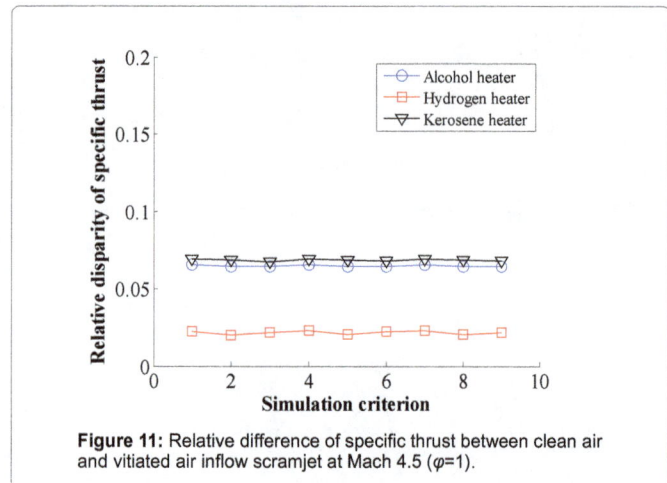

Figure 11: Relative difference of specific thrust between clean air and vitiated air inflow scramjet at Mach 4.5 ($\varphi=1$).

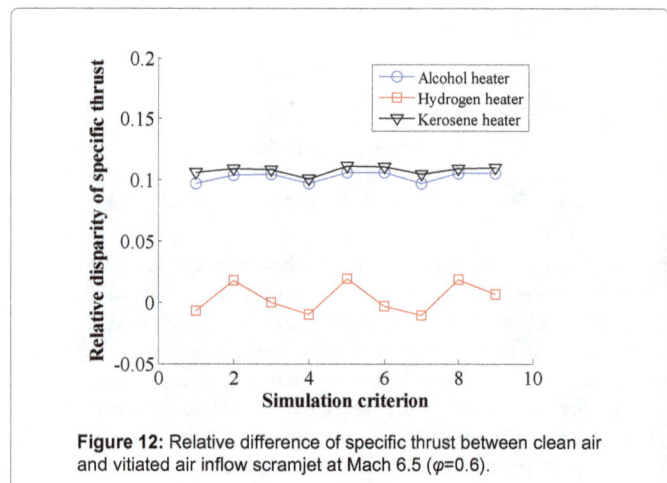

Figure 12: Relative difference of specific thrust between clean air and vitiated air inflow scramjet at Mach 6.5 ($\varphi=0.6$).

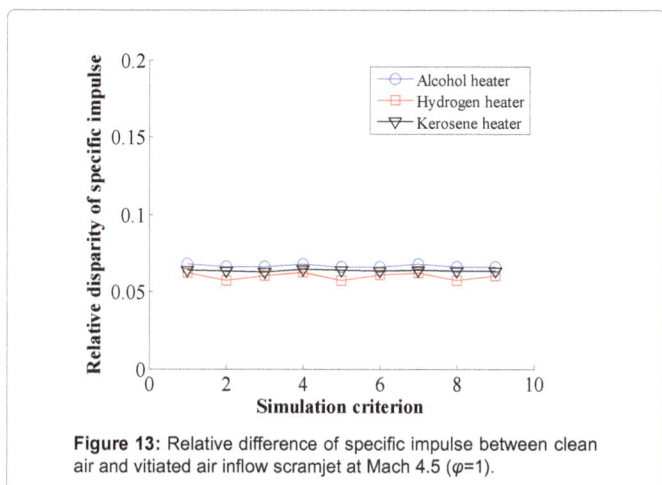

Figure 13: Relative difference of specific impulse between clean air and vitiated air inflow scramjet at Mach 4.5 (φ=1).

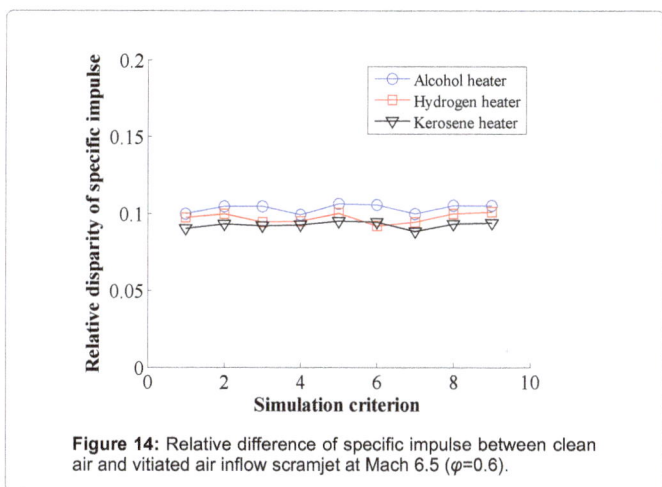

Figure 14: Relative difference of specific impulse between clean air and vitiated air inflow scramjet at Mach 6.5 (φ=0.6).

cycle analysis can validate and predict ground test results effectively and offer forecast when choosing simulation criterions and heating mediums. Therefore, this method will economize lots of costs and time. It should be noted that the theoretical calculations indeed cannot interpret all the authentic conditions in scramjet, it just provide a proper approach to predict the correct qualitative tendency. If we want to obtain the accurate quantitative effects, plenty of ground tests are inevitable.

Conclusions

The effects of contaminating species in vitiated air on scramjet were calculated by chemical equilibrium applications and theoretical analysis based on thermodynamic cycle with 9 simulation criterions. The method based on thermodynamic cycle analysis supply a quick, simple, and reliable approach to evaluate the scramjet performance related to vitiation effects. Properties of vitiated air heated by alcohol, hydrogen and kerosene combustion heater and scramjet performance (internal thrust, specific thrust, and specific impulse) were assessed at Mach 4.5 and 6.5 conditions, respectively, for the purpose of revealing the impact of combustion heating medium and finding out the optimum selection of simulation criterion. There are plenty of findings which can increase the understanding of vitiation effects on scramjet.

Specific heat and specific heat ratio of vitiated air are mainly determined by combustion heating medium, and also have minor

connection with temperature criterions. The specific heat of vitiated air is greater, but specific heat ratio of vitiated air is smaller than that of clean air. Moreover, the specific heat of vitiated air produced by hydrogen heater is the largest, followed by alcohol heater, then kerosene heater. Simulation criterions have great effect on temperature and pressure of vitiated air. What's more, the above differences are more obvious at Mach 6.5 relative to Mach 4.5.

The thrust and specific thrust of scramjet vitiated by hydrogen heater are greater than that of alcohol heater and kerosene heater at Mach 4.5 and 6.5 conditions. In addition, the scramjet thrust and specific impulse with vitiated air inflow is less than that of clean air. The performance of vitiated scramjet declines with the increasing Mach number. Simulating static pressure and dynamic pressure can obtain greater thrust compared to total pressure. The influence of pressure criterions on thrust is more remarkable than temperature criterions.

As the combustion efficiency was estimated by the help of references, further study will focus on the effects of chemical kinetic on combustion. In general, this method can calculate scramjet performance with all kinds of fuel and other simulation criterions and Mach number. It will be a significant guidance related to air vitiation effects on scramjet performance.

Nomenclature

c	velocity, m/s
Cp	specific heat, kJ/(kg·K)
f	fuel/air ratio
F	internal thrust, N
Fs	specific thrust, N·s/kg
Ht	total enthalpy, J/kg
Is	specific impulse, N·s/kg
$\dot{m}_a$	air mass flow rate, kg/s
$\dot{m}_f$	fuel mass flow rate, kg/s
Ma	Mach number
pd	dynamic pressure, kPa
ps	static pressure, kPa
pt	total pressure, kPa
Sa	stream thrust function, m/s
Tt	total temperature, K
Ts	static temperature, K
γ	specific heat ratio
φ	fuel equivalence ratio

Subscripts'

a	air
d	dynamic
f	fuel
s	static
t	total

Acknowledgments

This work was supported by China National Natural Science Foundation (No. 11372092).

References

1. Wang HB, Wang ZG, Sun MB, Qin N (2013) Large-Eddy/Reynolds-averaged

Navier–Stokes simulation of combustion oscillations in a cavity-based supersonic combustor. International Journal of Hydrogen Energy 38: 5918-5927.

2. Cao RF, Chang JT, Bao W, Guo ML, Qin J, et al. (2013) Analysis of combustion mode and operating route for hydrogen fueled scramjet engine. International Journal of Hydrogen Energy 38: 5928-5935.

3. Qin J, Zhang SL, Bao W, Zhang L, Zhou WX (2012) Effect of recooling cycle on performance of hydrogen fueled scramjet. International Journal of Hydrogen Energy 37: 18528-18536.

4. Qin J, Zhou WX, Bao W, Yu DR (2010) Thermodynamic analysis and parametric study of a closed Brayton cycle thermal management system for scramjet. International Journal of Hydrogen Energy 35: 356-364.

5. Powell ES, Stallings DW (1998)A review of test medium contamination effects on test article combustion processes.

6. Shuai-fan G, Wen-Yan S, Jian-ping L, Liang C (2013) Numerical Investigation of Effects of Vitiated Air on Scramjet Performance. Journal of Propulsion Technology 34: 493-498.

7. Edelman RB, Spabaccini LJ (1969) Analytical Investigation of the Effects of Vitiated Air Contamination on Combustion and Hypersonic Air Breathing Engine Ground Tests.

8. Goyne CP, McDaniel Jr. JC, Krauss RH, Whitehurst WB (2007) Test Gas Vitiation Effects in a Dual-Mode Scramjet Combustor. Journal of Propulsion and Power 23: 559-565.

9. Le J, Liu W, Song W, Xing J, Yang Y (2009) Experimental and Numerical Investigation of Air Vitiation Effects on Scramjet Test Performance.

10. Haw WL, Goyne CP, Rockwell RD, Krauss RH, McDaniel JC (2011) Experimental Study of Vitiation Effects on Scramjet Mode Transition. Journal of Propulsion and Power 27: 506-508.

11. Noda J, Masuya G, Tomioka S, Izumikawa M, Rockwell Jr. RD, et al. (2011) Comparison of Dual-Mode Combustor Performance with Various Heating Methods.

12. Mitani T, Hiraiwa T, Sato S, Tomioka S, Kanda T, et al. (1997) Comparison of Scramjet Engine Performance in Mach 6 Vitiated and Storage-Heated Air. Journal of Propulsion and Power 13: 635-642.

13. Rockwell RD, Goyne CP, Haw WL, KraussRH, McDaniel JC (2011) Experimental Study of Test-Medium Vitiation Effects on Dual-Mode Scramjet Performance. Journal of Propulsion and Power 27: 1135-1142.

14. Jian-ping L, Wen-yan S, Feiteng L, Liang C (2013) Numerical Investigation of H_2O/CO_2 Vitiation Effects on Kerosene-Fueled Supersonic Combustion. Journal of Propulsion Technology 34: 562-571.

15. Pellett GL, Bruno C, Chinitz W (2002) Review of Air Vitiation Effects on Scramjet Ignition and Flameholding Combustion Processes.

16. Tatman BJ, Rockwell RD,Goyne CP, McDaniel JC (2013) Experimental Study of Vitiation Effects on Flameholding in a Cavity Flameholder. Journal of Propulsion and Power 29: 417-423.

17. Tomioka S, Hiraiwa T, Kobayashi K, Izumikawa I, Kishida T, et al. (2007) Vitiation Effects on Scramjet Engine Performance in Mach 6 Flight Conditions. Journal of Propulsion and Power 23: 789-796.

18. Chemical Equilibrium with Applications, Glenn Research Center.

19. Heiser WH, Pratt DT (1994) Hypersonic Airbreathing Propulsion.

20. Srinivasan S, Erickson WD (1995) Interpretation of Vitiation Effects on Testing at Mach 7 Flight Conditions.

13

The Impact of Measurement Noise in GPA Diagnostic Analysis of a Gas Turbine Engine

Ntantis E*

School of Engineering, Cranfield University, Bedford, UK

Abstract

The performance diagnostic analysis of a gas turbine is accomplished by estimating a set of internal engine health parameters from available sensor measurements. No physical measuring instruments however can ever completely eliminate the presence of measurement uncertainties. Sensor measurements are often distorted by noise and bias leading to inaccurate estimation results. This paper explores the impact of measurement noise on Gas Turbine GPA analysis. The analysis is demonstrated with a test case where gas turbine performance simulation and diagnostics code TURBOMATCH is used to build a performance model of a model engine similar to Rolls-Royce Trent 500 turbofan engine, and carry out the diagnostic analysis with the presence of different levels of measurement noise. Conclusively, to improve the reliability of the diagnostic results, a statistical analysis of the data scattering caused by sensor uncertainties is made. The diagnostic tool used to deal with the statistical analysis of measurement noise impact is a model-based method utilizing a non-linear GPA.

Keywords: Measurement uncertainty; Measurement noise; Non-linear GPA analysis

Nomenclature

Acronyms: BPR: By-pass ratio; CC: Combustion chamber; DP: Design Point; FC: Mass flow capacity; FF: Fuel flow rate; HP: High pressure; HPC: High pressure compressor; HPT: High pressure turbine; IE: Isentropic efficiency; IP: Intermediate pressure; IPC: Intermediate pressure compressor; IPT: Intermediate pressure turbine; LP: Low pressure; LPC: Low pressure compressor; LPT: Low pressure turbine; N: Number of measured values; OPR: Overall pressure ratio; P: Total pressure; PCN: Relative rotational speed; PR: Pressure ratio; SLS: Sea level static; T: Temperature; TET: Turbine entry temperature

Notations: C1: Fan/Low pressure compressor; C2: Intermediate pressure compressor; C3: High pressure compressor; T1: High pressure turbine; T2: Intermediate pressure turbine; T3: Low pressure turbine

Subscripts: $\vec{z}$: Dependent parameter vector; $\vec{x}$: Independent (component) parameter vector; μ: Mean value; σ:Standard deviation; x_i: Measured (observed) value; $\bar{x}$: Average measurement

Introduction

The main gas path components of a gas turbine engine, namely compressor and turbine, are inherently reliable. However its operation under hostile environments, such as varying conditions of load, temperature and speeds, and the cycle sensitivity to component degradation, results into engine breakdowns and performance deterioration [1-4]. The deterioration of the gas path components cannot be prevented, and the engine performance always degrades increasingly with time. In an attempt to reduce the risk of such unwanted circumstances, commercial and military gas turbine users have engaged in some form of performance diagnostics.

Gas turbine performance diagnostics is a fairly mature methodology to accurately detect, isolate and assess the changes in engine module performance, engine system malfunctions and instrumentation problems from knowledge of measured parameters taken along the engine's gas path. Good estimates allow operators to make safe decisions, regarding the required maintenance actions. Different diagnostic approaches are adapted and developed, in order to restore the integrity and performance of the engine but one of the most popular is Gas Path Analysis (GPA), presented by Louis A. Urban [5].

GPA is a model-based mathematical technique that estimates individual modules and sensor performance shifts, from any specified set of engine measurable parameters and component characteristics, through the aero-thermodynamic relationships which exist between them [6]. The selection of sensors must be selected with great care since they will specify the level of confidence with which GPA will diagnose multiple component(s) faults. The optimal selection of instrumentation set is, therefore critical element in reducing the magnitude of prediction errors. Despite of the fact, sensors have good reputation on accurate measurements, measurement noise is inevitable, and there is a high possibility of affecting undesirably their reliability. The main reason is due to the harsh operating environment of gas turbine sensors such as high pressure and temperature, and large gradients. In many cases, the order of magnitude of the noise could be comparable to the variations in the measurements caused by an actual component fault [7].

In this research, the present of measurement noise is dealt with by processing a large number of readings and appropriate statistical techniques. The gas turbine diagnostic program used for this study is Gas Path Analysis (GPA) technique [8] developed at Cranfield University. The GPA has been applied to a gas turbine model engine, a civil high by-pass ratio turbofan engine, similar to Trent 500 manufactured by Rolls Royce plc. Non-linear GPA is the diagnostic tool that prefers to be used in this research because proved to take a significant advantage on the severe limitations of linear GPA models since it addresses the non-linear nature of the engine thermodynamic behavior.

*****Corresponding author:** Ntantis E, School of Engineering, Cranfield University, Bedford, UK, E-mail: entantis@gmail.com

Methodology

Gas path analysis

Gas Path Analysis (GPA), pioneered by *Urban* [9-11] is used to assess the condition of individual engine components, based on the aero-thermodynamic relationships that exist between the component and direct measurements of gas path parameters [12]. The theory behind this relationship can be summarized by: "*The presence of a primary gas-path physical fault induces change in the component characteristic that shows up a deviation of the measurable parameters from the baseline conditions*" [13]. Therefore, the purpose of the GPA is to detect, isolate and quantify the gas path components faults that have observable impacts on the measurable variables with the hope that will facilitate the subsequent isolation of the underlying physical fault. For a gas turbine engine, the mathematical relationship between dependent (engine component health parameters) and independent parameters (gas path measurements) is expressed analytically as [13,14]

$$\vec{z} = h(\vec{x}) \tag{1}$$

The assumption of linearity becomes increasingly false, when deteriorations cause the engine to operate further away from the condition for which the matrix was calculated [1]. Therefore, a non-linear GPA diagnostic technique is preferred instead of the linear technique at the present paper, due to the consideration of the non-linear nature of the engine thermodynamic behavior [1,10]. The non-linear GPA uses the *Newton-Raphson* iterative technique, where the linear GPA prediction process is applied iteratively until a converged solution is obtained [8].

Measurement uncertainty

The purpose of measurements is to numerically characterize the performance and condition of a gas turbine. Properly understanding of the data obtained from such measurements, is crucial to applying the knowledge thereby gained. It is important to note that, errors and flaws in measurements are always present and should never be stated with complete exactness. Real measuring indications can only be assumed to be made under perfect conditions; no measurement elaborate or precise or how often repeated, can ever completely eliminate any kind of uncertainty. A brief and generic definition of the *measurement error* is described in [15] as the amount by which a measured value differs from the true value. Therefore, the presence of uncertainties in measurements should be recognized as the starting point in a discussion of errors [16].

Measurement noise

In GPA diagnostic analysis of gas turbine engines, measurement noise exists in any measurement and its impact has to be taken into account as the measurement noise may affects the accuracy of GPA diagnostic results. Measurement noise in gas turbine measurements is a term to describe the value of measurement parameter centered on its average value; representing a band within which the true value of the measurement parameter is expected to lie [17]. Therefore, the values of the gas path parameters taken from the gas path measurement can only be seen as an approximation of the true values of those parameters. Suppose that the measurement noise distribution is of *Gaussian* type, the distribution of the measurement of a gas path parameter would be that shown in figure 1.

The accuracy of the measurements is determined by the standard deviation σ, supposing measurements are bias-free. The term standard deviation is used in quantifying measurement precision. The precision

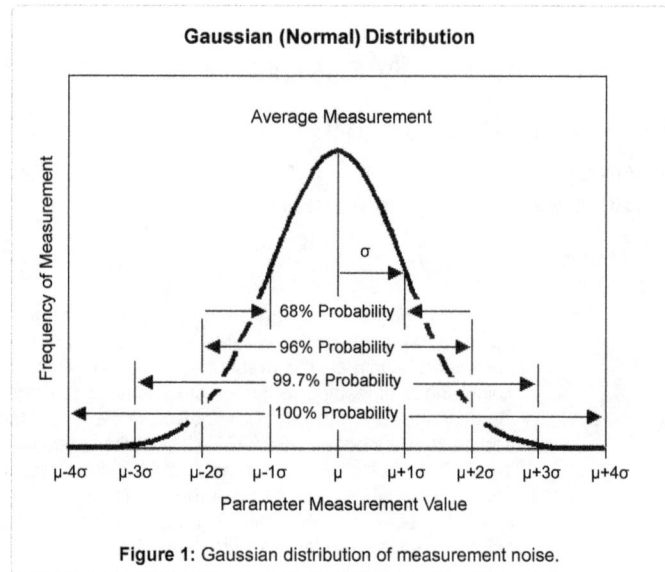

Figure 1: Gaussian distribution of measurement noise.

error is determined by taking N repeated measurements from the parameters of which can be approximated by the precision index and can be estimated as

$$\sigma = \sqrt{\frac{\sum \left(x_i - \bar{x}\right)^2}{N-1}} \tag{2}$$

Since the *Gaussian* or *Normal* distribution is a symmetric distribution, it has the property that a known percentage of all possible values of x lie within a certain number of standard deviation σ of the mean value. For example, around 68% of normally distributed observed samples lie within the interval $(\mu \pm 1\sigma)$, 96% within $(\mu \pm 2\sigma)$, and 99.7% within $(\mu \pm 3\sigma)$, as shown in Figure 1. Consequently, a large value of σ means that there is a lot of scattering in the measurements and a small value of σ reflects relatively less scatter.

Analysis procedure of noise impact

Besides measurement noise, other problems related measurement quality can be detected and derive from:

• Sensor accuracy, i.e. measurement bias.

• Uncertainties not directly related to measurements, but to ambient condition and gas turbine operating conditions.

In this research all the above uncertainties are assumed not exist and only measurement noise is being considered. The proposed methodology in analyzing the impact of measurement noise on gas turbine GPA diagnostics is divided into the following main steps:

• A model engine performance model is created with thermodynamic performance software and therefore all the true value of gas path measurements can be simulated.

• One set of gas path measurements were selected. Gas path measurements of the model engine at different engine health conditions were simulated.

• A large number of measurement samples with random measurement noise are simulated by superimposing the simulated measurements with different level of measurement noise.

• The simulated measurement samples are used as input to the GPA diagnostic system to predict engine degradation.

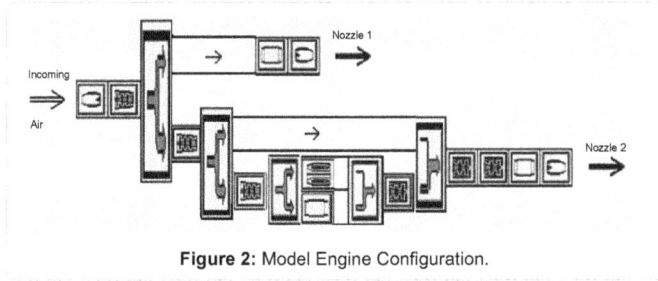

Figure 2: Model Engine Configuration.

- A statistical analysis of the predicted engine degradation with large number of measurement samples is carried out and the impact of measurement noise on the degradation prediction is analyzed.

Case Study

Model engine

The engine model selected for the noise impact analysis in this research is similar to Rolls Royce Trent 500, a three spool, high-bypass turbofan engine rated in 249 kN net thrust at sea level. In figure 2 there is a schematic representation of the model engine. The model engine was simulated using TURBOMATCH, a FORTRAN-based gas turbine simulator developed at Cranfield University [18]. Normally, it is appropriate to define DP of a gas turbine in cruise conditions because the aircraft spend most of the operational time at this situation. However, in the current research the DP was chosen at the take-off condition because the available open access databases for the performance parameters of this engine refer to the ground testing [19]. Therefore, the performance specifications of the engine at the DP are presented under the atmospheric conditions of SLS.

Performance Parameter		Value
Ambient Temperature	(K)	288
Inlet Mass Flow Rate	(kg/s)	879.5
BPR	(non-d)	7.5
TET	(K)	1600
Net Thrust	(kN)	249
LP Compressor / Fan PR	(non-d)	1.5
IP Compressor PR	(non-d)	5.9
HP Compressor PR	(non-d)	4.1
OPR	(non-d)	36.3

Instrumentation set selection

The success of any fault diagnosis technique depends critically on the sensor network, measuring the important observed parameters. An optimally designed sensor network for fault diagnosis should observe all the faults when they occur, and also distinguish between them to the maximum extent possible [20]. Therefore, the instrumentation set should be properly chosen for better detection of engine degradation because the quality of the diagnostic analysis relies on the quality of the measurements. Table 1 shows all the potential gas path measurements for the model engine and table 2 all the health parameters of potential degraded engine components. For the measurement set to be effective in the GPA diagnostic analysis, the chosen measurements should be sensitive to engine degradation and independent from each other. To assist the selection of measurements, a sensitivity analysis is carried out,

Symbol	Measurement Parameters	Unit
PCN1	C1 relative rotational speed	%
PCN2	C2 relative rotational speed	%
PCN3	C3 relative rotational speed	%
P3	C1 exit total pressure	atm
T3	C1 exit temperature	K
P5	C2 exit total pressure	atm
T5	C2 exit temperature	K
P7	C3 exit total pressure	atm
T7	C2 exit temperature	K
P10	CC exit total pressure Fuel	atm
FF	Fuel flow rate	kg/s
P14	T1 exit total pressure	atm
T14	T1 exit temperature	K
P16	T2 exit total pressure	atm
T16	T2 exit temperature	K
P17	T3 exit total pressure	atm
T17	T3 exit temperature	K

Table 1: Potential Gas Path Measurements.

Fault No.	Meaning	Health Parameter
1	C1 isentropic efficiency	IE_{C1}
2	C1 flow capacity	FC_{C1}
3	C2 isentropic efficiency	IE_{C2}
4	C2 flow capacity	FC_{C2}
5	C3 isentropic efficiency	IE_{C3}
6	C3 flow capacity	FC_{C3}
7	T1 isentropic efficiency	IE_{T1}
8	T1 flow capacity	FC_{T1}
9	T2 isentropic efficiency	IE_{T2}
10	T2 flow capacity	FC_{T2}
11	T3 isentropic efficiency	IE_{T3}
12	T3 flow capacity	FC_{T3}

Table 2: Health Parameters of potential degraded engine components.

where the response of all the potential gas path measurements, due to unit deviation of each of the engine component health parameters, are obtained, by using performance simulation of the model engine.

Figure 3 illustrates the plotted sensitivity of all potential gas path measurements against all the model engine health parameters and the number of faults taken from table 2. Based on the sensitivity of the measurements, a set of six measurement parameters is then selected for the diagnostic analysis of the model engine. The selected measurement set is:

Total pressure at the exit of IP and HP compressor   (P5, P7)

Total pressure at the exit of HP and IP turbine   (P14, P16)

LP and HP spool rotational speed   (PCN1, PCN3)

In an attempt to investigate the impact of measurement noise on performance diagnostics, engine component degradation is implemented into the model engine using TURBOMATCH software and gas path measurements are simulated. The implemented degradation is assumed unknown to the engine users and simulated measurements are used to predict the seeded fault. As the linear GPA is less effectively compared with its non-linear partner only the non-linear GPA technique is used in diagnostic analysis. Figure 4 presents in the form of histograms the differences of performance parameters with noisy and noise-free measurements. The example used to study the influence of measurement noise was an implemented fault of 2 percent drop in flow capacity and 2 percent loss in isentropic efficiency,

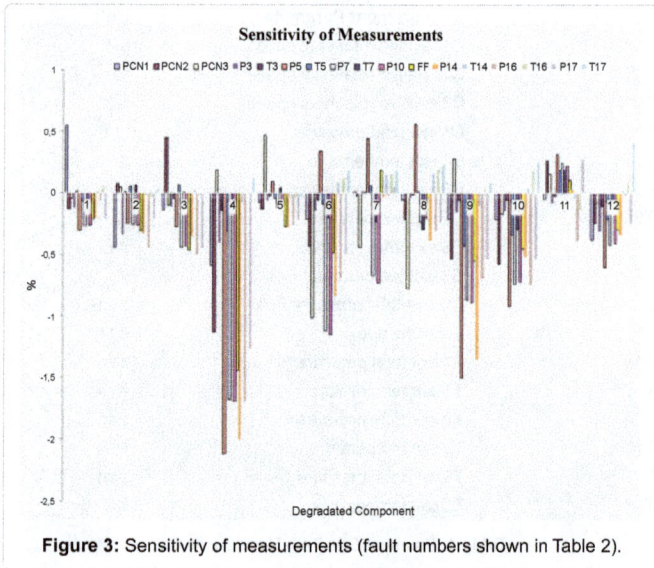

Figure 3: Sensitivity of measurements (fault numbers shown in Table 2).

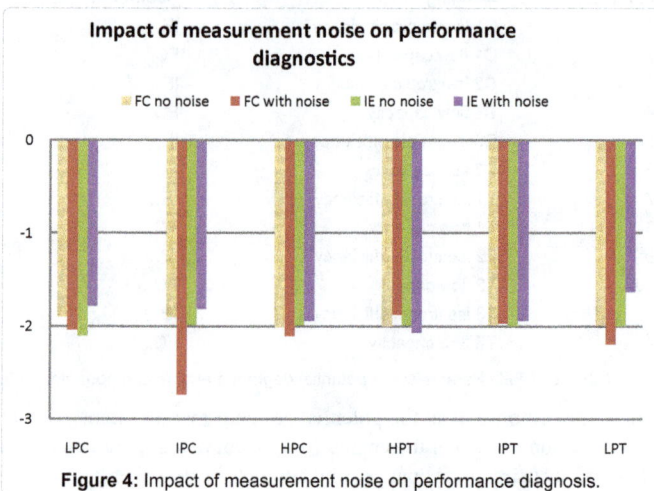

Figure 4: Impact of measurement noise on performance diagnosis.

		Number of Measurement Samples					
		Compressor			Turbine		
		LP/Fan	IP	HP	HP	IP	LP
Health Parameter	FC	3000	1500	4000	1500	2500	3000
	IE	2000	500	4000	2000	3000	500

Table 3: Number of optimum samples for each component fault case.

in both compressor and turbine of the engine. The accuracy of the non-linear GPA predictions with the influence of nominal noise level is about 0.2 percent in average except FC at IP compressor where its prediction error is over 0.8 percent. However, the noisy measurements do not seem to result in large prediction errors in the diagnosis of the examined degradation. Certainly, the impact of measurement noise on performance diagnostics is negative but it can consider being small and acceptable.

Measurement sample set selection

The impact of measurement noise was investigated by using selected set of measurements with a large number of samples and different levels of measurement noise. The problem arises is the number of samples that should be chosen to get meaningful statistic analysis results. After applying the non-linear GPA to the model engine with a varying sample

%	Standard Deviation %					
Noise Level	C1	C2	C3	T1	T2	T3
2σ/3	0.3	0.8	0.3	0.1	0.2	0.8
2σ/2	0.4	1.1	0.5	0.2	0.3	1.0
Std Dev. or 2σ	0.8	2.1	1.0	0.4	0.6	1.8
2x2σ	1.6	4.0	1.9	1.0	1.4	3.8
3x2σ	2.4	5.9	2.9	1.5	1.9	5.8

Table 4: Impact of measurement noise on flow capacity degradation predictions.

%	Standard Deviation %					
Noise Level	C1	C2	C3	T1	T2	T3
2σ/3	0.4	0.1	0.1	0.2	0.3	0.6
2σ/2	0.5	0.2	0.2	0.3	0.4	0.9
Std Dev.or 2σ	1.0	0.3	0.3	0.6	0.9	1.6
2x2σ	1.9	0.7	0.9	1.8	1.9	3.6
3x2σ	2.9	1.0	1.3	2.8	2.6	5.5

Table 5: Impact of measurement noise on isentropic efficiency degradation predictions.

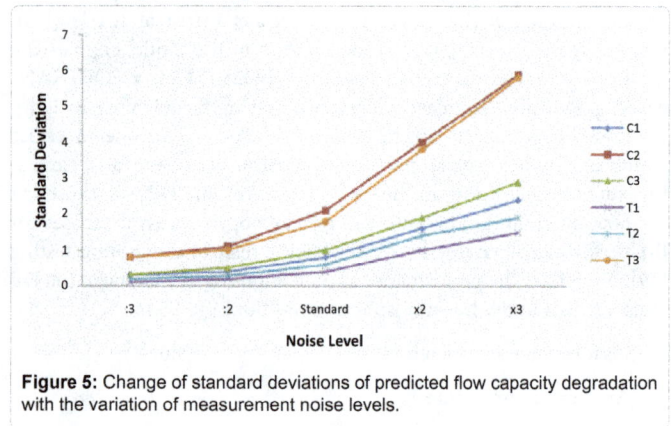

Figure 5: Change of standard deviations of predicted flow capacity degradation with the variation of measurement noise levels.

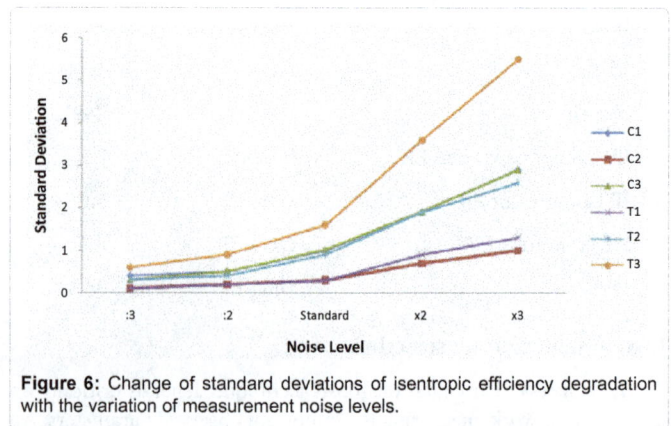

Figure 6: Change of standard deviations of isentropic efficiency degradation with the variation of measurement noise levels.

sizes (500 to 5000 samples), a fixed sample size is selected. Table 3 lists the optimum (minimum) measurement sets for each fault case, under the condition that the distribution of the health parameter on the selected set satisfies the *Gaussian distribution* conditions. The largest number of samples found to be 4000.

Statistical analysis of measurement noise impact

The difference between the predicted mean degradation and actual degradation in percentage indicates the accuracy of diagnostics. Tables 4 and 5, show the impact of measurement noise on the prediction accuracy of the model-based diagnosis of the gas turbine engine. For example, the standard deviation (2σ) in table 4 of the predicted C2 and T3 degradation are 2.1 and 1.8 percent. When noise level drops down to one third ($2\sigma/3$) the standard deviations seem to fall also down to 0.8 percent for both components. The same changes happen when measurement noise rises by i.e. three times ($3\times2\sigma$) and the standard deviations for the same components as mentioned before increase by 3.8 and 4 percent respectively. Similarly, in the case of isentropic efficiency drop presented in table 5, the standard deviation of the predicted degradation changes proportionally with the noise levels. Figures 5 and 6 plot results of table 4 and 5 respectively on graphs to demonstrate the impact of different level of measurement noise on the prediction accuracy of engine component diagnostic analysis.

The results also explain why the Gaussian distribution tends to be flatter as the measurement noise increases, making the scatter range wider. Flatter distributions lessen the possibility for an arbitrary measurement to be highly precise to the expected value, because the scatter around the average increases the precision error, which specifies the characteristics of measurement in greater detail. As precision refers to the level of measurement and exactness of description in the range of measurement uncertainty, the precision error increases the loss of information. More precise measurements have smaller uncertainties, therefore the negative impact of large standard deviations and measurement noise is the difficulty for any user to evaluate the condition of the engine, because the data from high distributions can mislead the diagnostic process due to high precision error.

Conclusion

The impact of measurement noise on engine component fault diagnosis was achieved in this paper. The impact of noisy measurements on the gas path components parameters analyzed and quantified where the measurement noise is assumed to exhibit *Gaussian distribution*. The statistical analysis shows that the variation of measurement noise levels has obvious impact on the prediction accuracy of the non-linear GPA gas path diagnostics and every measure should be taken to reduce the level of measurement noise in order to improve the accuracy of the diagnostic results.

References

1. Marinai L, Probert D, Singh R (2004) Prospects for aero gas-turbine diagnostics: a review. Appl Energ 79: 109-126.

2. Lakshminarasimha AN, Boyce MP, Meher-Homji CB (1994) Modeling and Analysis of Gas Turbine Performance Deterioration. J Eng Gas Turbines Power 116: 46-52.

3. Tabakoff W, Lakshminarasimha AN, Pasin M (1990) Simulation of Compressor Performance Deterioration Due to Erosion. J Turbomach 112: 78-83.

4. Ogaji SOT, Sampath S, Singh R, Probert SD (2002) Parameter selection for diagnosing a gas-turbine's performance-deterioration. Appl Energ 73: 25-46.

5. Urban LA (1969) Gas Turbine Engine Parameter Interrelationships. Hamilton Standard Division of United Aircraft Corporation, USA.

6. Volponi AJ (2003) Foundations of gas path analysis I & II. In Gas turbine condition monitoring & fault diagnosis. Editors Sieverding CH, Mathioudakis K.

7. Li YG (2004) Gas turbine condition monitoring & fault diagnosis. School of Engineering Lecturer Notes, Cranfield University, UK.

8. Escher PC (1995) Pythia: An object-orientated gas path analysis computer program for general applications. Cranfield University, UK.

9. Urban LA (1975) Parameter Selection for Multiple Fault Diagnostics of Gas Turbine Engines. J Eng Gas Turbines Power. 97: 225-230.

10. Li YG (2002) Performance-analysis-based gas turbine diagnostics: a review. Journal of Power and Energy 216: 363-377.

11. Saravanamuttoo HIH, MacIsaac BD (1983) Thermodynamic models for pipeline gas turbine diagnostics. J Eng Power-T Asme 105: 875-884.

12. Rolls Royce (1996) The jet engine. Technical Publications Department Edn Derby, England UK.

13. Scala S, Konrad M, Mason R, Skelton D (2003) Predicting the Performance of a Gas Turbine Engine Undergoing Compressor Blade Erosion. 39th AIAA/ASME/SAE/ASEE Joint Propulsion Conference and Exhibit, Alabama.

14. Urban LA (1973) Gas Path Analysis Applied to Turbine Engine Condition Monitoring. J Aircraft 10: 400-406.

15. Jermy M (2004) Uncertainty analysis and error propagation. Lecture Notes Cranfield University UK.

16. Bell S (1999) A Beginner's Guide to Uncertainty of Measurement. National Physical Laboratory, UK.

17. Abernethy RB, Thompson JW (1973) Handbook-Uncertainty in Gas Turbine Measurements. Arnold Engineering Development Center, USA.

18. MacMillan WL (1974) Development of a modular-type computer program for the calculation of gas turbine off-design performance. Cranfield University, UK.

19. Rolls Royce (1971) Technical description. Derby, England.

20. Bhushan M, Rengaswamy R (2000) Design of sensor location based on various fault diagnostic observability and reliability criteria. Comput Chem Eng 24: 735-741.

14

Effect of Fatigue Testing and Aquatic Environment on the Tensile Properties of Glass and Kevlar Fibers Reinforced Epoxy Composites

Menail Y[1], Abderrahim EL Mahi[2]* and Assarar M[3]

[1]*University of Badji Mokhtar, Sidi Ammar, LR3MI, BP 12, 23000, Annaba, Algeria*
[2]*University of Maine, LAUM, CNRS UMR 6613, Avenue Olivier Messiaen, 72085 Le Mans Cedex 9, France*
[3]*University of Reims Champagne-Ardenne, LISM, EA 4695, IUT de Troyes, 9 rue de Québec, 10026 Troyes Cedex, France*

Abstract

This paper presents the experimental results of the influence of water ageing after mechanical fatigue on glass-fiber composites, compared with Kevlar-fiber composites. The tested specimens were subjected to fatigue during various numbers of cycles (100 to 50000). After that, they were immersed into tap water and simulated seawater for different periods (4, 20 and 40 days). Next, the tensile tests were made on the unaged and aged samples in order to determine the evolution of the strength and stiffness under local interactions of the water absorption and fatigue. The obtained results showed that tensile characteristics were clearly affected by the immersion treatment and fatigue loading. In fact, the residual stiffness and residual strength decreased when the immersion time and cycle number of fatigue increased, indicating that the studied composites have experienced some forms of mechanical damage.

Keywords: Glass fiber; Kevlar fiber; Laminates; Fatigue; Damage mechanics; Accelerated aging

Introduction

Fiber reinforced polymer composite materials are widely used in several constructions (marine, aerospace and automobile, etc.) due to their various advantages: high stiffness to weight ratio, corrosion resistance, and low maintenance cost. Glass and Kevlar fibers are the most widely used to reinforce composite structures. Indeed, glass fiber has enjoyed widespread popularity to make reinforced composites in every field, because of their competitive cost and relatively good mechanical properties. Kevlar fibers widely used as reinforcement within several advanced composites, which were developed during the 1960s. Their high degree of toughness, associated with the failure mechanism of Kevlar, and damage tolerance promote good impact/ballistic performance. This is due to the low surface energy and the chemically inert surface of the Kevlar fiber, and consequently to the poor interfacial adhesion between fiber and matrix [1-4].

Generally, the mechanical properties of composite materials are well known by engineers, but there are still many concerns about their durability and their performance under severe environmental conditions [5-7]. In fact, with the use of composite materials in a warm and wet environment, the aggressive actions can appear under several aspects of biological, chemical and physical properties by altering materials and provoking a failure of the residual stiffness and residual strength. The deterioration of a composite material during a wet ageing is, in most cases, the results of a water absorption phenomenon depending on hygrometric and temperature. Indeed, water can penetrate into the composites by three main mechanisms: diffusion of the water through the matrix, capillary along fiber-matrix interface and percolating flow and storage of water in micro-cracks [8]. These diffusion mechanisms generally lead to the following damages in the composites: degradation by a hydrolysis reaction of unsaturated groups within the resin [9,10], interfacial fracture [11-14], debonding [15] and interlaminar toughness [16-20]. Beyond these considerations, it is well known that water absorption also affects the mechanical behavior of composite materials globally [21-27]. In spite of these investigations, few works studied the combined effect of damage mechanics and environmental ageing on the ultimate properties of composite materials.

The purpose of this paper is to describe the effect of the mechanical fatigue and accelerated ageing on two epoxy composite laminates reinforced with glass-fibers and Kevlar-fibers, respectively. From experimental procedures, the samples were subjected to artificial tap water and seawater at room temperature for various durations (4, 20 and 40 days) after fatigue during various numbers of cycles (100 to 50000). These investigations were done to evaluate the effect of the fatigue loading as well as the moisture uptake and the nature of the aquatic environment on the mechanical behavior of these two composites.

Experimental Procedure

Materials

Laminates were prepared by hand lay-up process from epoxy resin and unidirectional fabrics of weights 300 gm^{-2} for E-glass fibers and 170 gm^{-2} for Kevlar fibers. The first one, named GFRP, was composed of 4 layers of glass fiber with a volume fraction of fibers $Vf \sim 0.65$ and the second one, named KFRP, was constituted of 6 layers of Kevlar with $Vf \sim 0.42$. The choice of the number of layers was made in order to obtain plates with the same thickness of 1 mm. Plates of 110 cm length and 80 cm width were cured at room temperature with a pressure of 0.3 bar using a vacuum molding process for 6 hours, and then post-cured for 6 h at 80°C in an oven. Afterwards specimens were cut according to ASTM standard D3039 ($200 \times 20 \times 1$ mm), using a saw with diamond disc.

*Corresponding author: Abderrahim EL Mahi, University of Maine, LAUM, CNRS UMR 6613, Avenue Olivier Messiaen, 72085 Le Mans Cedex 9, France
E-mail: abderrahim.elmahi@univ-lemans.fr

Experimental procedure

The experimental protocol of studied composite materials is presented in 5 stages (Figure 1). The first stage is a phase of loading in static test, at a constant rate of 1 mm min^{-1}. The displacement is controlled up to 50% of displacement to the rupture in static tests.

The second stage is a phase of fatigue with a form of sinusoidal wave of 10 Hz frequency. A mean displacement *dmean* is maintained constant and the amplitude is equal to 10% of displacement at failure in static tests, which was imposed. The studied materials were subjected to ten cycles of fatigue between 100 and 50000.

The third stage is a phase of unloading of the specimens after being subjected to fatigue.

In the fourth stage, the samples tested in fatigue (stage 3) were immersed into tap water or artificial seawater (37 g/l) during various periods (4, 20 and 40 days). The choice of these periods was made to determine the evolution of the strength, stiffness and fatigue performance under local interactions of environmental condition and fatigue loading.

During the fifth phase, the samples will be tested in static in order to determine the influence of fatigue (stage 2) and ageing (stage 4), on the mechanical properties (Young's modulus *E*, ultimate strength and failure strain) of the samples in tensile tests. For each test and environmental condition, at least three specimens were submitted to tensile tests. The total number of tested specimens was 298 (Table 1) and only the mean values were taken into account.

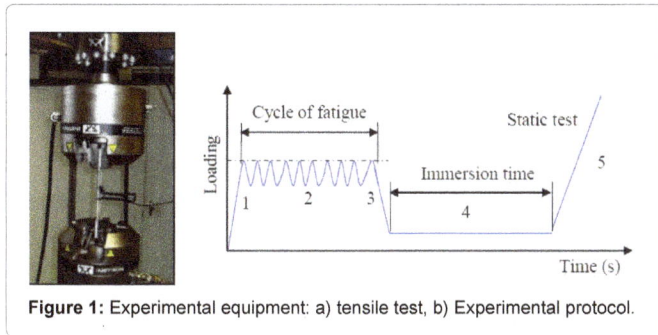

Figure 1: Experimental equipment: a) tensile test, b) Experimental protocol.

Results and Discussion

Static and fatigue test of unaged composites

In order to study the influence of fiber type on the static and fatigue behaviors of GFRP and KFRP composites, we present in Figures 2 and 3, the typical stress-strain curves and the loss of the load F_{max}/F_{0max} according to the number of cycles for both composites, respectively. The static behavior of these composites is quasi-linear until the failure. The value of tensile strength is around 380 MPa for the laminate with Glass-fiber, while it is 305 MPa for the laminate with Kevlar-fiber (Figure 2). This difference is related to the fibers volume fraction and fiber nature. The values of the longitudinal elastic modulus of the GFRP and the KFRP laminates are estimated as 16 GPa and 16.5 GPa, respectively (Figure 2).

For the fatigue test, we notice that the loss of the load F_{max}/F_{0max} (F_{0max} is the maximum load obtained in the first cycle) until the failure of the GFRP and KFRP composites occurs at three distinct stages (Figure 3). The first one is characterized by an initial loss in stiffness caused by the transverse ply cracking and some early fiber fractures. The second one is an intermediate stage representing a period of stable crack propagation in which stiffness reduction occurs as a result of formation

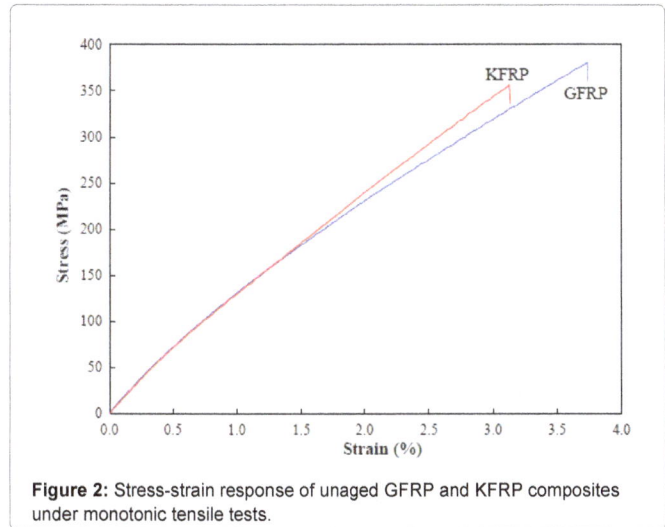

Figure 2: Stress-strain response of unaged GFRP and KFRP composites under monotonic tensile tests.

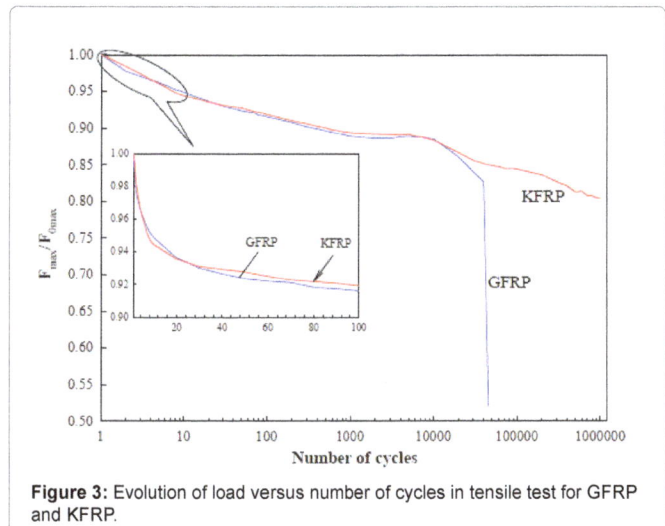

Figure 3: Evolution of load versus number of cycles in tensile test for GFRP and KFRP.

Aging conditions	Tap water								Sea water							
Immersion (days)	0		4		20		40		0		4		20		40	
K: Kevlar; G: Glass	G	K	G	K	G	K	G	K	G	K	G	K	G	K	G	K
Static	-	-	-	-	-	-	-	-	-	-	-	-	-	-	-	-
Testing — Fatigue (Cycles) — 100	3	3	3	3	3	3	3	3	3	3	3	3	3	3	3	3
1000	3	3	3	3	3	3	3	3	3	3	3	3	3	3	3	3
10000	3	3	3	3	3	3	3	3	3	3	3	3	3	3	3	3
500000	3	3	3	3	3	3	3	3	3	3	3	3	3	3	3	3
Total	298 samples															

Table 1: Summary of tests.

of additional matrix cracking, crack coupling along ply interface, and initiation of internal delamination. The last one is characterized by a rapid decrease in stiffness as a result of failure of fibers, including delamination and coalescence of all damage mechanisms resulting in a total failure of the laminate [28,29].

It should be noted that the first stage constitutes only 10% of the life expectancy while it corresponds to 80% of the damage rate (Figure 3). The GRFP composite reaches the failure before 50000 cycles of fatigue, whereas the rupture of the KFRP composite is not reached even at the end of millions of cycles.

In order to distinguish the different damages of these two composites due to the fatigue loading, the specimen surfaces were examined using an optical Microscope. After the test, small sections of the GFRP and KFRP composites were cut transversely to the beam axes 5 mm away from the failed center region.

Figure 4 shows representative fracture topographies taken from the GFRP and KFRP specimen after fatigue loading. The analysis of these pitchers shows that the mode of damage and failure strongly depends upon the fiber orientation. In fact, the first damage that occurs in the laminates is the transverse cracking, which induces local stress concentration at crack tips and consists in the formation of interlaminar cracks running parallel to the fibers in 90° layers due to matrix cracking (Figure 4). Transverse cracking is a progressive damage mechanism which develops with the increase of applied loading or cyclic fatigue numbers. These pitchers also show that dominant damage modes under fatigue are transverse cracks forming in the 90° plies and internal delamination at the 0/90° interface.

Static test after fatigue loading and ageing

Static tests were made on these two composites in order to determine the residual strength and residual stiffness as well as their evolution according to the numbers of cycles and ageing conditions. Figure 5 illustrates the stress-strain curves obtained for the GFRP and KFRP before and after each condition (immersion into water environment and fatigue cycles). The curves have been offset by 5% for clarity, with regard to strain. We observe that the two conditions performed after fatigue loading hardly affect the mechanical behavior of both composites which remains quasi-linear until failure. However, they induce significant variations in the mechanical properties. To better identify these variations, we present in Figures 6-9 the evolution

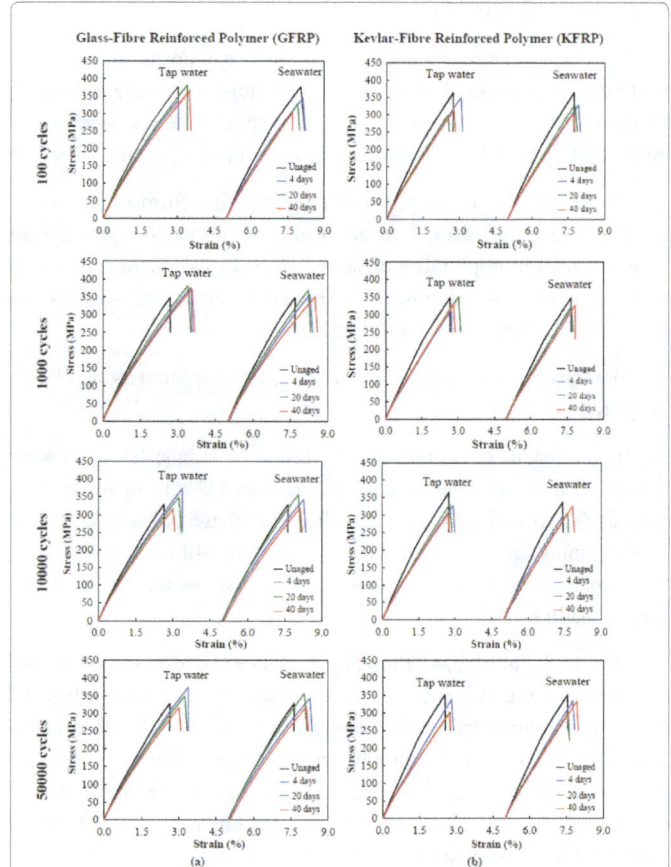

Figure 5: Stress-strain curves according to the immersion duration in tap water and seawater for various fatigue cycles: a) GFRP and b) KFRP.

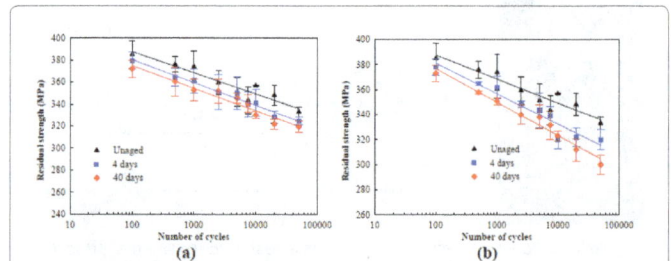

Figure 6: Evolution of residual strength according to the number of cycles for the GFRP: a) tap water and b) seawater.

of the residual strength (Figures 6-7) and residual stiffness (Figures 8 and 9) according to the number of cycles with a semi-logarithmic scale. These Figures show that mechanical fatigue induces a decrease in the residual strength (Figures 6 and 7) and residual stiffness (Figures 8 and 9) of both composites. We also note that these properties decrease with immersion time for each composite and these changes are amplified by the condition and duration of water ageing.

In order to compare the effect of fatigue and accelerated ageing on mechanical proprieties of both composites, Figure 10 presents the results obtained after fatigue of unaged and aged GFRP and KFRP composites during 40 days in seawater.

After 50000 cycles of fatigue, the residual strength decreases by about 15% and 10% for unaged GFRP and KFRP composites,

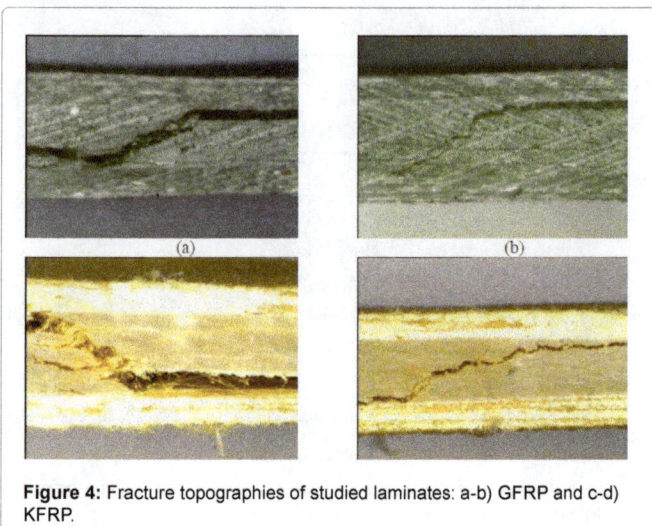

Figure 4: Fracture topographies of studied laminates: a-b) GFRP and c-d) KFRP.

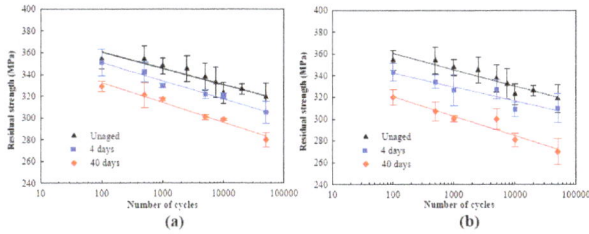

Figure 7: Evolution of residual strength according to the number of cycles for KFRP: a) tap water and b) seawater.

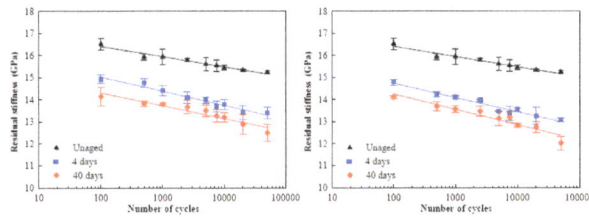

Figure 8: Evolution of residual stiffness according to the number of cycles for the GFRP: a) tap water and b) seawater.

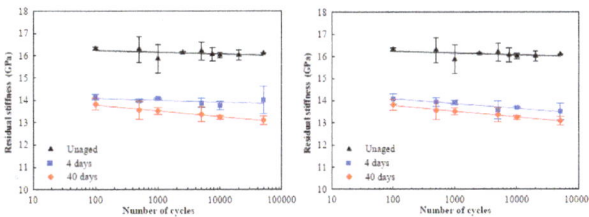

Figure 9: Evolution of residual stiffness according to the number of cycles for the KFRP: a) tap water and b) seawater.

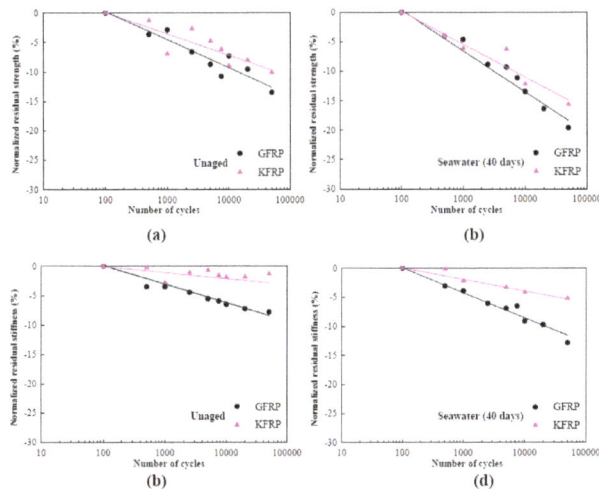

Figure 10: Comparison between mechanical properties of unaged and aged materials: a-b) residual strength and c-d) residual stiffness.

respectively (Figure 10a). On the other hand, its loss is about 20% for the GFRP and 15% for the KFRP after 40 days of immersion into

seawater (Figure 10b). For the residual stiffness, it is found to decrease by about 8% and 2% for the unaged GFRP and KFRP composites, respectively (Figure 10c), while its decrease is around 12% and 5% after 40 days of immersion into seawater (Figure 10d). These results show that the residual strength of GFRP composite is more affected by water ageing than this of KFRP composites. This difference can be attributed to several parameters such as the difference between volume fraction of glass fiber and Kevlar fiber composites, the interfacial bonds between fiber and hydrophobic polymer matrices and the morphology of each fiber.

Although the mechanical properties of GFRP and KFRP composites are affected by the ageing differently, we can conclude that the ageing and fatigue slightly modify the mechanical behavior of GFRP and KFRP composites, but they considerably affect the residual strength and residual stiffness of both these materials.

Conclusion

This study aimed at investigating on the influence of water ageing after fatigue on the mechanical proprieties of glass-fiber and Kevlar-fiber reinforced epoxy composite. The tested specimens were subjected to fatigue for during numbers of cycles (100 to 50000) and aged by immersion in tap water and simulated seawater for different lengths of time (4, 20 and 40 days). Then the composite specimens were subjected to the tensile tests in order to determine the evolutions of the strength and stiffness under local interactions of the water uptake and fatigue. The obtained results show that tensile proprieties are clearly affected by water ageing and fatigue loading, both for the shape of the stress-strain curves and for the values of mechanical properties. Consequently, the residual stiffness and residual strength decrease when immersion time increases, indicating that the material has experienced some forms of mechanical damage and irreversible chemical degradation. The decrease in stiffness and strength is explained by the degradation of fiber-matrix interface because of water absorption causing micro cracks of the matrix.

The static properties of GFRP and KFRP composites revealed that long periods of aging and high fatigue cycle numbers cause deterioration of their characteristics. In addition to that, the obtained results show that KFRP composite seems to be more suitable for the applied fatigue and water aging.

References

1. Lin TK, Kuo BH, Shyu SS, Hsiao SH (1999) Improvement of the adhesion of Kevlar fiber to bismaleimide resin by surface chemical modification. J Adhes Sci Technol 13: 545-560.

2. Lin TK, Wu SJ, Lai JG, Shyu SS (2000) The effect of chemical treatment on reinforcement/matrix interaction in Kevlar fiber/bismaleimide composites. Compos Sci Technol 60: 1873-1878.

3. Varelidis PC, Papakostopoulos DG, Pandazis CI, Papaspyrides CD (2000) Polyamide coated KevlarTM fabric in epoxy resin: mechanical properties and moisture absorption studies. Compos Part A Appl S 31: 549-558.

4. Park SJ, Seo MK, Ma TJ, Lee DR (2002) Effect of chemical treatment of Kevlar fibers on mechanical interfacial properties of composites. J Colloid Interface Sci 252: 249-255.

5. Ray BC, Rathore D (2014) Environmental damage and degradation of FRP composites: a review report. Polymer Composites.

6. Xian G, Li H, Su X (2012) Water absorption and hygrothermal ageing of ultraviolet cured glass-fiber reinforced acrylate composites. Polymer Composites 33: 1120e8.

7. Gutman EM, Soncino R (1995) Environmental effecton stress relaxation in polyester-fiberglass composite. Polymer Composites 16: 518-521 .

8. Tsenoglou CJ, Pavlidou S, Papaspyrides CD (2006) Evaluation of interfacial relaxation due to water absorption in fiber-polymer composites. Composites Science and Technology 66: 2855-2864.

9. Kootsooks A, Mouritz AP (2004) Seawater durability of glass and carbonpolymer composites. Compos Sci Technol 64: 1503-1511.

10. Alvarez VA, Vazquez A (2004) Effect of water sorption on the flexural properties of fully biodegradable composites. J Compos Mater 38: 1165-1181.

11. Kawagoe M, Doi Y, Fuwa N, Yasuda T, Takata K (2001) Effects of absorbed water on the interfacial fracture between two layers of unsaturated polyester and glass. J Mater Sci 36: 5161-5167.

12. Kootsookos A, Mouritz AP (2004) Seawater durability of glass and carbon polymer composites. Compos Sci Technol 64: 1503-1551.

13. Marais S, Metayer M, Nguyen TQ, Labbe M, Saiter JM (2000) Diffusion and permeation of water through unsaturated polyester resins - influence of resin curing. Eur Polym 36: 453-462.

14. Wood CA, Bradley WL (1997) Determination of the effect of seawater on the interfacial strength of an interlayer E glass/graphite/epoxy composite by in situ observation of transverse cracking in an environmental SEM. Compos Sci Technol 57: 1033-1043.

15. Krystyna L, Laurent G (2004)The effect of water immersion ageing on low velocity impact behaviour of woven aramid glass fibre/epoxy composites. Compos Sci Technol 64: 2271-2278.

16. Boukhoulda BF, Adda-Bedia E, Madani K (2006) The effect of fiber orientation angle in composite materials on moisture absorption and material degradation after hygrothermal ageing. Composite Structures 74: 406-418.

17. Hodzic A, Kim JK, Lowe AE, Stachurski ZH (2004) The effects of water aging on the interphase region and interlaminar fracture toughness in polymer-glass composites. Compos Sci Technol 64: 2185-2195.

18. Komai K, Minoshima K, Tanaka K (1997) Delamination induced by low velocity impact and influence of water absorption on delamination and CAI of FRPs. Trans Japan Soc Mech Eng 63: 1198-1204.

19. Akay M, Mun SKA, Stanley A (1997) Influence of moisture on the thermal and mechanical properties of autoclaved and oven cured Kevlar 49/Epoxy laminates. Compos Sci Technol 57: 565-571.

20. Hojo M, Ochiai S, Tanaka K (1995) Near threshold propagation of delamination fatigue cracks in unidirectional CF/PEEK laminates in air and in water. Mater Sci Res Int 1: 100-107.

21. Joliff Y, Rekik W, Belec L, Chailan JF (2014) Study of the moisture/stress effects on glass-fibre/epoxy composite and the impact of the interphase area. Composite Structures 108: 876-885.

22. Carra G, Carvelli V (2014) Ageing of pultruded glass fibre reinforced polymer composites exposed to combined environmental agents. Composite Structures 108: 1019-1026.

23. Ramezani DH (2014) Numerical and experimental investigation of hygro mechanical states of glass fiber reinforced polyester composites experienced by fbg sensors. Composite Structures 116: 38-47.

24. Roy XL, Arun K, Haibin N, Uday V (2012) A seawater tank approach to evaluate the dynamic failure and durability of E glass/vinyl ester marine composites. Composites: Part B 43: 2480-2486.

25. Visco AM, Campo N, Cianciafara P (2011) Comparison of seawater absorption properties of thermoset resins based composites Composites: Part A 42: 123-130.

26. Poodts E, Minak G, Zucchelli A (2013) Impact of seawater on the quasi static and fatigue flexural properties of GFRP. Composite Structures 97: 222-230.

27. Menail Y, El Mahi A, Assarar M, Redjel B, Kondratas A (2009)The effects of water aging on the mechanical properties of glass fiber and Kevlar fiber epoxy composite materials. Mechanika 2: 28-32.

28. Talreja R (1999) Damage mechanics and fatigue life assessment of composite materials. International Journal of Damage Mechanics 8: 339-354.

29. Muc A (2000) Design of composite structures under cyclic loads. Composites Structures 76: 211-218.

15

Experiences of Using a Mobile RFID-Based Triage System

Jokela Jorma[1]*, Laapotti Heli[2], Engblom Janne[3] and Harkke Ville[3]

[1]Laurea Simulated Hospital, Laurea University of Applied Sciences, Hyvinkää, Finland
[2]Päijät-Häme Social and Health Care Group, Centre for Prehospital Care and Emergency Medicine Lahti, Finland
[3]Turku School of Economics, University of Turku, Finland

Abstract

A number of triage support systems which use Radio Frequency Identification (RFID) have been introduced in recent years. This paper will focus on one mobile triage system; known as "mTriage" The purpose of this paper was to determine the applicability of Radio Frequency communication (RFID) technology and a "mobile triage" system in a simulated multicasualty situation by examining the system's performance during a military winter exercise in Finland year 2009. This paper focuses on the medical personnel's opinion on this matter, answering the question: Are the medical personnel who use the system in the field satisfied with its performance.

Several field medics were asked to complete a questionnaire. The results of the evaluation were mainly positive. Conclusion was that mobile triage has potential to contribute to the management of mass casualty situations.

Keywords: Triage; Radio frequency identification; Mobile technology and simulation

Introduction

Triage is used daily by the emergency and health care workers around the world. It is a very important tool when processing and categorising casualties. In mass casualty events, those with severe, life-threatening injuries may receive a lower priority than those with more survivable injuries. [1]. According to Szul [2] triage is an attempt to make order during chaos and to make an overwhelming situation manageable. There is no standardised system of triage and several are in use throughout the world. The most common classification uses the four-colour code system: red signals high priority, yellow for medium priority, green is used for ambulatory patients and black for deceased [1].

This paper sets out to examine the applicability of a digital triage marking system for mass casualty situations. We present the findings of a field test conducted during a military field exercise. This paper is organized as follows: First we introduce the existing methods for triage tagging, including the mTriage system that is under study here. Then we define the test setting and methodology and present the results of the tests. In the final chapter we discuss the implications of the findings for managing mass casualty situations and needs for further research.

All prehospital triage systems have different documentation methods. Primitive types are based on paper tags, and others on triage tags with Radio Frequency Identification (RFID) or other wireless "intelligent" tags. Most of the current mass casualty triage systems still rely on a paper triage tags on which rescue and medical workers write the casualties' triage status and limited medical information [3,4]. There are different methods for marking the paper tags, either by writing on them or by making tears on designated spots. Regardless of the type the paper tags have serious limitations. The space for recording medical information is very limited. The "tear-off" triage tags allow only unidirectional changes in casualty's condition. Paper triage tags are not weather proof and are easily destroyed [5]. According to Gao et al. [6] paper triage tags are difficult for responders to update the triage color at the designation easily, tags have little room for manually write the vital signs and complaints and reading tags can be difficult because casualty information recorded is poorly written. Baker [7] wrote how important is to retriage casualties at every medical facility in their travel and how triage priority may also be altered by new findings.

Castren et al. [8] states that triage tags should show who gave treatment to casualties, who made decisions about what kind of treatment of care to give, which unit transported the casualties and what treatment facility were used. This is something that paper tags cannot always do; because there is not enough room to write to the required information.

Mobile systems of triage have recently been introduced. Newer methods are using triage tags with Radio Frequency Identification (RFID) or other wireless "intelligent" tags for casualty triage and tracking. These systems promise to produce accurate and on time information on victims status and triage class in the field [3]. Medical command must coordinate the number of casualties and their needs with the known availability of resources, such as on-scene medical personnel, ambulance locations, and medical capacities. Real-time information is also critical to determining the appropriate patient destination, the type of injuries and the capabilities of the receiving facilities [5, 9].

According to the researchers working on the Finnish project, getting relevant information about casualty location, numbers and categorizations to the command center has taken up to 24 hours in previous exercises.

Triage tags with Radio Frequency Identification (RFID) or other wireless "intelligent" tags might be the future of systems for both identification and tracking the casualties [7]. Real-time information is critical to overall management of field medical care [3].

Chan et al. [4] write that information technology used in

---

*Corresponding author: Jokela Jorma, Laurea Simulated Hospital, Laurea University of Applied Sciences, Hyvinkää, Finland, E-mail: Jorma.Jokela@laurea.fi

emergency health care should be easy to use. Effective training should also be provided. Lack of adequate training has prevented deployment of many systems [10]. Patient tracking devices would need to be small, durable and rugged enough to withstand environmental and manmade elements.

Background

A number of types of triage have been introduced in recent years. Technology has been combined with triage through the use of barcodes, tag readers, passive RFID tags, hand-held computers, and geolocation to collect data about the mass casualty events [11]. Rapid triage with flexible data management is vital in response to emergency care. RFID technology is making its way into emergency health care to enhance emergency data management. This article will focus on a RFID triage system, mTriage.

mTriage

The mTriage system is based on RFID technology, Nokia Field Force Management Solution using Near Field Communication (NFC) and WM-data (is part of CGI) mTriage software. The medical personnel on the field used the triage system via a Nokia 5140i phone and RFID mTriage tags. The system is specifically designed to triage and track casualties in the battlefield [12].

Medical personnel have their own RFID tags-(B) and casualties their own personal tags-(A); one per person. Instead of writing- and hand-collecting, the triage information is transmitted to the system via nokia 5140i phones, used by medical personnel. When casualties transported forward to medical facilities, the RFID tags carry the triage information [12].

Casualties are classified to four triage categories; immediate, delayed, minor and dead. Triage category at the tag is possible to change if needed by the medic or the medical personnel at the medical facilities [12].

Methods

The purpose of this paper is to determine the applicability of Radio Frequency Identification (RFID) technology and commercial cellular networks to provide an online triage system for handling mass casualty situation. The system was tested during a military exercise in actual field conditions. The users are field medics who received the standard Finnish military medical training.

The usability of the new triage system is evaluated with a standard post-test questionnaire, with 19 questions regarding the field medics' subjective confidence in the personal use, general use and applicability of the system. Free comments are allowed to be made at the end of the questionnaire [12,13].

The questionnaire is divided into five groups, each one dealing with a different aspect of the user experience: overall reactions to 1) reading the tag, 2) technology, 3) time consumption, 4) triage and 5) training.

Field medics in the exercise are given a personal RFID tag and a Nokia 5140i phone with integrated RFID reader/writer and mTriage software from a division of Logica.

Several medical facilities and 5 evacuation vehicles are equipped with the RFID tags and readers as well [12].

The mock casualties include 130 randomly selected conscripts, each tagged with an injury card (including triage Tag) of their injury [12].

During the study in the field, medics treat casualties with paper triage tags or RFID-tags.

Results

The test took place during the Finnish military's Pyry 2006 exercise. Sub-arctic weather conditions gave the Finnish Defence Forces good testing fields to test medical technology [12].

All 10 field medics complete the entire questionnaire [12]. The field medics had various professional backgrounds. One is a physician, two have hospital background, four have finished high school, and two have graduated from a vocational school and one with unknown background. Since we were especially interested in the reactions of the users who are not familiar with the RFID technology, none of these field medics had RFID-triage experience before. Before the study started users were briefly taught the basic concepts needed to operate the system. They had 10 minutes of training how to operate the equipment. Medics also had change to ask questions if they had any difficulties while using the equipment on the scene. The staff from the Finnish National Center for Military Medicine were on the scene to provide support if there were any difficulties [12].

After users complete their task, the questionnaire collects their subjective opinion concerning the usability of the system. As earlier mentioned the questionnaire is divided into five groups, each one dealing with a different aspect of the user experience. Here we take a closer look at these groups.

Reading the tag: While using the equipment in the field medics felt that that RFID-tag reading was not difficult and was mostly successful. Most of the users strongly disagreed and disagreed that reading the RFID-tag would be difficult and only one had neutral. When asked if the RFID-tag reading was slow; nine of the field medics disagreed only one agreed to that opinion. All medics agreed that RFID- tag reading was mostly successful.

Technology: Reader operation was mostly problem free and the reader operation was not difficult. All users agreed the reader operation was mostly problem free. While using the RFID-reader all field medics disagreed or strongly disagreed that reader operation was difficult.

Questionnaire responses show also that field medics found the application easy to use and stable. All field medics found the application easy to use. They also found application stable.

Time: Five of the users felt that RFID-patients required more time than normal patients, one had neutral opinion and four disagreed that RFID-patients required more time. Normal patient is here referred as a patient with paper triage tag.

Six of the field medics disagree that RFID-triage took longer because of technology. Four agreed that RFID technology influenced they triage timxe, and RFID-patient triage took longer because of the technology. Field medics found RFID-patient triage to be quick and reliable, all of the users agreed to that opinion (Figure 1).

Triage: Seven of the user disagree that they viewed RFID-patients differently to normal patients, two were neutral and one agree. Field medics did not found RFID –patient triage labour intensive. Eight of them disagreed that RFID-patient triage was more labour intensive and two were neutral in their opinion. To question "normal patient triage was even more labour intensive" four of the users had neutral opinion, four disagreed and two agreed.

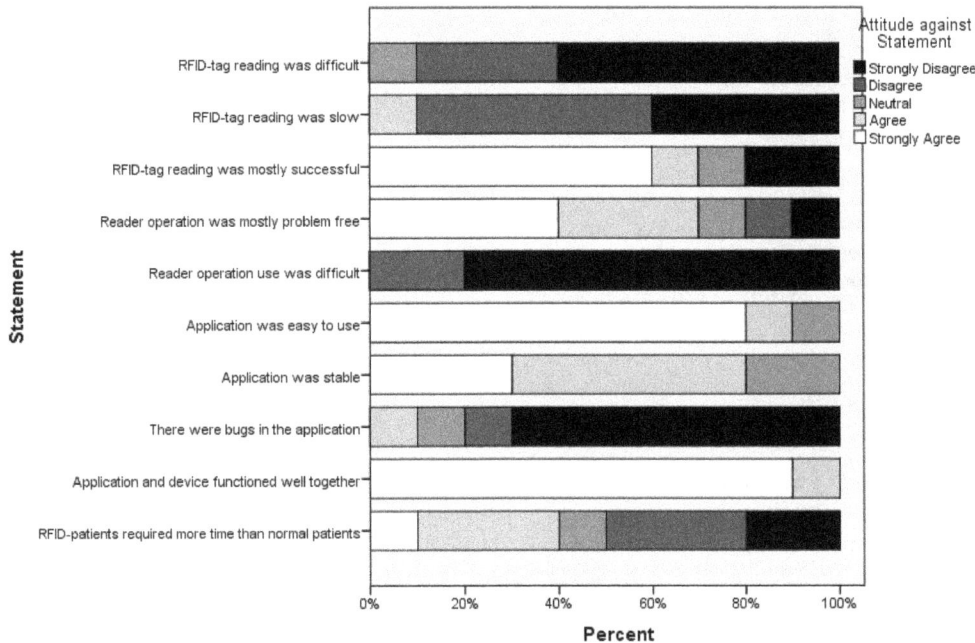

Figure 1: The attitudes towards statements about tag reading, technology operation and time.

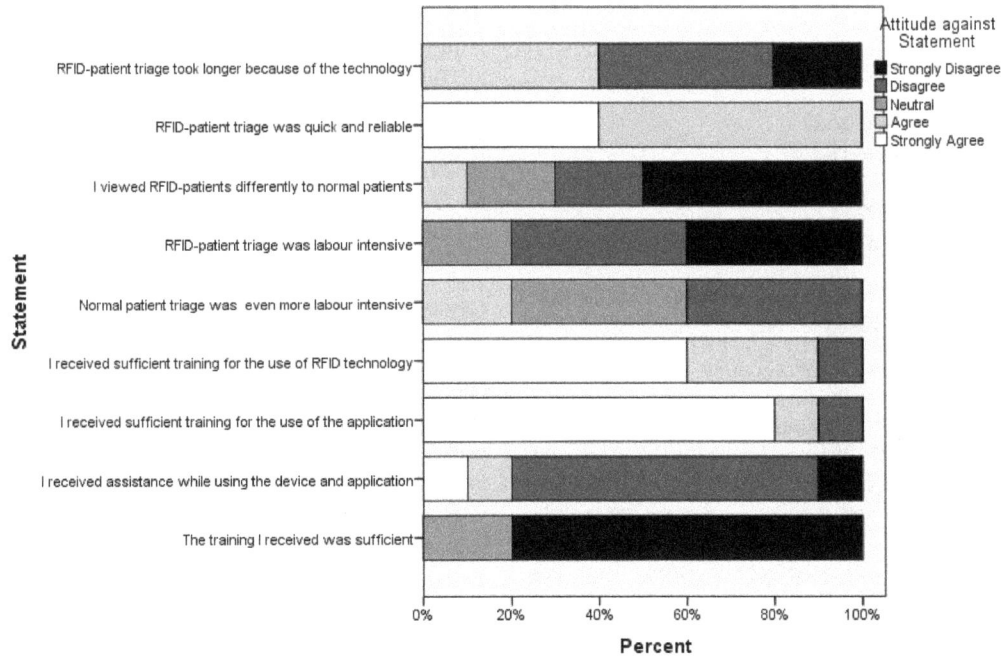

Figure 2: The attitudes towards statements on triage and received training.

Training: Most field medics agreed that they received sufficient training for the use of the RFID-technology and the use of the application. Almost every medics agreed that they received sufficient training for the use of the RFID-technology.

While using the application, nine medics strongly agreed that they received sufficient training and one disagreed. Eight of the users disagreed that they have received assistance while using the device and application and two agreed that they have received assistance while using the system (Figure 2).

Statistical Dependence Between Statement Questions

Statistical dependence i.e. correlations between statement questions were examined using explorative factor analysis. In (Table 1) varimax-rotated factor loadings (i.e. correlations between statements and factors;

	Factor					
	1	2	3	4	5	6
I received sufficient training for the use of the application	,940					
I received assistance while using the device and application	-,935					
I received sufficient training for the use of RFID technology	,932					
I viewed RFID-patients differently to normal patients	-,684					
Application was stable	,643					
RFID-tag reading was mostly successful		-,905				
Reader operation was mostly problem free		-,835				
RFID-patients required more time than normal patients		,815				
RFID-patient triage was quick and reliable		-,637				
RFID-tag reading was difficult			,919			
Application was easy to use			-,910			
The training I received was sufficient			,657			
Application and device functioned well together				-,938		
There were bugs in the application				,929		
Reader operation use was difficult				,851		
RFID-patient triage took longer because of the technology					,909	
Normal patient triage was even more labour intensive					-,752	
RFID-patient triage was labour intensive						,744
RFID-tag reading was slow						-,605

Mineigen criteria were used to define the number of factors (i.e. principal components) as 6. Factor loadings which were smaller than 0.5 or greater than -0.5 are not presented in the table
Negative factor loadings with the factor means positive correlations with each other

Table 1: Factor loading of descriptive factor analysis.

	Cluster	
	1	2
I received sufficient training for the use of the application	2	1
RFID-tag reading was mostly successful	1	4
RFID-tag reading was difficult	5	4
Application and device functioned well together	1	1
RFID-patient triage took longer because of the technology	4	3
RFID-patient triage was labour intensive	5	4

Table 2: Final cluster centers.

Factors can be interpret as groups of statement variables) are presented. Principal component method was used to obtain the estimates.

Statement question "I received sufficient training for the use of the application" had a very high factor loading with factor 1. It means it had a high correlation with other statements within this factor as well. In practice respondents with positive attitude against this statement usually also had positive attitude against "I received sufficient training for the use of RFID technology" and "Application was stable" as well. On the other hand respondents in these cases tended to have negative attitudes more often with "I received assistance while using the device and application" and "I viewed RFID-patients differently to normal patients". These first five statements didn't correlate highly with other statements in the questionnaire.

In the second factor "RFID-tag reading was mostly successful", "Reader operation was mostly problem free" and "RFID-patient triage was quick and reliable" had a high positive correlation with each other. They all correlated negatively with "RFID-patients required more time than normal patients".

The third factor consisted of three statements: "RFID-tag reading was difficult", "Application was easy to use" and "The training I received was sufficient" of which the second had a high negative correlation with other two.

The last three factors had three, two and two statements with high positive or negative correlations. This result indicates that instead of 19 statements only six could be used to measure opinions against these features experiences of Mobile RFID-Based Triage Systems. Although it is difficult to name the factors (groups of statements) they had some common features. For example some statements in the first factor had something to do with training. The second factor may be called "No problem" factor. The fourth factor with three statements may be called the "Flaws" factor. Six factors explained 91.4% of the total variance of the statement variables which justified the grouping the statements into six groups and also indicated high model fit with the data.

Grouping the respondents

Cluster analysis was used to cluster respondents with similar opinions against statements. Statements used in the cluster analysis were selected based on the factor analysis. Statements with the highest loading with the factors were used as a representative variable of the factors. In that sense all 19 statements were used in the cluster analysis. In (Table 2) final cluster centers (i.e. cluster means of statements using scale 1=Strongly agree, 5=Strongly disagree) are given. Number of clusters was chosen as two. This was done because of small sample size.

The first cluster (n=6) respondents tended to agree "I received sufficient training for the use of the application", "RFID-tag reading was mostly successful" and "Application and device functioned well together" and agree with other three statements. On the other hand respondents of the second cluster (n=4) tended to agree with "I received sufficient training for the use of the application" and "Application and device functioned well together". Average attitude against "RFID-patient triage took longer because of the technology" was neutral. Attitudes against "RFID-tag reading was mostly successful", "RFID-tag reading was difficult" and "RFID-patient triage was labour intensive" were typically negative.

Conclusion

This paper evaluated the implementation of the mobile triage in a military field exercise.

This system has many benefits over the current paper-based paper triage. The medics whom were unfamiliar to RFID triage quickly learned how to use it, and found it be easy-to-use. Where this kind of equipment is used in disaster relief efforts, equipment should be quick and easy to repair and hands-on training should be routine [10]. The training on site was effective and helped the use of the equipment.

While using the mTriage in the field medics did not find triage to be labour intensive, it was quick and reliable. The users in the field felt that the application and the device functioned well together while they were performing the casualty triage. The mTriage system also made it possible to change casualties triage category when needed, which is often done at the secondary triage point or during the transportation.

Mobile triage has the potential to contribute to the management of mass casualty situations; it also has potential for improving the quality of medical care. New information technologies, such as mTriage, will improve triage and patient tracking on the field. The mTriage has potential not only for the military medicine use in the future. The system could also be adapted without any difficulties by the civilian sector. It could be used for management of mass casualty disasters; such as earthquakes, storms and mass casualty incidents.

All new technologies have limitations as well as capabilities. The RFID tags would need some sort of human readable element (similar to the color coding on traditional triage cards) to simplify the work of transport and field personnel who are not equipped with RFID readers [13].

Due to nature of emergencies more studies need to be done on user satisfaction to evaluate the usability of the RFID triage in the field.

References

1. The Sphere Project (2011) Humanitarian Charter and Minimum Standards in Disaster Response. Belmont Press Ltd, United Kingdom.

2. Szul AC (2004) Emergency War Surgery 3rd U.S. revision. Border Institute. Walter Reed Army Medical Center. Department of Defense Washington, DC.

3. Killeen JP, Chan TC, Buono CC, Griswold WG, Lenert LA (2006) Patient Assessment and Documentation during Mass Casualty Incidents. AMIA.

4. Chan TC, Killeen J, Griswold W, Lenert L (2004) Information Technology and Emergency Medical Care during Disasters. ACAD EMERG MED 11: 11.

5. Plischke M, Wolf K H, Lison T, Pretschner D P (1999) Telemedical support of prehospital emergency care in mass casualty incidents. Eur J Med Res 4: 394-398.

6. Gao T, Hauenstein LK, Alm A, Crawford D, Sims CK, et al. (2006) Vital signs monitoring and patient Tracking over a Wireless network. Johns Hopkins ApL TechnicAL DigesT (2006) 27: 1.

7. Baker M (2007) Creating Order from Chaos: Part 1: Triage, Initial Care, and Tactical Considerations in Mass Casualty and Disaster Response. Military Medicine 172: 232-236.

8. Castrén M, Ekman S, Martikainen M Sahi T, Söder J (2006) Suuronnettomuusopas. Duodecim.

9. Teich JM, Wagner MM, Mackenzie C F, Schafer K O (2002) The informatics response in disaster, terrorism, and war. J Am Med Inform Assoc 9: 97-104.

10. Garshnek V, Burkle FM (1999) Applications of Telemedicine and Telecommunications to Disaster edicine: Historical and Future Perspectives. J Am Med Inform Assoc 6: 26-37.

11. Massey T, Gao T, Welsh M, Sharp JH, Sarrafzadeh M (2006) The Design of a Decentralized Electronic Triage System AMIA.

12. Finnish Defence Forces (2006) Material from Pyry 2006 -exercise. Centre for Military Medicine, Finnish Defence Forces.

13. Jokela J, Rådestad M, Gryth D, Nilsson H, Rüter A, et al. (2012) Increased Situation Awareness in Major Incidents-Radio Frequency Identification (RFID) Technique: A Promising Tool. Prehosp Disaster Med 27: 81-87.

16

Generic Framework for Multi-Disciplinary Trajectory Optimization of Aircraft and Power Plant Integrated Systems

Rukshan Navaratne[1]*, Marco Tessaro[1], Weiqun Gu[1], Vishal Sethi[1], Pericles Pilidis[1], Roberto Sabatini[2] and David Zammit-Mangion[2]

[1]Department of Power and Propulsion, Cranfield University, Bedfordshire, UK
[2]Department of Aerospace Engineering, Cranfield University, Bedfordshire, UK

Abstract

Engineering improvements, technology enhancements and advanced operations have an important role to play in reducing aviation fuel consumption and environmental emissions. Currently several organizations worldwide are focusing their efforts towards large collaborative projects whose main objective is to identify the best technologies or routes to reduce the environmental impact and fuel efficiency of aircraft operations. The paper describes the capability of a multi-disciplinary optimization framework named GATAC (Green Aircraft Trajectories under ATM Constrains) developed as part of the Clean Sky project to identify the potential cleaner and quieter aircraft trajectories.

The main objective of the framework is to integrate a set of specific models and perform multi-objective optimization of flight trajectories according to predetermined operational and environmental constraints. The models considered for this study include the Aircraft Performance Model, Engine Performance Simulation Model and the Gaseous Emissions Model. The paper, further discusses the results of a test case to demonstrate trade-offs between fuel consumption, flight time and NOx emissions that the trajectory optimization activity achieves at a primary level. It thereby forms the basis of a complete reference base-line trajectory which will be used to determine more accurate environmental gains that can be expected through optimization with the integration of more models within the framework in the future.

Keywords: GATAC; Aircraft performance model; Engine Performance simulation Model; NOx emissions

Introduction

The air transport industry today is paying a lot of attention to growing public concern about the environmental issues of air pollution, noise and climate change. The past decade has witnessed rapid changes both in the regulations for controlling emissions and in the technologies used to meet these regulations. Considering the critical nature of the problem regarding the environmental footprint of aviation several organizations worldwide have focused their efforts through large collaborative projects such as Clean Sky Joint Technical Initiative (JTI). Clean Sky is a European public private partnership between the aeronautical industry and the European Commission. It will advance the demonstration, integration and validation of different technologies making a major step towards the achievement of the environmental goals set by ACARE (Advisory Council for Aeronautics Research in Europe). The ACARE Vision 2020 and associated Strategic Research Agendas (SRAs) have successfully steered European aeronautics research in recent years by setting the objectives of reducing CO_2 by 50%, NO_x by 80% and Noise by 50% compared to year 2000 [1]. Ability to meet these challenges only is possible with a strong commitment to the vigorous evolution of technologies and achieving new breakthroughs. Over the last few years several alternatives have been proposed and most of them are long term solutions such as changing the aircraft and engine configurations and architectures. Hence all the manufacturers have started focusing and developing their strategies along the other possible options. The management of trajectory and mission is one of the key identified solutions found in achieving the above set goals and is a measure that can readily be implemented.

In order to truly understand the optimized environmental friendly trajectories it is necessary to simultaneously consider the combined effects of aircraft performance, propulsion system and engine performance, environmental emissions, noise and flying trajectories. GATAC (Green Aircraft Trajectories under ATM Constrains) is a multi-disciplinary optimization frame work which is being collaboratively developed to achieve the above requirement by Cranfield University and other partners as part of the Systems for Green Operations - Integrated Technical Demonstrator (SGO-ITD) under the Clean Sky Joint Technical Initiative [1].

The Gatac Environment

This section presents an overview of the main features and capabilities of the GATAC multi-disciplinary optimization framework. It can be considered as a state-of the-art optimization framework with optimizers and simulation models to perform multi-objective optimization of flight trajectories under Air Traffic Management (ATM) constraints. The top level structure of the GATAC framework is shown in Figure 1.

The framework consists of, the GATAC Core, Model Suite, Graphical User Interface (GUI) and Post-Processing Suite. It interacts with a suite of models as configured at set-up time. The GATAC core is the core engine of the interaction framework and provides the connectivity between the various models. It also provides for the organization of an evaluation process (within the Evaluation Handler) and includes functionalities such as parameter stores, data parsing, translation

*Corresponding author: Rukshan Navaratne, Department of Power and Propulsion, Cranfield University, Bedfordshire, MK43 0AL England, UK
E-mail: r.navaratne@cranfield.ac.uk

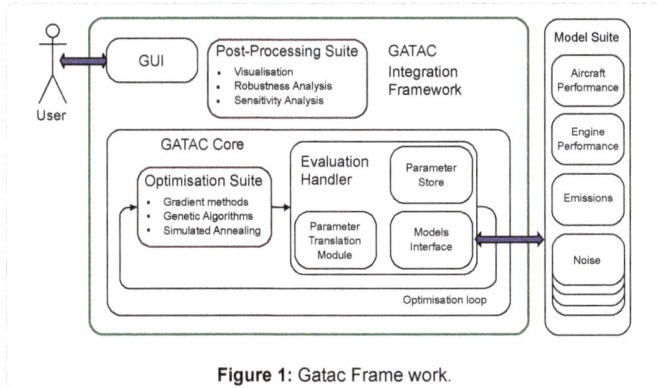

Figure 1: Gatac Frame work.

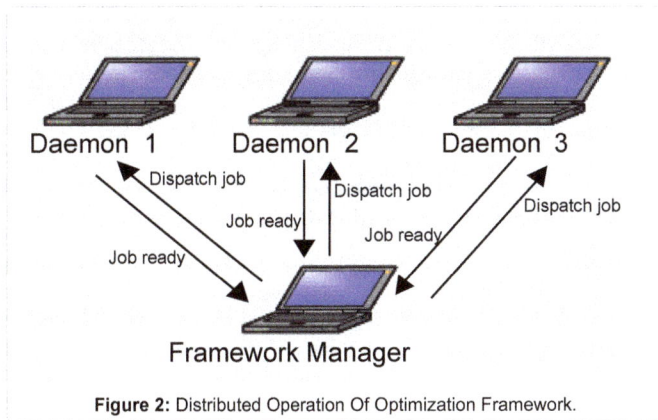

Figure 2: Distributed Operation Of Optimization Framework.

function and interfacing with models. It also supports the repeated calling of sets of models to enable trajectories to be evaluated step by step with number of steps being defined by the user at set-up time. The core, therefore, is programmable as the user sets-up the problem at hand within the Evaluation Handler by defining connectivity between models and any data translation and other similar functions. This can be done either directly using a purposely defined domain specific language or graphically via GUI. In this way, the user effectively defines (formulates) the optimization problem. The optimization process takes place in the GATAC Core, which accesses an optimization function chosen from a suite by the user [2,3].

A key feature of GATAC is that, he user can select any algorithm from the optimization suite without the need to modify the problem formulation because; the framework caters for normalization of data. Indeed, the algorithm in the optimization suite are designed to handle normalize variable parameters. The normalized parameters are then de-normalized by the integration framework as specified by the user before being input to the evaluation handler. Similarly the data that are output from the evaluation handler are again normalized before being input to the optimizer to close the optimization loop (Figure 1).

As the data exchanged between the optimization core and the models need to be defined according to the input and output data of each model and module. GATAC caters for the automatic definition of data structures by means of a dictionary. The automatic definition is carried out by GATAC at set-up time according to the output and input variables of the specific models and modules invoked in the problem definition. These data structures then enable the correct data transfer between the models and modules.

The GATAC can be run either on a single stand-alone machine or a distributed system with multiple computers (Figure 2). In the

latter case the model suite is replicated on a number of different machines, on which a daemon will be running in the background. The daemon is even-triggered and instructed to run particular models by the Framework Manager, where the GATAC core resides. When its particular job is complete, the relevant daemon will return the results to the GATAC core. In this way, the core maintains full control of the optimization process. Data exchange between the GATAC core and the daemons is achieved through Ethernet LAN connectivity between the respective computers. The model suite is distributed on a single machine or different machines acting as hosts. The data exchange between components carried out through Ethernet LAN. The Figure 2 illustrates the architecture and operating network of the GATAC distributed system [2,3].

The Nsgamo Genetic Optimiser

The NSGAMO (Non Dominated Sorting Genetic Algorithm Multi-Objective) is one of the genetic based optimizers incorporated in the GATAC framework. This optimizer is able to perform optimization of two objectives without or with constraints. Figure 3 shows the sequence of steps of the NSGAMO genetic algorithm.

According to the flowchart, at the first step an initial population of the test cases (candidate trajectories) is created randomly. The size of the initial population determined by the product of the prescribed population size with an initialization factor (>=1) A larger initial population size increases the probability of the optimizer converging to the global optimum point but slows down the optimization process. The optimizer then sends all the cases to the GATAC framework for the evaluation handler to evaluate and return the results (optimization objective) to the optimizer. On receipt of the results, the optimizer performs fitness evaluation on the data (i.e. qualifies the population). As optimum point is identified on the first generation, a second generation population is created and the process repeated. The process is repeated until convergence criteria are met (either a maximum number of generations will have been generated and evaluated or Pareto convergence will have been reached). In order to reduce the computational time of subsequent generation is reduced to

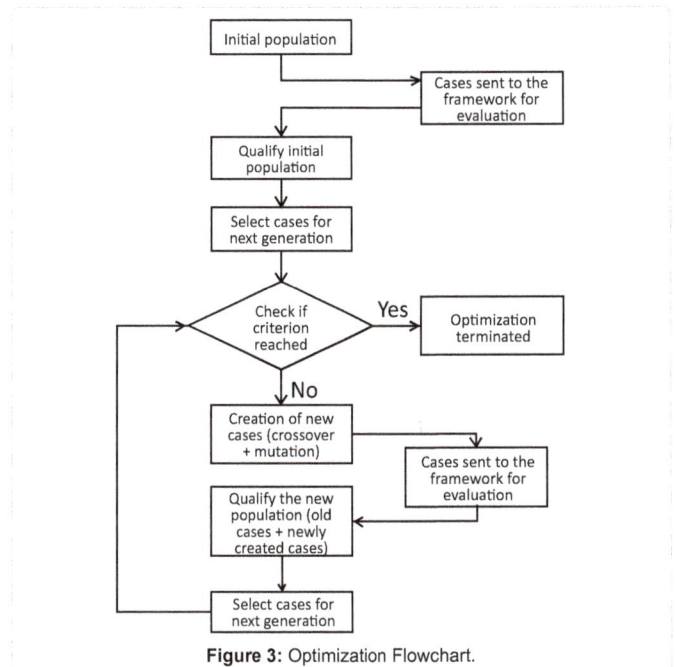

Figure 3: Optimization Flowchart.

prescribed population size. To achieve this only the best solution s of the previous population are selected to generate the next generation. New generations are created using different methods such as stochastic universal sampling, random selection and genetic operators (crossover and mutation). In the case of single objective optimization the result is the best-case while for a multi-objective optimization, the final result is a Pareto Front [3].

The implementation of the NSGAMO algorithm allows, for via a text file, the user definition of the various parameters associated with the optimization, which include population size, optimization method, mutation and crossover ratio, selection method and type of mutation and crossover and other parameters. A detailed description of the testing and benchmarking of the optimizer performance is presented in reference [4].

Simulation Models Engine Model

Engine model

The engine model developed for this study is based on Trent 895 which is a 3-spool high by-pass ratio turbofan engine with separate exhausts. The engine model is designated as CUHBR (Cranfield University High By-Pass Ratio) and was modeled using data available from public domain and making educated engineering assumptions where necessary. This engine has been selected to power the long-range aircraft which has been used to develop the aircraft performance model. The engine model has been developed and simulated using TURBOMATCH which is an in-house gas-turbine performance simulation and diagnostics software developed at Cranfield University [5]. The tool is used to model the design point of the engine and study its off-design performance. TURBOMATCH is a fully modular engine cycle simulator that can perform design point, off-design, steady state,

and transient conditions as well as degraded performance analysis of gas turbines.

The TURBOMATCH engine model is assembled from a collection of existing interconnected elements called 'Bricks'. Individual bricks are controlled by a numerical solver and represent the thermodynamic equivalent of gas turbine components including; intake, fan, compressor, combustion chamber, turbine, duct, and nozzle. Bricks are called up to model the architecture of the gas turbine and a numerical solver is used to solve the mass and energy balances between the interconnected bricks. TURBOMATCH also allows for the modeling of different fuels, extraction of bleed air and the shaft power off-takes, cooling air, component degradation, reheating, or sequential combustion etc. The outputs from the tool include the calculation of the overall performance of the engine in terms of gross and net thrust, fuel flow, Specific Fuel Consumption (SFC) as well as the thermodynamic parameters and gas properties at the inlet and outlet of each component. Detailed operational parameters such as efficiency, rotational speed, power required/power delivered, surge margin in case of the fan and compressor or thrust coefficients in the case of nozzles, are also provided. For the purpose of this study, the engine is modeled by developing a representative input file that represents the configuration of the CUHBR engine. Figure 4 is a schematic of the CUHBR TURBOMATCH model.

As shown in the figure, the LP turbine drives the fan. Similarly, IP turbine and HP turbine drive the IP compressor and HP compressor respectively. It has been assumed that part of the air is bleed from the HP compressor to cool the HP turbine and no cooling air bleeds for the IP turbine and LP turbines. The secondary air system has been largely simplified and handling bleed has not been considered.

The engine design point has been selected at maximum rated thrust during take-off under ambient International Standard Atmospheric (ISA) conditions at sea level, and the engine mass flow, bypass ratio and overall pressure ratio have been obtained from the public domain. An iterative trial and error process has been required in order to match the performance of the engine model with the reference engine performance data at DP as well as cruise phase. At the design point assumptions are made with regards to the pressure ratio split between the compressors, component efficiencies, surge margin, cooling mass flows, duct/intake and burner pressure losses, burner efficiency, as well as bleed and power off-takes. The fan pressure ratio is iterated and optimized for the maximum thrust and minimum SFC. The Turbine Entry Temperature (TET) of the cycle is iterated until the calculated

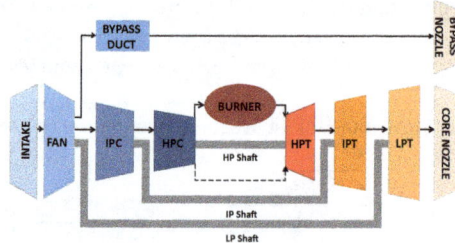

Figure 4: Engine Model Sketch.

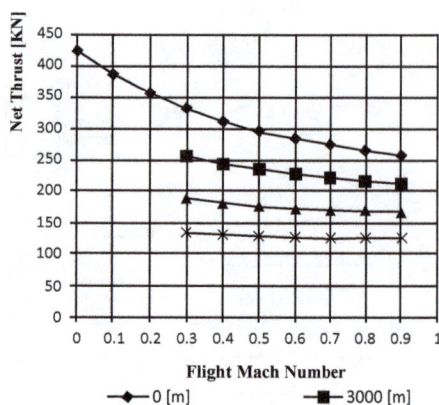

Figure 5: Net Thrust As A Function Of Altitude And Flight Mach Number (For Constant Tet= 1771K).

Parameter	VALUE	UNIT
Engine Mass Flow	1,208	Kg/s
Overall Pressure Ratio	41.7	
Bypass Ratio	5.8	
Fan Pressure Ratio	1.81	
Ipc Pressure Ratio	4.79	
Hpc Pressure Ratio	4.79	
Fan Efficiency	89.5	%
Ipc Efficiency	88	%
Hpc Efficiency	88	%
Combustor Efficiency	99.9	%
Combustional Fractional Pressure Loss	5	%
Turbine Inlet Temperature	1,771	K
Hpt Cooling Flow	13	%
Hpt Efficiency	89	%
Ipt Efficiency	90	%
Lpt Efficiency	91	%

Table 1: Cuhbr Engine Performance At Design Point [Take-Off Is A Sls Condition].

	TAKE-OFF			CRUISE		
	T895	CUHBR	%	T895	CUHBR	%
W (Kg/S)	1,208	1,208	0.00	-	441.0	-
FPR	1.81	1.81	0.00	-	1.66	-
BPR	5.8	5.8	0.04	-	6.551	-
PR	41.6	41.6	0.00	-	32.66	-
TET[K]	-	1771	-	-	1383	-
Wf [Kg/S]	-	4.19	-	-	0.972	-
SFC [Mg/Ns]	-	9.90	-	-	16.20	2.14
Fn [Kn]	422.6	422.8	0.04	-	60.01	-0.06
Wc [Kg/S]	178.2	177.6	-0.35	-	4.00	-

Table 2: Real Engine Vs Engine Model Performance.

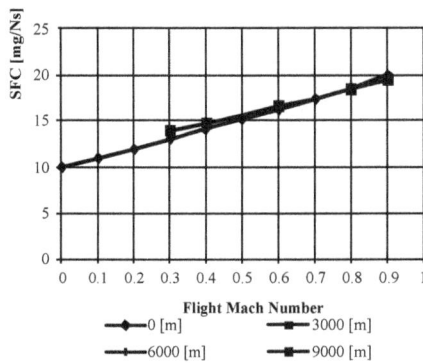

Figure 6: Sfc Vs Flight Mach Number And Altitude For Constant Tet.

thrust and the fuel consumption match the values found in the public domain with marginal difference. Being the differences in percentage about 2% the engine model can be considered verified for the scope of this project. The performance comparison of the CUHPR engine and summary of component specifications are shown in the Table 1 and 2 respectively.

With the fixed design point, a series of Off-Design (OD) performance simulations have been performed in order to simulate the effects of ambient temperature, altitude, flight Mach number and TET on net thrust and SFC as a further model verification process. The Figure 5 and 6 shows the variation of net thrust and specific fuel consumption for different flight Mach numbers at different altitudes under OD performance.

As the flight velocity increases the performance of the engine is influenced by three main factors: momentum drag, ram compression and ram temperature rise. The momentum drag rises with the flight speed with a consequence reduction of the momentum imparted to the air by the engine. Therefore, the net thrust, which is defined as the difference between gross thrust and intake momentum drag, drops with the rising of flight Mach number. The second effect is the ram compression and it has a double effect. Firstly, it increases the nozzle pressure ratio and therefore the net thrust. Secondly, it raises the inlet pressure and thus air density along with mass flow. The last effect is the ram temperature rise, which produces an increment of air temperature at fan inlet. This leads, at constant shaft speed, to a decrement of non-dimensional power setting and hence thermal efficiency. The momentum drag and the ram compression are generally the main effects. At low speed, momentum drag is the main effect and the net thrust drops quickly with the rising of flight speed. Since Mach number is less than 0.3 the effects of temperature rise and am compression are

small. At higher Mach number the effect of compressor rise starts to be important and, as it is possible to observe, the gradual decrease in net thrust.

As shown in Figure 5 the net thrust decreases when the altitude increases with a constant Mach number. When the altitude increases the air density drops, leading to a reduction of mass flow and hence net thrust. The reduction of air density does not alter the non-dimensional power setting of the engine. Moreover, in the troposphere the reduction of net thrust due to the drop of the air density is in partly offset by the positive effect of the decrement in ambient temperature. Indeed, the ambient temperature falls linearly in the troposphere, from 15°C at sea level to -56°C at the top of the Troposphere at 11 km. At constant shaft speed, when the temperature drops the non-dimensional power setting raises leading to an increment in pressure ratio therefore in net-thrust. Figure 6 shows how the increases with flight Mach number. In order to fly at faster speed more fuel is required. The increment of the fuel flow overcomes the decreasing on net thrust and rises with Mach number.

In Figure 7 and 8 are the effects of ambient temperature and on net thrust and are shown. At low TET give rise to low thermal efficiency and jet velocity which create a high propulsive efficiency which resulted in high SFC. Similarly at high TET leads to give high thermal efficiency and high jet velocity which result in low propulsive efficiency. The figure 7 shows the best compromise between thermal efficiency and propulsive efficiency for several ISA deviations.

With the variation of ambient temperature there are two main effects that have to be considered [6]. The first effect is well described

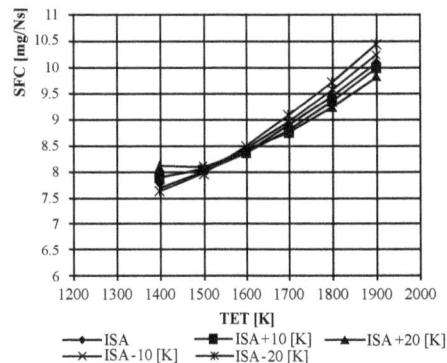

Figure 7: Sfc Vs Tet And Ambient Temperature At Sls Conditions.

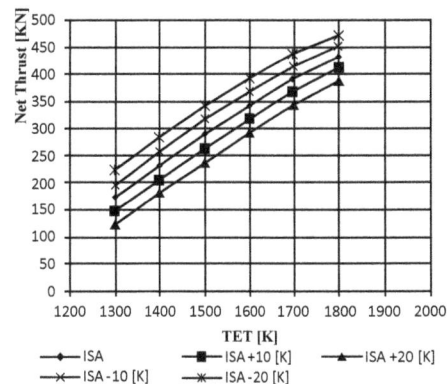

Figure 8: Net Thrustvs Tet and Ambient Temperature at Sls Conditions.

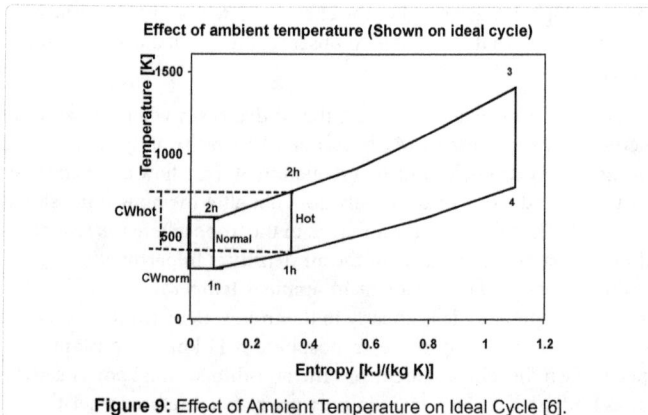

Figure 9: Effect of Ambient Temperature on Ideal Cycle [6].

in Figure 9 whereas considering an ideal cycle between fixed values of overall pressure ratio is shown.

Figure 9 shows that on a 'hot day' the compressor work will be greater than in a 'normal day'. This is due to the fact that compressing hot air requires more work. However, the turbine's work is not affected by ambient temperature because in this case the overall pressure ratio does not change. Consequently on a hot day, the difference between turbine work and compressor work will be less than the normal day therefore the net thrusts will decrease. Also this effect is reflected with a shift in compressor operating point with a variation in ambient temperature. This is due to the fact that, the non-dimensional rotational speed $\frac{N}{\sqrt{T}}$ depends on shaft speed and temperature. Assuming constant the rotational speed of the shaft, in a hot day the ambient temperature increases and hence the non-dimensional rotational speed decreases. Therefore, the operating point will move to the left and downwards so the pressure ratio and the non-dimensional mass flow will decrease. The opposite will occur in a cold day. For constant ambient temperature, with the increment of the pressure ratio and net thrust increase. For constant with the rising of ambient temperature the net thrust drops. Vice versa, with the decrement of ambient temperature the net thrust rises.

Aircraft Performance Model

The software that has been used to simulate the integrated aircraft-engine performance is called HERMES. It has been developed at Cranfield University in order to assess the potential benefit of adopting new aircrafts, engine concepts and technologies [7]. The aircraft model is capable to simulate the performance of different types of aircrafts, from a baseline aircraft to an advanced one for a given civil mission. The software consists of six different modules;

a. Input data

b. Mission profile module

c. Atmospheric module

d. Engine module

e. Aerodynamic module

f. Aircraft performance module

The aircraft model computation starts reading the required input data from an input file. As described below, these data regard the general arrangement of the aircraft and mission profile. Some of these data are usually available from the public domain or defined by the user. The user has to specify as an input the MTOW and the weight of the

payload. Moreover, the user has to set either the fuel load or the mission range. In the first case HERMES will compute the mission range whilst in the second case HERMES will assess the required amount of fuel to complete the mission. In the case that the user has set the initial amount of fuel, the value of the range will be considered as an initial guess and will not influence the resulting values.

The range is calculated iteratively. In each iteration process the fuel required for a trial distance is computed. As soon as the total fuel is consumed the convergence is achieved. This is obtained calculating the trial OEW by subtracting the assessed total fuel, which is given by the sum of mission fuel plus reserve, and the payload from the MTOW. The trial OEW is then compared with the OEW set up in the input file and the difference is used to redefine the distance and the time spent at the cruise. The convergence is achieved when the difference of OEW is within 0.1 %.

Input data module

The input data required for the aircraft model are information regarding the geometry, configuration and the required performance of the aircraft. These input data are used by the aircraft performance and aerodynamic modules to calculate the performance and aerodynamics characteristics of the aircraft.

Mission profile module

The mission profile is subdivided into different phases. The overall mission profile is defined by the user and is used by the aircraft performance module to compute the distance, fuel and time for each the each segment in which the mission is subdivide. In addition, TURBOMATCH refers to the mission profile in order to calculate the engine performance.

Atmospheric module

The atmospheric conditions for a given Mach number and altitude have a great influence on the aircraft and engine performance. Therefore, the atmospheric module calculates the ISA conditions both in the lower atmosphere and stratosphere. Moreover, the user has the possibility to alter the temperature from ISA standard values to simulate non-standard conditions.

Engine data module

The performance of the engine greatly influences the aircraft performance. The engine data usually includes, maximum take-off thrust: required to assess the length, fuel and time required for the take-off; maximum climb thrust and SFC: required to compute fuel consumed, horizontal distance covered, rate of climb and time to climb; cruise and descent performance.

Aerodynamic module

Calculates the aerodynamic performance of the aircraft for the given flight conditions. The module elaborates the information regarding the mission profile, the aircraft and aerodynamics properties in order to compute the drag characteristics in form of drag polar profile and drag coefficients.

The drag polar can be always expressed using two main components of drag; one is dependent on lift and the other independent on lift. Therefore, the total drag coefficient can be expressed using the following equation:

$$C_D = C_{D0} + C_{DI}$$

The term C_{DO} is the zero lift drag coefficient and is a constant while C_{DI} is the lift dependent drag or induced drag coefficient and it can be expressed as follows:

$$C_{DI} = K.C_L^2$$

Where K is called lift dependent factor. Combining the previous two expressions it is possible to write the well-known drag coefficient expression:

$$C_D = C_{D0} + K.C_L^2 \quad C_D = C_{D0} + K.C_L^2$$

The calculation of the zero lift drag coefficient is performed using the component build-up method, which has the following general expression [5]:

$$(C_{DO}) = \frac{\Sigma(Cf_c, \varphi_c, Q_c, S_{Wet_c}}{S_{ref}}$$

The flat-plate skin friction coefficient Cf and the form factor φ, which estimates the pressure drag due to viscous effects, are used to assess the subsonic profile drag of a particular component. The factor is Q used in order to take into account the effects of the interference drag on the component. As highlight in the previous equation, the product of the wetted surface of the component, S_{Wet} and Cf, φ and Q allows to calculate the total drag on the component, c. Using this method it is possible to calculate the drag arising from several components such as fuselage, tail plane, fins, nacelle, outer wings and engine pylons. It also allows the estimation of miscellaneous drag arising from deployed flaps, landing gear and trim conditions. The coefficient C_{DO} is then calculated dividing the total drag by the reference area S_{ref}, which is the plan wing area.

The lift induced drag is estimated using the following equation [8]:

$$C_{DI} = \left[\left(\frac{C_1}{C_2.\mu.AR} \right) + C_3 + C_4.C_{DO} \right].C_L^2$$

Where the coefficients C_1 and C_2 are a function of the wing aspect ratio and taper ratio and are used to take into account the wing plan form geometry. The coefficient C_3 and C_4 are used to account the non-optimum wing twist and viscous effect respectively.

Aircraft performance module

Information from the other modules is passed to the aircraft performance module. In turn, the aircraft performance module computes the overall performance of the aircraft for each segment in which the entire mission is divided.

Typical outputs include: fuel consumption, distance covered, mission duration, engine thrust and SFC for the whole, mission and for each flight segment.

The calculations of the climb rely on the rate of climb, which is defined as the ratio between the change in height and the time assuming zero wind velocity:

$$RateofC\lim b = \frac{dh}{dt} = V_{Vertical}$$

During the calculation of the rate of climb appropriate acceleration factors are included in order to take into account the following cases:

• When the aircraft is climbing in the stratosphere, the ambient temperature reduces, thus at constant Mach number the airspeed decreases because the speed of sound drops.

• During a climb at constant equivalent air speed the true air speed is increasing because the air density drops with altitude.

Therefore, the time required to flight from an altitude h_1 to an altitude h_2 is given by the following expression:

$$t_i = \int_{h_2}^{h_1} \frac{1}{V_{Vertical}} dh$$

The flight distance, the time and the fuel consumed (thus aircraft weight) at the end of each segment are a function of the rate of limb. In turn the rate of climb relies on thrust, drag and mass of the aircraft. The integration of the previous equation is therefore complicated to compute because of numerous interrelationships regarding the variables involved. For this reason the model used an iterative procedure with the estimation of the weight and the time at the end of each segment, which are then used to compute the correct values.

The total time of the climb phase is given by the sum of the time of each segment:

$$Total\,Time = \sum_{i=1}^{n} t_1$$

Using the climb speed and gradient it is possible to work out the horizontal distance covered by the aircraft:

$$TotalDis\tan ce = \sum_{i=1}^{n} t_i V_i .cos(\gamma_i)$$

Similarly, the total fuel is computed by summing the fuel burnt in each interval.

The calculation of the flight range is a function of the engine and aircraft parameters and the available quantity of fuel. In order to derive the equations implemented in the aircraft performance module it is necessary to define some fundamental variables that are involved.

Firstly, for an aircraft in horizontal, steady state flight at constant true airspeed V, the engine thrust has to be equal so the aerodynamic drag. Therefore it is possible to write:

$$F = D = \frac{L}{E} = \frac{W}{E}$$

Where F is the engines thrust, D and L are the drag and lift of the aircraft respectively W is the aircraft weight and E is the aerodynamic efficiency, which is defined as the ratio between the lift and the drag. Considering the definition of lift and the drag forces, the aerodynamic efficiency can be expressed as follows:

$$E = \frac{D}{L} = \frac{\frac{1}{2}.\rho.C_L.V^2.S_{Ref}}{\frac{1}{2}.\rho.C_D.V^2.S_{Ref}} = \frac{C_L}{C_D}$$

Where ρ is the density of the air and S_{Ref} is the wing plan area. The specific fuel consumption, SFC, of an aircraft powered by a turbojet or turbofan engine is defined as the ratio of fuel flow (Q) per specific thrust (F_s):

$$SFC = \frac{Q}{F_S}$$

The specific range r_a is defined as the flight distance dR per unit of fuel consumed so:

$$r_a = \frac{dR}{dm} = \frac{V.E}{SFC.W}$$

The integration of the above equation leads to compute the total cruise range:

$$Range = -\int_{m_1}^{m_2} r_a.dm = -\int_{m_1}^{m_2} \frac{V.E}{SFC.m.g}.dm$$

Where m_1 and m_2 are the initial and final mass of the aircraft during the cruise. In order to be able to integrate the range equation it is necessary to express the variables V, E and SFC as a function of the aircraft mass (m).

Three different flight schedules can be chosen which lead to three different sets of assumptions for the variables:

- Cruise at constant altitude, SFC and lift coefficient;
- Cruise at constant airspeed, SFC and lift coefficient;
- Cruise at constant altitude, airspeed and SFC

Usually the second option is used. Therefore the lift coefficient, expressed as following:

$$C_L = \frac{W}{\frac{1}{2}.\rho.V^2.S_{Ref}}$$

Has to remain constant. This allows concluding that the ratio of the aircraft weight to the air density has to remain constant. During the cruise the fuel is consumed thus the weight of the aircraft decreases and the density has to decrease accordingly. This can be achieved allowing the aircraft to climb. At the same time, with the decrement in air density the thrust will decrease and it can be assumed that the true air speed is constant.

In practice the aircraft are not allowed to climb during cruise by the air traffic control so airlines adopt a stepped climb procedure. For each segment at constant altitude, it is possible rewrite the range equation considering constant SFC airspeed and lift coefficient:

$$Range = -\frac{V.E}{sfc.g}\int_{m_1}^{m_2} \frac{1}{m}.dm = \frac{V.E}{sfc.g} in\left(\frac{m_1}{m_2}\right)$$

This equation of range is known as Brequet equation. A variant to solve the basic range equation is using a numerical integration dividing the mass variation in intervals:

$$Range = \sum_{i=1}^{n} \frac{V_i}{Q_i}.\left(m_i - m_{i-1}\right)$$

In HERMES a similar approach of the integrated range method has been implemented. However, instead of dividing the mass into intervals and then calculating fuel consumed and range, the cruise is split into intervals of time.

Weight variation during the time for an aircraft flying horizontally is given by the following equation:

$$\frac{dW}{dt} = -\frac{sfc.W}{E}$$

and integrating the previous equation respect to the time:

$$FuelConsumed = \int_{1}^{2} sfc\frac{W}{E}.dt\, b$$

onsidering n intervals the total fuel consumed can be worked out as follows:

$$FuelConsumed = \sum_{i=1}^{n} \frac{sfc.W.\left(t_i - t_{i-1}\right)}{E}$$

At each interval the range is equal to:

$$dR = V_{Cruise}.dt$$

The total range is the sum of the flight range for each interval. As pointed out above, in order to improve the overall efficiency of the aircraft the airliners allow performing what it is known as *step climb cruise*. Similarly, the user can subdivide the cruise into intervals and specify for each interval different flight Mach numbers and altitudes.

During the descent phase the drag of the aircraft is greater than the thrust produced by the engine leading the aircraft to glide. The descent starts at cruise Mach number and it reduced till the 250 knots at sea level. The calculation of the descent phase is similar to the climb calculation presented above (Figure 10). The user has to set up in the mission profile input file the different intervals of the descent phase. Using an iterative method the flight time, rate of descent and horizontal distance covered are assessed.

The aircraft model developed in this project is designated as LRACM (Long Range Aircraft Model). The LRACM model is based on the performance data of a typical twin-engined turbofan long range civil aircraft LRACPD (Long Range Aircraft Public Domain) available in

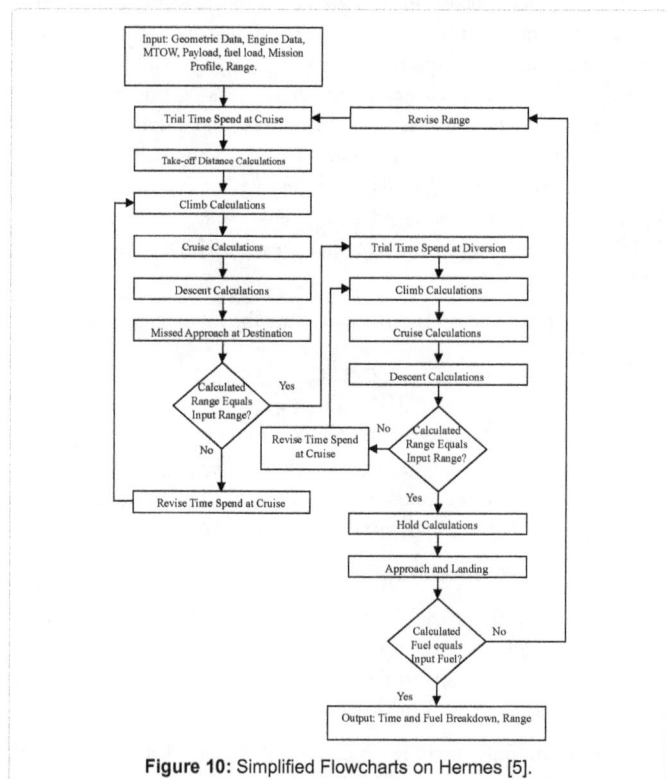

Figure 10: Simplified Flowcharts on Hermes [5].

PARAMETER	VALUE	UNIT
PAASENGERS	301	
ENGINES	Trent 895	
MAXIMUM TAXI WEIGHT	298,460	Kg
MAXIMUM TAKE-OFF WEIGHT	297,550	Kg
MAXIMUM PAYLOAD WEIGHT	59,430	Kg
MAXIMUM LANDING WEIGHT	213,180	Kg
MAXIMUM ZERO FUEL WEIGHT	199,580	Kg
OPERATING EMPTY WEIGHT	141,880	Kg
FUEL CAPACITY	171,170	Kg
Cruise Mach Number	0.84	

Table 3: Performance Summary of Lracpd.

	PARAMETERS	VALUE	UNIT
FUELSAGE	LENGTH	63	M
	DIAMETER	6	M
WING	AREA	428	M²
	ASPECT RATIO	9	
TAIL PLANE	AREA	101	M²
	ASPECT RATIO	5	
	TAPER RATIO	0.3	
FIN	AREA	53	M²
	ASPECT RATIO	5	
	TAPER RATIO	46	
ENGINE	DIAMETER	3	M
	LENGTH	7	M

Table 4: Aircraft and Engine Model Geometry.

public domain. As mentioned before, in order to configure the aircraft model it is required to setup an input file with several parameters, including, aircraft geometry, configuration, mission profile and weight breakdown.

Table 3 reports the main performance parameters regarding the LRACPD available from the public domain.

Regarding the geometry of the aircraft and the engine in some of the required information is listed. Some of the parameters are not available in the literature therefore they have been assumed.

The accuracy of the aircraft model has been verified against published data using the payload-range diagram. In this diagram the aircraft range is plotted against the payload (Table 4). There are usually three-baseline aircraft configurations, including:

- Maximum Payload range;
- Maximum economic range;
- Ferry range.

In the maximum payload range the aircraft take-off with both maximum take-off weight and maximum payload weight. Therefore, the amount of fuel is given by the following equation:

$$MTOW = OEW + PW_{Max} + FW$$
$$\rightarrow FW = MTOW - OEW - PW_{Max}$$
$$= 297,550 - 141,880 - 59,430$$
$$= 96,240 kg$$

In the maximum economic range, similarly to the previous case aircraft take-off with its maximum take-off weight, but this time with the maximum amount of fuel. Therefore the range will increase. The amount of carried payload is given by:

$$MTOW = OEW + FW_{Max} + PW$$
$$\rightarrow PW = MTOW - OEW - FW_{Max}$$
$$= 297,550 - 141,880 - 137,520$$
$$= 18,150 kg$$

Regarding the ferry range, the aircraft take-off with no payload and with the maximum amount of fuel. The ferry range is the maximum range of the aircraft. The take-off weight is given by the following equation:

$$TOW = OEW + PW_{Max} = 297,550 + 141,880 = 493,350 \text{ kg}$$

Considering that the initial amount of fuel was known for each mission using HERMES the flight range has been calculated and compared with published data.

Figure 11 shows the comparison between the payload range diagram of LRACPD and the aircraft model. Due to the lack of more information a step cruise from 10,000 to 11,000 meters was assumed.

In Table 5 shows the values of the payload and fuel weight are reported for each mission along with the difference between the range of the real aircraft and the model.

Emissions Prediction Model

The emission prediction model used in this work is the P3T3 empirical correlation model which has been integrated as part of Cranfield University HEPHASTUS emission prediction tool. This model estimates the level of emissions at altitude using a correlation with the emissions measured at ground level. This methodology is straightforward. Firstly, during the certification test of the engine the emission indices are measured. These indices are subsequently corrected to take into account the variation of altitude and flight speed. In order to do that, it is necessary to know the combustion parameters for the operating conditions at both ground level and altitude. These parameters are: burner inlet pressure and temperature, fuel and air ratio and fuel flow. In addition the model takes into account the variation of humidity from the sea level to altitude. The model is capable of predicting all the emissions and in this paper main focus given to the NOx emissions only. Detailed model layout shown is shown in Figure 12.

The engine tests results published by ICAO the level of emissions and other main parameters are measured for different engine operating conditions. These conditions are:

1. Take-off: full power (maximum level of thrust);

2. Climb out: 85% of take-off thrust;

3. Approach: 30% of Take-off thrust;

4. Idle: 7% of Take-off thrust.

$EINO_x$ measurements at ground level are plotted for different combustor inlet temperatures. Moreover, as explained above, in order to calculate the emissions at certain flight altitude and speed, the

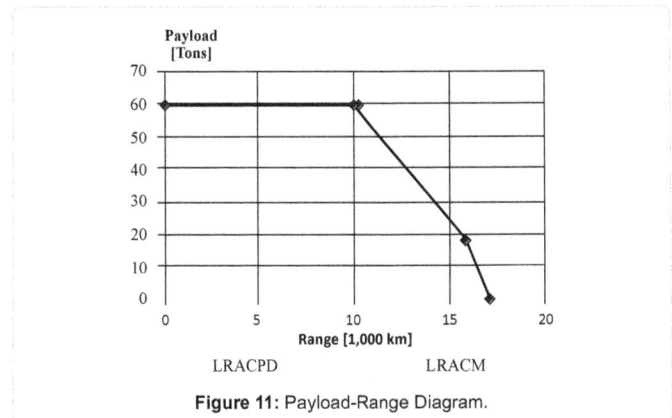

Figure 11: Payload-Range Diagram.

	Maxim Payload	Economy	Ferry	Unit
Payload Weight	59,430	18,150	0	Kg
Fuel Weight	96,240	137,520	137,520	Kg
LRACM	10,200	15,890	17,060	Km
LRACPD	10,010	15,870	17,190	Km
Difference	1.82	0.15	-0.74	%

Table 5: Aircraft Model Validation.

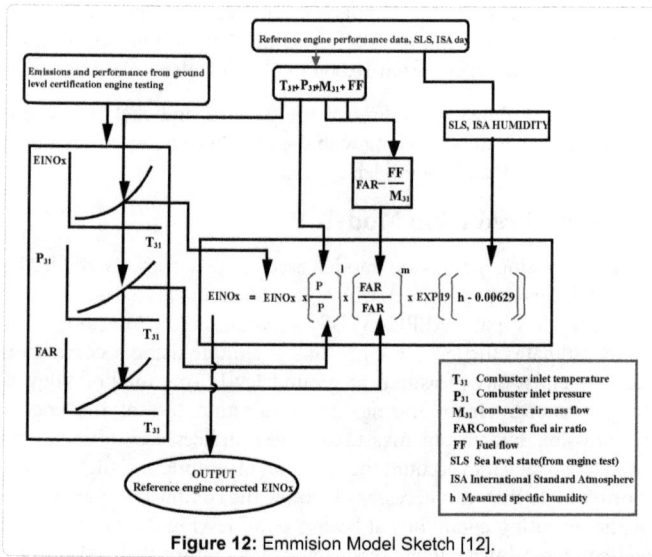

Figure 12: Emmision Model Sketch [12].

combustor inlet temperature, inlet pressure and air mass flow have to be known. Even if these values are not measured during the ICAO tests they can be assessed using gas turbine performance simulation software. At this point, similarly to $EINO_x$, burner inlet pressure and FAR are plotted for different burner inlet temperatures. Then, using the combustor inlet temperature at altitude it is possible to obtain the respective value of $EINO_x$ at ground level from the specific plot. This value of $EINO_x$ is then corrected for taking into account the differences in FAR and inlet combustor pressure between ground level and altitude. The values of exponent and establish the severity of $EINO_x$ correction. Finally, a correction for the humidity influence is also taken into account. Having calculated the value of $EINO_x$, the emitted NO_x in kilograms is given by:

$$NO_x = (FF.time).EINO_x$$

Where FF, is the fuel flow in [kg/s], and the time in seconds.

The variation of humidity change with the altitude relative to ISA sea level is taken into consideration. The correction increases with increasing altitude. If the measurement of $EINO_x$ at sea level has been done in a day with a high level of relative humidity, say 60%, the correction with the altitude will increase $EINO_x$ by around 12.5%. At typical cruise altitude the error by choosing different curves of relative humidity is small because the air is dry. ICAO suggests using 60% of relative humidity for calculations [9]. Engine manufactures during the years have gathered a large amount of data from engine testing, which have facilitated in defining the pressure coefficient to be set in the model. In the rig tests the combustor inlet conditions are varied independently in order to establish their relative effect on NO_x formation. The value of pressure exponent is commonly in the range between 0.3 and 0.5 in typical cruise condition [10]. This value varies as a function of the combustor type, operating conditions and measurement variability. An average value of 0.4 is normally used for all civil aircraft engines. Regarding the FAR, the data from the engine manufacturers shows that during the cruise the FAR is 10% richer than at ground level with constant combustor inlet temperature.

The main advantage of using the P3T3 model relative to other emissions models such as the physics based stirred reactor model is the low computational time required because it is based on empirical correlations. The required computational time is a key feature for a model that has to be used in aircraft multi-objectives trajectory

optimization study considering the large amount of calculations involved.

Emission Prediction Model Setup

The file used to setup the engine emission model requires information about engine emissions, the combustor inlet pressure and temperature, the fuel flow and the fuel/air ratio for the four operating conditions at ground level.

In the ICAO database it is possible to find only data regarding the emissions indices. Therefore, the values of combustor inlet temperature and pressure, fuel flow and fuel/air ratio have to be assessed using an engine simulation tool. TURBOMATCH has been used for this work. Table 6 indicates the relevant data available from the ICAO engine database:

As it is possible to notice from the Table 5 only the fuel flow and the emissions indices are available along with the power setting of each mission phase, a series of off-design simulations has been carried out using TURBOMATCH in order to find the other necessary performance data of the engine. In the OD section of the engine model input file the value TET for each phase of the mission will be taken to match the values of the fuel flow with ICAO database. In Table 7 compare the fuel flows of the engine model under different flight phases with the public domain data available in ICAO. Then, the values of pressure and temperature at the burner inlet and the fuel/air ratio have been set in the emission model input file.

Aircraft Trajectory Optimization

In this study the entire flight profile has been divided in to three main phases: climb, cruise and descent. Three parameters have been used to define the flight trajectory: aircraft speed (M, TAS and EAS), flight altitude and mission range. The mission range has been kept constant for the all optimization studies. Therefore the study has been mainly focused on the trajectory optimization between two-fixed destinations. The climb and cruise phases are simply defined using 18 points and the cruise Mach number.

Mode	Power Setting [% TOT]	Fuel Flow [Kg/S]	EIHC [G/Kg]	CO [G/Kg]	Nox [G/Kg]
Take-Off	100	4.03	0.02	0.27	47.79
Climb Out	75	3.19	0	0.19	34.29
Approach	30	1.05	0	0.54	11.39
Idle	7	0.33	0.89	14.71	5.11

Table 6: Icao Engine Emmisions Data Bank.

Mode	ICAO Fuel Flow [Kg/S]	CUHBR Fuel Flow [Kg/S]	Difference [%]
Idle	0.33	0.33	0.27
Approach	1.05	1.03	-1.63
Climb Out	3.19	3.10	-2.84
Take-Off	4.03	3.86	-4.17

Table 7: Emmision Model Setup–Sea Level Isa Condition.

Design Variable	Min Value	Max Value	Unit
Climb Altitude 1	1,000	2,400	M
Climb Altitude 2	2,700	4,400	M
Climb Altitude 3	4,800	6,400	M
Climb Altitude 4	7,000	8,400	M
Climb Altitude 5	9,000	11,000	M
Cruise Mach Number	0.75	0.85	

Table 8: Trajectory Designs Variables and Their Boundaries.

However, in the optimization process only six design variables have been considered in order to reduce the required computational time. These design variables are: five values of altitude and the cruise Mach number. The first four values of altitude are used to define the climb trajectory whilst the last altitude point defines the cruise altitude, which is constant for the entire cruise. The other points are computed by interpolation between two consecutive design variables maintaining constant increment in altitude.

For each design variable a boundary has been set to ensure that the resulting optimized trajectories were both feasible and with constant rising climb altitude. These boundaries can be considered as explicit constraints since they are directly applied to the design variables. Table 8 shows the limitation values for each design variable. A gap in altitude between two consecutive variables has been considered in order to guarantee a constant increment in altitude.

Speed (EAS) during climb was fixed with the aircraft performance input file to 250 knots for the first two climb segment and 320 knots for the three subsequent climb segments. Moreover, climb and descent phases are flown at fixed power setting. For both phases maximum power setting is selected, i.e. maximum thrust at maximum TET permitted in the given flight phase. According to Laskaridis et al. [8] a common method to climb is at constant EAS.

As it is shown in Table 8 the maximum allowable altitude has been limited to 11,000 meters. This limitation is related to the fact that, as altitude increases, the Reynolds number falls because the ratio density to absolute viscosity drops. At certain altitude, the Reynolds number will fall below a critical value of 105   and the flows about the blades of compressors and turbines will start to separate. This situation leads to two main consequences: (a) The flow is not deflected as much as before thus the compressors and turbines power drops leading a reduction of thrust; (b) SFC increases because of the increment in losses associated with the turbulent wakes that, in turn, cause a reduction of compressors and turbines efficiencies.

According to Pilidis [6] at an altitude of 11,000 meters, for a large turbofan, the effect due to the drop of Reynolds number lead a reduction about 2 % of thrust. In the engine performance model adopted in this work the effect of Reynolds number is not taken into account. Therefore a limitation of 11,000 meters has been considered in order to obtain more realistic results.

The descent trajectory starts at cruise altitude and speed and it has been divided into 10 segments. For each segment the flight speed has been chosen as it is stated in Table 9.

Multi Disciplinary Optimization of Aircraft Trajectories

The overall optimization running sequence is shown in Figure 13. At

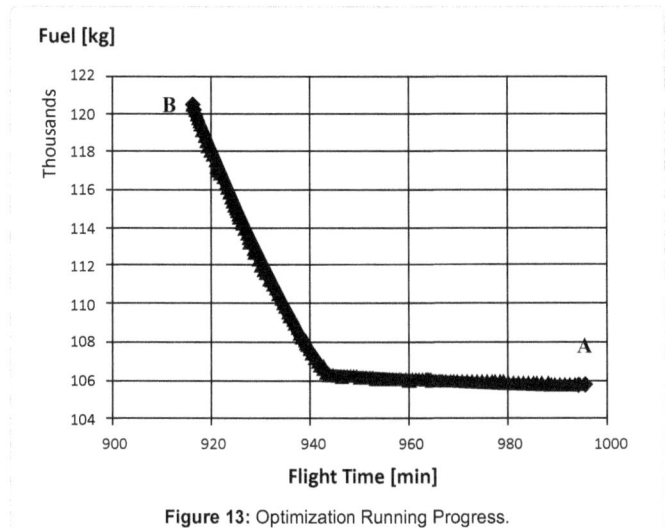

Figure 13: Optimization Running Progress.

the start of the optimization process the optimizer (GATAC) generates the first set of design variables. In this work the design variables are five altitude points and the cruise Mach number. The first four altitude points describe the climb trajectory whilst the last point corresponds to the altitude of the cruise.

The design variables are written in the input file of the aircraft model along with all the other required parameters. The following step consists in the execution of the aircraft model. As already explained, the aircraft model also requires the specifications about the aircraft and engine performance data.

The execution of the aircraft model generates two output files. The first file regards the aircraft performance. The results include mission duration, fuel consumption and distance covered for the whole mission and for each flight segment. The second file contains the performance of the engine for each phase of the mission. SFC, thrust, shaft speed as well as engine temperatures and pressures can be found in this file.

Information regarding the mission profile, fuel consumed and engine performances during the mission are used to generate the input file for the emission model. In addition, the emission model required other data regarding combustion specification and engine emission. These data are read from a specific input file. The emission prediction model computes the values of NO_x, CO, UHC emitted during the mission in kilograms. Then, the output data are read by GATAC and, based on these values; a new generation of design variables are created. This process is repeated till the integrated optimisation criteria are satisfied.

Assumptions and Considerations

A number of assumptions have been required in the present study, including:

1. A procedure is implemented in the aircraft model to ensure that each point, which defines the climb trajectory, has a higher altitude of the previous one.

2. The cruise phase is flown at constant altitude and constant flight Mach number;

3. The climb phase is flown at constant power setting. This means that the profile generated for every range and altitude selected is nearly the same;

Power Setting During Descent	TAS	Unit
Idle	223	Knots
Idle	221	Knots
Idle	202	Knots
Idle	195	Knots
Idle	183	Knots
Idle	164	Knots
Idle	150	Knots
Approach	140	Knots
Approach	135	Knots
Approach	135	Knots

Table 9: Descent Phase Power Setting and Tas.

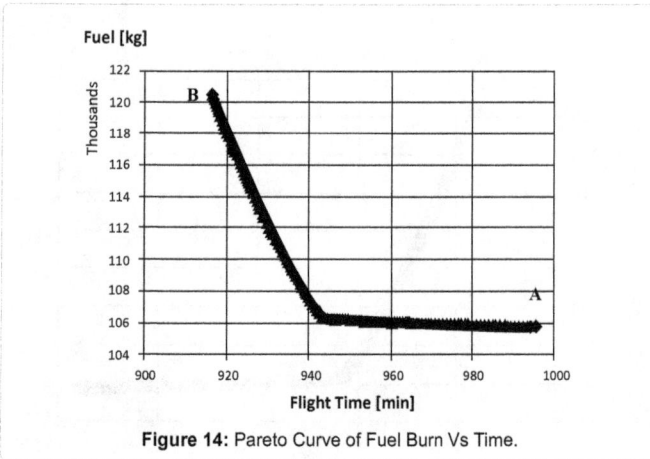

Figure 14: Pareto Curve of Fuel Burn Vs Time.

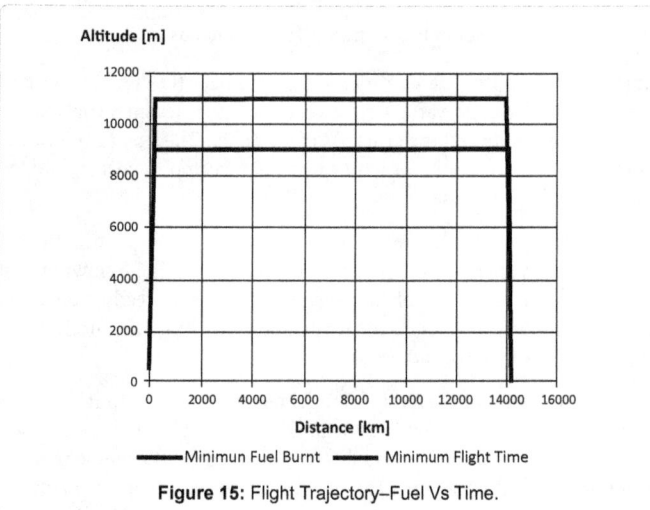

Figure 15: Flight Trajectory–Fuel Vs Time.

4. The descent phase not taken for the optimization process.

5. The user cannot choose arbitrary descent profile and it is automatically calculated by the HERMES code by interpolation between the cruise altitude and the landing altitude. Therefore, the descent profile is a function of the cruise altitude only;

6. The continuity of the flight speed has not been guaranteed between the cruise and the descent phase. This could be cause variation of flight speed between cruise and descent phase;

7. For simplicity, taxi phases, take-off and landing have not been included in the mission profile and hence in the overall calculation and optimization. The consequence is that the total flight range considered by the optimizer comprises only climb, cruise and descent. The climb phase starts at 475 meters of altitude whilst the descent phase terminates at sea level altitude;

8. Although in HERMES it is possible to take into account the flight diversion mission, it has not been considered in this work;

9. A deviation of +3 degrees has been assumed respect ISA conditions for the entire mission.

Following section presents the different optimization studies carried out in the GATAC framework. In each the aircraft flight trajectory which has been optimized keeping the aircraft and engine configurations unchanged including the payload equivalent to 301

passengers. As stated in the above section the design variables utilized are associated with flight altitude during the climb and cruise phases and aircraft flight Mach number during cruise. The trade-offs of conflicting objectives such as flight time, fuel burnt and NO_x emitted have been considered under each case study (Table 10).

Case 1: fuel burnt vs flight time

This optimization study has been carried out for two conflicting objectives: minimum fuel burnt and minimum flight time. No other constrains were applied. Figure 14 illustrates the Pareto front obtained with the GATAC NSGAMO optimizer [11]. The mission range was set equal to 14,195 kilometers.

The Pareto curves were generated from 100 and 300 generations with series of points, where each point represents a trajectory, with its combination of design variables (altitudes and Mach number). For each point of the Pareto curve it is impossible to minimize further any objective from points given in the Pareto front. In Figure 14 the point A corresponds to the trajectory of minimum fuel consumed and the point B refers to the minimum flight time trajectories. The remaining points are other intermediate solutions. In the case of a trade-off between two conflicting objectives there is no a unique optimum solution or trajectory but the solution consists of a series of optimum trajectories with their unique combinations of time and fuel.

The two trajectories lead to important differences in terms of flight time, fuel burnt and emissions. This was expected considering the trade-off between minimum fuel and minimum flight time. The two trajectories differ of 8.63% (79 min) in flight time and about 12.27% (14,790 kg) in fuel burnt. Less fuel burnt means less emission of NO_x and CO_2. While the emission of CO_2 is directly related with fuel burnt, the relation between NO_x and fuel is different, as it will be described in the following sections. However, the fuel-optimized trajectory leads to higher emissions during the descent phase than the time-optimized trajectory. Considering that the aircraft is flying at higher altitude in the

Parameter	Minimum Fuel	Minimum Time	Unit
Altitude 1	2,200	2,200	M
Altitude 2	2,700	2,700	M
Altitude 3	4,802	4,800	M
Altitude 4	7,000	8,220	M
Altitude 5	11,000	9,000	M
Mach Number	0.804	0.85	

Table 10: Variables Corresponding To Optimum Fuel and Time Trajectories.

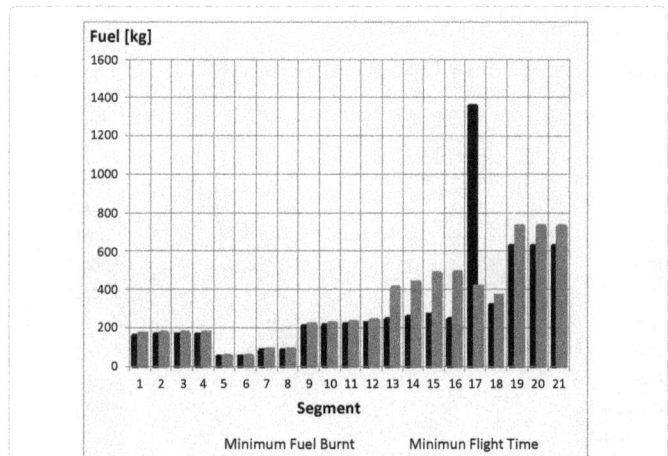

Figure 16: Fuel Consumptions–Fuel Vs Time.

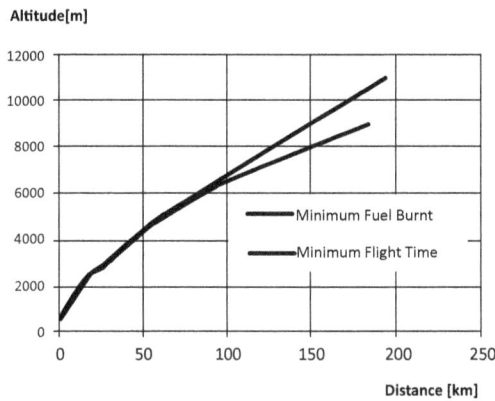

Figure 17: Flight Trajectories during Climb-Fuel Vs Time.

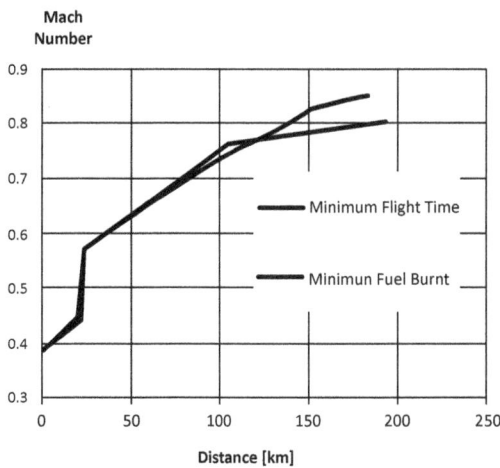

Figure 18: Mach Number During Climb- Fuel Vs Time.

fuel-optimized trajectory the descent phase will be longer. Moreover, differently from the climb phase, the descent phase is fixed and it is not part of the optimization process. The two trajectories are shown in Figure 15 based on the selected design variables in Table 11.

In order to minimize the fuel burnt the optimizer suggests a solution where the aircraft flies the cruise phase at highest possible altitude 11,000 meters and Mach number equal to 0.804.

Generally, decreasing the speed and increasing the altitude lead to a decrement in drag and therefore in required thrust. This lower thrust requirement, in turn, means lower engine power setting along with lower fuel burnt. However, other important aspects have to be considered. A reduction in flight speed means more flight time with a negative effect on the fuel burnt. Therefore the optimizer has to assess the best compromise. In this case the resulting cruise Mach number for minimum fuel burnt trajectory is 0.804, which is higher than the minimum allowed value that was set to 0.75.

Moreover, in order to reach as quickly as possible the highest allowable altitude an increment of engine thrust and power setting is also required. It is interesting to notice in Figure 16 how the optimizer for the fuel-optimized trajectory proposes segment 17 affording a much greater fuel consumption respect to the time-optimized trajectory. This is done in order to gain height as quickly as possible, which leads to lower fuel consumption for the following segments. In Figure 16 it is worth noting that the climb phase comprises the first 18 segments and

the following segments represent the cruise phase.

The minimum flight time means maximum TAS. Therefore in order to minimize the time during the cruise the aircraft flies at the highest Mach number permitted, which is 0.85. The optimizer suggests also flying at lowest altitude. This is correct since the speed of sound is highest at sea level along with TAS. Moreover, as already explained, the thrust increases with the decreasing in altitude because of the air density. The altitude profile of the two climb trajectories is shown in Figure 17

Again, it is possible to notice that the climb gradient of the fuel-optimized trajectory is greater than the time-optimized one. The aircraft has to accelerate as faster as possible in order to gain height. The acceleration to gain height for the fuel-optimized trajectory is well shown in Figure 18. The step in the flight Mach number is due to the passage from 250 knots to 320 knots of EAS during the climb phase as described above.

Case 2: minimum flight time vs minimum nox emitted

The next study that has been carried out regards the optimization of two conflicting objectives: minimum flight time and minimum

Parameter	Unit	Min Nox	Min Time	Diff [%]
Flight Time	Min	1,065	916	13.96
Fuel Burnt	Tons	106.74	120.53	-12.91
Total Nox	Tons	1.46	1.97	-35.07
Nox Climb	Tons	0.14	0.16	-9.50
Nox Cruise	Tons	1.31	1.81	-38.11
Nox Descent	Tons	6.6	5.7	13.89
Total CO2	Tons	337.44	380.69	-12.82
CO2 Climb	Tons	16.94	18.33	-8.16
CO2 Cruise	Tons	318.46	360.68	-13.26
CO2 Descent	Tons	2.03	1.69	17.15

Table 11: Results Comparison-Time Vs Nox.

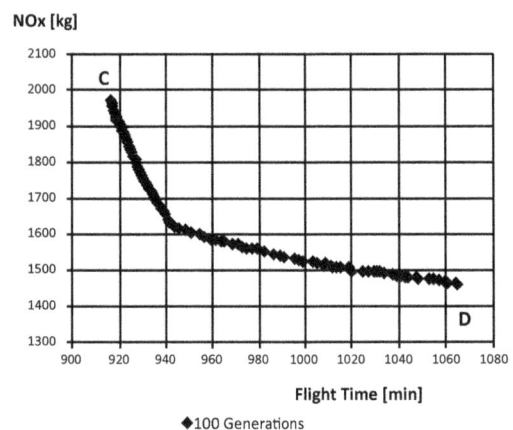

Figure 19: Pareto Curve-Time Vs Nox.

Parameter	Minimum Nox	Minimum Time	Unit
Altitude 1	2,200	2,200	M
Altitude 2	2,700	2,700	M
Altitude 3	5,105	5,167	M
Altitude 4	7,000	8,400	M
Altitude 5	11,000	9,000	M
Mach Number	0.75	0.85	

Table 12: Variables Corresponding To Optimum Nox and Time Trajectories.

Principles of Aircraft Engineering

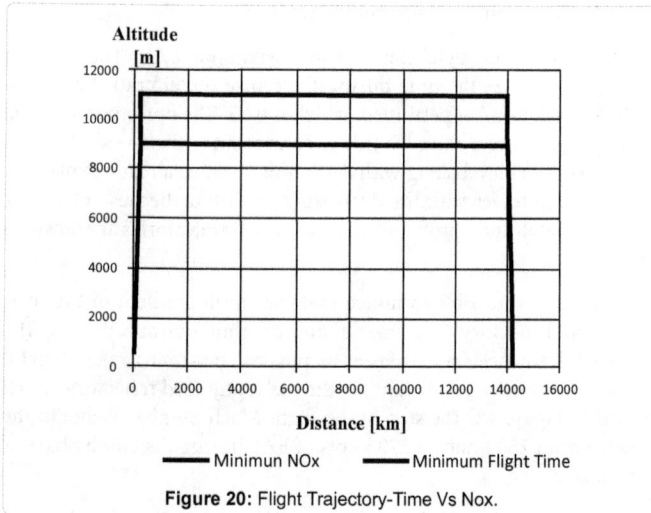

Figure 20: Flight Trajectory-Time Vs Nox.

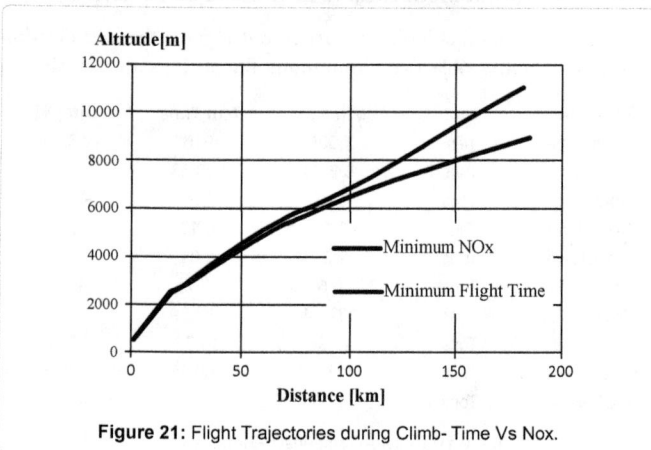

Figure 21: Flight Trajectories during Climb- Time Vs Nox.

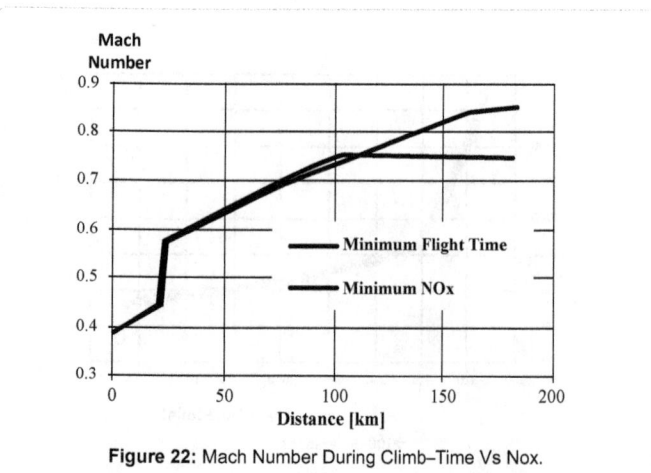

Figure 22: Mach Number During Climb–Time Vs Nox.

NO_x emitted. It is important bearing in mind that optimizing for fuel consumption is different from optimizing for NO_x emissions. Similarly to the previous case, the trade-off of the two objectives leads to the characteristic of the Pareto curve shown in Figure 19.

The point C and D represent the minimum flight time and minimum NOx trajectories respectively. A comparison of the results considering Time and NO_x optimized trajectories is shown in Table 11.

The minimum flight time trajectory has exactly the same values

of flight time, fuel burnt and emissions already calculated in Case1. Regarding the NO_x-optimized trajectory, the total NO_x emitted is about 1,457 kilograms and the reduction is up to 35.07% compared with the time-optimized trajectory. In the fuel optimized trajectory the total NO_x emitted is 1,519 kilograms. The flight time, for the NO_x-optimized trajectory is 1,065 minutes while for the fuel-optimized trajectory is 995 minutes. Considering that the two cruise altitudes are the same for minimum NO_x and fuel, in order to reduce the NO_x emission the optimizer suggests flying at a lower speed than the fuel-optimized trajectory with an increment in the flight time. Figure 20 shows the flight path for the two optimized trajectories.

Considering that the time-optimized trajectory is exactly the same trajectory described in case one further consideration is required at this point. Regarding the NO_x-optimized trajectory, it is interesting to notice that the optimizer suggests a solution where the aircraft flies at lowest and highest allowable Mach number and cruise altitude respectively. Highest altitude and lowest speed lead to minimize the engine thrust requirement. In turn, low thrust requirement means low TET, which results in low NO_x emission.

In Figure 21 and Figure 22 shows the altitudes Mach number variation of the two climb trajectories against the flight distance.

It is evident how, similarly to the fuel-optimized trajectory, in order to minimize the NO_x emission the optimizer suggests to reach the highest admissible altitude as faster as possible.

It is important to consider the fact that large amount of NO_x is produced at TET values in the region of 1700 ~ 1800 K and it increases exponentially with TET. In this respect, the trajectory optimized for flight time produced a large amount of NO_x. One of the main reasons for that, besides the fuel burnt, is the high value of TET required for the thrust. Considering only the cruise phase, the time-optimized trajectory emitted about 409 kilograms of more NO_x respect to the NO_x-optimized trajectory.

Conclusion

The multi-disciplinary optimization framework has been implemented in order to investigate the potential of greener trajectories as future possible solutions for the reduction of aircraft environmental impact. The optimization framework comprises three different simulation models: engine model, aircraft model and emissions prediction model and a GA based NSGAMO optimizer.

The multi-objectives optimization studies have been carried out in GATAC frame work focusing on minimization of conflicting objectives, such as fuel burnt versus flight time and NO_x versus flight time for long range trajectories.

In the first optimization study a long-range mission of 14,195 kilometers has been considered. The results show a difference of 8.63% (79 min) in flight time and about 12.27% (14,790 kg) in fuel burnt between the fuel-optimized and time-optimized trajectories. In order to minimize the flight time the optimizer suggests a solution where the aircraft has to fly at minimum allowable altitude and maximum flight Mach number. On the other hand, the flight trajectory that minimized the fuel burnt is one in which the aircraft has to fly at maximum permissible altitude. The cruise Mach number that minimizes the fuel consumed does not correspond to the minimum allowed Mach number but it is a result of a compromise between fuel flow (power setting) and flight time.

Regarding the minimization of NO_x emissions the results show that the aircraft has to fly at the highest allowable altitude and optimization is the minimization of NO_x emissions the optimizer tends to reduce the power setting (TET) and all other aspects become secondary.

This preliminary application clearly shows the capabilities of GATAC framework as an optimization tool in obtaining optimum solutions at multidisciplinary level. However, more research efforts needed in enhancing the spectrum of the capability of the tool, for an example, use different optimizers with different objective functions; integration of different models such as

Noise, Lifing, & engine degradation etc. introducing practical ATM constrains and changing the current 2-D optimization to 3-D and 4-D.

In order to achieve the ACARE 2020 targets by reducing the impact of aviation on the environment in the short term, introduction of changes in aircraft operational procedures and rules under optimum trajectories is a promising solution. In conclusion, the implemented framework proved to be a valuable tool for identifying the characteristic features of aircraft trajectories with a minimum environmental impact and other operational consequences.

Acknowledgements

The research project has received funding from the European Union's Seventh Framework Program (FP7/2007-2013) for the Clean Sky Joint Technology Initiative under grant agreement n⁰ CJSU-GAMSGO-2008-001. The authors are also grateful to EADS Innovation Works, Thales Avionics, Airbus France and the GSAF (Green Systems for Aircraft Foundation) members of the Technical University Delft (Faculty of Aeronautical Engineering), National Aerospace Laboratory NLR, and particularly Matthew Xuereb, Matthew Sammut and Kenneth Chircop of The University of Malta.

References

1. Clean Sky JTI (Joint Technology Initiative).

2. Chircop K, Xuereb M, Zammit-Mangion D, Cachia E (2010) A Generic Framework for Multi-Parameter Optimization of Flight Trajectories, 27th International Congress of the Aeronautical Sciences (ICAS), France 2010.

3. Gu W, Navaratne R, Quaglia D, Yu Y, Chircop K, et al. (2012) Towards the Development of a Multi-disciplinary Trajectory optimization Tool-GATAC, GT2012-69862, Proceedings of ASME Turbo Expo, Copenhagen, Denmark.

4. Pervier H, Nalianda D, Espi R, Sethi V, Pilidis P, et al. (2011) Application of Genetic Algorithm for Preliminary Trajectory optimization. SAE Int J Aerospace 4: 973-987.

5. Pachidis V (2004) Gas Turbine Simulation - PYTHIA workshop guide, Part 1 and 11. ASME/IGTI proceedings aero engine life management conference, London.

6. Pilidis P, Palmer JR (2010) Gas Turbine Theory and Performance, Course Notes, School of Engineering, Cranfield, UK.

7. Hermes V5 and TmatchCalls V3 User Manual (2009) Department of Power and Propulsion, School of Engineering, Cranfield University UK.

8. Laskaridis P, Pilidis P, Kotsiopoulos P (2005) An integrated engine-aircraft performance platform for assessing new technologies in aeronautics. In proceedings of the 17th international symposium on air breathing engines (ISABE), Munich.

9. Jenkinson LR, Simpkin P, Rhodes D (1999) Civil jet aircraft design (Arnold London).

10. ICAO Engine Emissions, available at: http://www.caa.co.uk/default.aspx?catid=702.

11. Deb K, Agrawal S, Pratap A, Meyarivan T (2002) A Fast and Elitist Multi-objective Genetic Algorithm: NSGA-II. IEEE Transactions on Evolutionary Computation.

17

Computation Study of Oblique Shock Wave-Vortex Iinteraction in Supersonic External Flows

Ahmad Sedaghat* and Mohammad Amin Aghahosaini

Department of Mechanical Engineering, Isfahan University of Technology, Isfahan, 84156-83111, I. R. of Iran

Abstract

Interaction of a supersonic vortex with an oblique shock wave is an important phenomenon because of strong acoustic levels and its adverse effect in aerodynamic performance in high speed flows. In this paper, a mathematical model for the vortex is introduced which predicts experimentally generated vortices accurately. Then, the shock wave-vortex interaction was numerically studied using a finite-volume TVD scheme for solving Euler equations. From the results, two interaction regimes are recognized, namely weak and strong regimes. It is shown that vortex breakdown is possible in the case of strong interaction which can significantly alter aerodynamic characteristics of flyers or acoustic field. Results for two case studies are presented here with the freestream Mach numbers of 2.5 and 2.9. The effects of different parameters in vortex breakdown are investigated. It is found that the smaller vortex core sizes with higher intensity may result in stronger interaction at higher Mach numbers leading to vortex breakdown.

Keywords: Supersonic flow; Burgers vortex; Shock wave; Reflected shock; Vortex breakdown

Introduction

Shock wave-vortex interaction that is generated in external flows and internal flows (inlets, nozzles, and combustors) is a complex and often unsteady phenomenon. For analyzing external flow around supersonic aircraft with wings of large sweep and small aspect ratio this classical problem in theoretical gas dynamics is very important. Generation of a vortex sheet near the leading edge of the wing and its subsequent rolling into two isolated vortices is the important feature of such flows. These vortices interact with shock waves formed on the aircraft surface elements. Two isolated vortices are usually formed in the flow around a delta wing at subsonic speeds at some incidence angles [1,2]. These vortex structures can enter the engine inlet of a supersonic aircraft. The flow-rate, drag, and other parameters of this propulsion element significantly depend on the interaction mode of the vortex and the shock wave that always formed ahead of a supersonic inlet engine. Such situations can lead to problems for a supersonic aircraft. The vortex can interacts with the aircraft surface in supersonic external flow, which significantly alters the lifting and moment characteristics of wings and other aircraft elements. Loss of stability and controllability of the aircraft is the result of changes in the force characteristics.

One of the usages of interaction of a vortex with a shock wave can also arise in the study of fuel and oxidizer mixing in combustion chambers of hypersonic flying vehicles with air-breathing engines [3]. In such vehicles, there is supersonic flow regime in the combustor causing complications in fuel and oxidizer mixing.

The experimental and numerical works on shock-vortex interactions are extensively presented by a number of researchers. Gnani et al. [4] have studied shock diffraction around sharp and curved splitters at Mach speeds of 1.31 and 1.59 using Schlieren photography. Their study suggests that the flow field evolves more rapidly with stronger structures for the higher incoming Mach number. Quinn et al. [5] have made some measurements of global pressure filed generated by shock wave diffraction using paints for their CFD validations at two Mach numbers of 1.28 and 1.55. They concluded that the strong interaction at Mach number 1.55 observed numerically may not supported by their experimental observations. Sun and Takayama [6] have made a numerical simulation of laminar flows of vortical

structures in shock diffraction and explained factors responsible for the deviation of laminar solutions from experimental results. Chatterjee [7] has numerically studied shock wave deformation in shock-vortex interaction. He has introduced a simple model to explain shock structure formation and its dependence on the strengths of the interacting vortex and shock wave. Ellzey et al. [8] have performed a computational study based on a High Resolution Shock Capturing scheme of the interaction of a planar shock wave with a cylindrical vortex. They observed a severe reorganization of the flow field and acoustic levels in the downstream region reflected mainly due to the strength of shock wave. Abate and Shyy [9] have studied the dynamic of a normal shock wave within a tube leading to vortical flows. Their calculations identified a multiple physical mechanisms included shock–strain rate interaction, baroclinic effect, vorticity generation, and different aspects of viscous dissipation as observed experimentally. Zare-Behtash and Kontis [10] have conducted experimental study to examine the interaction of shocks and vortices with a two-dimensional ejector. The incident shock Mach numbers of 1.34, 1.54, and 1.66, were studied. They concluded that the induced flow is unsteady and dependent on the degree of compressibility of the initial shock wave.

The aim of present study is to model high Mach number shock-vortex interactions and to assess robustness of the developed finite-volume TVD scheme for solving compressible flow equations for such complex problems. The other goal is to assess the levels of acoustic pressure at high supersonic Mach numbers of 2.5 and 2.9.

The action of a strong external disturbance on the vortex propagating in the flow results in the so-called vortex explosion

---

*Corresponding author: Ahmad Sedaghat, Department of Mechanical Engineering, Ishafan University of Technology, Ishafan, Iran
E-mail: sedaghat@cc.iut.ac.ir

or vortex breakdown, as was shown in the experimental works of [11,12]. This phenomenon was observed both in incompressible flows containing vortices and in compressible subsonic and supersonic flows. The meaning of term "vortex explosion" is formation of a point (or a surface) of total stagnation of the flow in the region of interaction of the vortex and the strong disturbance also the formation of a reverse flow region near the vortex centerline. The problem of vortex explosion has not been ultimately solved, even in the case of an incompressible fluid [11]. Various experiments for determining vortex-breakdown conditions were done for an incompressible fluid. Zatoloka et al. [13] were obtained important results in the experiments for supersonic velocities. They assumed that the mechanisms of vortex explosion and boundary-layer separation are similar. This idea was developed by Glotov [14]. Origination of unsteady oscillatory regimes in the region of interaction of the vortex and the shock wave was observed in [14,15]. Quantitative experimental results were obtained in [12,16]. Vortex-core radius, intensity of vortex and free-stream Mach number was used as parameters determining the interaction mode, and data on the structure of interaction regimes were obtained. In addition to the information given above, investigations [1] show formation of supersonic vortices on the back side of a delta wing and in the wake behind a complex-shaped body. Supersonic circular components of velocity and a wake-type vortex were studied experimentally [12,16].

If the vortex intersects the shock wave, two interactions types are possible: 1) interaction of the vortex with the shock wave perpendicular to the vortex axis; 2) interaction of the vortex with the shock wave inclined to the vortex axis. The first type of interaction was experimentally studied in [12,15,16], and the second type was considered in [14,17]. It was shown in these works that, in the first case of interaction, the vortex break down, and a reverse flow region with flow unsteadiness appears. There are a few numerical works in high supersonic where the interaction of the vortex and the shock wave is very strong causing numerical instabilities. Some vortex-shock wave interaction regimes with vortex explosion were observed in some works [18,19], whereas no vortex breakdown was obtained in other works [20]. Therefore, it is necessary to develop a mathematical model for predicting experimentally generated vortices in a complex (often unsteady) flow types such as interaction of a vortex with an oblique shock wave. There is little works to study on interaction of a vortex with an oblique shock wave and its reflection. Zhang et al. [21] has studied an oblique shock wave interaction with two vortices at sonic to slightly supersonic flows. Mahesh [22] has also reported an analytical model on the onset of vortex breakdown for several cases.

The objective of the present work is numerical simulation of various types of interaction of a vortex with an oblique shock wave (and its reflection from surface) by means of unsteady two-dimensional Euler equations. A mathematical model of the vortex is developed. Calculation results of interaction of the vortex with oblique shock waves are presented.

Formulation of the Problem

A supersonic perfect-gas flow is considered with a vortex entering to the flow at some upstream position. The velocities in the vortex core and in the cocurrent flow are assumed to be supersonic. An oblique shock wave is inclined at a certain angle to the path of the vortex. This shock hits a flat surface representing a flyer surface at supersonic flows and reflects so as to generate a weaker reflected shock wave. To simplify problem formulation, we assume that the shock-wave generator is outside the computational domain (Figure 1). Direction of the vortex is parallel to the direction of horizontal inflow. The intensity of the

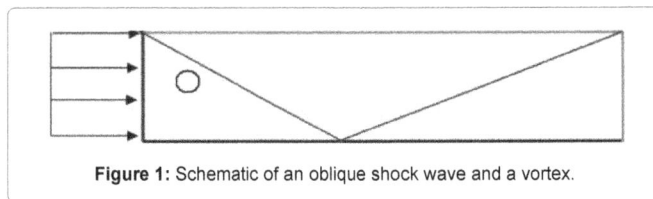

Figure 1: Schematic of an oblique shock wave and a vortex.

vortex is characterized by vortex circulation Γ_0 and the vortex-core radius R_C are the vortex governing parameters. The values of Γ_0 and R_C are prescribed to the flow and then left to interact with the flow and the oblique shock waves. In determining properties on the shock wave, it is assumed that the angle of the wedge generating the shock wave is prescribed, and thus, the angle of inclination of the shock wave generated is known. The flow properties at the upper boundary are specified from the property relations for oblique shock wave. A reflective wall boundary condition and an outflow boundary condition implemented [23].

Glotov [14] has shown that the beginning of vortex-core breakdown on the shock wave is independent of viscosity and is mainly determined by the relative component of velocity in the vortex core and by the angle of inclination of the shock wave. Therefore, the inviscid ideal-gas flow is assumed and the unsteady Euler equations are solved numerically in this work. The effects of viscosity is little on the interaction zone and the main features of the flow structure can be obtained by implementing a more proper vortex model which is introduced in this study.

Vortex model

It is necessary to construct a mathematical model of the vortex to investigate interaction of the vortex and shock waves. It is desirable to have a vortex insulated from the external cocurrent flow (at least in the initial step). The structure of a vortex formed behind a diamond-shaped body exposed to a supersonic flow was experimentally studied in [15,16]. The results showed that the structure of the vortex being formed is close to the structure of the Burgers vortex. In the present work, the Burgers vortex model is used that satisfy experimental dependences of the velocity in the vortex center on the vortex radius and flow velocity at infinity [12]. The distribution of quantities across the vortex is similar to the distribution of properties in the classical Burgers vortex; a linear distribution of velocity is set in the vortex core and velocity decreases exponentially with distance from the vortex core. The tangential velocity V_θ in the Burgers vortex is written as

$$V_\theta = \frac{\Gamma_0}{r}(1 - \exp[-(r/R_C)^2]), \qquad (1)$$

where Γ_0 is the vortex circulation and R_C is the vortex-core radius. The velocity components u and v in Cartesian coordinate system (x,y) may be expressed as

$$u = V_\theta \sin\theta, \quad v = -V_\theta \cos\theta. \qquad (2)$$

The pressure distribution in the vortex is calculated from the conservation of momentum in radial direction expressed by

$$\frac{dp}{dr} = \frac{\rho V_\theta^2}{r}, \qquad (3)$$

where the radial component of velocity is assumed to be zero. In calculating density in the vortex, it was assumed that the total temperatures in the vortex and in the cocurrent flow are identical. This assumption is confirmed by the experimental data of [12], which show that the ratio of total temperatures in the vortex and in the cocurrent flow is within 0.95-1.05. The differential equation (3) for the pressure across the vortex is then simplified using equation (1) and assuming $H_0 = H_{0\infty} = constant$, i.e. the total enthalpy to be constant, leads to [23]

$$\ln p = \Gamma_0 \left[\frac{-1}{2r^2} + \frac{2Exp\left(-r^2/R_C^2\right) - Exp\left(-2r^2/R_C^2\right)}{2r^2} + \frac{Ei\left(1, 2r^2/R_C^2\right) - Ei\left(1, -r^2/R_C^2\right)}{R_C^2} \right] \quad (4)$$

where $Ei(a, z)$ is the Exponential integral of the arguments inside and it can be expressed by

$$Ei\left(a, z\right) = z^{(a-1)}\Gamma\left(1-a, z\right) \quad (5)$$

where Γ is the Gamma function.

Figure 2 shows the distributions of pressure and tangential velocity along the radius of the vortex modeled by the differential equation derived above. The calculations were performed with the following parameters: $M_\infty = 2.5 - 2.9$, $\Gamma_0 = 2.5 - 12.5$, $R_C = 0.05 - 0.15$, and ratio of specific heats $\gamma = 1.4$.

In Figure 2, the pressure is normalized to γM_∞^2, the velocity V_θ to the velocity of sound in the flow at infinity, and r to the size of the computational domain along the y axis (equal to 1.0 m). It follows from Figure 2 that the tangential component of velocity increases with increasing radius, reaches a maximum at the vortex-core boundary, and then exponentially decreases. The static pressure in the vortex core changes significantly weaker and is approximately 10% lower than the pressure in the ambient flow.

Table 1 summarizes and compares experimentally generated vortices with the present analytical results with the maximum error

Property	Experiment [12]	Present	Error (%)
R_C (mm)	5.5	5.5	0.0
M_∞	1.64	1.643	0.183
$M_{\theta max}$	0.58	0.62	6.9
$(p/p_\infty)/\gamma M_\infty^2$	0.41	0.411	0.24

Table 1: The characteristics of experimental vortex compared with present analytical solution.

bound of 6.9% for the maximum angular Mach number.

The governing euler equations

The equations of fluid flow without viscous forces, body forces, heat conduction or energy sources are referred to as the Euler equations. The Euler equations in two-dimensional Cartesian coordinates and in conservation-law form can be written as

$$\frac{\partial U}{\partial t} + \frac{\partial F(U)}{\partial x} + \frac{\partial G(U)}{\partial y} = 0 \quad (6)$$

where

$$U = \begin{bmatrix} \rho \\ \rho u \\ \rho v \\ e \end{bmatrix}, \quad F = \begin{bmatrix} \rho u \\ p + \rho u^2 \\ \rho uv \\ (e+p)u \end{bmatrix}, \quad G = \begin{bmatrix} \rho v \\ \rho uv \\ p + \rho v^2 \\ (e+p)v \end{bmatrix} \quad (7)$$

The primitive variables are the density ρ, the velocity components u and v, and the pressure p. The total energy per unit volume e, is related to p by,

$$p = (\gamma - 1)\left[e - \frac{1}{2}\rho(u^2 + v^2) \right] \quad (8)$$

where γ is the ratio of specific heats ($\gamma = 1.4$ for air).

Acoustic field

For spherical sonic waves, the sound intensity is given by:

$$I = \frac{1}{r^2} \cdot \frac{(p^2)_{av}}{\rho \cdot c} \quad (9)$$

where ρ is the density of the medium, c is the sound speed, and $(p^2)_{av}$ is the square of mean sonic pressure for r=1. Sound intensity level (SIL) is generally used to express acoustic filed in decibel as

$$I = 10 \cdot \log_{10}(I/I_{ref}) \quad \text{dB(SIL)} \quad (10)$$

where $I_{ref} = 10^{-12}$ W/m^2 is a reference value.

Numerical TVD algorithm

A total variation diminishing (TVD) method was introduced

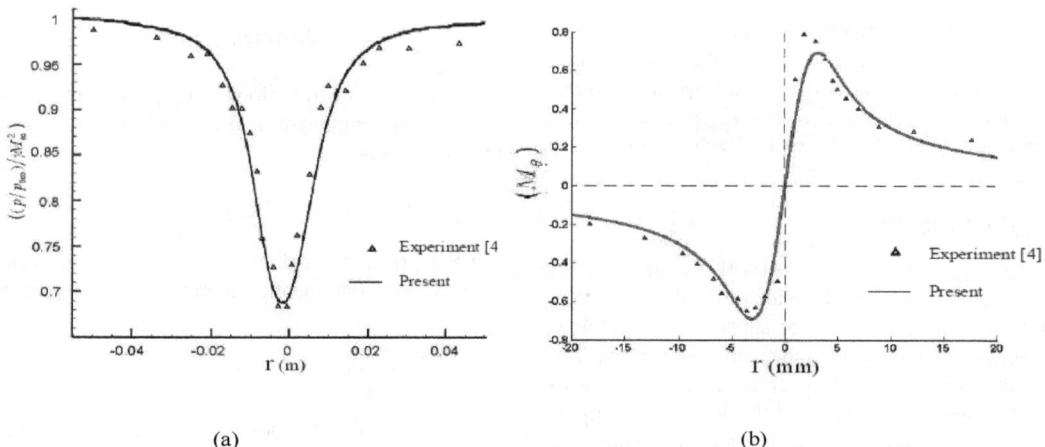

Figure 2: Distributions of (a) pressure and (b) tangential component of velocity across the vortex.

by Harten [24,25] and later generalized by Yee [26-28] for solving compressible fluid flows. This is a class of finite-difference high resolution shock capturing technique for discretising and solving the governing fluid flow equations in conservation law forms.

The TVD approach is then modified to be of finite-volume type by Sedaghat [29,30]. In this approach, the governing equation (6) is integrated over a finite-volume ABCD as follows:

$$\iint_{ABCD}(\frac{\partial \mathbf{U}}{\partial t}+\frac{\partial \mathbf{F}}{\partial x}+\frac{\partial \mathbf{G}}{\partial y})dA = 0 \tag{11}$$

By implementing the Green theorem, one may arrive at:

$$\frac{d}{dt}\int_V \mathbf{U}\,dV + \int_{ABCD}\mathbf{H.n}ds = 0 \tag{12}$$

Where s is a unit vector along any faces of a control volume ABCD and n is a unit vector normal to the faces, as shown in Figure 3.

The vector $\mathbf{H}=(\mathbf{F},\mathbf{G})$ represents the fluxes in (8) and is expressed in two dimensional flows by

$$\mathbf{H.n}ds = \mathbf{F}dy - \mathbf{G}dx \tag{13}$$

The governing fluid flow equation (12) then can be approximated as

$$\frac{d}{dt}\mathbf{U}+\frac{1}{A}\sum(\Delta y\mathbf{F}-\Delta x\mathbf{G})=0 \tag{14}$$

where is discretized as

$$\mathbf{U}_{i,j}^{n+1} = \mathbf{U}_{i,j}^n - \frac{\Delta t_{i,j}}{A_{i,j}}[\Delta y_{AB}\mathbf{F}_{AB}+\Delta y_{BC}\mathbf{F}_{BC}+\Delta y_{CD}\mathbf{F}_{CD}+\Delta y_{DA}\mathbf{F}_{DA}]$$
$$- \frac{\Delta t_{i,j}}{A_{i,j}}[\Delta x_{AB}\mathbf{G}_{AB}+\Delta x_{Bc}\mathbf{G}_{BC}+\Delta x_{CD}\mathbf{G}_{CD}+\Delta x_{DA}\mathbf{G}_{DA}] \tag{15}$$

Where $A_{i,j}$ is the area of the quadrilateral ABCD, and expressions for some of terms in (15) may be given as

$$A_{i,j}=\frac{1}{2}|\Delta x_{AC}\Delta y_{DB}-\Delta x_{DB}\Delta y_{AC}|$$
$$x_A=\frac{1}{4}(x_{i-1,j-1}+x_{i,j-1}+x_{i,j}+x_{i-1,j}),y_A=\frac{1}{4}(y_{i-1,j-1}+y_{i,j-1}+y_{i,j}y_{i-1,j})$$
$$\Delta y_{AB}=y_B-y_A \quad,\quad \Delta x_{AB}=x_B-x_A$$
$$\mathbf{F}_{AB}=\frac{1}{2}(\mathbf{F}_{i,j-1}+\mathbf{F}_{i,j}+\mathbf{R}_{i,j-1/2}\mathbf{\Phi}_{i,j-1/2})\quad,\quad \mathbf{G}_{AB}=\frac{1}{2}(\mathbf{G}_{i,j-1}+\mathbf{G}_{i,j}+\mathbf{R}_{i,j-1/2}\mathbf{\Phi}_{i,j-1/2}) \tag{16}$$

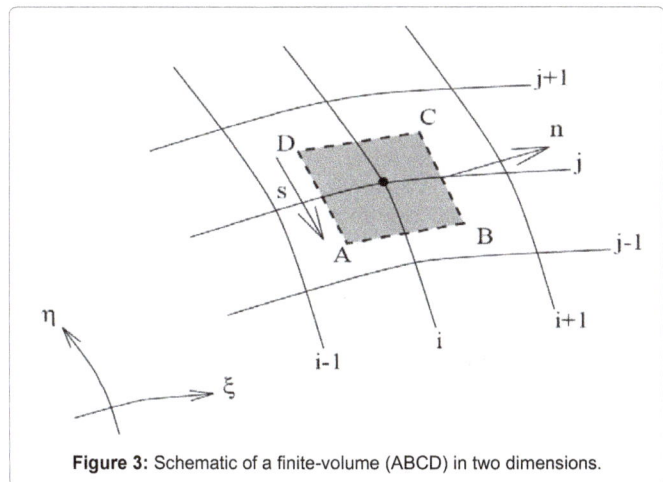

Figure 3: Schematic of a finite-volume (ABCD) in two dimensions.

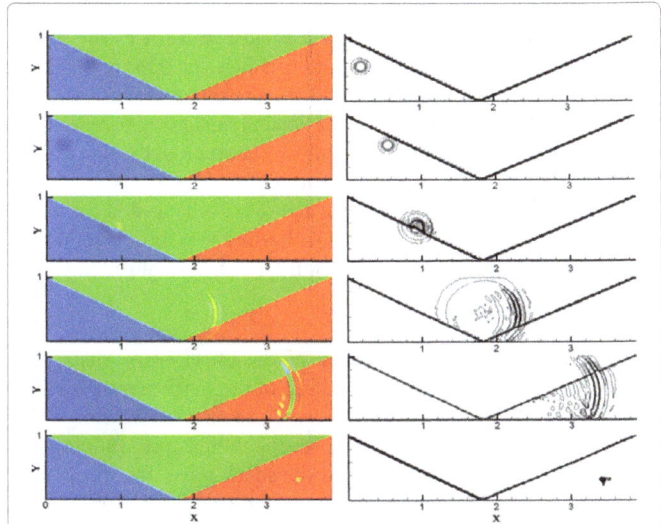

Figure 4: Isolines of pressure in the case of weak interaction $(M_\infty = 2.5,\ R_C = 0.1,\ \Gamma_0 = 2.5)$.

More expressions can be obtained similarly for the rest of terms in equation (16). The TVD terms associated with $\mathbf{R}_{i,j}\mathbf{\Phi}_{i,j}$ represents numerical nonlinear artificial viscosity terms to maintain numerical stability when solving Euler equations. Derivation of this term and the corresponding symmetric and upwind limiter functions are expressed in detail in Sedaghat [29,30].

Computational Results

Numerical simulation of the flow structure in the case of interaction of a vortex with an oblique shock wave was performed for a range of free-stream Mach numbers $M_\infty = 2.5 - 2.9$. It was assumed that the stream wise direction of the vortex axis is parallel to the direction of the x axis. It was also assumed that the direction of the flow velocity vector ahead of the shock-wave front coincides with the direction of the x axis. The variable properties were the circulation Γ_0 and vortex-core radius R_C. The computations were performed for the following geometric dimensions of the domain: length of 4.1 m and width of 1.0 m. The geometric definition of the incoming vortex was also assumed to be unchanged. The experimental and numerical investigations of vortex-shock wave interaction show that a drastic increase along the vortex (vortex breakdown) in the region of its interaction with the shock wave depends mainly on velocity deficit. Depending on velocity deficit, the interaction structure was essentially different. Our computational results revealed two modes of interaction of the vortex with shock waves (weak and strong) depending on the combination of the governing properties M_∞, Γ_0, R_C.

Weak interaction

Figure 4 shows a typical example of the weak mode of interaction of the vortex and an oblique shock wave. The flow structure is shown in this figure. The computation was performed for the following flow parameters: $M_\infty = 2.5,\ R_C = 0.1,\ \Gamma_0 = 2.5$. The angle of the wedge generating the shock wave is 10.94°. It follows from Figure 4 that the vortex passes through the shock wave and, interacting with the reflected one from the wall surface, turns at a certain angle; the axis of symmetry of the vortex becomes parallel to the velocity vector behind the shock wave (actually, the vortex propagates along the wedge generating the shock wave). It is found that the vortex remains almost undistorted when passing through the shock wave. The vortex shape behind the shock-wave front changes weakly and is close to a circle. The shock-wave shape does not change significantly either. The slope of the shock wave also remains almost unchanged. As a result of weak interaction, the flow in the entire computational domain remains supersonic. Thus, weak interaction is characterized by weak distortion of the shock-wave front, minimum change in the vortex structure, and supersonic velocity in the entire flow region. The flow structure is unchanged in time. This agrees with the experimental data of [12], which also show that the flow structure in the case of weak interaction remains unchanged in time.

Strong interaction

In this case, extreme changes in the shock-wave shape and properties of the incoming vortex ahead of the shock wave, especially behind the reflected shock are observed. Because of the low values of the total pressure, Mach number in the vortex core, and angle of inclination of the velocity vector to the shock wave, the shock-wave front and the flow structure in the interaction region ahead of the front become substantially different. Figure 5 shows the flow structure in the case of strong interaction of the vortex with the shock wave for $M_\infty = 2.5$, $R_C = 0.1$, $\Gamma_0 = 10$.

In this case, the vortex breakdown does not occur but the vortex oscillates and constitutes two counter rotating vortices which again combined to make a single vortex as shown in Figure 6. This was also experimentally observed [12].

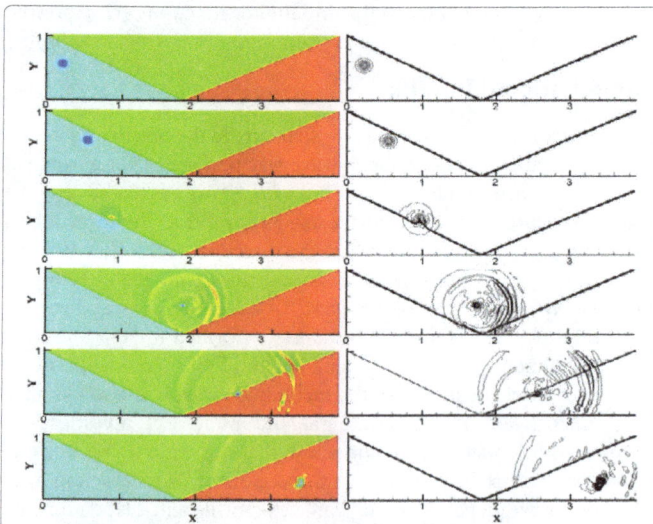

Figure 5: Contours of pressure in the case of strong interaction $(M_\infty = 2.5,\ R_C = 0.1,\ \Gamma_0 = 10)$.

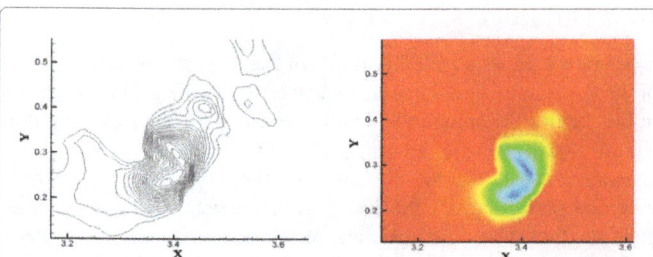

Figure 6: Core vortex division into two counter rotating vortices in the case of strong interaction $(M_\infty = 2.5,\ R_C = 0.1,\ \Gamma_0 = 10)$.

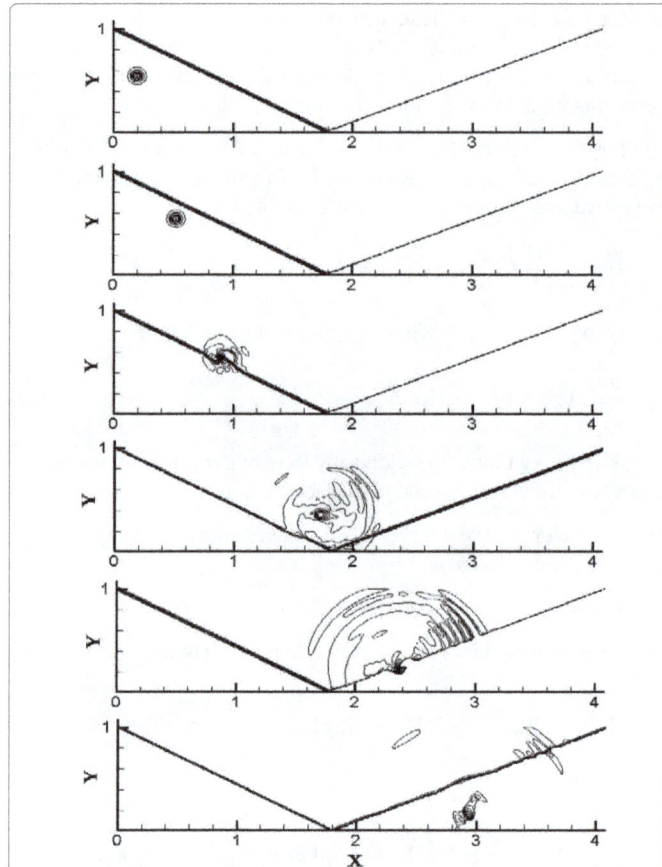

Figure 7: Contours of pressure in the case of strong interaction $(M_\infty = 2.9,\ R_C = 0.07,\ \Gamma_0 = 12.5)$.

Figure 7 shows the flow structure in the case of strong interaction of the vortex with the shock wave leading to vortex breakdown for the case $M_\infty = 2.9$, $R_C = 0.07$, $\Gamma_0 = 12.5$, and $\theta = 29°$. In this case, extreme changes in the shock-wave shape and properties of the incoming vortex ahead of the shock wave, especially behind the reflected shock are observed. Because of the low values of the total pressure, Mach number in the vortex core, and angle of inclination of the velocity vector to the shock wave, the shock-wave front and the flow structure in the interaction region ahead of the front become substantially different as sketched in Figure 8. In the case of strong interaction, the shape of the shock-wave front in the interference region differs from the straight line. A local zone of subsonic recirculation flow is formed, which is located upstream of the initial position of the shock-wave front at a certain distance depending on the governing properties.

Recirculation flow regions are visible ahead of the main shock wave and its reflection in Figure 9. It follows from the computation results that the flow in these regions is subsonic. A shock wave is formed around the recirculation region. In the case of interaction of the vortex and the shock wave, there is a closed subsonic recirculation flow region near the centerline of vortex. Between the normal shock and recirculation region, there is a point with zero flow velocity at the vortex centerline. The angle of inclination of the shock wave formed around the recirculation flow region to the vortex axis is variable, since the Mach number in the vortex core increases significantly in the direction away from its axis (at the centerline, the shock-wave front is perpendicular to the vortex axis). With distance from the centerline, the slope of the shock-wave front decreases.

Thus, in the case of strong interaction, there arises a subsonic reverse flow region ahead of the front of the main shock wave. In its structure, this region is similar to the separated flow region arising, for instance, in the viscous supersonic flow around a compression corner. The emergence of the reverse flow region leads to significant expansion of the vortex cross section. In all examined regimes of strong interaction, a subsonic reverse flow region was observed, which was separated from the supersonic flow region by a slip surface over the entire perimeter. The size of the reverse flow region depends on the shock-wave strength, i.e., on the angle of the wedge generating the shock wave. The length of the local subsonic flow region can serve, to a certain extent, as a characteristic of the strong interaction mode. Similar to weak interaction, rarefaction regions appear behind the shock wave in the case of strong interaction. This leads to a decrease in pressure on the wedge surface as compared to the pressure arising in a uniform supersonic flow around the wedge; this decrease is more significant in the case of strong interaction. Thus, the pressure on the wedge in the case of strong interaction is lower than that in the case of weak interaction.

The following features of the flow structure behind the recirculation flow region can be noted. One can see the forming of the Mach structure and Mach and Regular reflection structure in Figure 9. The presence of two vortices is confirmed by a significant decrease in density in them and by the change in the direction of the velocity vector over the vortex perimeter. The vortices rotate in the opposite directions. With increasing x coordinate (i.e., with distance from the shock-wave front), the area occupied by the subsonic flow in the vortex decreases.

The results of the weak shock wave–vortex interaction (M=2.5)

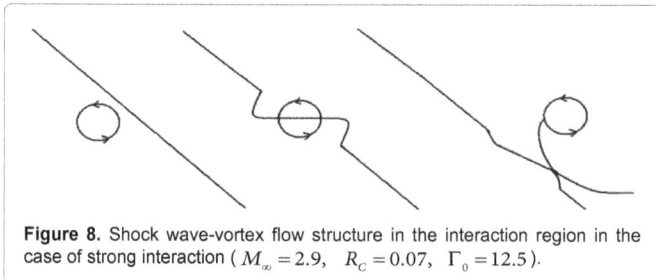

Figure 8. Shock wave-vortex flow structure in the interaction region in the case of strong interaction ($M_\infty = 2.9$, $R_C = 0.07$, $\Gamma_0 = 12.5$).

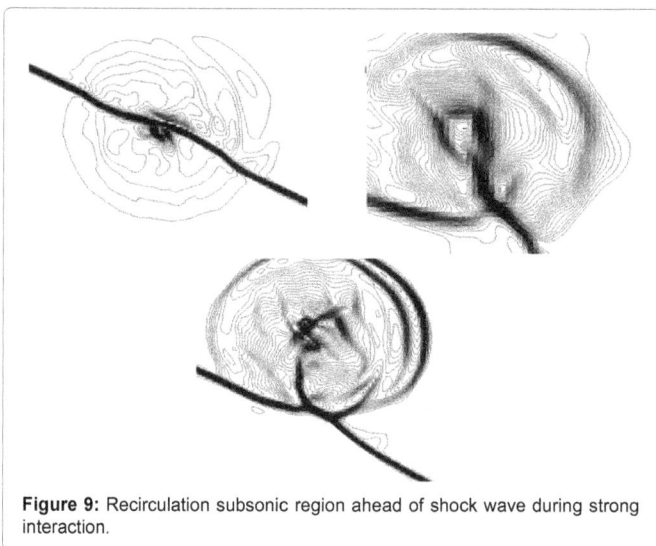

Figure 9: Recirculation subsonic region ahead of shock wave during strong interaction.

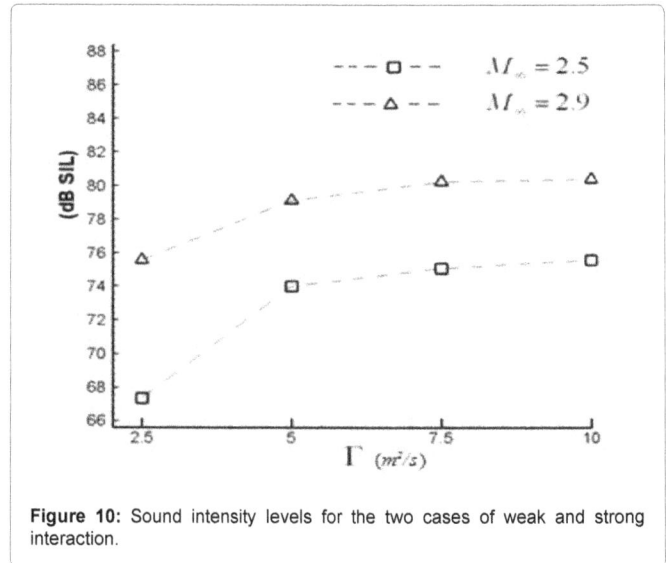

Figure 10: Sound intensity levels for the two cases of weak and strong interaction.

and the strong interaction (M=2.9) are compared in terms of Sound Intensity Level (SIL) in decibel in Figure 10 for different values of vortex strength (Γ) which shows higher acoustic strength at higher Mach number.

In Figure 10, we see the effect of vortex strength Γ_0 and the Mach number on the acoustic field of a vortex interaction with an oblique shock wave. The computational results show that vortex strength become less effective compared with Mach number at higher vortex strength. At higher vortex intensities, the sound intensity levels remain nearly unchanged.

Conclusions

The numerical result of the interaction of a vortex with an oblique shock wave is presented here. A mathematical model of the vortex is introduced which describes vortices generated experimentally. Two types of interaction, namely weak and strong are identified. It is shown that vortex breakdown is possible in the case of strong interaction. The effect of vortex breakdown is characterized by a significant change in its structure: a subsonic recirculation region appears and the vortex diameter substantially increases. It is also demonstrated that the intensity and the size of vortex may be responsible for destabilizing effects when is hit by an oblique shock wave. It is observed that at smaller vortex intensities, the free stream Mach number may play a crucial role on the sound intensity levels.

References

1. Brodetsky MD, Krause E, Nikiforov SB, Pavlov AA, Kharitonov AM, et al. (2001) Evolution of vortex structures on the leeward side of a delta wing. J Appl Mech Tech Phys 42: 243-254.

2. Lugovtsov BA (1999) Asymptotic behavior of the far region of turbulent wake vortices. J Appl Mech Tech Phys 40: 198-207.

3. Che Idris A, Saad MR, Zare-Behtash H, Kontis K (2014) Luminescent measurement systems for the investigation of a scramjet inlet-isolator. Sensors 14: 6606-6632.

4. Gnani F, Lo KH, Zare-Behtash H, Kontis K (2014) Experimental investigation on shock wave diffraction over sharp and curved splitters. Acta Astronautica 99: 143.152.

5. Quinn MK, Kontis K (2013) Pressure-sensitive paint measurements of transient shock phenomena. Sensors 13: 4404-4427.

6. Sun M, Takayama K (2003) A note on numerical simulation of vortical structures in shock diffraction. Shock Waves 13: 25-32.

7. Chatterjee A (1999) Shock wave deformation in shock-vortex interactions. Shock Waves 9: 95-105.

8. Ellzey JL, Henneke MR, Picone JM, Oran ES (1995) The interaction of a shock with a vortex: Shock distortion and the production of acoustic waves. Physics of Fluids 7: 172-184.

9. Abate G, Shyy W (2002) Dynamic structure of confined shocks undergoing sudden expansion. Progress in Aerospace Sciences 38: 23-42.

10. Zare-Behtash H, Kontis K (2009) Compressible flow structures interaction with a two-dimensional ejector: A cold-flow study. Journal of Propulsion and Power 25: 707-716.

11. Leibovich S (1983) Vortex stability and breakdown: survey and extention. AIAA J 22: 1192-1206.

12. Delery JM (1994) Aspects of vortex breakdown. In: Progress in Aerospace Sciences, Pergamon Press, Oxford, 30: 1-59.

13. Zatoloka VV, Ivanyushkin AK, Nikolaev AV (1975) Interference of vortices with shock waves in the inlet. Vortex breakdown. Uch. Zap. TsAGI, 6: 134-138.

14. Glotov GF (1989) Interference of a vortex core with shock waves in free flow and nonisobaric jets. Uch. Zap. TsAGI, 20: 21-32.

15. Kalkhoran IM (1994) Vortex distortion during vortex-surface interaction in a Mach 3 stream. AIAA J 32: 123-129.

16. Cattafesta LN, Settles G (1992) Experiments on shock vortex interaction.

17. Smart MK, Kalkhoran I (1995) Effect of shock strength on oblique shock-wave vortex interaction. AIAA J 33: 2137-2143.

18. Nedungadi A, Lewis MJ (1996) Computational study of the flowfields associated with oblique shock vortex interactions. AIAA J 34: 2545-2553.

19. Rizzetta DP (1996) Numerical investigation of supersonic wing-tip vortices. AIAA J 34: 1203-1208.

20. Corpening G, Anderson JD (19989) Numerical solutions to three-dimensional shock wave-vortex interaction at hypersonic speeds. AIAA J.

21. Zhang S, Zhang YT, Shuc CW (2006) Interaction of an oblique shock wave with a pair of parallel vortices: Shock dynamics and mechanism of sound generation. Physics of Fluids 18: 1-21.

22. Mahesh K (1996) A model for the onset of breakdown in an axisymmetric compressible vortex. Phys. Fluids 8: 3338-3345.

23. Mohammad Amin Aghahosaini (2011) Simulation of Shock Wave-Vortex Interaction Emphasizing TVD Schemes.

24. Harten A (1983) A High Resolution Scheme For The Computation Of Weak Solutions of Hyperbolic Conservation Laws. J Comp Phys 49: 357-393.

25. Harten A (1984) On A Class of High Resolution Total-Variation-Stable Finite Difference Schemes. SIAM J. Numer. Anal 21: 1-23.

26. Yee HC, Warming RF (1983) Implicit Total Variation Diminishing (TVD) Schemes for Steady-State Calculations. AIAA Paper No.83-1902, Proc. Of the AIAA 6th Computational Fluid Dynamics Conference, Danvers, Mass.

27. Yee HC, Kutler P (1983) Application of second-order Accurate Total Variation Diminishing (TVD) Schemes to the Euler Equations in General Geometries. NASA TM-85845, Ames Research Centre, Moffett Field, CA.

28. Yee HC, Klopfer GH, Montagne JL (1990) High Resolution Shock-Capturing Schemes for Invisid and Viscous Hypersonic Flows. J Comp Phys 88: 31-61.

29. Sedaghat A (1993) Comparative study of High Resolution Shock-Capturing Schemes For external supersonic flows.

30. Sedaghat A (1997) A Finite Volume TVD Approach to Transonic Flow Computations.

18

Kite Sky Anchor Analysis for Drone Launching System

Rossett M*

Rowan University, USA

Abstract

A system was developed to launch unmanned aerial vehicles (UAV) near cruising altitude by utilizing a kite based system. A kite was flown in the sky asking as an anchor in the sky. A UAV was brought up the tether and released. The tension, tether angles, and other flight characteristics of the kite were simulated and field tested. This data was used to determine the feasibility of the overall project.

Keywords: Field testing; Kite selection; Simulation

Introduction

Unmanned aerial vehicles, also known as UAVs or drones, are becoming favorable assets in multiple fields including, but not limited to, military defense, ecology research, environmental research, aerial photography, agricultural crop surveillance, search and rescue, and small package delivery [1,2]. With this growth comes opportunity for development and improvement. One major issue facing UAVs is flight range. A system was developed to remotely raise small UAVs into the sky then launch them near cruising altitude. Current methods of launching UAVs normally utilize a catapult type system launching the UAV from the ground [3,4]. By rising the UAVs into the sky via this external system onboard energy that would have been require to bring the UAV to cruising altitude would be conserved.

This effectively increases the UAV's range and flight time. This system used a kite flown attached to a tether. The kite would act as an anchor in the sky. A robotic shuttle system would then climb the tether and release the UAV. For this system to properly work a proper kite would have to be selected and its flight properties analyzed.

Kite

Kite requirements

The function of the kite was to act as a sky anchor, creating a fixed point in the sky. This kite was tethered to the ground with a rope. The kite created tension in the rope. This tension allowed the shuttle system to have the ability to climb the rope into the sky in order to release the UAV. The required tension to lift the UAV and shuttle is dependent on the angle of the tether. Once the shuttle was attached onto the tether, two angles in the line were created as shown in Figure 1. The "shuttle angle" is the angle created between the ground and shuttle. The "kite angle" is the angle created between the shuttle and kite. The kite angle is always greater than the shuttle angle.

Since the kite tension is the only controllable design parameter, finding the minimum required kite tension is necessary to select an appropriate kite. The shuttle is acting as a connection point. This creates two separate tensions in the tether: the kite tension and shuttle tension. A free body diagram of the shuttle is shown in Figure 2. The kite tension can be found from a force balance, shown in Equation Set 1. The factors contributing to the kite tension are kite angle, shuttle angle, shuttle weight, and shuttle tension. The shuttle tension is a reacting force that is interdependent with the kite tension. The tether angles themselves are geometrical parameters determined by multiple factors including: length of tether, location of shuttle, and kite tension. With all of these variable codependent factors, creating a practical model

for estimation is difficult. There are more variables than governing equations thus creating many possible solutions. To create a design requirement estimation, some reasonable assumptions were made. The mass of the shuttle was estimated to be 6 lbf. The shuttle and kite angles were assumed to be 25 and 40 degrees respectively from our experience of the experiments. The kite tension correlating to these requirements is 21 lbf.

Kite selection

A series of kite styles were researched and tested. The first kite investigated was the "Liquid Force Spectrum II", a double lined dynamic kite.

The "Liquid Force Spectrum II" was originally designed for Kite surfing, a sport in which a kite drags a rider on a board across the water. It could produce reasonable tension, however, had a major drawback: the nature of the kite itself. The kite, by design, continuously moves with a figure "8" pattern in the air. This creates the need for a complicated control system to continuously steer the kite. More importantly, launching the UAV from a moving, non-stable, system would be nearly impossible and lead to many potential issues. Due to this, a static style of kite had to be chosen. After some additional research, the "Flow Form 4.0", was chosen to be the successor to the "Liquid Force Spectrum II". The "Flow Form 4.0" was marketed as a "stable flyer" that is "often used for aerial photography" [5,6]. This made it an ideal choice. This kite was a single line kite with a compact size of 2.2 meters by 2.6 meters.

Field testing

The kite was field tested three times in open field environments. Each time between 100 and 150 feet of tether was let out. The tether angle relative to the ground, tension in tether, and wind speed were observed. The tension was measured using a digital fish scale attached to the end of the tether, while the angle was measured by taking a digital photograph of the tether. The photo was then uploaded to a computer and the angle was found. The wind speed was obtained

---

*Corresponding author: Rossett M, Associate professor, Rowan University, USA
E-mail: rosset94@students.rowan.edu

from the National Weather Service's website, which provided current weather data for each location [7].

Due to the nature of wind, wind speed is not truly constant. To obtain a constant wind speed in an ideal open field environment, an altitude of at least 274 m or 900 ft would have to be obtained [8]. This is possible but impractical for most applications. This means the kite cannot be truly static. The average angle of the tether relative to the ground, average tension in the tether, and National Weather Service's wind speeds were recorded to account for these continuous fluctuations. The results of this can be seen in Table 1.

Simulation

A 3D Solid Works model of the kite was created, shown in Figure 3. Using Solid Works Flow Simulation the kite model was evaluated using finite element analysis. The kite was simulated at multiple wind speeds with three different angles of attack. The simulations were run using a mesh setting of 4. The resulting forces in the X and Y direction on the kite were determined. From these forces the resulting tension and angle of the tether could be determined using a simple force balance because it is a static system. The forces in balance are X force on kite (Kx), Y force on kite (Ky), tension in tether (T), and weight of tether. Fifty feet of tether used in our system was weighed at 0.16 pounds, correlating to 0.0032 pounds per foot. Three hundred feet of this tether would weigh less than one pound. Since the mass of this tether was minimal, its weight was assumed to be negligible. Therefor we can make the tension in the tether the only reaction force on the kite. This is demonstrated in Figure 4 and Equation Set 2.

$$T = \sqrt{K_x^2 + K_y^2}$$

$$\phi = \tan^{-1}\left(\frac{K_y}{K_x}\right)$$

Equation Set 2: Tether Tension Equation

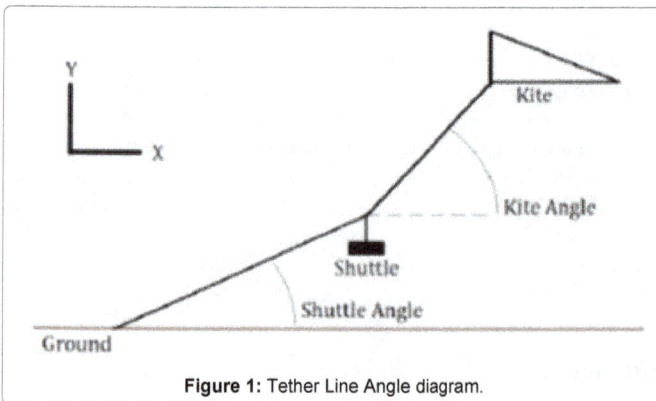

Figure 1: Tether Line Angle diagram.

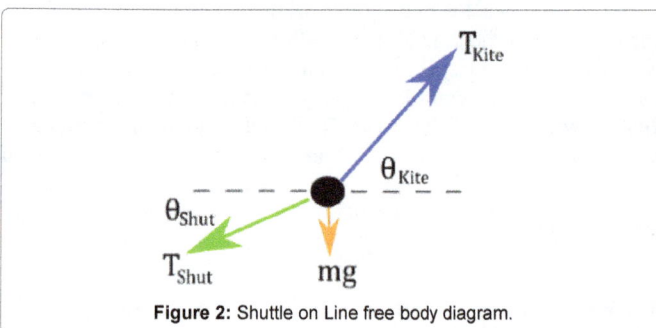

Figure 2: Shuttle on Line free body diagram.

Test date	Wind speed	Avg TetherTension (± 2)	Avg TetherAngle (± 5)
Feb 10, 2015	15 MPH	16 lbf	40 Deg
April 2, 2015	19 MPH	20 lbf	45 Deg
April 4, 2015	16 MPH	16 lbf	45 Deg

Table 1: Kite performance data.

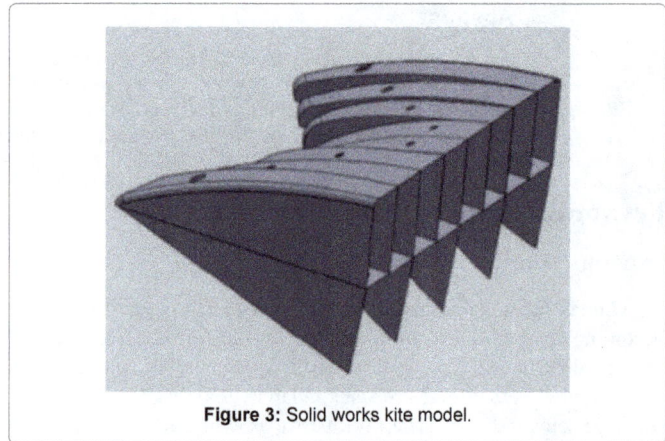

Figure 3: Solid works kite model.

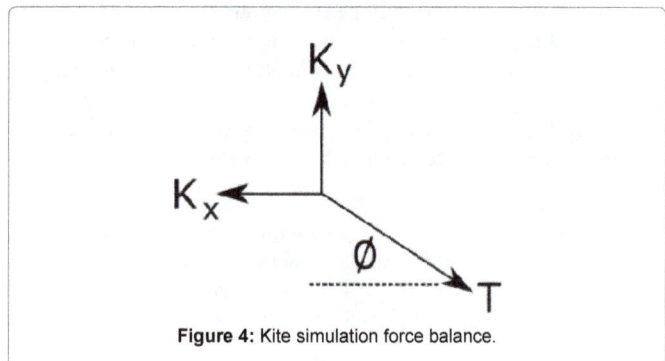

Figure 4: Kite simulation force balance.

The model's default wind attack angle is shown in Figure 5. This angle needed to be altered to simulate real world conditions. As observed during field testing, the orientation of the kite during flight was approximately 20 degrees (± 5 degrees) off of the model's default orientation, shown in Figure 6. To account for this, the simulation was run at angles of attack of 15, 20, and 25 degrees.

The resulting data collected from the simulation can be seen in Table 2. The simulated model data seemed to line up to the field test data at an angle of attack between 15 and 20 degrees. This was close to the orientation angle the kite was observed flying at. An example of this was the Feb 10th, 2015 field test. During that test, a wind speed of 15 MPH resulted in 16 lbf of tension (± 2) at a tether angle of 40 degrees (± 5). According to the simulation, a 15 MPH wind should produce a tension of between 14.14 and 16.02 lbf at a tether angle of between 43.9 and 49.1 degrees. This confirmed that the angle of attack of the kite was between 15 to 20 degrees.

The simulated tether angle was consistent at any given angle of attack regardless of wind speed. This is because the lift to drag ratio at any given angle of attack was dependent on shape and not wind speed, making the lift to drag ratio constant. The tension however increased with wind speed. The tension of the tether was plotted against wind speed for each of the three angles of attack. A second order polynomial trend line was then fitted to each data set. The trend line equation and coefficient of determination (R^2) for each was calculated. This data is shown in Figure 7.

This trend line equation can be used reliably to estimate tension in the tether at given wind speeds without the use of complicated and time consuming CFD simulations.

System Integrated Testing

The kite was field tested with the complete system incorporating the shuttle climbing the tether. The kite itself produced enough tension in the tether to support the shuttle however, variation in tension is caused the overall system to fail. The tension in the line would increase or decrease by a few pounds, due to slight changes in the wind. This caused the tether to become either taut or relaxed. These slight decreases in tension caused the shuttle on the tether to drop downwards, while slight increases in tension caused the shuttle to be pulled vertically upwards. This is shown in Figure 8. As the tension increased, the difference between the "kite angle" and "shuttle angle" decreased. This pulled the shuttle vertically up. The same effect occurred in the opposite direction when the tension decreased. This oscillating disturbance caused the

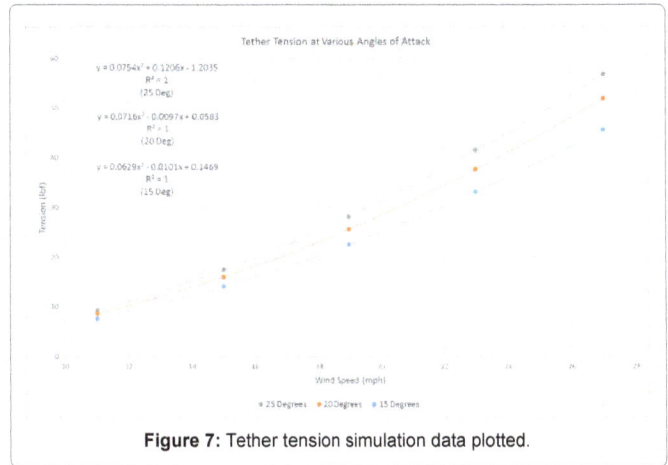

Figure 5: Model default angle of attack orientation.

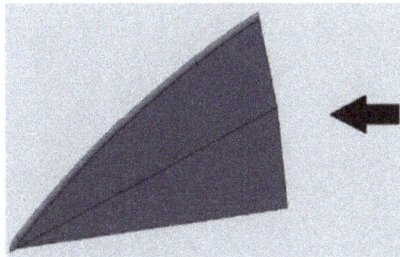

Figure 6: Model modified angle of attack orientation.

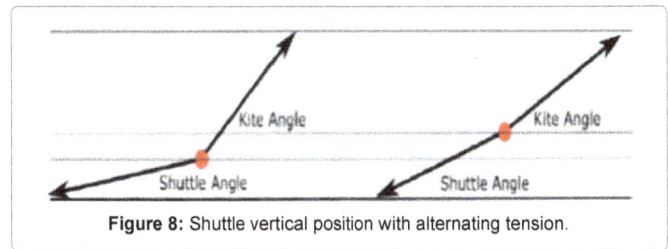

Figure 7: Tether tension simulation data plotted.

Figure 8: Shuttle vertical position with alternating tension.

shuttle to bounce around on the line. This made it impossible to safely release the UAV with the shuttle flailing around. During one test, the buildup of this disturbance even caused the flailing shuttle to pull the kite down from the sky.

The disturbance can be quantified by the change in height of the shuttle. The height of the shuttle can be determined from the shuttle angle and length of tether from the ground to the shuttle. The largest deflection occurs when the shuttle is located in the center of the tether. To account for the worst case, the length of tether to shuttle can be considered half the total tether length. This is shown in Equation Set 3.

$$h = \frac{L_{total}}{2} * \sin(\theta_{shut})$$

$$\Delta h = h_{t+1} - h_t$$

Equation Set 3: Disturbance Shuttle Height

The shuttle angle can be found either by field test; or calculated from a known kite tension, shuttle weight, and kite angle. This equation was found by rearranging the previous equation for kite tension and solving for the shuttle angle. This is shown in Equation Set 4.

$$\theta_{shut} = \tan^{-1}\left(\tan\theta_{kite} - \frac{mg}{T} * \frac{1}{\cos\theta_{kite}}\right)$$

Equation Set 4: Shuttle Angle Formula

Conclusion and Future Work

The disturbance in the line caused by the changing wind speed made it impossible to keep the shuttle stable enough to safely release a UAV. To minimize disturbance, the tension in the kite tether would have to be increased. The same amount of change in tension would create a lesser disturbance with a higher kite tension. This means a tension drop from 25 lbf to 20 lbf in the line would create more of a disturbance than a tension drop from 40 lbf to 35 lbf. To obtain this

Angle (Deg)	Wind Speed (mph)	X (lbf)	Y (lbf)	Tension (lbf)	TetherAngle (Deg)
15	11	5.5	5.3	7.64	43.94
15	15	10.2	9.8	14.14	43.85
15	19	16.3	15.7	22.63	43.93
15	23	23.9	23	33.17	43.9
15	27	33	31.6	45.69	43.76
20	11	56.6	6.5	8.62	48.95
20	15	10.5	12.1	16.02	49.05
20	19	16.87	19.44	25.74	49.05
20	23	24.7	28.5	37.71	49.09
20	27	34.1	39.27	52.01	49.03
25	11	5.67	7.33	9.27	52.28
25	15	10.5	14.05	17.54	53.23
25	19	16.9	22.58	28.2	53.19
25	23	24.89	33.3	41.57	53.22
25	27	34.17	45.54	56.93	53.12

Table 2: Kite simulation data.

higher tension, the kite would have to be flown higher. This would require adding an anchoring system and possibly upgrading to a higher tension rope to account for the stronger winds.

This brings up the question of how this system is stable enough for aerial photography. During kite aerial photography, the camera is attached to a stabilizing rig with pan and tilt. This rig is then attached to the tether close to the kite itself [9]. Since the rig is always close to the kite, which is acting as an anchor point, it sees minimal deflection as the tension varies. This is unlike the UAV system which climbs up the entire tether experiencing the worst deflection in the middle of the line between the kite and ground. This can be thought of as a beam supported by a fixed point at both ends. The most deflection would be caused by adding a weight in the center of the beam while the minimal amount of deflection would be caused by adding that same weight on the end of the beam directly above the anchor point. Future research would focus on overcoming the disturbance issue. Flying the kite at higher altitudes should be the first option to test. Another possibility is to modify the project to incorporate a blimp or balloon style of sky anchor instead of a kite. These, while still affected by wind, would be less sensitive to wind changes. This could assist in minimizing the disturbance. This style of system would be simpler overall and easier to launch compared to the kite system. Similar technology is already being used to gather weather data using weather balloons [10]. If the kite aspect is retained instead of investigating a balloon style sky anchor, the system could be integrated with an autonomous flight control system to fly the kite. There is current researching into developing these types of systems [11,12]. Research into damping the motion of the tether or adding damping to the counterweight stability system are other possible areas of future development. A possibly solution is developing an active tether tension control system. This additional mechanism would pull or release tether to maintain constant tension in the tether line to minimize disturbance.

References

1. Ogden LE (2013) Drone Ecology BioScience 63: 776-776.

2. Cook M (2015) The 10 Most Innovative Drone Applications Today.

3. Fahlstrom P, Gleason T (2012) Introduction to UAV Systems. Hoboken Wiley.

4. Hindle P (2015) UAVs Unleashed. Microwave Journal 56: 18-22.

5. (2015) Flow Form 4.0 Lifter Sled.

6. (2015) Great selection of sutton flow form kites ideal for power to haul your favorite line laundry FunWithWind Kites.

7. (2015) National Weather Service.

8. Chen W (1997) Handbook of Structural Engineering. Boca Raton Fla CRC Press.

9. Aber JS, Marzolff I, Ries JB (2010) Small-format aerial photography. Amsterdam Elsevier Science.

10. Crane R (1990) Sampling the Weather in the Upper Atmosphere by Ballon American Association of Physics Teachers 28: 182.

11. McGarey P, Saripalli S (2013) Auto kite. Journal of Intelligent & Robotic Systems 74: 363-370.

12. Erhard M, Strauch H (2013) Control of Towing Kites for Seagoing Vessels IEEE Transactions on Control Systems Technology 21: 1629-1640.

19

Characterization of Physical and Structural Properties of Aluminium Carbide Powder: Impact of Biofield Treatment

Mahendra Kumar Trivedi[1], Rama Mohan Tallapragada[1], Alice Branton[1], Dahryn Trivedi[1], Gopal Nayak[1], Omprakash Latiyal[2] and Snehasis Jana[2]*

[1]Trivedi Global Inc, 10624 S Eastern Avenue Suite A-969, Henderson, NV 89052, USA
[2]Trivedi Science Research Laboratory Pvt. Ltd, Hall-A, Chinar Mega Mall, Chinar Fortune City, Hoshangabad Rd., Bhopal-462026, Madhya Pradesh, India

Abstract

Aluminium carbide (Al_4C_3) has gained extensive attention due to its abrasive and creep resistance properties. Aim of the present study was to evaluate the impact of biofield treatment on physical and structural properties of Al_4C_3 powder. The Al_4C_3 powder was divided into two parts *i.e.* control and treated. Control part was remained as untreated and treated part received biofield treatment. Subsequently, control and treated Al_4C_3 samples were characterized using X-ray diffraction (XRD), surface area analyser and Fourier transform infrared spectroscopy (FT-IR). XRD data revealed that lattice parameter and unit cell volume of treated Al_4C_3 samples were increased by 0.33 and 0.66% respectively, as compared to control. The density of treated Al_4C_3 samples was reduced upto 0.65% as compared to control. In addition, the molecular weight and crystallite size of treated Al_4C_3 samples were increased upto 0.66 and 249.53% respectively as compared to control. Furthermore, surface area of treated Al_4C_3 sample was increased by 5% as compared to control. The FT-IR spectra revealed no significant change in absorption peaks of treated Al_4C_3 samples as compared to control. Thus, XRD and surface area results suggest that biofield treatment has substantially altered the physical and structural properties of treated Al_4C_3 powder.

Keywords: Biofield treatment; Aluminium carbide powder; X-ray diffraction; Fourier transform infrared spectroscopy; Surface area

Introduction

Aluminium carbide (Al_4C_3) is known for its abrasive and creep resistance properties. Generally, it is produced by reaction of aluminium with carbon in electric arc furnace [1]. Al_4C_3 plays a major role in production of some important structures such as diamond related structures, nanostructure carbons, and growth of diamonds on boron nitride etc. In addition, Al_4C_3 react with water under high pressure and generates methane [2]. Moreover, Al_4C_3 particles are used as fine dispersion in aluminium alloy to strengthen the material. In aluminium matrix, Al_4C_3 particles increase the creep resistance, especially with silicon carbide, which is widely utilizing in automobile and aircraft industries [3]. In order to improve the creep resistance of material, its crystal structure and crystallite size plays an important role. Furthermore, Al_4C_3 is also used as an abrasive material in cutting tools, where its crystallite size plays a crucial role. After considering the vast importance of Al_4C_3 in several industries, authors wish to investigate an approach that could be beneficial to modify the physical and structural properties of Al_4C_3 powder.

Energy is considered as the ability to do work, which interrelates with matter as $E=mc^2$ (Einstein's famous equation). The energy can effectively interact with any matter at a distance and cause action. In addition, energy also exists with various fields such as electric, magnetic etc. Furthermore, researchers have confirmed that bio magnetic fields are present around the human body, which have been evidenced by electromyography (EMG), Electrocardiography (ECG) and Electroencephalogram (EEG) [4]. Scientists have postulated that it is due to the flow of bioelectricity (generated from heart, brain functions or due to the motion of charged particles such as protons, electrons, and ions) in the human body. As per the basic fundamental law in physics, when an electrical signal passes through any material, a magnetic field is generated in the surrounding space [5]. Due to this, a human has ability to harness the energy from environment/universe and can transmit into any object (living or non-living) around the Globe. The

object(s) always receive the energy and responded into useful way that is called biofield energy. This process is termed as biofield treatment. These healing treatments suggest their mechanism upon modulating patient-environmental energy fields [6]. The National Center for Complementary and Alternative Medicine (NCCAM) considered this biofield treatment (therapy) in subcategory of energy therapies [7]. Furthermore, Mr. Trivedi's unique biofield treatment is known as Trivedi Effect®. Mr.Trivedi's biofield treatment has substantially altered the physical, structural and atomic characteristic in various metals [8-10] and ceramics [11,12]. Additionally, the influence of biofield treatment was significantly studied in the field of microbiology [13,14], biotechnology [15,16], and agriculture [17-19]. Recently, it was reported that biofield treatment had increased the particle size by six fold and enhanced the crystallite size by two fold in zinc powder [20]. Our group previously reported that biofield treatment has substantial altered the atomic, structural and physical properties in silicon carbide [21] and carbon allotropes [22]. Based on the outstanding results achieved by biofield treatment on metals and ceramics, an attempt was made to evaluate the effect of biofield treatment on physical and structural properties of Al_4C_3 powder.

*Corresponding author: Dr. Snehasis Jana, Trivedi Science Research Laboratory Pvt. Ltd , Hall-A, Chinar Mega Mall, Chinar Fortune City, Hoshangabad Rd, Bhopal-462026, Madhya Pradesh, India
E-mail: publication@trivedisrl.com

Experimental

The Al_4C_3 powder was purchased from Sigma Aldrich, India. The sample was equally divided into two parts, considered as control and treated. Treated group was in sealed pack and handed over to Mr. Trivedi for biofield treatment under laboratory condition. Mr. Trivedi provided the biofield treatment through his energy transmission process to the treated group without touching the sample. The control and treated samples were characterized using X-ray Diffraction (XRD), surface area analyzer, and Fourier Transform Infrared Spectroscopy (FT-IR).

X-Ray Diffraction Study

XRD analysis of control and treated Al_4C_3 powder was carried out on Phillips, Holland PW 1710 X-ray diffractometer system, which had a copper anode with nickel filter. The radiation of wavelength used by the XRD system was 1.54056 Å. The data obtained from this XRD were in the form of a chart of 2θ vs. intensity and a detailed table containing peak intensity counts, d value (Å), peak width (θ°), relative intensity (%) etc. Additionally, PowderX software was used to calculate lattice parameter and unit cell volume of Al_4C_3 powder samples. Weight of the unit cell was calculated as, molecular weight multiplied by the number of atoms present in a unit cell. Density of the unit cell was computed as follows:

density =mass of unit cell/volume of unit cell

The crystallite size (G) was calculated by using formula: G=kλ/(bCosθ),

Here, λ is the wavelength of radiation used, b is full width half maximum (FWHM) and k is the equipment constant (0.94). Furthermore, the percent change in the lattice parameter was calculated using following equation:

$$\% \text{ change in lattice parameter} = \frac{\left[A_{Treated} - A_{Control}\right]}{A_{Control}} \times 100$$

where $A_{treated}$ and $A_{control}$ are the lattice parameter of treated and control samples respectively. Similarly, the percent change in all other parameters such as unit cell volume, density, molecular weight, and crystallite size were calculated.

Surface Area Analysis

The surface area was measured by the surface area analyser, Smart SORB 90 based on Brunauer–Emmett–Teller (BET), which had a detection range of 0.20–1000 m^2/g. Percent changes in surface area were calculated using following equation:

$$\% \text{ change in surface area} = \frac{\left[S_{Treated} - S_{Control}\right]}{S_{Control}} \times 100$$

Where, $S_{control}$ and $S_{treated}$ are the surface area of control and treated samples respectively.

FT-IR Spectroscopy

To see the impact of biofield treatment at bonding level in Al_4C_3, the FT-IR analysis of control and treated Al_4C_3 samples was carried out. For FT-IR analysis, Shimadzu, Fourier transform infrared (FT-IR) spectrometer with frequency range of 300-4000 cm^{-1} was used.

Results and Discussion

X-ray diffraction (XRD)

XRD analysis results of control and treated Al_4C_3 samples are

Group	Lattice parameter (Å)	Unit cell volume (×10⁻²² cm³)	Density (g/cc)	Molecular weight (g/mol)	Crystallite size (nm)
Control	3.3350	2.4012	3.013	145.234	81.56
Treated, T1	3.3446	2.4149	2.996	146.064	142.59
Treated, T2	3.3429	2.4124	2.999	145.915	190.07
Treated, T3	3.3455	2.4162	2.994	146.143	285.08
Treated, T4	3.3460	2.4169	2.993	146.187	190.03

Table 1: X-ray diffraction analysis of aluminium carbide powder.

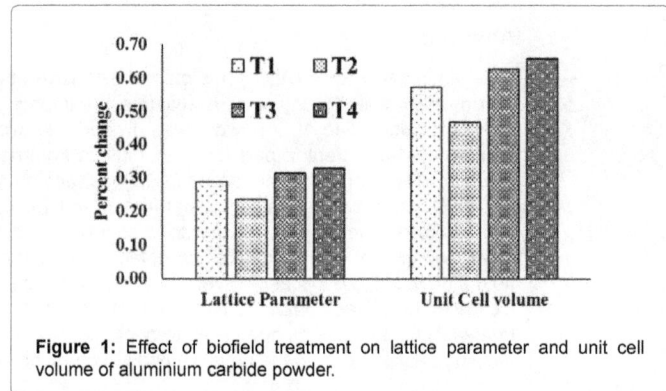

Figure 1: Effect of biofield treatment on lattice parameter and unit cell volume of aluminium carbide powder.

illustrated in Table 1 and Figures 1-3. Data showed that the lattice parameter of unit cell was increased by 0.29, 0.24, 0.31, and 0.33% in treated Al_4C_3 samples T_1, T_2, T_3, and T_4, respectively as compared to control [23]. The change in lattice parameter is also known as lattice strain (ε), which is related to stress (σ) by following equation:

$$\sigma = Y\varepsilon$$

Where, Y is Young's Modulus

In above equation, negative and positive lattice strain indicates the compressive and tensile stress respectively. Thus, positive strain found in treated Al_4C_3 sample suggests that biofield treatment might induce tensile stress, which probably stretched the unit cell lattice parameter. Our group previously reported that biofield treatment has altered the lattice parameter in silicon carbide powder [21]. In addition, the lattice strain less than 0.2% is considered as elastic strain, while more than 0.2% is referred as plastic strain [24]. Thus, the positive lattice strain (>0.2%) in treated Al_4C_3 indicates that biofield treatment probably induced plastic strain. Furthermore, the unit cell volume was increased by 0.57, 0.47, 0.63, and 0.66% in treated Al_4C_3 samples T_1, T_2, T_3, and T_4, respectively as compared to control (Figure 1). Data also showed that density was reduced by 0.57, 0.47, 0.62, and 0.65% in treated Al_4C_3 samples T_1, T_2, T_3, and T_4, respectively as compared to control. Contrarily, the molecular weight of treated Al_4C_3 was increased from 145.23 g/mol (control) to 146.06, 145.91, 146.14, and 146.18 g/mol in T_1, T_2, T_3, and T_4 respectively. It suggest that molecular weight was increased by 0.57, 0.47, 0.63, and 0.66% in treated Al_4C_3 samples T_1, T_2, T_3, and T_4, respectively as compared to control (Figure 2). This could be possible if number of protons and neutrons altered after biofield treatment. Thus, it is hypothesized that a weak reversible nuclear level reaction including neutrons-protons and neutrinos might occurred in treated Al_4C_3 powders after biofield treatment [25]. It is already reported that biofield treatment has significantly altered the atomic weight and density in silicon dioxide, zirconia [26], and silicon carbide [21]. Besides this, the crystallite size of control and treated Al_4C_3 powder were computed using Scherrer formula and calculated result are presented in Table 1. Data showed that the crystallite size was

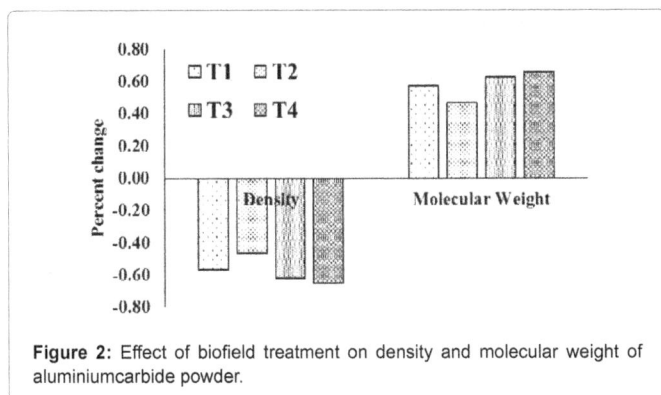

Figure 2: Effect of biofield treatment on density and molecular weight of aluminiumcarbide powder.

	Surface Area (m²/g)		Percent change
	Control	Treated	
	1.60	1.68	5.0

Table 2: Surface area analysis of aluminium carbide powder.

increased from 81.56 nm (control) to 142.59, 190.07, 285.08, and 190.03 nm in treated Al_4C_3 samples T_1, T_2, T_3, and T_4, respectively. It suggests that crystallite size of treated Al_4C_3 powder was significantly increased by 74.83, 133.04, 249.53, and 133.0% in treated Al_4C_3 samples T_1, T_2, T_3, and T_4, respectively as compared to control (Figure 3). Previously, our group reported that biofield treatment has increased the crystallite size in antimony powder [27]. Al_4C_3 is utilized in aluminium matrix and silicon carbide to increase the creep resistance. Furthermore, Coble proposed that the strain rate in a material is inversely proportional to crystallite size as given below [28]:

$$\frac{d\varepsilon}{dt} = \frac{\sigma}{G^3} D_{gb} e^{\frac{Q_{Coble}}{RT}}$$

where σ is the applied stress, G is crystallite size, Dgb is diffusion coefficient in grain boundary, Q_{Coble} is activation energy for coble creep, R is gas constant, and T is temperature. Coble equation suggests that the strain rate decreases as increase in crystallite size (G) at constant temperature and stress for a given material. Further, the reduction in strain rate increases the creep resistance of a material. In Coble-creep, vacancies and atoms diffused along crystallite boundaries to elongate the crystallite along stress axis to deform the material. Thus, the increase in crystallite size in Al_4C_3 reduced the crystallite boundaries, which prevents the movement of vacancies along boundaries [29,30]. Shah et al. demonstrated that the creep resistance of metal-carbide was improved after heat treatment due to increase in crystallite size. The increase in crystallite size leads to stabilize the grain boundaries and thus improves creep resistance [31]. In addition, it was demonstrated that grain boundary sliding via slip dominates the creep process in case of finer crystallite size as compared to coarser [32]. Hence, the higher crystallite size found in treated Al_4C_3 indicates that creep resistance probably enhanced after biofield treatment as compared to control. Therefore, XRD data suggest that biofield treatment has significantly altered the atomic and structural properties in Al_4C_3.

Surface area analysis

Surface area analysis of Al_4C_3 powder is presented in Table 2. Data exhibited that surface area of treated Al_4C_3 powder was increased from 1.60 m²/g (control) to 1.68 m²/g after biofield treatment. This indicates that surface area of treated Al_4C_3 powder was slightly increased by 5.0% as compared to control. Our group previously reported that biofield

treatment has significantly reduced the particle size and increased the surface area in zirconium oxide [26]. Thus, it is assumed that the increase of surface area in treated Al_4C_3, possibly due to particle size reduction after biofield treatment. The existence of internal strains in treated Al_4C_3 was evidenced by XRD data (Figure 1), which might induce fractures in particles and reduced size. Hence, it is concludes that biofield treatment has altered the physical characteristics of Al_4C_3 powder as compared to control.

FT-IR analysis

FT-IR spectra of control and treated Al_4C_3 samples are illustrated in Figure 4. In control Al_4C_3 samples absorption peaks were observed at 499, 609, 711, and 785 cm⁻¹, which could be due to Al-C bonding vibrations. The control data is well supported by literature data [33]. The treated Al_4C_3 also showed similar absorption peaks at 499, 609, 709, and 785 cm⁻¹, which could be assigned Al-C bonding vibrations. Furthermore, peaks observed at 1490 and 1440 cm⁻¹ in control and treated Al_4C_3 respectively, could be due to moisture absorption. In addition, the peaks observed at 2358 and 2395 cm⁻¹ in control and treated Al_4C_3 respectively, could be due to CO_2 absorption by samples. Thus, FT-IR data suggest that no significant change was observed in absorption peaks of treated Al_4C_3 as compared to control.

Conclusion

Biofield treatment showed an increased lattice parameter and unit cell volume of treated Al_4C_3 samples upto 0.33 and 0.66% respectively,

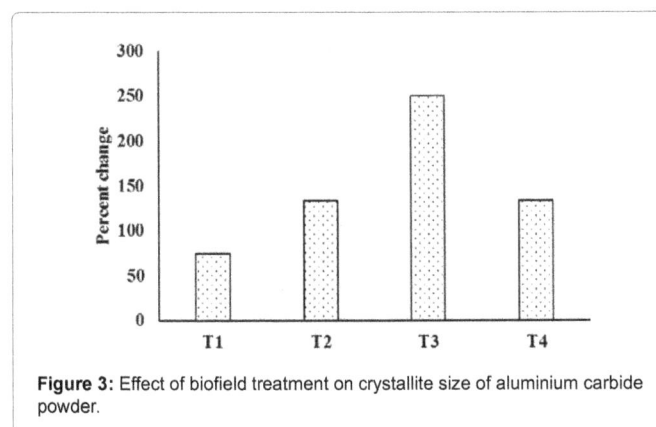

Figure 3: Effect of biofield treatment on crystallite size of aluminium carbide powder.

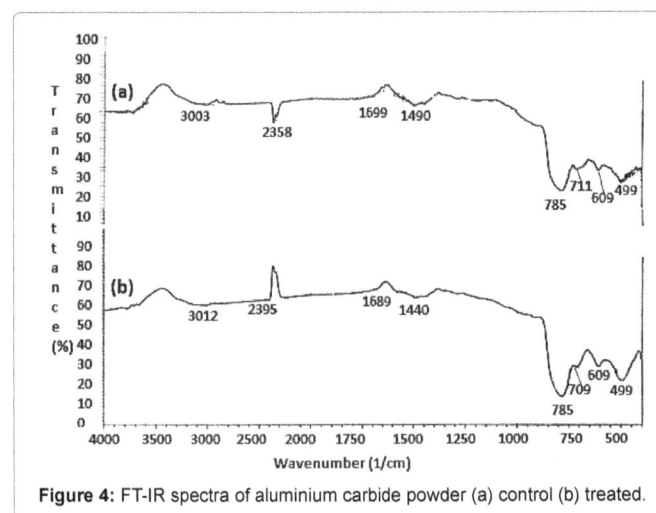

Figure 4: FT-IR spectra of aluminium carbide powder (a) control (b) treated.

as compared to control. It may be due to tensile stress, which probably generated in treated Al_4C_3 samples after biofield treatment. In addition, the molecular weight was increased upto 0.66% in treated Al_4C_3 samples as compared to control. It is hypothesized that biofield treatment may induce nuclear level reaction, which resulted into increase of molecular weight in treated Al_4C_3 sample. Besides, the crystallite size of treated Al_4C_3 samples was significantly increased upto 285.08 nm from 81.56 nm (in control). The increase in crystallite size could improve the creep resistance and abrasive properties of treated Al_4C_3 samples. Furthermore, the surface area was increased by 5% in treated Al_4C_3 samples as compared to control. It could be due to alteration of shape/ size of Al_4C_3 particles after biofield treatment. However, no significant change was observed in absorption peaks in FT-IR spectra of treated Al_4C_3 as compared to control. Therefore, based on above outcomes of XRD and surface area analysis, it is assumed that treated Al_4C_3 with high creep resistance could be more useful in automobile and aircraft manufacturing industries.

Acknowledgement

Authors gratefully acknowledged to Dr. Cheng Dong of NLSC, Institute of Physics, and Chinese academy of Sciences for providing the facilities to use PowderX software for analyzing XRD data. Authors also would like to thank Trivedi science, Trivedi master wellness and Trivedi testimonials for their support during the work.

References

1. Greenwood NN, Earnshaw A (1997) Chemistry of the Elements. Butterworth-Heinemann.

2. Ji C, Ma Y, Chyu MC, Knudson R, Zhu H (2009) X-ray diffraction study of aluminum carbide powder to 50 GPa. J App Phys 106: 083511.

3. Zhu SJ, Peng LM, Zhou Q, Ma ZY, Kucharova K, et al. (1998) Creep behaviour of aluminium strengthened by fine aluminium carbide particles and reinforced by silicon carbide particulates DS Al-SiC/Al4C3 composites. Mater Sci Eng A 268: 236-245.

4. Zahra M, Farsi M (2009) Biofield therapies: Biophysical basis and biological regulations. Complement Ther Clin Pract 15: 35-37.

5. Maxwell JC (1865) A dynamical theory of the electromagnetic field. Phil Trans R Soc Lond 155: 459-512.

6. Aldridge D (1991) Spirituality healing and medicine. Br J Gen Pract 41: 425-427.

7. Hok J, Tishelman C, Ploner A, Forss A, Falkenberg T (2008) Mapping patterns of complementary and alternative medicine use in cancer: An explorative cross-sectional study of individuals with reported positive exceptional experiences. BMC Complement Altern Med 8: 48.

8. Trivedi MK, Patil S, Tallapragada RM (2012) Thought intervention through bio field changing metal powder characteristics experiments on powder characteristics at a PM plant. Future Control and Automation LNEE 173: 247-252.

9. Trivedi MK, Patil S, Tallapragada RM (2015) Effect of biofield treatment on the physical and thermal characteristics of aluminium powders. Ind Eng Manage 4: 151.

10. Trivedi MK, Patil S, Tallapragada RM (2013) Effect of biofield treatment on the physical and thermal characteristics of silicon, tin and lead powders. J Material Sci Eng 2: 125.

11. Trivedi MK, Patil S, Tallapragada RM (2013) Effect of biofield treatment on the physical and thermal characteristics of vanadium pentoxide powder. J Material Sci Eng S11: 001.

12. Trivedi MK, Nayak G, Patil S, Tallapragada RM, Latiyal O (2015) Studies of the atomic and crystalline characteristics of ceramic oxide nano powders after bio field treatment. Ind Eng Manage 4: 161.

13. Trivedi MK, Patil S, Shettigar H, Gangwar M, Jana S (2015) Antimicrobial sensitivity pattern of Pseudomonas fluorescens after biofield treatment. J Infect Dis Ther 3: 222.

14. Trivedi MK, Patil S, Shettigar H, Bairwa K, Jana S (2015) Phenotypic and biotypic characterization of Klebsiella oxytoca: An impact of biofield treatment. J Microb Biochem Technol 7: 203-206.

15. Patil S, Nayak GB, Barve SS, Tembe RP, Khan RR (2012) Impact of biofield treatment on growth and anatomical characteristics of Pogostemon cablin (Benth). Biotechnology 11: 154-162.

16. Nayak G, Altekar N (2015) Effect of biofield treatment on plant growth and adaptation. J Environ Health Sci 1: 1-9.

17. Shinde V, Sances F, Patil S, Spence A (2012) Impact of biofield treatment on growth and yield of lettuce and tomato. Aust J Basic and Appl Sci 6: 100-105.

18. Lenssen AW (2013) Biofield and fungicide seed treatment influences on soybean productivity, seed quality and weed community. Agricultural Journal 8: 138-143.

19. Sances F, Flora E, Patil S, Spence A, Shinde V (2013) Impact of biofield treatment on ginseng and organic blueberry yield. Agrivita J Agric Sci 35.

20. Trivedi MK, Tallapragada RM (2008) A transcendental to changing metal powder characteristics. Met Powder Rep 63: 22-28, 31.

21. Trivedi MK, Nayak G, Tallapragada RM, Patil S, Latiyal O, et al. (2015) Effect of biofield treatment on structural and morphological properties of silicon carbide. J Powder Metall Min 4: 1.

22. Trivedi MK, Tallapragada RM (2009) Effect of super consciousness external energy on atomic, crystalline and powder characteristics of carbon allotrope powders. Mater Res Innov 13: 473-480.

23. Soboyejo W (2002) Mechanical properties of engineered materials. CRC press.

24. Daymond MR, Bourke MAM, Dreele RBV, Clausen B, Lorentzen T (1997) Use of Rietveld refinement for elastic macrostrain determination and for evaluation of plastic strain history from diffraction spectra. J Appl Phys 82: 1554-1562.

25. Narlikar JV (1993) Introduction to cosmology. (2nd edn), Jones and Bartlett Inc Cambridge University Press.

26. Trivedi MK, Patil S, Tallapragada RM (2014) Atomic crystalline and powder characteristics of treated zirconia and silica powders. J Material Sci Eng 3: 144

27. Dhabade VV, Tallapragada RM, Trivedi MK (2009) Effect of external energy on atomic, crystalline and powder characteristics of antimony and bismuth powders. Bull Mater Sci 32: 471-479.

28. Chawla KK (1999) Mechanical Behavior of Materials. (1st edn) Prentice Hall.

29. Lu L, Sui ML, Lu K (2000) Superplastic extensibility of nanocrystalline copper at room temperature. Science 287: 1463-1466.

30. Yamakova V, Wolfa D, Phillpota SR, Gleiterb H (2002) Grain-boundary diffusion creep in nanocrystalline palladium by molecular-dynamics simulation. Acta Materialia 50: 61-73.

31. Sha JJ, Nozawa T, Park JS, Katoh Y, Kohyama A (2004) Effect of heat treatment on the tensile strength and creep resistance of advanced SiC fibers. J Nucl Mater 329: 592-596.

32. Sherby OD, Taleff EM (2002) Influence of grain size solute atoms and second-phase particles on creep behavior of polycrystalline solids. Mater Sci Eng A322: 89-99.

33. Meyer FD, Hillebrecht H (1998) Synthesis and crystal structures of ternary phases in the system Al/C/N. European Crystallographic Meeting. Praha, Czech Republic.

20

Designing and Fly Testing a Long Endurance Solar Unmanned Air Vehicle

Harasani W*

Faculty of Engineering, King Abdul Aziz University, Saudi Arabia

Abstract

The scope of the present work was to design, build and fly test a solar UAV, Sun Falcon 2 for long endurance day and night flight operations. A software program was written to design the UAV with appropriate aerodynamic attributes, power requirements and other flight mission constraints to keep the vehicle airborne for multi day and night operations. More specifically the design called for an least 12 hours of endurance during the day with solar panels deployed to absorb sufficient daylight energy to top up the on board batteries for the complete subsequent night flight mission. With Sun radiation levels averaging at about 6003 W/m2 during the Saudi day it was not too difficult to conform to multi day and night design requirements. The prototype Sun Falcon 2 has already been built and flight tested with satisfactory performance records satisfying the design criteria.

Keywords: Unmanned Air Vehicle; Fly testing; Designing

Introduction

This article reports on the second leg of the collaboration between the students and the staff of the Tokai and King Abdul Aziz universities tasked to design long duration UAVs. The present work, thus is followed on from the first design of Sun Falcon 1[1] which was successfully designed and flight tested at both locations Tokai University and King Abdul Aziz University. While solar powered aircraft are not unknown to the industry especially with the famous Gossamer [2] version of vehicles flying ever higher, heavier and longer to confirm the viability of solar power as a reliable means to power airborne vehicles. Elsewhere German Akaflieg group coming up with such large sized UAV's as the VELA2, NACRE and AMPAIR have proven that a variety of different UAV configurations can be flown but few as pointed out in Reference 1 and by Noth and Siegwart [3] UAVs have demonstrated the continuous day/night capability using the solar power.

Design Procedure

The overall design procedure demands an accurate weight estimation which can be supported comprehensively by the configuration aerodynamics throughout the complete itinerary of the flight mission. An even more demanding challenge is the adequate supply of the power dispensation especially during the sundown hours. The methodology is heavily based on the principle that the on board batteries would be sufficiently charged during the day operation by the on board solar panels to cover the power requirements during the night hours. Obviously there is an iterative process which optimises the weight against the aerodynamic loads and stability as well as the available power demands [4].

In terms of the actual design and performance specifications, the Sun Falcon 2 was estimated to have a 200 g weight with a continuous flight capability lasting at least 5 days (120 hours) with a cruise velocity of 30 km/h. It will have a climb rate of about 2 m/s operating by an electrical motor powered by a battery replenishable by solar panels. It will take-off in a normal fashion from an appropriate ground terrain and remain airborne at an altitude of about 500 m. A typical flight mission would then require the UAV to climb to a maximum height of 500 m, remain airborne continuously for 120 hours loiter at that altitude and ten descend to a prescribed location [5]. The final design will be subject to the safety regulations of the European Aviation Safety Agency and Certification Safety of Very Light Airplanes (CS-VLA).

Power Requirements

Power required to maintain uniform flight in cruise, is one of the most important parameters in sustained flight over a long period of time. This parameter in turn is dependent upon the aerodynamic performance of the vehicle, particularly the drag,

$$D = \frac{\rho * V^2 * C_D * S}{2} \tag{1}$$

Where

$$C_D = C_{Di} + C_{Do} \tag{2}$$

$$C_D = C_{Di} + C_{Do} \tag{3}$$

$$C_{Di} = \frac{C_L^2}{e * AR * \pi} \tag{4}$$

$$C_{Do} = 0.455 \times Log(Re)^2 \tag{5}$$

Where is the Oswald's efficiency factor and Re is the Reynold's number

According to the induced drag and viscous drag calculation, the total drag can be obtained by equation 6

$$D = \frac{\rho * V^2 * (Cd_0 \times \text{Total Flat Plat Area}) + (Cd_i \times S)}{2} \tag{6}$$

Since the required thrust is equal to the drag, the require power can be calculated from equation 7

$$P = \frac{\text{Thrust} * V}{\eta_m * \eta_p} \tag{7}$$

***Corresponding author:** Harasani W, Faculty of Engineering, King Abdul Aziz University, Jeddah, Saudi Arabia, E-mail: wharasani@kau.edu.sa

Weight Estimates

Along with the aerodynamic performance of the vehicle the weight of the components of vehicle must be obtained with extreme accuracy. The total weight of the vehicle, which must in all cases be

less than the aerodynamic lift $L = \dfrac{\rho V^2 C_L S}{2}$ is

given by the equation:

Total Weight = Weight of Batteries + Payload + Weight of Solar Cells + Weight of Airframe (7)

The weight of solar cells and the weight of airframe will be calculated from the following two equations:

Weight of Solar Cells [kg] = Solar Module Area $* 0.8$ (8)

Weight of Air frame [kg] = $\dfrac{0.2 * S^{1.55} * AR^{1.3}}{9.81}$ (9)

Another important feature of solar based flight is to have an accurate estimate of the energy generated from the solar cells which must exceed the energy requirements of the vehicle.

Energy Requirements

Energy needed is calculated from adding the energy needed from motor and energy needed to charge th-e batteries.

Energy needed at Day[w_h] = Energy motar [W. h]+ Energy to charge battery [W_h]

And the power required can be calculated from the equation:

Where energy needed for motars and energy to Charge battery can be calculated by the equation

Energy to mortar = Required power [w]$*$ Day time [h]   (10)

And the power required can be calculated from the equation :

Energy to charge battery [W_h] =

$\dfrac{\text{Capacity of Battery} [mAh] * \text{volt}[v] * \text{Number of Batteries}}{1000}$ (11)

The length of day the day time is assumed to be 12 hours in this analysis, and the number of batteries can be obtained from were 13 in the next section. From the batteries specifications in Table 1 the capacity of battery and volt can be found.

The above energy required has to be balanced against the energy available from the solar cells. The energy generated by solar cell is given by equation 12:

Energy generated by Solar, cell [Wh] = $\eta * \Omega_{rad} \left[\dfrac{W.hr}{m^2}\right] * S[m^2]$ (12)

Where Ωr is the radiation vector obtained from the Jeddah Met Office and S is the wing area and

is the solar cell efficiency.

Where minimum capacity and charging current can be found from the battery specifications and the number of batteries required can be obtained from equation 13

Number of Batteries $= \dfrac{\text{Total energy at night} [w_h]}{\text{Single Battery Energy} [w_h]}$ (13)

The total energy and single battery energy can be calculated from

Description	Amount	Unit
Wing span length	7.5	[m]
Aspect ratio	19.66	--
Total amount of global radiation	6006	[W.h/m²]
Chord at root	0.4	[m]
Cruise speed	33.4	[km/h]
Altitude	500	[m]
Efficiency of the propeller	0.8	--
Efficiency of the motor	0.85	--
Efficiency of the solar module	0.23	--
Predicted Weigh of the airframe	5	[kg]
Predicted Weight of the solar module	2.06	[kg]
Predicted Weight of the battery (Lithium ion)	3.69	[kg]
Predicted Weight of the plane	10.86	[kg]
Power during cruise	70	[W]

Table 1: Sun Falcon 2 Design Parameters.

equation 14 and 15 respectively:

Energy needed at $[W_h]$ = Night Time $[hr] *$ Power Need $[W]$ (14)

Single Battery Energy $[W_h] = \dfrac{\text{Minimum Capacity} [mAh] * \text{Voltage} [V]}{1000}$ (15)

The time needed to charge batteries is given by the following equation:

Time to charge battery $[h] = \dfrac{(\text{min capacity} [mAh]) * \text{number of battery} / 1000}{\text{Charging current} [A]}$ (16)

Figure 1 displays an EXCEL based SUNDOME flow chart diagram, in which the individual compartments are updated as the iterative design procedure is advanced to converge towards the final design. It has the ability to iterate between configuration aerodynamics, weight, energy and power requirements as well as the critical time to energy absorption from the daylight operations. The mission parameters are introduced into the mission specification module which feeds such information towards the aerodynamics module which uses such basic performance coefficients as the lift, drag and configuration geometry to arrive at the power and energy generations and other motor specifications. This information is in turn used in the power plant design to arrive at the solar cell and battery weight, area and power requirements. A final decision module interrogates whether the available energy and weight quantities satisfy the appropriate constraints and meet the critical time needed to replenish the battery charge for night operation in a repetitive manner. If the constraints are not satisfied than the frame geometry in terms of the aspect ratio and airframe weight is updated to repeat the convergence iteration.

Typical configuration geometry, aerodynamics, mission specification, and other energy and power requirements as well as battery and solar power requirements at any instant as they are updated during successive design iterations is shown in the Figure 2.

The configuration which was used to provide various geometry, weight and other aerodynamic characteristics is shown in Figure 3. The wing aerofoil section is based on an S8037 airfoil without a fuselage having an inverted V tail configuration supported by a tail boom which is extended forward for an appropriate c.g location. Detailed geometry

Given from mission specification:
Velocity 35 [km/hr] about 8.3 [m/s]
Payload 0.12 g + 5 Servo+ Auto polit
Altitude 500 m
Endurance time to fly 5 days

Sizing the motor [power]
calculate the total drag
Total drag 85 % = induced drag + viscous drag
force needed for motor = factor * total drag
D [N] = 1/2 ρ V² A C_D
Power needed for motor [watt] = force needed for motor[N] * V [m/s]
Power Req. for motor [watt]= 1/2 ρ V³ A C_D
power RC = need
out put:
power of the motor [watts]
weight of the motor [kg]
Energy generated [W.h]

Assumed
Span [m]
AR

Battery specifications

Power calculations
Energy needed to sustain night and day flight
out put:
Weight of the batteries[kg]
Solar cell area [m2]
Weight of the cells in [kg]
Energy needed [Wh]

Airframe weight [kg]

Airframe = (0.2 b $^{3.18}$ AR $^{-0.88}$)/g

Calculate weights and time to charge batteries

NO — Energy needed < Energy generated
Weight < Lift
time for charging battery < Day time

YES

The design is fezabile, and Predicted weights

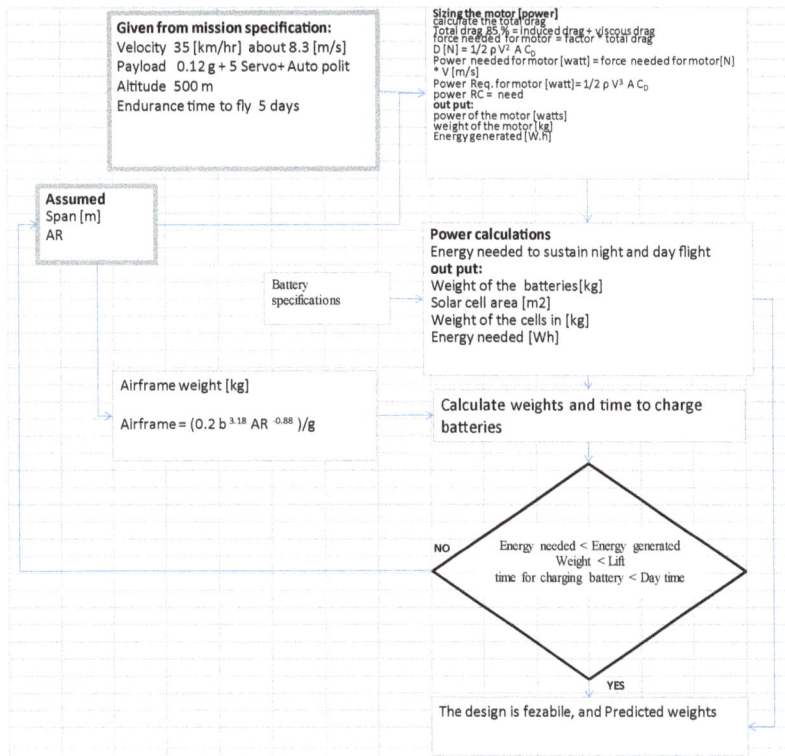

Figure 1: Sundowm flow chart.

Mission	
Cruising Velocity [km/hr]	33.48
Cruising Altitude H [m]	500.00
Payload [Kg] (Camera and Data Transmitter)	0.12
Rate of Climb R/C [m/s]	2.00
day time time [h]	12.00

Configuration	
Wing Span [m]	7.50
AR	19.66

Given Data	
Cruising CL	1.00
Oswald Efficiency Factor e (for High Mounted Wing)	0.80
Viscosity [Pa*s]	0.0000182
Sea Level Temperature To [K]	288.00
Gas Constant R [J/(K*Kg)]	287.00
Decrease Rate of Temperature a [K/m]	0.0065
Sea Level Pressure Po [Pa]	101325.00
Sea Level Density po [kg/m^3]	1.226
Efficiency of Motor(0.85) * Propeller(0.8)	0.680
Solar Cell Efficiency	0.23
Minimum Average Daily Horizontal Radiation [W*hr/m^2]	4300.00
Battery Weight [kg/(W*hr)]	0.005
Margin Time of Battery Usage [hrm]	0.014

Calculated Parameter	
Temperature T [K]	284.75
Pressure P [Pa]	95454.86
Density ρ [kg/m^3)	1.17
Cruising Velocity [m/s]	9.30
Wing Area S [m^2]	2.86
Chord Length c[m]	0.38
Reynolds Number Re	2.277E+05
Total Flat Plate Area [m^2] {Wing, Fuselage, Tail, and others (=2.6*Wing Area)}	7.44
Induced Drag Coefficient Cdi	0.020
Flat Plate Turbulent Viscous Drag Coefficient Cd0	0.00599
Total Drag [N]	5.18
Solar Module Area [m^2]	2.58
Night time [h]	12.00

Airframe	
Weight of Airframe [kg] (Including Weight of Motor, Controller, Propeller, Electric Circuit, Solar Cell, and Actuator)	5.00

Motor and Propeller	
Required Thrust T [N]	5.18
Required needed Power [W]	70.78
Weight of Motor [kg]	0.8
Charging current [A]	20

Stalling Speed	
Maximum CL	1.00
Stalling Speed at Maximum CL and Sea Level [m/s]	1.69
Airfoil: S8037	

predicted Weights	
weight of solar cells [kg]	2.060
weight of batteries [kg]	3.69
total weight [kg]	10.86

Battery specifications	
Nominal Capacity min [mAh]	2950
Nominal Capacity Typical [mAh]	3100
Approx. Weight [g]	45.5
Nominal Voltage [V]	3.6
Energy for a single battery, [W.h]	10.62

Night flight	
Night flight hours [h]	12.00
Energy total needed [W.h]	850
Number of Batteries needed	81
Weight of batteries [kg]	3.6855

Day flight	
Solar Energy per [m2, Wh/m2]	989
Energy needed from motor [W.h]	850
Energy from battery needed to charge [W.h]	903.96
Energy needed from solar [W.h]	1753.96
Solar cells area from Energy needed [m2]	1.7735

Energy Balance	
Energy generated from solar power [W.h]	2546.7
Energy needed < Energy generated	OK

Weight Balance	
Total Weight W [kg]	10.86
Lift L [N]	144.52
Lift L [Kg]	15
Lift must > weight predicted	OK

Time Balance	
time to charge battery [hr]	11.9475
time to charge < day time	OK

Figure 2: Unit Excel Sheet.

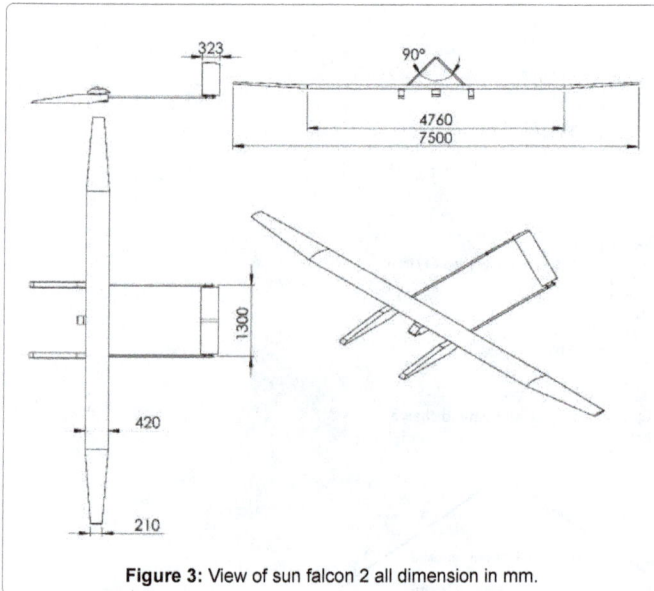

Figure 3: View of sun falcon 2 all dimension in mm.

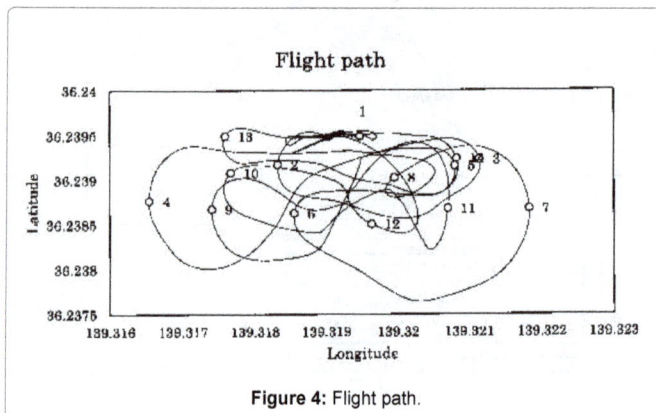

Figure 4: Flight path.

and other aerodynamic features of the final design are included in Table 1.

Loiter Control

The UAV was controlled to loiter within a specified airspace defined say by a number of prescribed boundary points identified by points 1 to 13 in the flight path as shown in Figure 4.

The stability and control feed-back system on board the UAV would manipulate the control surfaces to provide just the correct incremental acceleration at each step to advance it towards the next point on the trajectory. The incremental acceleration as discussed by Park et al. is obtained from the relationship:

$$\alpha_{scmd} = 2\frac{V^2}{L_1}\sin\eta$$

With reference to the Figure 5.

The incremental distance to next location on the point is repeated step by step until the antire trajectory is completed bring the UAV back to the point of the origin. This entire flight path can be repeated many times over to maintain the craft within the permissible airspace for the entire duration of the many days of flight.

Figure 6 shows the photograph of the SUN Falcon 2 during its

augrational flight in May 2015. The model was first flown under battery power and the UAV remained airborne for at least 30 minutes. It was controlled to fly within a specified airspace. After at least 30 minutes of flight, the UAV was flown singularly under power scooped up from the sunlight. It was able to fly without difficulty from the reserved power.

Conclusion

Sun Falcon 2 has been designed from the lessons learnt from the successful designs of Sun Falcon 1. Suitable temperature models have been used to assess the functions of the solar cells and their inevitable impact on the power /unit area distribution and the weight estimates. Meticulous design procedures with fast turnaround times, were devised to arrive at the most optimum design for the multi day operation of the Sun Falcon 2. The flight of the prototype Sun Falcon 2 demonstrated the successful design strategies adopted for the continuous flight vehicles. The first tests albeit for short duration of time demonstrated that the UAV could operate from the power reservoir recovered from the recharging of batteries from the power cells. The first flights were accomplished at Tokai University, Japan where the energy recovery source from a maritime climate is not as supportive as the more appropriate Jeddah desert climate.

Acknowledgment

This project was funded by the Deanship of Scientific Research (DSR) King Abdulaziz University, Jeddah, under the grand No. (431/009), the authors, therefore, acknowledge with thanks DSR technical and financial support, furthermore the

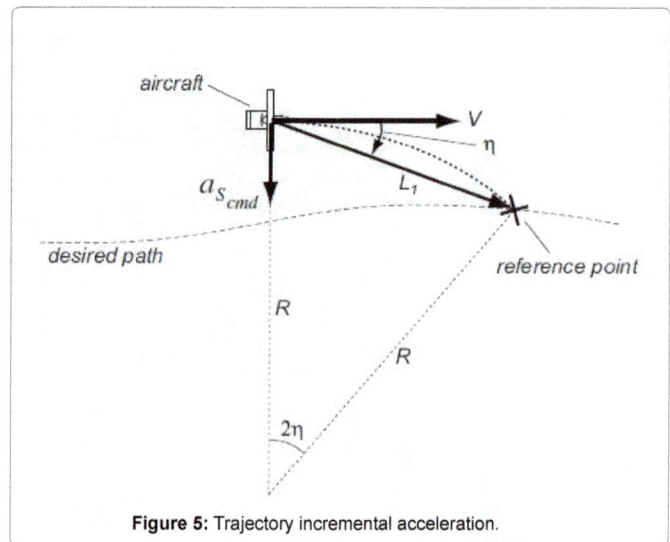

Figure 5: Trajectory incremental acceleration.

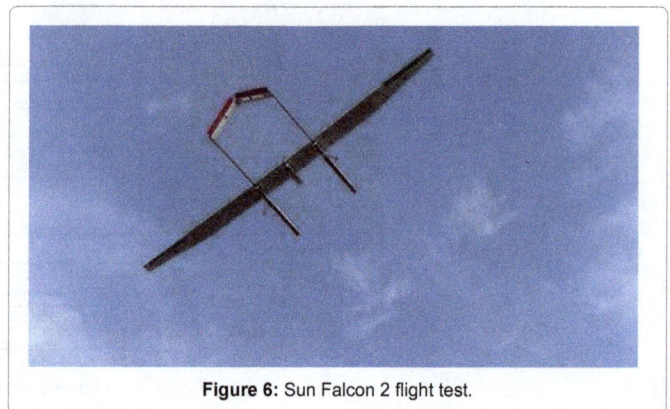

Figure 6: Sun Falcon 2 flight test.

authors would like to express their gratitude and appreciation to Tokai University for their technical help and support.

References

1. (2014) NASA Armstrong Fact Sheet. Solar Power Research.

2. Harasani W, Khalid , Arai N, Fakuda K, Hiroaka K (2014) Initial Design and Wing Aerodynamic Analysis of a Solar Powered UAV. The Aeronautical Journal.

3. Noth A, Siegwart R (2006) Design of Solar Powered Airplanes. Swiss Federal Institute of Technology.

4. Lyons CA, Boeren AP, Giguere P, Gopalarathnam A, Selig MS (1995) Summary of low speed airfoil data. University of Illionois at Urbana Champaign.

5. Park S, Deysr J, How P (2004) A New Nonlinear Guidance Logic for Trajectory Tracking AIAA - 4900.

21

Green Engines: Possible Damages by Firing Alternative Fuels and Protection

Hong Z[1], Cao G[2] and Chen WR[2]*

[1]*AVIC Commercial Aircraft Engine Co., Ltd., R & D Center, 3998 S. Lianhua Road, Shanghai Minhang, 201108, China*
[2]*AVIC Commercial Aircraft Engine Manufacturing Co., Ltd., 77 Hongyin Road, Shanghai Lingang, 201306, China*

Abstract

With the rising cost of fossil fuels along with greenhouse gas emission such as NO_x and CO_x, use of alternative fuels such as syngas and biofuels is intense interesting, and in the meantime using ceramic matrix composites that eliminate the need of film cooling in combustors, vanes and other hot section components to improve the efficiency of gas turbine engine and reduce the NO_x and CO_x emission becomes increasingly attractive for green engines. However, the alternative fuels have an increased hydrogen/carbon ratio; in turn during combustion it produces more water vapor than the conventional jet fuels. The increased water vapor level will have an impact on the protective oxide scale developed on the gas turbine hot section components, particularly on those made of SiC/SiC ceramic matrix composites (CMC), leading to an accelerated degradation of the turbine components. In addition, some alternative fuels derived from biomass may contain alkali elements such as potassium, sodium and calcium, as well as chlorine, sulfur and/or phosphorus, which may result in possible corrosion of the hot section components in gas turbines, leading to premature failure during service. This paper will review some of the alternative fuels and their combustion products, the possible damages to gas turbine hot section components, as well as some potential protective coatings that may mitigate such damage.

Keywords: Gas turbine; Alternative fuel; Ceramic matrix composite; Environmental barrier coating

Introduction

With the rising cost of fossil fuels (oil and natural gas) along with greenhouse gas (GHG) emission such as NO_x and CO_x, use of alternative fuels such as synthesis gas (syngas) and biofuels becomes increasingly attractive [1-3]. Syngas is a gas mixture that contains hydrogen and carbon monoxide generated by the gasification of a carbon containing fuel, such as coal or municipal waste [4]. Biofuels are produced from recently living organisms, most often referring to plants or plant-derived materials [5]. While syngas will increase the water vapor content of the combustion gas, the biofuels and their combustion products mainly contain alkali elements such as potassium, sodium and calcium, as well as chlorine, sulfur and/or phosphorus [6-10]. These may accelerate the degradation process of the hot section components of gas turbine engines such as combustor liners, nozzle guide vanes and turbine blades, leading to premature failure during service [11-13]. This paper will review some of the alternative fuels and their combustion products, the possible damages to the gas turbine hot section components by the combustion products, as well as some potential protective coatings to mitigate such damage.

Properties of Jet Fuels and Some Alternative Fuels

The rising costs of fuel and potential environmental benefits, along with an increasing desire to enhance the security of fuel supply have driven feasibility and viability assessment studies of alternative renewable fuels for commercial aviation applications. Among those studies fuels derived from biomass or synthesis from coal and natural gas via the Fischer-Tropsch (F-T) process [14] are of particular interest, with other alternative biomass based fuels, e.g. fatty acid methyl ester (FAME), also being considered. The properties of some jet fuels and fuel blends are listed in Table 1 [15]. It is noticeable that the hydrogen to carbon (H/C) ratio increases with the increase of biofuels volume.

The overall particulate matter (PM) number emissions over the International Civil Aviation Organization (ICAO) Landing Takeoff

(LTO) Cycle are reduced when burning the candidate alternative fuels using a CFM56-7B commercial jet engine [15], and the results are shown in Table 2. It is believed that both fuel aromatic content and H/C ratio will influence PM emissions.

For commercial jet engines, sulfur from the fossil fuel is generally limited to 0.3%; however, as reported in literatures, some biofuels may contain alkali elements such as potassium, sodium and calcium, as well as chlorine, sulfur and/or phosphorus [6-10].

Possible Damages to the Gas Turbine Engine Components by Burning Alternative Fuels

As the alternative fuels have higher H/C ratio, during combustion it will produce increased amount of water vapor than the conventional jet fuel. As a result, the increased water vapor level may have an impact to the hot section components in gas turbine engines, such as combustor liners, nozzle guide vanes and turbine blades.

In modern aircraft engines, hot section components are often protected by oxidation/hot corrosion resistant metallic coatings or thermal barrier coating (TBC) systems, composed of an yttria-stabilized-zirconia (YSZ) ceramic topcoat and a metallic bond coat deposited on the superalloy substrate [16]. The YSZ topcoat provides heat insulation to the components, while the metallic bond coat

***Corresponding author:** Chen WR, AVIC Commercial Aircraft Engine Manufacturing Co., Ltd., 77 Hongyin Road, Shanghai Lingang, 201306, China
E-mail: wrchenca@yahoo.com

provides adhesion between the YSZ topcoat and superalloy substrate as well as oxidation and hot corrosion protections. The blades of high pressure turbines (HPTs) that withstand most severe thermal cycling condition are usually protected by the state-of-the-art electron beam physical vapor deposition (EB-PVD) produced TBC system, which has a chemical vapor deposition (CVD) produced Pt-modified NiAl bond coat Figure 1.

β type Pt-modified NiAl coatings, produced by platinum electro-plating followed by CVD aluminizing, has excellent oxidation resistance and Type I hot corrosion resistance, and has a higher temperature capability than the conventional MCrAlY type oxidation and hot corrosion resistant coatings. However, a recent study shows that it degrades faster in water vapor environment by depleting Al from the β-NiAl phase at high temperatures Figure 2 [17], leading to a reduction of lifetime. Such degradation proceeds via the reaction between water vapor and the oxide scale developed on the surface of Pt-NiAl coating, $Al_2O_3+3H_2O= 2 Al (OH)_3(g)$. Influence of water vapor on another type of oxidation/hot corrosion resistant coating, i.e. the MCrAlY type coatings, is barely reported.

In gas turbine engines, metallic components that experience high temperature exposure are protected by TBCs, along with internal cooling through the cooling channels, which will also generate NOx emission. The use of ceramic or ceramic matrix composite (CMC) components becomes increasingly attractive because of the elimination of the need for film cooling, with the candidate ceramic materials being SiC, Si_3N_4, SiC/SiC CMC, and oxide/oxide CMCs. However, silicon based ceramic materials will be oxidized to form a SiO_2 scale on the component surface upon thermal exposure, which will react with water vapor at high temperatures, leading to volatilization of SiO_2 Figure 3 [18], also, SiC may react with water vapor directly. The reactions can be described as

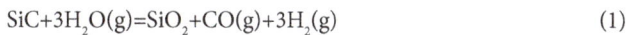

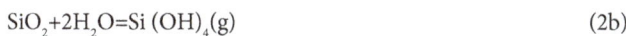

$$SiC+3H_2O(g)=SiO_2+CO(g)+3H_2(g) \qquad (1)$$

$$SiC+3/2O_2(g)= SiO_2+CO(g) \qquad (2a)$$

$$SiO_2+2H_2O=Si (OH)_4(g) \qquad (2b)$$

During combustion of biofuels that derived from biomass and municipal waste, it may produce some products containing alkali elements such as potassium, sodium and calcium, as well as chlorine, sulfur and/or phosphorus [6-10]. Although alkali chlorides are highly detrimental to the metallic components by reacting with the protective oxide scale at around 600°C [19], because gas turbines operates at very high temperatures, impact of alkali chlorides to gas turbine hot section components may not be as significant as sulfates in Type I hot corrosion [12]:

$$2NaCl (g)+SO_3(g)+ H_2O(g)=Na_2SO_4(l)+2 HCl(g) \qquad (3a)$$

$$Al_2O_3+Na_2O(in Na_2SO_4)=2 NaAl O_2 \qquad (3b)$$

For the CMC based components, molten sodium sulfate also partially decomposes and dissolves the protective silica scale [20-23].

Potential Protective Coatings to Prevent Damage from Water Vapor

Thus, the major problem in gas turbine engines burning alternative fuels is apparently water vapor formed during combustion, as a result of the high H/C ratio of the fuel. Water vapor consequently reacts with the protective oxide scales that developed on the component surfaces, making the oxide scales less protective and therefore leading to a faster degradation of the protective coating or substrate material, compared

to in the dry oxidation environment. Environmental barrier coatings (EBCs) to protect gas turbine components from water vapor attack at high temperatures are under developing, among which Ta_2O_5-based EBCs appear to be effective in inhibiting water vapor induced volatilization of SiO_2 scale Figure 4 [24]; however, Ta_2O_5-based coatings are not sufficient to act as a stand-alone EBC.

	Fuels and fuel blends					
	Jet-A	Jet-A1	20% FAME / 80% Jet-A1	40% FAME / 60% Jet-A1	50% F-T / 50% Jet-A1	100% F-T
Fuel identifier	Jet A	Jet A1	20% FAME	40% FAME	50% F-T	100% F-T
Specific gravity @ 15°C	0.803	0.797	0.808	0.825	0.776	0.755
Kinematic viscosity @ -20°C (mm²/s)	5.78	4.27	4.74	5.62	4.4	4.65
EI CO_2 (g/kg fuel burned)	3155	3155	3045	2942	3127	3100
Heat of combustion (kJ/kg)	43302	43300	42000	40300	43600	44100
H/C ratio	1.92	1.92	1.94	1.94	2.04	2.17
Aromatic content (vol.%)	12.3	18.5	14.8	11.1	9.25	< 0.2
Oxygen content (wt.%)	0	0	3.4	6.6	0	0

Table 1: Properties of fuels and fuel blends.

	Alternative fuel			
	20% FAME	40% FAME	50% F-T	100% F-T
PM number reduction	22% ± 7%	35% ± 6%	34% ± 7%	52% ± 4%
PM mass reduction	20% ± 8%	52% ± 5%	39% ± 7%	62% ± 7%

Table 2: Overall PM number and mass reductions over the LTO cycle achieved using candidate alternative fuels

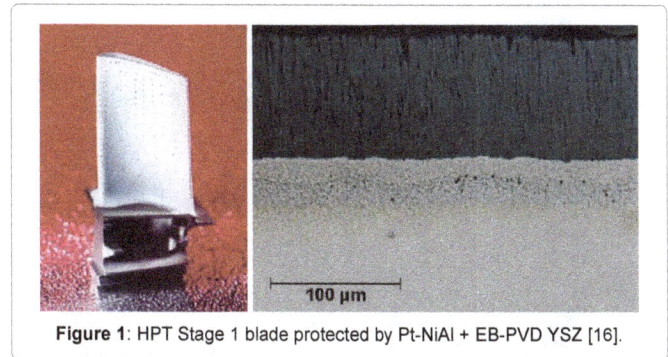

Figure 1: HPT Stage 1 blade protected by Pt-NiAl + EB-PVD YSZ [16].

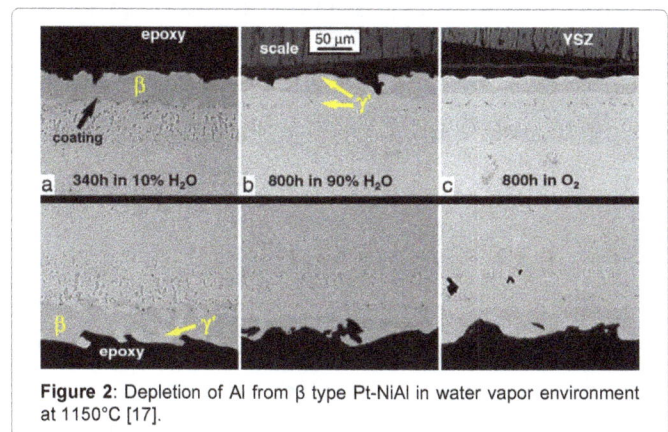

Figure 2: Depletion of Al from β type Pt-NiAl in water vapor environment at 1150°C [17].

The EBCs to protect SiC/SiC CMC from water vapor attack have to be very dense, because any open pores and/or cracks in the EBCs are openings for water vapor to penetrate to the SiC/SiC matrix material. A multilayered Si/mullite+BSAS (barium-strontium-aluminum-silicate)/ScSiO$_5$ EBC for SiC/SiC CMC vane is shown in Figure 5 [25]. The Si layer is applied to get a strong bond to the SiC/SiC matrix material, whereas the mullite+BSAS layer is applied to improve the crack resistance of the EBC layer. As the cross section shows, the EBC layer is nonporous, which prevents water vapor penetration.

Moreover, the design surface temperature for the CMC components is expected to exceed the BSAS thermal stability limit, thus, in addition to increase the EBC thickness, materials having high temperature capability and low thermal conductivity are required. In the meantime, environmental protection of the CMC substrate from the recession and hot corrosion will still be needed. One of the approaches to "building" a high-temperature T/EBC system would consist of an added high-temperature ceramic layer, for example, zirconia-based, on top of the three-layer BSAS type EBC system Figure 6 [26]. A transition layer with an intermediate coefficient of thermal expansion (CTE) will be needed to accommodate the CTE mismatch between the BSAS layer (about 5 ppm/$^\circ$C) and the top ceramic layer (about 10 ppm/$^\circ$C in the case of stabilized zirconia). This approach will result in a five-layer

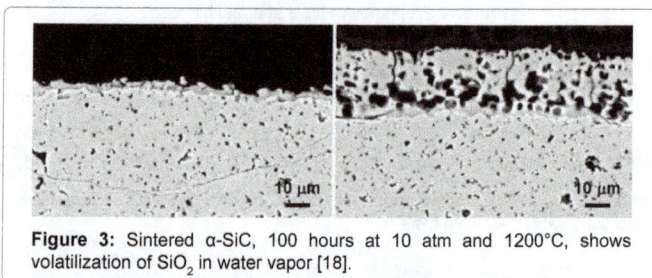

Figure 3: Sintered α-SiC, 100 hours at 10 atm and 1200°C, shows volatilization of SiO$_2$ in water vapor [18].

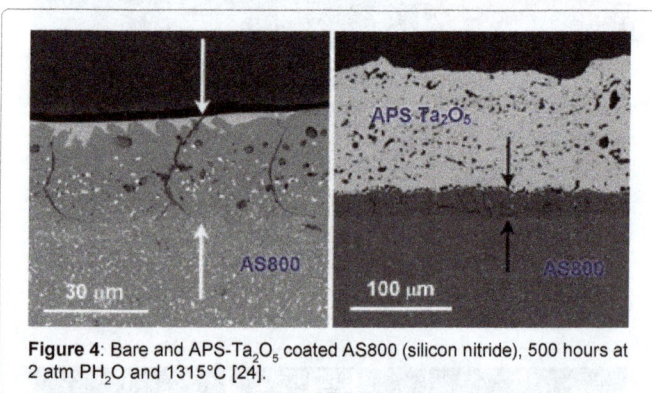

Figure 4: Bare and APS-Ta$_2$O$_5$ coated AS800 (silicon nitride), 500 hours at 2 atm PH$_2$O and 1315°C [24].

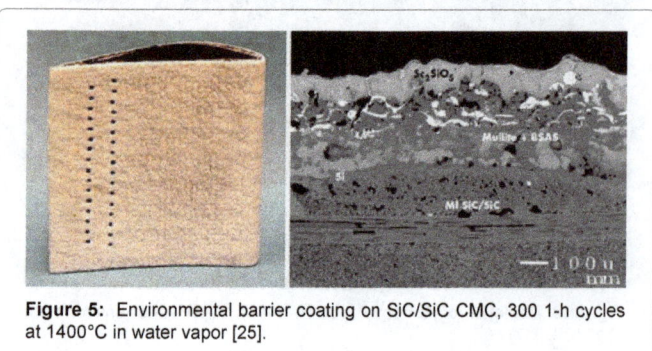

Figure 5: Environmental barrier coating on SiC/SiC CMC, 300 1-h cycles at 1400°C in water vapor [25].

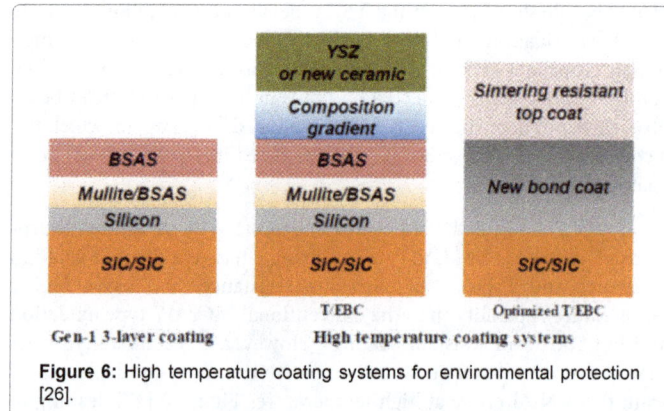

Figure 6: High temperature coating systems for environmental protection [26].

coating system which presents challenges in terms of processing, meeting thickness requirements, and cost. Therefore, new types of materials that could replace the three-layer EBC system, provide the environmental protection, and serve as a bond coat for the top ceramic layer are currently under development.

Concluding Remarks

Use of alternative fuels such as syngas/biofuels and ceramic matrix composites can help to reduce greenhouse gas emission such as NO$_x$ and CO$_x$; however, water vapor and possible molten salts formed during alternative fuel combustion may damage gas turbine hot section components by reacting with the protective oxide scales that developed on the component surfaces. Water vapour results in volatilization of the protective SiO$_2$ scale on the CMC surface leading to faster degradation of the CMC; however, influences of water vapor on oxidation and hot corrosion resistant metallic coatings, such as Pt modified NiAl and MCrAlY coatings, are not well explored, which deserve further and detailed investigation.

In gas turbine engines, environmental barrier coatings protect engine components from the volatilization and the resulting recession caused by water vapour and molten salt, but coating structure and deposition process need to be optimized. Moreover, thermal/environmental barrier coating systems merit continuous development to meet the requirement of high design temperature for the gas turbine hot section components. Furthermore, simplicity and affordability of the T/EBC systems warrant consideration.

Acknowledgement

This work is supported by the Science and Technology Commission of Shanghai Municipality Fundamental Research Program #12DJ1400400, R&D Programs #13521101100 and #14XD1424000.

References

1. Daggett D, Hadaller O, Hendricks R (2006) Walther, Alternative Fuels and Their Potential Impact on Aviation. NASA/TM-2006-214365, Cleveland OH.

2. Hileman JI, Stratton RW, Donohoo PE (2010) Energy Content and Alternative Jet Fuel Viability. Journal of Propulsion and Power 26: 1184-1196.

3. Blakey S, Rye L, Wilson CW (2011) Aviation gas turbine alternative fuels: A Review. The Proceedings of the Combustion Institute 33: 2863-2885.

4. Syngas.

5. Biofuel.

6. Nordin A (1994) Chemical elemental characteristics of biomass fuels Biomass and Bioenergy 6: 339-347.

7. Miles TR, Baxter LL, Bryers RW, Jenkins BM, Oden LL (1996) Boiler deposits from firing biomass fuels. Biomass and Bioenergy 10: 125-138.

8. Jensen PA, Stenholm M, Hald P (1997) Deposit investigation in straw-fired boilers. Energy and Fuels 11:1048-1055.

9. Christensen KA, Stenholm M, Livbjerg H (1998) The formation of submicron aerosol particles HCl and SO2 in straw-fired boilers. Journal of Aerosol Science. 29: 421-444.

10. Öhman M, Nordin A (2000) Bed agglomeration characteristics during fluidized bed combustion of biomass fuels Energy and Fuels 14: 169-178.

11. McCreath CG (1976) The role of evaporation of ingested sea salts in hot corrosion mechanisms of gas turbines Transactions of the Institute of Marine Engineers 88: 145-149.

12. Pettit FS, Meier GH (1984) Oxidation and hot corrosion of superalloys in Superalloys eds. M Gell CS Kartovich RH Bricknel WB Kent JF Radovich The Metallurgical Society of AIME Warrendale PA: 651-687.

13. Meier GH (1989) A review of advances in high-temperature corrosion. Materials Science and Engineering A 120: 1-11.

14. Fischer-Tropsch process

15. Lobo P, Hagen DE, Whitefield PD (2011) Comparison of PM Emissions from a Commercial Jet Engine Burning Conventional Biomass and Fischer-Tropsch Fuels, Environmental Science and Technology 45: 10744-10749.

16. Stolle R. Conventional and advanced coatings for turbine airfoils. MTU Aero Engines.

17. Pint BA, Garner GW, Lowe TM, Haynes JA, Zhang Y (2011) Effect of increased water vapor levels on TBC lifetime with Pt-containing bond coatings Surface & Coatings Technology 206: 1566-1570.

18. Tortorelli PF, More KL (2000) Oxidation of SiC-based materials at high water-vapor pressure. UT Materials Science and Engineering Seminar.

19. Pettersson C, Svensson JE, Johansson LG (2006) Corrosivity of KCl(g) at temperatures above its dew point-Initial stages of the high temperature corrosion of alloy Sanicro 28 at 600°C Materials Science Forum 522-523; 539-546.

20. Jacobson NS (1993) Corrosion of Silicon-Based Ceramics in Combustion Environments, Journal of the American Ceramic Society 76: 3-28.

21. Nickel KG (1997) Corrosion of non-oxide ceramics. Ceramics International 23: 127-133.

22. Kang NL, Robert AM (1996) Development and environmental durability of mullite and mullite/YSZ dual layer coatings for SiC and Si3N4 ceramics. Surface and Coatings Technology 86-87; 142-148.

23. Jacobson NS, Opila EJ, Lee KN (2001) Oxidation and Corrosion of Ceramics and Ceramic Matrix Composites. Current Opinion in Solid State and Materials Science 5: 301-309.

24. Schenk B (2002) Environmental Barrier Coating Development Challenges U.S. DOE EBC Workshop Nashville TN November: 6-7.

25. (2004) NASA Environmental Barrier Coatings (EBCs) for Ceramic Gas Turbine Components, LEW-17275-1.

26. Spitsberg I, Steibel J (2004) Thermal and Environmental Barrier Coatings for SiC/SiC CMCs in Aircraft Engine Applications. International Journal of Applied Ceramic Technology 1.

22

Hydromagnetic Nanofluid Flow in the Presence of Radiation and Heat Generation/Absorption Past an Exponential Stretching Sheet with Slip Boundary Conditions Using HAM

Shagaiya Daniel Y*

Department of Mathematical Sciences, Kaduna State University, Kaduna, Nigeria

Abstract

This paper considers the theoretical problem of hydrodynamic and slips boundary conditions over an exponential stretching sheet in the presence of radiation and heat generation/absorption. Similarity solutions are obtained from the governing boundary layer equations for different various of slip parameters, exponential parameter, magnetic field parameter, radiation parameter, heat source parameter, thermophoretic parameter and porosity parameter. The resulting couple system of equations which is highly nonlinear ordinary differential equations are solved semi-analytically using homotopy analysis method (HAM). Numerical results are obtained for non-dimensional governing parameters on skin friction, heat and mass transfer coefficient in the presence of suction. Comparison with published results seen in literature is in perfect agreement.

Keywords: HAM; Radiation; Hydromagnetic; Exponential parameter; Slip boundary condition; Porosity

Introduction

The boundary layer flows play a vital role in different areas of fluid mechanics as results of motion of a viscous fluid close to the surface. This stems from different fields of engineering and metallurgical applications such as wire drawing, hot rolling, plastic and metal extrusion, glass fiber production, paper production and crystal growing.

Flow and heat transfer features past a stretching sheet have important role in industrial applications, for example, extrusion of polymer sheet from a die. In the process of manufacturing sheets, the melt process from a slit and is thereafter stretched. The rates of cooling and stretching have a great influence on the quality of the finished product with desired features. This processes which involved cooling of a molten liquid through drawing into a cooling system. At first is the cooling liquid used then thereafter is the rate of stretching are the processes involved. Weak electrical conductivity of non-Newtonian fluid can be chosen for cooling fluid as their flow and after mentioned the regulated the heat transfer rate through some external medium. Highest rate of stretching is vital as results in sudden solidification then destroying the properties expected from the final product.

After the pioneering study of Hayat et al., [1] a great number of research works on a stretching sheet have been investigated and published by considering different governing parameters. The problem of radiative magnetohydrodynamic flow of Jeffrey nanofluid by an exponentially stretching sheet was resolved by Hussain et al., [2]. Hayat et al., [3] considered flow of viscoelastic fluid using thermal radiation and convective boundary condition on three-dimensional mixed convection fluid flow. Madhu et al., [4] presented an investigation on hydromagnetic mixed convection stagnation point flow of a power-law by using Non-Newtonian nanofluid close to a stretching surrounding with radiation and heat source/sink. Hayat et al., [5] worked on influence of inclined magnetic field in flow of third grade fluid flow in the presence of variable thermal conductivity. Effects of Newtonian heating on hydrodynamic flow of couple stress Nanofluid in the presence of viscous dissipation and Joule heating was discussed by Ramzan [6]. Hayat et al., [7] investigated temperature and concentration stratification influence on mixed

convection flow of an Oldroyd-B fluid in the presence of thermal radiation and chemical reaction. Nield et al., [8] worked on external natural convection through the porous media. Magnetohydrodynamic mixed convection on stagnation point fluid flow of a power law Non-Newtonian Nanofluid closed to a stretching surrounding medium in the presence of radiation and heat source was presented by Madhu and Naikoti [9]. Narayana et al., [10] used Numerical techniques to resolve the problem of magnetohydrodynamic heat and mass transfer of a Jeffrey fluid flow past a stretching sheet in the presence of chemical reaction and thermal radiation. Hayat et al., [11] investigated on the problem of three-dimensional flow past an exponentially stretching sheet medium in porous medium, chemical reaction and heat source/sink. Convection heat and mass transfer in hydromagnetic mixed convection flow of Jeffrey nanofluid past a radially stretching surface in the presence of thermal radiation was presented by Ashraf et al., [12]. Lakshmi and Suryanarayana [13] studied radiation on mixed convection flow of a Non-Newtonian nanofluid past a Non-linearly stretching sheet in the presence of heat source/sink. Hayat et al., [14] investigated Joule heating and thermal radiation in flow of third grade fluid over a radiative surface. Similarity solution to three dimensional boundary layer flow of second grade nanofluid flow over a stretching surface in the presence of thermal radiation and heat source/sink was presented by Hayat et al., [15].

Khan et al., [16] discussed on three dimensional of Oldroyd-B nanofluid flow closed to stretching surface in the presence of heat generation/absorption. An extension was made by Mabood et al.,

*Corresponding author: Shagaiya Daniel Y, Department of Mathematical Sciences, Kaduna State University, Kaduna, Nigeria
E-mail: shagaiya12@gmail.com

[17] to hydromagnetic flow past a non-isothermal stretching sheet with heat generation/absorption and transpiration using approximate analytical modeling. Effects of convective heat and mass transfer in hydromagnetic fluid flow of a nanofluid were discussed by Shehzad et al., [18]. Time dependent hydromagnetic Nano-Second Grade fluid flow was investigated by Ramzan and Muhammad [19] using induced by permeable vertical stretching sheet. Lin et al., [20] investigated hydromagnetic pseudo-plastic nanofluid using finite thin film on unsteady flow and heat transfer over stretching sheet with internal heat generation. An extension was made by Hayat et al., [21] to Jeffrey fluid flow over a bidirectional stretching sheet in the presence of source/sink.

Swapna et al., [22] used finite element analysis on radiative mixed convection magneto-micropolar fluid flow in a Darcian porous material in presence of variable viscosity and convective surface condition. Influence of thermophoresis and heat generation/absorption of hydromagnetic flow in the presence of oscillatory stretching sheet and chemically reactive species was presented by Sheikh and Zaheer [23]. Hayat et al., [24] presented a work on boundary layer flow of Maxwell nanofluid using semi-analytically method of solution (HAM).

Slip flow and hydromagnetic of nanofluid past an exponentially stretching sheet with permeable material in the presence of heat generation/absorption was discussed by Ranga Rao et al., [25]. Shen et al., [26] work on hydromagnetic mixed convection slip flow closed to a stagnation point flow with nonlinearly vertical stretching materials. Rashidi et al., [27] investigated on Double Diffusive hydromagnetic mixed convection slip flow.

The aim of the paper is to study theoretically the slip boundary conditions in velocity, temperature and concentration on the boundary layer flow and heat transfer of a nanofluid. Aftermentioned considered the no-slip thermal and solutal boundary condition. Effects of slip conditions are vital for fluids that shows wall slip seen in polymer solutions, suspensions, emulsions, foams, etc. Nanofluid exhibiting slips are vital in different technological areas of application such as in the polishing of internal cavities and artificial heart values. But in some instance where no-slip boundary condition may not be applicable, we may be force to look into slip boundary condition. In this case, this study fulfills this gap. The study investigated magnetohydrodynamic nanofluid flow in the presence of radiation and heat generation/absorption over an exponential stretching sheet with velocity, thermal, and solutal slip boundary condition. Moreso, the combined influences of exponential parameter, magnetic field parameter, radiation parameter, heat source parameter, thermophoresis parameter, and porosity parameter on heat transfer and boundary layer flow due to nanofluid are studied.

Mathematical Formulation

Consider two-dimensional steady, electrically conducting, incompressible viscous flow of a nanofluid with dissipative past an exponentially stretching sheet in permeable medium. Chosen a fluid stagnation-point flow been the origin of the coordinate system such that the x-axis is flat (direction) and y=0. The fluid flow surrounding is confined to y-axis is normal to the surface. In the x-axis we have two equal and opposite force on the stretching flat sheet such that the wall is continuous stretching along the initial fixed direction. The permeable medium with non-uniform permeability k is taken with a variable magnetic field B(x) is chosen along y direction. A variable heat source/sink Q (x), influence of suction/injection and thermophoretic are incorporated. The continuous stretching surface of the temperature and the nanoparticle fraction are said to have constant value T_w and

C_w respectively. For large value of y, the temperature and nanoparticle fraction are deemed to have constant values T_∞ and C respectively. The boundary layer governing equations of the fluid flow and temperature subjected to Boussinesq approximations are given as follows:

$$\frac{\partial u}{\partial x} + \frac{\partial u}{\partial y} = 0 \tag{1}$$

$$u\frac{\partial u}{\partial x} + v\frac{\partial u}{\partial y} = \frac{1}{\rho_{nf}}\left[\mu_{nf}\frac{\partial^2 u}{\partial y^2} + g(\rho\beta)_{nf}(T - T_\infty) - \sigma B^2(x)u - \frac{v_f}{k}u\right] \tag{2}$$

$$u\frac{\partial T}{\partial x} + v\frac{\partial T}{\partial y} = \alpha_{nf}\frac{\partial^2 T}{\partial y^2} - \frac{1}{(\rho c_p)_{nf}}\frac{\partial q_r}{\partial y} + \frac{Q(x)}{(\rho c_p)_{nf}}(T - T_\infty) + \frac{\mu_{nf}}{(\rho c_p)_{nf}}\left(\frac{\partial u}{\partial y}\right)^2 \tag{3}$$

$$u\frac{\partial C}{\partial x} + v\frac{\partial C}{\partial y} = D_m\frac{\partial^2 C}{\partial y^2} + \frac{kv_f}{T_r}\frac{\partial T}{\partial y}\frac{\partial C}{\partial y} - C\frac{\partial C}{\partial y} \tag{4}$$

The boundary conditions for the problem are given as follows:

$$u = U_w + L\frac{\partial u}{\partial y}, v = V_w, \quad T = T_w + K_1\frac{\partial T}{\partial y}, \quad C = C_w + K_2\frac{\partial C}{\partial y} \text{ at } y = 0$$

$$u \to U_\infty, \quad T \to T_\infty, \quad C \to C_\infty \text{ as } \to \infty \tag{5}$$

where $\alpha_{nf} = \frac{k}{(c\rho)_{nf}}$, $u = U_w(x) = U_0 e^{Nx/L_g}$ is the surface velocity and $V_w(x) = v_0 e^{Nx/2L_g}$ is the velocity at the surface. The velocity components along x and y-axis are u and v respectively. v is the kinematic viscosity, T is the temperature inside the boundary layer, $(c\rho)_{nf}$ effective heat capacity of the nanoparticle, ρ is the density, $V_w(x) > 0$ is the suction and $V_w(x) < 0$ represents injection on the permeable surface. The radiative heat flux q_r under Rosseland approximation is given by

$$q_r = -\frac{4\sigma_1}{3k}\frac{\partial T^4}{\partial y} \tag{6}$$

Here σ_1 and k are the Stefan-Boltzmann constant and mean absorption coefficient, respectively. Let the temperature difference in the flow be sufficiently small in such a way that the term T^4 will be expressed as linear function of temperature. By using Taylor series about a free stream temperature T_∞ is represented as

$$T^4 = T_\infty^4 + 4T_\infty^3(T - T_\infty) + 6T_\infty^3(T - T_\infty)^2 + \dots \tag{7}$$

By neglecting higher-order terms in equation (7) beyond the first order in $(T-T_\infty)$ we have:

$$T^4 \cong 4T_\infty^3 T - 3T_\infty^4 \tag{8}$$

Substituting equation (8) in equation (6) we get:

$$q_r = -\frac{16T_\infty^3 \sigma_1}{3k}\frac{\partial T}{\partial y} \tag{9}$$

We now defined similarity transformation variables as follows:

$$\eta = y\sqrt{\frac{U_0}{2v_{nf}Le^{Nx/2L_g}}}, \quad u = U_0 e^{Nx/2L_g}f'(\eta), \quad C = C_0 e^{Nx/2L_g}\beta(\eta)$$

$$T = T_w = T_\infty + T_0 e^{Nx/2L_g}\theta(\eta), v = -N\sqrt{v_{nf}U_0/2L_g e^{Nx/2L_g}}\{f(\eta) + \eta f'(\eta)\} \tag{10}$$

The equation of continuity is satisfied if a stream function $\Psi(x, y)$ is defined as:

$$u = \frac{\partial \psi}{\partial y}, \quad v = -\frac{\partial \psi}{\partial x} \tag{11}$$

Using the similarity transformation variables, the governing partial differential equations (1)-(4) are now transformed to ordinary differential equation as presented:

$$f''' + Nff'' - 2Nf'^2 + Gr\theta - (M+K)f' = 0 \tag{12}$$

$$\left(\frac{1}{Pr} + R\right)\theta'' - 4Nf'\theta + Nf\theta' + Ec(f'')^2 + Q_H\theta = 0 \tag{13}$$

$$\beta'' - NSc(4f'\beta - f\beta') - \frac{Sc\tau}{\theta}(\theta'\beta' + \theta'') = 0 \tag{14}$$

With boundary conditions:

$$f(0) = s, \quad f'(0)1 + \gamma f'(0), \quad \theta(0) = 1 + \delta\theta'(0), \quad \beta(0) = 1 + \alpha\beta'(0) \text{ at} \eta = 0$$

$$f'(\infty) = \theta(\infty) = \beta(\infty) = 0 \quad \text{as } \eta \to \infty \tag{15}$$

Where the governing parameters are given as follows:

$$Gr = \frac{2L_g\beta_{nf}T_0}{U_0^2}, \quad M = \frac{2L_g\sigma B_0^2}{\rho_{nf}U_0}, \quad K = \frac{2L_g v_{nf}}{k_0 U_0}, \quad Pr = \frac{v_{nf}}{\alpha_{nf}}, \quad R = \frac{16\sigma_1 T_\infty^3}{3k(\mu c_p)_{nf}}$$

$$Ec = \frac{U_0^2}{T_0(c_p)_{nf}}, \quad Q_H = \frac{2L_g Q_0}{(\rho c_p)_{nf}U_0}, \quad Sc = \frac{v_{nf}}{D_m}, \quad \tau = -\frac{k(T-T_\infty)}{T_r}, \quad s = -V_w(x)\sqrt{v_{nf}U_0/2L_g}$$

$$\gamma = L\sqrt{\frac{U_0}{v}}, \qquad \delta = K_1\sqrt{\frac{U_0}{v}}, \qquad \alpha = K_2\sqrt{\frac{U_0}{v}} \tag{16}$$

Where Gr is the thermal Grashof number, M is the Hartmann number, K is the porosity parameter, Pr is the Prandtl number, R is the radiation parameter, Ec is the Eckert number, Q_H is the internal heat source/sink, Sc is the Schmidt number, is the thermophoretic parameter, s is the permeability of the porous surface ($s>0$ is suction and $s<0$ is injection). γ, δ, α are velocity, thermal and concentration slip parameters, respectively. f', θ & β are the dimensionless velocity, temperature and nanoparticle concentration respectively and Prime represents differentiation with respect to η (similarity variable). The physical quantities of interest in our problem are the local skin friction coefficient c_f, the local Nusselt number Nu_x and the local Sherwood number sh_x, Reynolds number Re_x are defined as:

$$c_f = \frac{\tau_w}{\rho_{nf}U_w^2}, \qquad Nu_x = \frac{xq_w}{k(T_w - T_\infty)}, \qquad sh_x = \frac{xh_m}{D_m(\beta_w - \beta_\infty)} \tag{17}$$

Where the mass heat flux h_m and wall heat flux q_w are presented as:

$$h_m = -D_m\left(\frac{\partial\beta}{\partial y}\right)_{y=0}, \qquad q_w = -k\left(\frac{\partial T}{\partial y}\right)_{y=0} \tag{18}$$

From the above equations, we have:

$$c_f\sqrt{Re_x} = -f''(0), \qquad \frac{Nu_x}{Re_x^{1/2}} = -(1+R)\theta'(0), \qquad \frac{sh_x}{Re_x^{1/2}} = -h'(0)s \tag{19}$$

Homotopy Analysis Method (HAM)

In this section, we will apply HAM an idea from Liao [28-34] to solve equations (12)-(15). We choose the set of base functions { $\eta^k e^{(-n\eta)}; k \geq 0, n \geq 0$ is an integer} to approximate the unknown functions $f(\eta)$, $\theta(\eta)$ and $\beta(\eta)$ respectively, as

$$f(\eta) = f_0(\eta) + \sum_{i=1}^{+\infty} f_i(\eta), \quad \theta(\eta) = \theta_0(\eta) + \sum_{i=1}^{+\infty}\theta_i(\eta), \quad \beta(\eta) = \beta_0(\eta) + \sum_{i=1}^{+\infty}\beta_i(\eta) \tag{20}$$

where

$$f_0(\eta) = s\eta + \frac{1}{1+\gamma}(s-1)(e^{-k\eta} - 1), \theta_0(\eta) = \frac{1}{1+\delta}e^{-\eta}, \beta_0(\eta) = \frac{1}{1+\alpha}e^{-\eta} \tag{21}$$

The auxiliary linear operators are selected as:

$$\mathcal{L}_f[f] = \frac{\partial^3 f}{\partial\eta^3} + \frac{\partial^2 f}{\partial\eta^2} \tag{22}$$

$$\mathcal{L}_\theta[\theta] = \frac{\partial^2\theta}{\partial\eta^2} - \theta \tag{23}$$

$$\mathcal{L}_\beta[\beta] = \frac{\partial^2\beta}{\partial\eta^2} - \beta, \tag{24}$$

are taken to be the initial guess approximations, with the property:

$$\mathcal{L}_f[c_1 + c_2\eta + c_3\exp(-\eta)] = 0 \tag{25}$$

$$\mathcal{L}_\theta[c_4\exp(-\eta) + c_5\exp(\eta)] = 0 \tag{26}$$

$$\mathcal{L}_\beta[c_6\exp(-\eta) + c_7\exp(\eta)] = 0 \tag{27}$$

Here c_i (where $i = 1,2,3,4,5,6,7$) are they constants. Base on equations (17) and (18), the non-linear operators

$$\mathbb{1}_f[\hat{f}(\eta;q), \hat{\theta}(\eta;q)] = \frac{\partial^3 f(\eta;q)}{\partial\eta^3} + Nf(\eta;q)\frac{\partial^2 f(\eta;q)}{\partial\eta^2} - 2N\left[\frac{\partial f(\eta;q)}{\partial\eta}\right]^2 + Gr\theta(\eta;q) - (M+K)\frac{\partial f(\eta;q)}{\partial\eta} = 0 \tag{28}$$

$$\mathbb{1}_\theta[\hat{f}(\eta;q), \hat{\theta}(\eta;q)] = \left(\frac{1}{Pr} + R\right)\frac{\partial^2\theta(\eta;q)}{\partial\eta^2} - 4N\theta(\eta;q)\frac{\partial f(\eta;q)}{\partial\eta} + Nf(\eta;q)\frac{\partial\theta(\eta;q)}{\partial\eta} + Ec\left(\frac{\partial f(\eta;q)}{\partial\eta}\right)^2 + Q_H\theta(\eta;q) = 0 \tag{29}$$

$$\mathbb{1}_\beta[\hat{f}(\eta;q), \hat{\theta}(\eta;q), \hat{\beta}(\eta;q)] = \frac{\partial^2\beta(\eta;q)}{\partial\eta^2} - NSc\left(4\frac{\partial f(\eta;q)}{\partial\eta}\beta(\eta;q) - f(\eta;q)\frac{\partial\beta(\eta;q)}{\partial\eta}\right) - \frac{Sc\tau}{\theta(\eta;q)}\left(\frac{\partial\theta(\eta;q)}{\partial\eta}\frac{\partial\beta(\eta;q)}{\partial\eta} + \frac{\partial^2\theta(\eta;q)}{\partial\eta^2}\right) = 0 \tag{30}$$

Where $q\dot{\in}[0,1]$ is an embedding parameter, and $\hat{f}(\eta;q), \hat{\theta}(\eta;q)$ and $\hat{\beta}(\eta;q)$ are kind of mapping function for $f(\eta)$, $\theta(\eta)$, and $\beta(\eta)$ respectively. From the operators, we can construct the zeroth-order deformation equations as:

$$(1-q)\mathcal{L}_f[\hat{f}(\eta;q) - f_0(x)] = qh\mathrm{N}_f[\hat{f}(\eta;q)] \tag{31}$$

$$(1-q)\mathcal{L}_\theta[\hat{\theta}(\eta;q) - \theta_0(x)] = qh\mathrm{N}_\theta[\hat{f}(\eta;q), \hat{\theta}(\eta;q), \hat{\beta}(\eta;q)] \tag{32}$$

$$(1-q)\mathcal{L}_\beta[\hat{\beta}(\eta;q) - \beta_0(x)] = qh\mathrm{N}_\beta[\hat{f}(\eta;q), \hat{\theta}(\eta;q), \hat{\beta}(\eta;q)] \tag{33}$$

Where h is an auxiliary non-zero parameter. The boundary conditions for equations (31)-(33) are presented as:

$$\left.\frac{\partial\hat{f}(\eta;q)}{\partial\eta}\right|_{\eta=0} = 1 + \gamma\left.\frac{\partial^2\hat{f}(\eta;q)}{\partial\eta^2}\right|_{\eta=0}, \quad \left.\frac{\partial\hat{\theta}(\eta;q)}{\partial\eta}\right|_{\eta=0} = 1 + \delta\left.\frac{\partial\hat{\theta}(\eta;q)}{\partial\eta}\right|_{\eta=0}$$

$$\left.\frac{\partial\hat{\beta}(\eta;q)}{\partial\eta}\right|_{\eta=0} = 1 + \alpha\left.\frac{\partial\hat{\beta}(\eta;q)}{\partial\eta}\right|_{\eta=0}, \left.\hat{f}(\eta;q)\right|_{\eta=0} = s \tag{34}$$

$$\left.\frac{\partial\hat{f}(\eta;q)}{\partial\eta}\right|_{\eta=\infty} = 0, \quad \left.\hat{\theta}(\eta;q)\right|_{\eta=\infty} = 0, \quad \left.\hat{\beta}(\eta;q)\right|_{\eta=\infty} = 0 \tag{35}$$

Clearly, when $q=0$ and $q=1$, the above zeroth-order deformation equations have the following solutions.

$$\hat{f}(\eta;0) = f_0(\eta), \qquad \hat{\theta}(\eta;0) = \theta_0(\eta), \qquad \hat{\beta}(\eta;0) = \beta_0(\eta) \tag{36}$$

$$\hat{f}(\eta;1) = f(\eta), \qquad \hat{\theta}(\eta;1) = \theta(\eta), \quad \hat{\beta}(\eta;1) = \beta(\eta) \tag{37}$$

When q increases from 0 to 1, and $\hat{f}(\eta;q)$, $\hat{\theta}(\eta;q)$ and $\hat{\beta}(\eta;q)$ vary from $f_0(\eta)$, $\theta_0(\eta)$ and $\beta_0(\eta)$ to $f(\eta)$, $\theta(\eta)$ and $\beta(\eta)$. From Taylor's theorem and expanding equation (36), we obtained

$$\hat{f}(\eta;0) = f_0(\eta) + \sum_{m=1}^{+\infty} f_m(\eta)q^m \tag{38}$$

$$\hat{\theta}(\eta;0) = \theta_0(\eta) + \sum_{m=1}^{+\infty}\theta_m(\eta)q^m \tag{39}$$

Hydromagnetic Nanofluid Flow in the Presence of Radiation and Heat Generation/Absorption Past...

147

$$\hat{\beta}(\eta;0) = \beta_0(\eta) + \sum_{m=1}^{+\infty} \beta_m(\eta) q^m \qquad (40)$$

where

$$f_m(x) = \frac{1}{m!}\frac{\partial^m \hat{f}(\eta;q)}{\partial q^m}|_{q=0}, \quad \theta_m(x) = \frac{1}{m!}\frac{\partial^m \hat{\theta}(\eta;q)}{\partial q^m}|_{q=0}, \quad \beta_m(x) = \frac{1}{m!}\frac{\partial^m \hat{\beta}(\eta;q)}{\partial q^m}|_{q=0} \quad (41)$$

The convergence of the series solutions (38)-(40) depend upon the choice of auxiliary parameter h. Assume that h is chosen such that the series solution (31)-(33) are convergent at $q=0$ & $q=1$, then due to equations (38)-(40). The i th-order deformation equations, we differentiate equations (31)-(33) up to m times with respect to q and divide by $m!$ and then set $q=0$. The resulting deformation equation at the m th-order are

$$\mathcal{L}_f\left[f_m(\eta) - \chi_m f_{m-1}(\eta)\right] = h R_{f,m}(\eta) \qquad (42)$$

$$\mathcal{L}_\theta\left[\theta_m(\eta) - \chi_m \theta_{m-1}(\eta)\right] = h R_{\theta,m}(\eta) \qquad (43)$$

$$\mathcal{L}_\beta\left[\beta_m(\eta) - \chi_m \beta_{m-1}(\eta)\right] = h R_{\beta,m}(\eta) \qquad (44)$$

with the following boundary conditions

$$f_m(0) = 0, \qquad f'_m(0) = 0, \qquad f'_m(\infty) = 0 \quad (45)$$

$$\theta_m(0) = 0, \qquad \theta_m(\infty) = 0 \quad (46)$$

$$\beta_m(0) = 0, \qquad \beta_m(\infty) = 0 \quad (47)$$

where

$$R_{f,m}(\eta) = f'''_{i-1}(\eta) + N\sum_{k=0}^{i-1} f_{i-1}(\eta) f'_k(\eta) - 2N\sum_{k=0}^{i-1}\sum_{j=0}^{k}\left(f'_{i-1-k}(\eta) f'_{k-1}(\eta)\right) + Gr\theta_{i-1}(\eta) - (M+K)f_{i-1}(\eta) \quad (48)$$

$$R_{\theta,m}(\eta) = \left(\frac{1}{Pr} + R\right)\theta''_{i-1}(\eta) - 4N\theta_{i-1}(\eta) f'_{i-1}(\eta) + N\theta'_{i-1}(\eta) f_{i-1}(\eta) + Ec f'^2_{i-1}(\eta) + Q_H\theta_{i-1}(\eta) \quad (49)$$

$$R_{\beta,m}(\eta) = \beta''_{i-1}(\eta) - NSc\left(f'_{i-1}(\eta)\beta_{i-1}(\eta) - f_{i-1}(\eta)\beta'_{i-1}(\eta)\right) - \frac{Sc\tau}{\theta_{i-1}(\eta)}\left(\theta'_{i-1}(\eta)\beta_{i-1}(\eta) + \theta'_{i-1}(\eta)\right) \quad (50)$$

and

$$\chi_m = \begin{cases} 0, & m \le 1 \\ \\ 1, & m > 1 \end{cases} \qquad (51)$$

The general solutions of equations (39)-(44) which can be represent as follows:

$$f_m(\eta) - \chi_m f_{m-1}(\eta) = f^*_m(\eta) + C_1^m + C_2^m\eta + C_3^m\exp(-\eta) \qquad (52)$$

$$\theta_m(\eta) - \chi_m \theta_{m-1}(\eta) = \theta^*_m(\eta) + C_4^m\exp(-\eta) + C_5^m\exp(\eta) \qquad (53)$$

$$\beta_m(\eta) - \chi_m \beta_{m-1}(\eta) = \beta^*_m(\eta) + C_6^m\exp(-\eta) + C_7^m\exp(\eta) \qquad (54)$$

Assume $f^*_m(\eta)$, $\theta^*_m(\eta)$ and $\beta^*_m(\eta)$ represent the particular solutions of equations (31)-(33), one after the other as C_i^m $(i=1,2,3,...)$ the computation, can easily be compute by symbolic computation software such as Mathematica, Maple, Matlab etc for $m=1,2,3...$ in that order. Now by computation using homotopy analysis method (HAM) the problems containing equations (42)-(44) see Hayat et al., [31]. Therefore, the Figures 1 and 2 are shown to determine the values of auxiliary parameters h_f, h and h_β for the convergent solutions. It was found that the admissible convergent ranges are $-2.8 \le h_f \le 2.8$ and $-2.2 \le (h_\theta, h_\beta) \le 1.5$ respectively. It obvious the solutions converges in the whole region of η ($0 < \eta < \infty$) for hf, h_θ and $h_\beta = -0.5$

Results and Discussion

The graphical and numerical results representative of velocity,

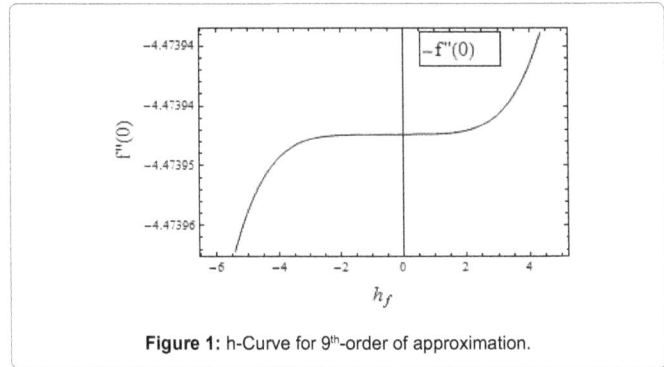

Figure 1: h-Curve for 9th-order of approximation.

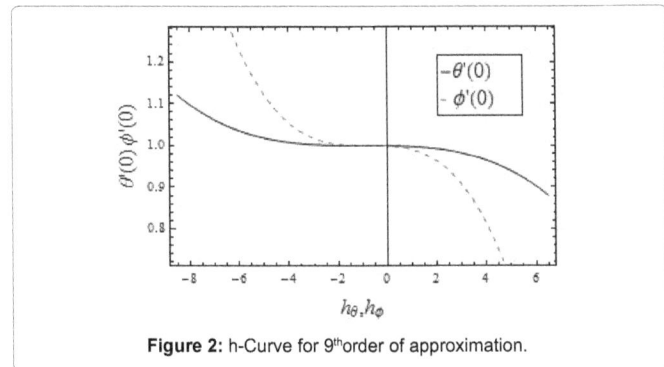

Figure 2: h-Curve for 9th order of approximation.

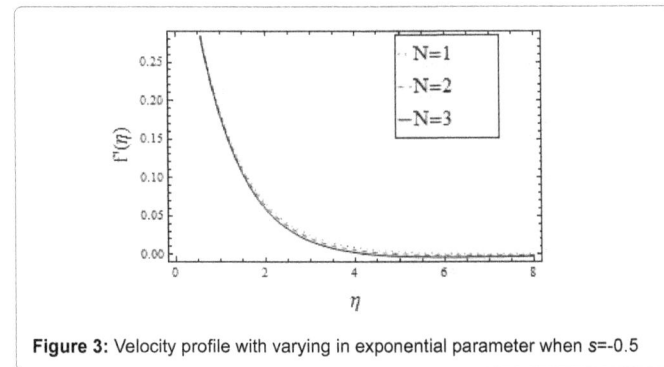

Figure 3: Velocity profile with varying in exponential parameter when s=-0.5

temperature, concentration, skin friction, heat and mass transfer coefficients, for various values of the thermophysical parameters controlling the dynamics nanofluid in the flow region. In Table 1 below shows the numerical values in order to validate the new computation scheme of our studied using homotopy analysis method (HAM). The results for special case of the current study (i.e permeable flat surface s=0.5 are compared with the work of Sandeep and Sulochana [35]. It was found that the theoretical results are in perfect agreement with the one published in literature. This gives us the confidence on our new approach scheme to the numerical results seen and reported subsequently. Figure 3 shows the effects of exponential parameter on the velocity profile when (s<0) injection. For different values, the nanofluid velocity decreases from the fixed plate surface and hence attains the free stream value which statisfied the boundary conditions. Generally, the momentum boundary layer thickness decreases with increases in the fluid exponential parameter at the plate surface. For suction see Figure 4. Similar trend happened, but is more pronounced in injection(s<0). Figure 5 illustrates the influence of exponential parameter on temperature profile. The nanofluid temperature gets to the highest at the plate surface and decreases greatly to zero free

Pr	N	M	K	R	Q_H	T	Gr	Sc	Sandeep, Sulochana [35]			Present Result		
									f"(0)	θ'(0)	B'(0)	f"(0)	θ'(0)	B'(0)
1									-2.0552323	2.390596	2.560407	-2.552322	2.390597	2.560407
	1								-2.551039	2.307011	2.554629	-2.551039	2.307011	2.554628
		1							-2.626792	2.304187	2.552583	-2.626791	2.304187	2.552583
			0.5						-2.551039	2.307011	2.554629	-2.551039	2.307011	2.554629
				0.5					-2.551039	2.307011	2.554629	-2.551039	2.307011	2.554629
					0.5				-2.551039	2.307011	2.554629	-2.551039	2.307011	2.554629
						0.1			-2.551039	2.307011	2.469642	-2.551039	2.307011	2.469643
							1		-2.551039	2.307011	2.554629	-2.551039	2.307011	2.554629
								0.2	-2.551039	2.307011	2.184415	-2.551039	2.307011	2.184414

Table 1: Comparison test results. Values of the skin friction, heat and mass transfer coefficients in present of suction.

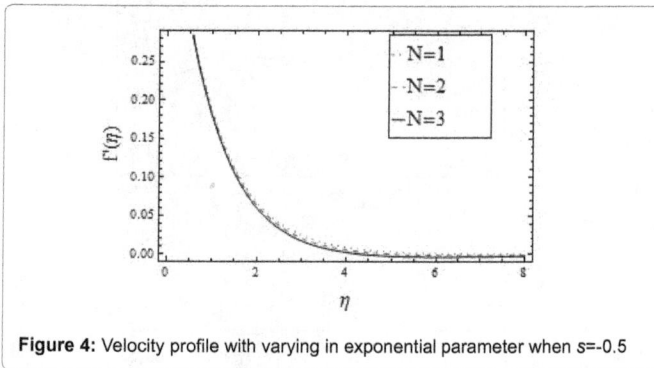

Figure 4: Velocity profile with varying in exponential parameter when s=-0.5

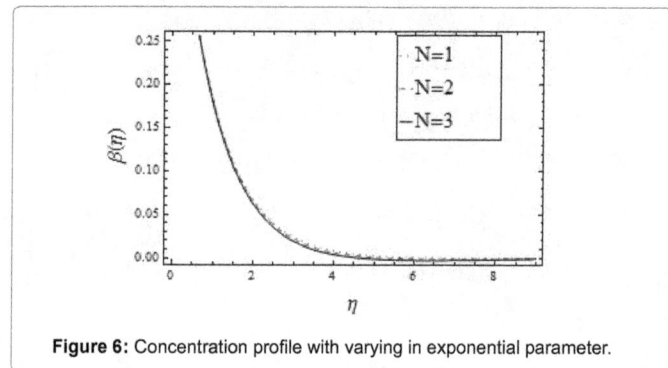

Figure 6: Concentration profile with varying in exponential parameter.

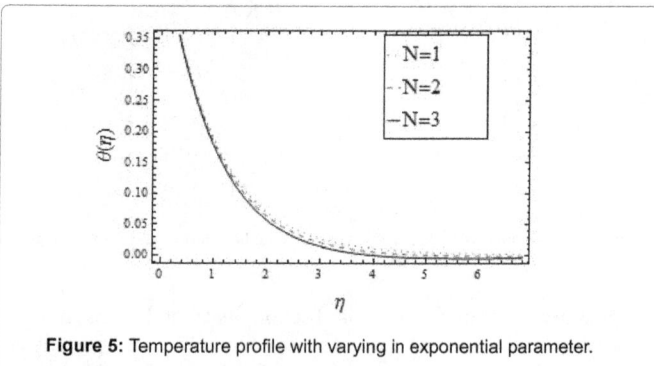

Figure 5: Temperature profile with varying in exponential parameter.

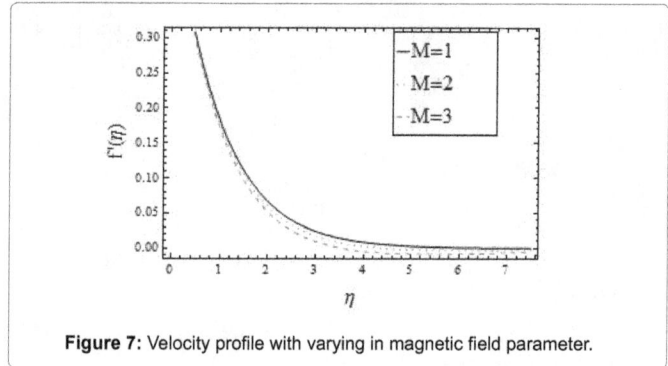

Figure 7: Velocity profile with varying in magnetic field parameter.

stream value satisfying the boundary conditions. An increase in fluid exponential parameter yields a decrease in the plate surface temperature and the thermal boundary layer thickness. Notice that the thermal boundary layer thickens with exponential parameter leading to an increasing in nanofluid temperature. Figure 6, shows the effects of exponential parameter on the concentration profile. It is observed that an increase in the exponential parameter decreases the concentration profile. The exponential parameter decreases the concentration boundary layer thickness.

The influence of magnetic field parameter on the velocity profile is shown in Figure 7. It was observed that the velocity profile depreciate monotonically as η increases. At any point on the boundary layer the velocity of a nanofluid increases with an increase in magnetic field parameter. The reason is that application of the magnetic field on the fluid flow region creates a Lorentz force which retards the fluid speed and as a consequence the velocity of the fluid within the domain of

the boundary layer increases. The increases in the boundary layer thickness also increase with an increase on the strength of the applied magnetic field. Hence the surface velocity of the stretching sheet can be controlled by controlling the strength of the applied magnetic field. In Figure 8, shows the effects of radiation parameter on the temperature profile. We found that the influence of radiation parameter is directly proportional to the radiation parameter. It is noticed that an increase of the radiation parameter increases the temperature profile within the thermal boundary layer. The thickness of the thermal boundary layer thickness decreases with an increase of the radiation influence. In Figure 9, describe the influence of heat source (Q_H >0) in the boundary layer on the temperature profile. The presence of a heat source in the thermal boundary layer generates energy which causes the temperature of the nanofluid to increases. This increases in temperature yields an increase in the nanofluid fluid.

Figure 10 shows the effect of thermophoretic parameter on the

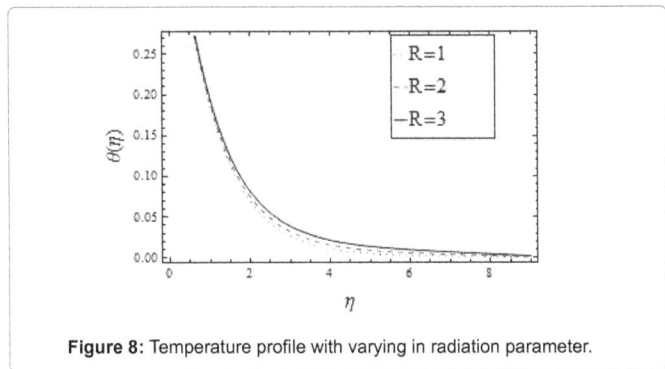

Figure 8: Temperature profile with varying in radiation parameter.

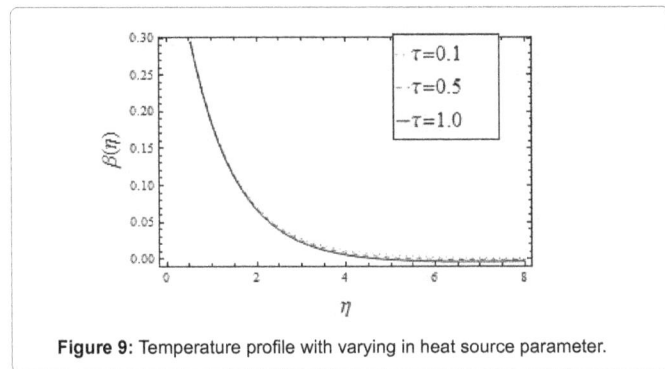

Figure 9: Temperature profile with varying in heat source parameter.

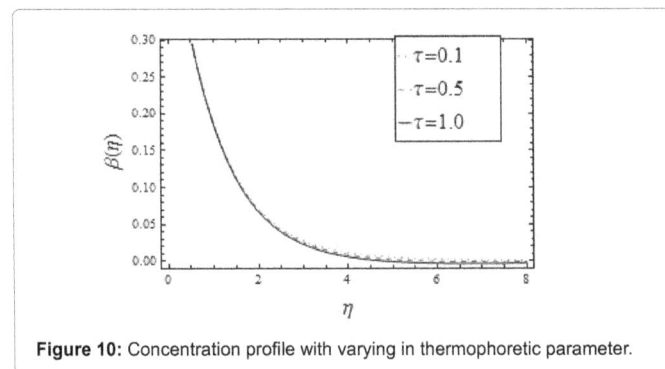

Figure 10: Concentration profile with varying in thermophoretic parameter.

velocity slip parameter, the nanofluid speed decreases monotonically. The reason is that the slip condition at the plate the speed of the fluid closed to the plate is positive values and the thickness of the momentum boundary layer decreases. As the slip parameter increases in magnitude it permits less fluid to slip past the plate, the flow gets decelerated for distance adjacent to the plate, the distance far away from the plate opposite the behavior. A similar trend occurred in temperature profile see Figure13. It was observed that for varying temperature with respect to thermal slip parameter. The profile shows that the wall temperature and thermal boundary layer thickness depreciate when the values of the thermal slip parameter increases. Also in case of concentration profile see Figure 14 the concentration profile and the concentration boundary layer thickness decreases when the values of the solutal slip parameter is increases. It decreases with an increase in solutal slip parameter.

Conclusion

The similarity solution in this work was considered to the problem of two dimensional flow of a steady laminar viscous, incompressible nanofluid over a permeable continuous stretching sheet in the

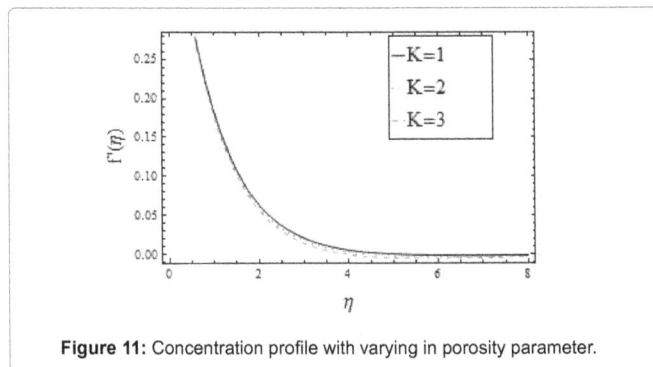

Figure 11: Concentration profile with varying in porosity parameter.

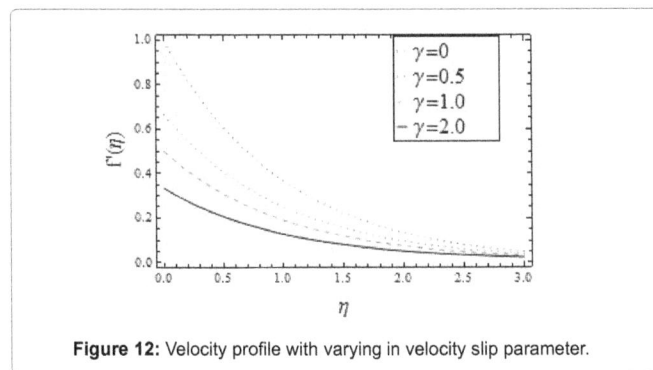

Figure 12: Velocity profile with varying in velocity slip parameter.

concentration profile. It is noticed that both the plate surface and the nanofluid concentration increase when thermophoretic parameter increased. This results to an increase in concentration boundary layer thickness. As the value of thermophoretic parameter increases, the intensity of heat generation on the plate surface increases, the intensity of heat generation on the plate surface increases, which results to an increasing rate of convective heat transfer from the nanofluid on the surface of the plate. The rise in nanofluid concentration can be attributed to nanoparticle interaction connected to increasing thermophoresis. In Figure 11, we can see the influence of porosity parameter on the concentration profile. It was found that porosity parameter on the nanofluid concentration profile rise as the concentration profile increases. The reason is that increase in porosity increase the porous layer region and increase the concentration boundary layer thickness. It is seen also that, it's generates the internal heat to the nanofluid flow, which will increase the concentration boundary layers.

Figure 12 shows the effects of velocity slip parameter on the velocity profile. We observed that the velocity profile decreases as the values of velocity slip parameter increases. With an increasing values of the

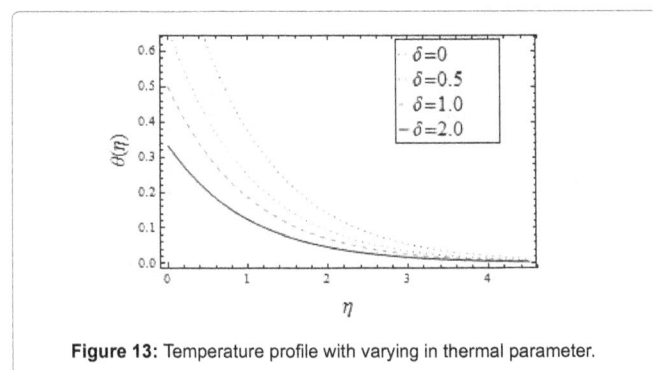

Figure 13: Temperature profile with varying in thermal parameter.

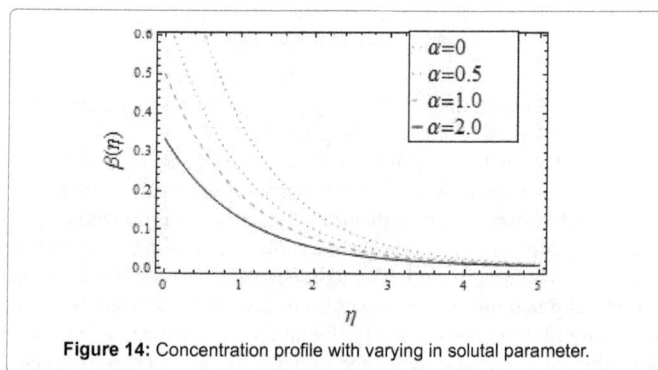

Figure 14: Concentration profile with varying in solutal parameter.

presence of heat generation/absorption with velocity, thermal and solutal slip boundary conditions is studied semi-analytically. Using similarity transformations the involved partial differential equations of the problem are transformed into nonlinear ordinary differential equations and then solve by homotopy analysis method (HAM), by using MATHEMATICA 9.0. In our study the following conclusions can be drawn:

1. The velocity, temperature and concentration profiles decrease with an increase in exponential parameter.

2. The momentum boundary layer thickness of the velocity profile increase with an increase in magnetic field parameter.

3. The thermal boundary layer thickness of the temperature of the temperature profile increase with an increase in radiation parameter.

4. The thickness of temperature boundary layer increases with an increases in heat source parameter.

5. The thermophoretic parameter decreases with an increase in the concentration profile.

6. Increase in porosity parameter decreases the momentum boundary layer thickness and the velocity profile.

7. An increase in velocity, thermal and solutal slip parameters decreases the momentum, thermal and concentration boundary layer thickness and also their respective profiles.

References

1. Hayat T, Shehzad SA, Qasim M, Asghar S (2014)Threedimensional stretched flow via convective boundary condition and heat generation/absorption. International Journal of Numerical Methods for Heat & Fluid Flow 24: 342-358.

2. Hussain T, Shehzad SA, Hayat T, Alsaedi A, Al-Solamy F et al. (2014) Radiative hydromagnetic flow of Jeffrey nanofluid by an exponentially stretching sheet e103719.

3. Hayat T, Ashraf MB, Alsulami HH, Alhuthali MS (2014) Three-dimensional mixed convection flow of viscoelastic fluid with thermal radiation and convective conditions. Plos one 9.

4. Rashidi MM, Kavyani N, Abelman S, Uddin MJ, Freidoonimehr N (2014) Double Diffusive Magnetohydrodynamic (MHD) Mixed Convective Slip Flow along a Radiating Moving Vertical Flat Plate with Convective Boundary Condition.

5. Hayat T, Shafiq A, Alsaedi A, Asghar S (2015) Effect of inclined magnetic field in flow of third grade fluid with variable thermal conductivity. AIP Advances 5: 087108.

6. Ramzan M (2015) Influence of Newtonian Heating on Three Dimensional MHD Flow of Couple Stress Nanofluid with Viscous Dissipation and Joule Heating: e0124699.

7. Hayat T, Muhammad T, Shehzad SA, Alsaedi A (2015) Temperature and Concentration Stratification Effects in Mixed Convection Flow of an Oldroyd-B Fluid with Thermal Radiation and Chemical Reaction. PloS one 10: e0127646.

8. Nield Donald A, Bejan A (2013) External Natural Convection. In Convection in Porous Media Springer New York: 145-220.

9. Macha M, Kishan N (2015) Magnetohydrodynamic Mixed Convection Stagnation-Point Flow of a Power-Law Non-Newtonian Nanofluid towards a Stretching Surface with Radiation and Heat Source/Sink. Journal of Fluids 14.

10. Satya Narayana PV, Harish Babu D (2015) Numerical study of MHD heat and mass transfer of a Jeffrey fluid over a stretching sheet with chemical reaction and thermal radiation. Journal of the Taiwan Institute of Chemical Engineers.

11. Hayat T, Muhammad T, Shehzad SA, Alsaedi A (2015) Soret and Dufour effects in three-dimensional flow over an exponentially stretching surface with porous medium, chemical reaction and heat source/sink. International Journal of Numerical Methods for Heat & Fluid Flow 25: 762-781.

12. Bilal Ashraf M, Hayat T, Alsaedi A, Shehzad SA (2015) Convective heat and mass transfer in MHD mixed convection flow of Jeffrey nanofluid over a radially stretching surface with thermal radiation. Journal of Central South University 22: 1114-1123.

13. Vijaya Lakshmi S, Suryanarayana Reddy M (2013) Effect of Radiation on Mixed Convection Flow of a Non-Newtonian Nanofluid over a Non-Linearly Stretching Sheet with Heat Source/Sink. International Journal of Modern Eng. Research 3: 2675-2696.

14. Hayat T, Shafiq A, Alsaedi A (2014) Effect of Joule heating and thermal radiation in flow of third grade fluid over radiative surface. Plos one 9.

15. Hayat T, Muhammad T, Shehzad SA, Alsaedi A (2015) Similarity solution to three dimensional boundary layer flow of second grade nanofluid past a stretching surface with thermal radiation and heat source/sink. AIP Advances 5: 017107.

16. Azeem Khan W, Khan M, Malik R (2014) Three-dimensional flow of an Oldroyd-B nanofluid towards stretching surface with heat generation/absorption.

17. Mabood F, Khan WA, Ismail AIM (2015) Approximate analytical modeling of heat and mass transfer in hydromagnetic flow over a non-isothermal stretched surface with heat generation/absorption and transpiration. Journal of the Taiwan Institute of Chemical Engineers.

18. Shehzad SA, Hayat T, Alsaedi A (2015) Influence of convective heat and mass conditions in MHD flow of nanofluid. Bulletin of the Polish Academy of Sciences Technical Sciences 63: 465-474.

19. Ramzan M , Bilal M (2015) Time Dependent MHD Nano-Second Grade Fluid Flow Induced by Permeable Vertical Sheet with Mixed Convection and Thermal Radiation.

20. Lin Y, Zheng L, Zhang X, Ma L, Chen G (2015) MHD pseudo-plastic nanofluid unsteady flow and heat transfer in a finite thin film over stretching surface with internal heat generation. International Journal of Heat and Mass Transfer 84: 903-911.

21. Hayat T, Shehzad SA, Alsaedi A (2014) Three-dimensional flow of Jeffrey fluid over a bidirectional stretching surface with heat source/sink. Journal of Aerospace Engineering 27: 04014007.

22. Swapna G, Kumar L, Anwar Bég O, Singh B (2014) Finite Element Analysis of Radiative Mixed Convection Magneto-Micropolar Flow in a Darcian Porous Medium with Variable Viscosity and Convective Surface Condition. Heat Transfer Asian Research.

23. Sheikh M, Abbas Z (2015) Effects of thermophoresis and heat generation/absorption on MHD flow due to an oscillatory stretching sheet with chemically reactive species. Journal of Magnetism and Magnetic Materials 396: 204-213.

24. Hayat T, Muhammad T, Shehzad SA, Alsaedi A (2015) Three-dimensional boundary layer flow of Maxwell nanofluid mathematical model. Applied Mathematics and Mechanics 1-16.

25. Ranga Rao T, Gangadhar K, Sundar Raju HB, Subba Rao VM (2014) Slip Flow and Magneto-NANOFLUID over an Exponentially Stretching Permeable Sheet with Heat Generation/Absorption.

26. Shen M, Wang F, Chen H (2015) MHD mixed convection slip flow near a stagnation-point on a nonlinearly vertical stretching sheet. Boundary Value Problems 1-15.

27. Rashidi MM, Kavyani N, Abelman S, Uddin MJ, Freidoonimehr N (2014) Double Diffusive Magnetohydrodynamic (MHD) Mixed Convective Slip Flow.

28. Dinarvand S, Rashidi MM (2010) A reliable treatment of a homotopy analysis method for two-dimensional viscous flow in a rectangular domain bounded by two moving porous walls. Nonlinear Analysis Real World Applications 11: 1502-1512.

29. Rashidi MM, Freidoonimehr N, Hosseini A, Anwar Bég O, Hung TK (2014) Homotopy simulation of nanofluid dynamics from a non-linearly stretching isothermal permeable sheet with transpiration. Meccanica 49: 469-482.

30. Ziabakhsh Z , Domairry G, Ghazizadeh HR (2009) Analytical solution of the stagnation-point flow in a porous medium by using the homotopy analysis method. Journal of the Taiwan Institute of Chemical Engineers 40: 91-97.

31. Hayat T, Rahila Naz, Sajid M (2010) On the homotopy solution for Poiseuille flow of a fourth grade fluid. Communications in Nonlinear Science and Numerical Simulation 15: 581-589.

32. Sohouli AR, Domairry D, Famouri M, Mohsenzadeh A (2008) Analytical solution of natural convection of Darcian fluid about a vertical full cone embedded in porous media prescribed wall temperature by means of HAM. International Communications in Heat and Mass Transfer 35: 1380-1384.

33. Liao S (2003) Beyond perturbation introduction to the homotopy analysis method. CRC press.

34. Liao S, Tan Y (2007) A General Approach to obtain Series Solutions of Nonlinear Differential Equations. Studies in Applied Mathematics 119: 297-354.

35. Sandeep N, Sulochana C (2015) Dual solutions of radiative MHD nanofluid flow over an exponentially stretching sheet with heat generation. Appl Nanaosci 1-9.

23

Analysis of Various Surface Roughness Parameters of Low Modulus Aerospace Materials Using Speckle Photography

Retheesh R*, Samuel B, Radhakrishnan P, Nampoori VPN and Mujeeb A

International School of Photonics CUSAT, Kochi 682022, India

Abstract

Surface roughness is a principal measure of product quality of the materials used for sophisticated applications. The present paper reports an experimental investigation of various parameters related to the surface roughness of the materials used for the aerospace applications by an inexpensive non-invasive technique known as speckle photography. Conventional non-destructive testing techniques such as X-Ray imaging, ultrasound scanning, microwave testing etc, have certain limitations on testing the surface characteristics of low modulus aerospace vehicle components. This paper explains characterization of surface roughness by analyzing certain roughness parameters such as intensity distribution, size distribution and contrast ratio of different low modulus materials used in aerospace applications. The optical setup employed for the work is simple and uses an inexpensive digital camera for grabbing the specimens under different static conditions.

Keywords: Laser speckle photography; Surface roughness; Aerospace applications; Non-destructive testing

Introduction

Measurement methods based on the speckle phenomenon have become increasingly important in recent years [1]. Speckle metrology encompasses a range of techniques that are utilized for non-contact, full-field inspection of diffusely reflecting surfaces [2]. Speckle techniques rely on the illumination of region of interest on a diffuse surface with an expanded laser beam to produce a speckle pattern. The principal speckle techniques such as speckle photography, speckle interferometry and speckle shearography have already been established to be emerging optical NDT techniques for structural inspection [3]. Out of the three techniques, speckle photography is the simplest scheme to set up and operate in speckle metrology [4]. In this technique, the speckle patterns of the specimens are recorded by use of a CCD camera and are then analyzed by digital image processing algorithms to obtain parameters that describe the surface roughness [5]. The surface roughness can be regarded as a variation of the surface height and there exists a copious number of parameters to characterize the surface roughness that must be considered [6]. This paper illustrates the utilization speckle photography for surface roughness assessment of certain low modulus rubberized constituents used for aerospace applications. Since aerospace technology demands stringent quality requirement, the constituents utilized are to be accessed by different means. The conventional methods like x-ray imaging, ultrasound scanning, acoustic and microwave analysis have certain restraints in giving the surface roughness properties of low modulus materials. Generally, x-ray does not give good surface images of rubberized compounds. The low modulus materials completely absorb acoustic waves, making ultrasound imaging not feasible. The microwave is not suitable for these specimens due to its high wavelength compared to laser sources used in optics experiments [7]. The proposed speckle photography experimental setup can be easily configured with computer system and further analysis can be performed using image processing software. The specklegrams of the specimen are grabbed under different static conditions. Since the specklegrams and its characteristics are different for each surface, one or more of the attributes of specklegrams such as contrast ratio, size distribution, intensity distribution and correlation between the different areas may change. In the present experiment, the properties such as contrast ratio, intensity distribution and

size distribution have been studied for three different low modulus materials and are compared to a glass surface.

Theory

The basic characteristic of the speckle pattern is its randomness. This randomness is caused by the surface roughness because the phase of the light scattered will vary from point to point in proportion to the local surface height [8]. Since speckle patterns are produced by the roughness of a surface, it is reasonable to assume that the statistics of the speckle patterns produced might be used to provide a measure of this roughness [9]. At this juncture, the statistical properties of image speckle patterns in relation to the surface roughness of objects have been studied in depth with the condition that the surface, comprising of several independent scatterers, constructs a complex Gaussian random amplitude distribution at the image surface [10]. Goodman communicates a comprehensive treatment of the statistical distribution of speckle patterns originating from various rough surfaces [11]. The amplitude of the intensity field at a given observation point (x,y) consists of a multitude of de-phased contributions from different elementary scattering areas on the rough surface. Consequently, at a specific distance from a source, a group of plane wave fronts propagating in space is denoted by a random phasor sum. A Random phasor sum is statistically represented as [12].

$$A = Ae^{j\phi} = \frac{1}{\sqrt{N}} \sum_{n=1}^{N} a_n = \frac{1}{\sqrt{N}} \sum_{n=1}^{N} a_n e^{j\theta n} \qquad (1)$$

This phasor sum can be considered as a random walk in the complex plane where 'N' represents the number of phasor components in the

---

*Corresponding author: Retheesh R, International School of Photonics CUSAT, Kochi 682022, India, E-mail: wharasani@kau.edu.sa

random walk, 'A' is the resultant phasor, 'A' is the length and 'θ' is the phase of the resultant phasor. Further, a_n is the nth component phasor in the sum. The scaling factor $1/\sqrt{N}$ is introduced to preserve the finite second moment of the sum even when a_n approaches infinity. Finally, a_n represents the length and represents the phase of a_n.

The joint probability density function P for the real and imaginary parts of the resultant phasor is given by

$$P_{R,J}\{R, J\} = \frac{1}{2\pi\sigma^2}\exp\left(-\frac{R^2 + J^2}{2\sigma^2}\right) \tag{2}$$

Where $\sigma 2 = \sigma_R^2 = \sigma_J^2$ denotes variance of real (R) and imaginary (J) parts of the resultant phasor. Hence, the marginal statistics of the length A turn out to be

$$p_A(A) = \int_{-\pi}^{\pi} p_{A,\theta}(A,\theta)d\theta = \frac{A}{\sigma^2}\exp\left\{-\frac{A}{2\sigma^2}\right\} \tag{3}$$

This is known as the Rayleigh density function and its typical plot for A ≥ 0, is shown in Figure 1 The marginal statistics of the phase θ is found by

$$P_\theta(\theta) = \frac{1}{2\pi} \tag{4}$$

for $-\pi \leq \theta \leq \pi$. Here it is used that the integral of the Rayleigh density function must be unity.

For a speckle pattern, there are a large number of contributing amplitudes. So, the result obtained for random walks with large number of independent steps is used for plotting the Rayleigh probability function.

Applying the transformation laws, for a sum of complex field amplitudes with uniformly distributed phases, Goodman and others have shown that the intensity I, displays a negative exponential probability density function.

$$p_I(I) = \frac{\sqrt{I}}{\sigma^2}\exp\left\{-\frac{I}{2\sigma^2}\right\}\frac{1}{2\sqrt{I}} = \frac{1}{\sigma^2}\exp\left\{-\frac{I}{2\sigma^2}\right\} \tag{5}$$

for I ≥ 0, equation (5) can be modified to obtain a fully developed speckle

$$p_I(I) = \frac{1}{I}\exp\left\{-\frac{I}{2I}\right\} \tag{6}$$

Thus the magnitude of the resultant speckle field follows a Rayleigh probability distribution, the phases are uniformly distributed between [−π, π] and the speckle intensities distribution can be described by a negative exponential distribution as shown in Figure 2.

Experimental Setup

The experimental arrangement is shown schematically in the Figure 3. The specimens under investigation are coherently illuminated by means of a 10 mw He-Ne laser operating at 632.8 nm.

To begin with, the laser beam is expanded by a beam expander optics to illuminate the mirror M1. The beam thus deflects off from the mirror M1 passes through a spatial filter arrangement. Another mirror M2 guides the spatially filtered beam towards the test specimen. The object beam from the each sample is steered on a 9.1 megapixel digital camera (Sony DSC-H50) where resultant speckle patterns are captured. These patterns are then transferred to a computer for further processing of the images. The specimens (1-3) used for this current experiment are rubberized diffusing materials used for aerospace applications whilst the specimen (4) is a well scratched glass plate. The entire experimental setup has been arranged on a vibration isolation table. Trial experiments

are conducted before recording the specklegrams for achieving better stability of the table within the vibration isolation environment and for achieving the steady state focusing. A program has been scripted using MATLAB and C for processing the specklegrams and to analyze the surface roughness properties such as intensity distribution, contrast ratio and size distribution. Several trial exposures are taken for grabbing specklegrams of the specimens with different aperture sizes such as f/2, f/4, f/5, f/8 at a shutter speed of 1/20 Sec. It has been spotted that the images obtained with an aperture size f/4 has given high contrast specklegrams for further analysis and are shown in Figure 4.

Results and Analysis

An experiment has been performed in order to investigate the effect of intensity distribution, contrast ratio and size distribution of the speckle on surface roughness assessment.

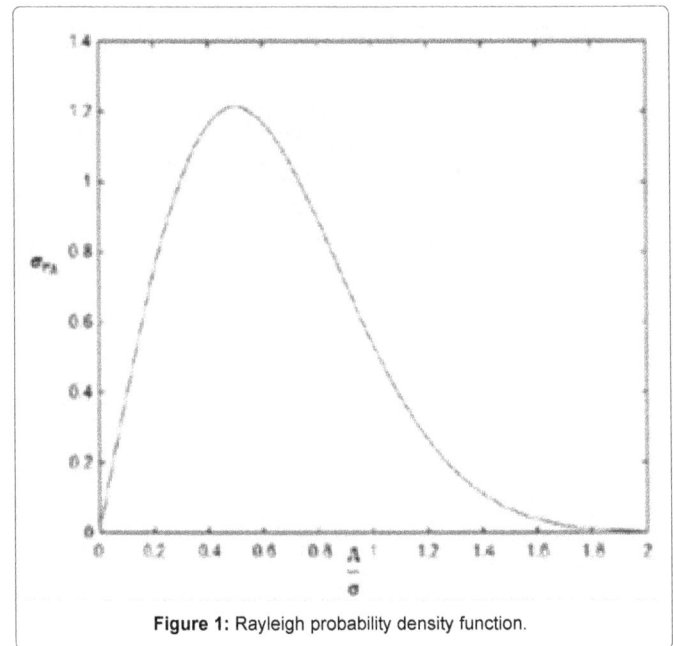

Figure 1: Rayleigh probability density function.

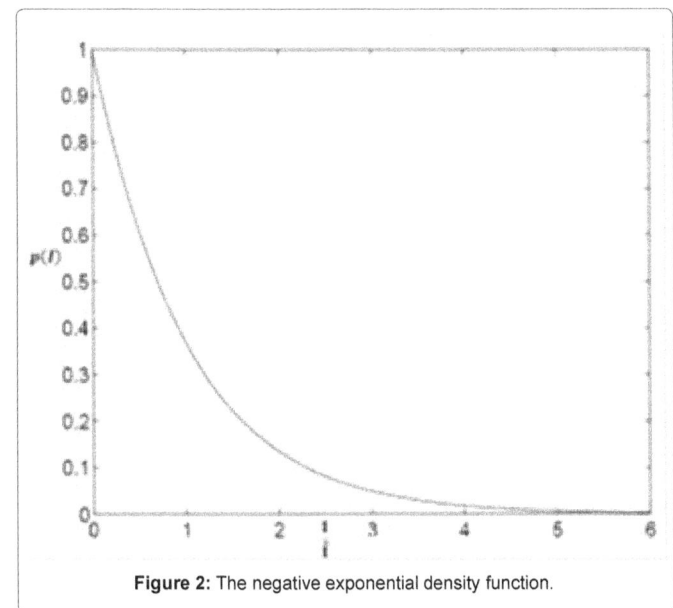

Figure 2: The negative exponential density function.

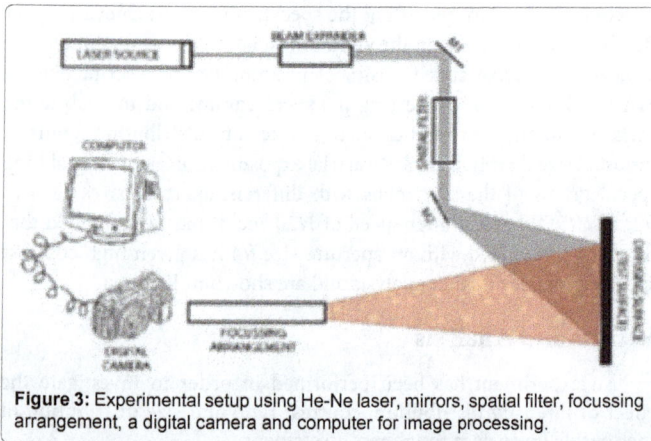

Figure 3: Experimental setup using He-Ne laser, mirrors, spatial filter, focussing arrangement, a digital camera and computer for image processing.

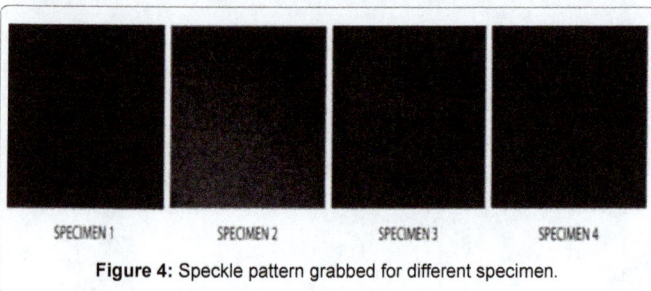

SPECIMEN 1   SPECIMEN 2   SPECIMEN 3   SPECIMEN 4

Figure 4: Speckle pattern grabbed for different specimen.

Intensity distribution

The intensity distribution of the specklegram will be random, with a normal distribution [13]. The scattered intensity distribution produced by the four specimens is curve fitted with a matlab Gaussian peak model. The Gaussian model is provided by the equation (6) where **a** corresponds to amplitude, **b** is the centroid location, **c** is related to peak width and **n** is the number of peaks to fit the random independent variable **x**.

$$f(x) = a_1 \exp\left(-\left(\frac{x-b_1}{c_1}\right)^2\right) + a_2 \exp\left(-\left(\frac{x-b_2}{c_2}\right)^2\right) \qquad (7)$$

The intensity distribution of the speckle pattern generated by the specimens is fitted to the Gaussian model as shown in Figures 5-9. A comparison of fitting curves for all the specimens is presented in Figure 9.

Thus, the intensity distribution of the specklegram is normally distributed at random. The fitting shows that it is almost the same as Rayleigh distribution and in agreement with the theoretical results. The analysis of the specimens used leads to a fact that though the distribution remains Gaussian, the distribution parameters change for all the four specimens and that can be exploited to contribute more information about the type of material.

Contrast ratio

The intensity statistics of the speckle pattern can be well explained by the contrast ratio measurements [14]. The speckle contrast ratio, the ratio of maximum to minimum intensities, is a scale of the speckle intensity variation [15-19]. Since maximum intensity possible for a monochrome pattern is 256, contrast ratio will be between 0 and 256. The noise resulting from the recording instruments is eliminated by trial experiments as they may cause a difference in the actual minimum and maximum point.

The contrast ratio analysis has been performed on all the specimens

and is shown in Table 1. The contrast ratio of the specklegram depends on the roughness of the surface on which specular reflection is happening. When roughness of the surfaces increases, the contrast ratio will also be increased.

Size distribution

The size of the speckle has also been considered for the present study. The size of the individual speckle depends on the illumination area, the wavelength of illuminating light, etc. In order to calculate the size distribution, the obtained grayscale specklegram is converted to black and white image using a mean threshold value and then the size of the white spots is measured. The equation used for the best fit for the present size distribution has been selected as

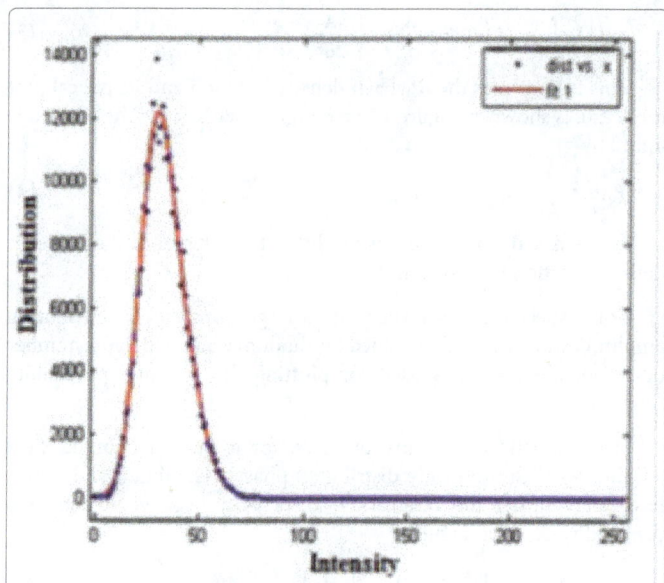

Figure 5: Intensity distribution of a specklegram, fitted to a gaussian function for specimen-1.

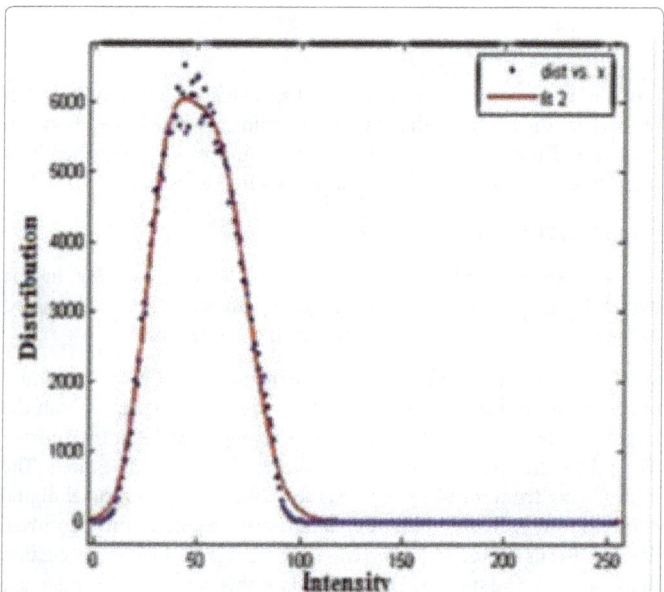

Figure 6: Intensity distribution of a specklegram, fitted to a gaussian function for specimen-2.

Specimen	Contrast Ratio
Specimen 1 (Roughest Surface)	79
Specimen 2 (Medium roughness)	14. 14
Specimen 3 (Least roughness)	7. 54
Specimen 4 (Scratched Mirror)	58

Table 1: Contrast ratio for different specimen.

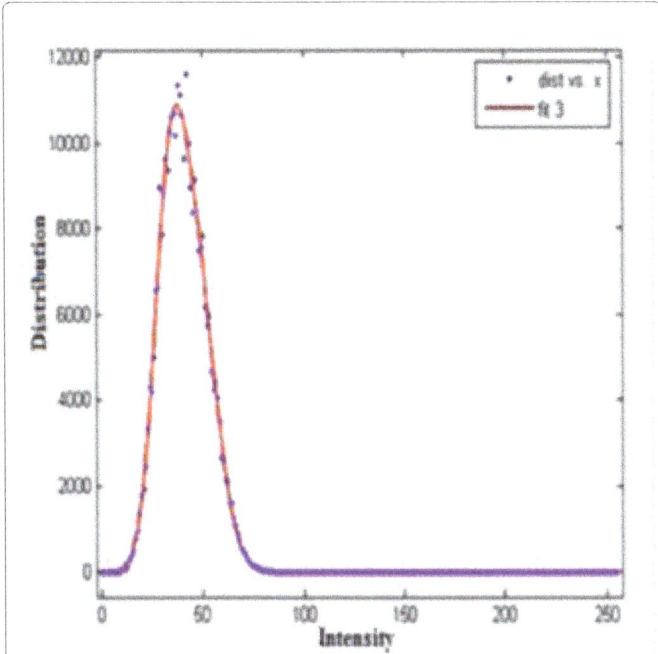

Figure 7: Intensity distribution of a specklegram, fitted to a gaussian function for specimen-3.

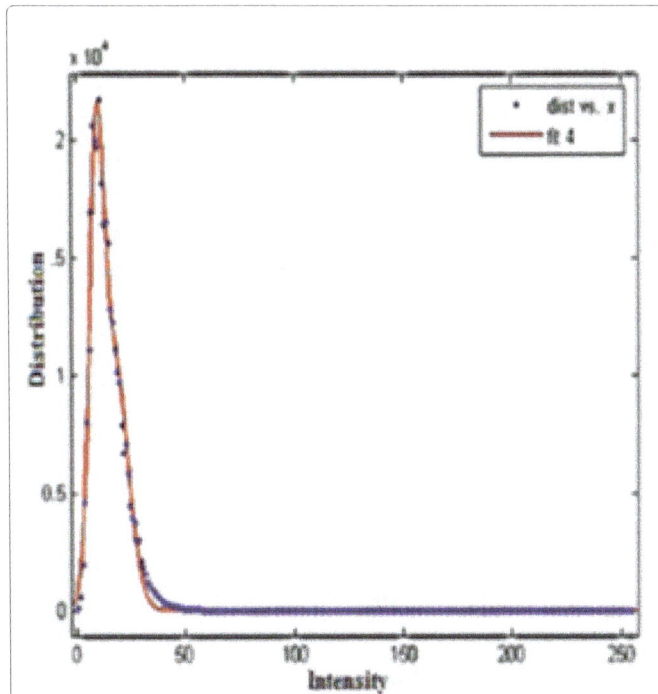

Figure 8: Intensity distribution of a specklegram, fitted to a gaussian function for specimen-4.

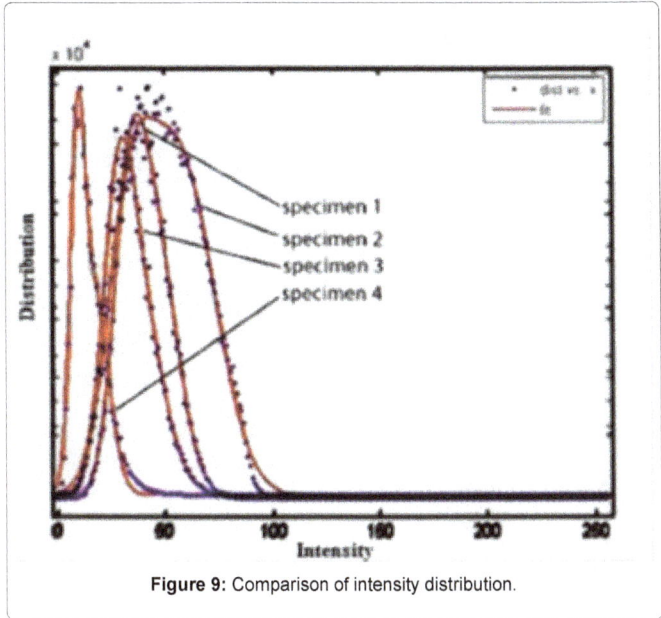

Figure 9: Comparison of intensity distribution.

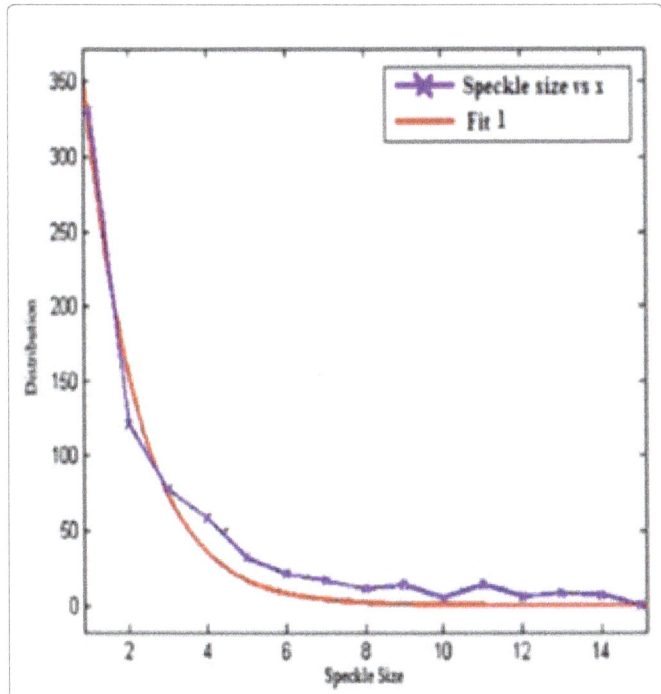

Figure 10: Speckle size distribution of a specklegram, fitted to an exponential function specimen 1.

$$f(x) = a \, exp(bx)$$

The fitting obtained for the size distribution of the specklegrams (Figures 10-14) is exponential and in agreement with Figure **2**.

Even though the size distribution remains exponentially decreasing for all the surfaces, it can be seen that there are slight changes in the fitting parameters. It should be pointed out that for the low modulus materials, the peak size can be seen as 350 units whereas for the Glass the peak size is around 600 units. Thus, the change in material is observed to be producing the change in the size of the speckle. Thus, materials can also be characterized using the maximum speckle size that is obtained using a particular wavelength.

Conclusion

The present work exploited speckle photography as a method to characterize different surface roughness properties of materials using an indigenous low cost method. The intensity distribution of the

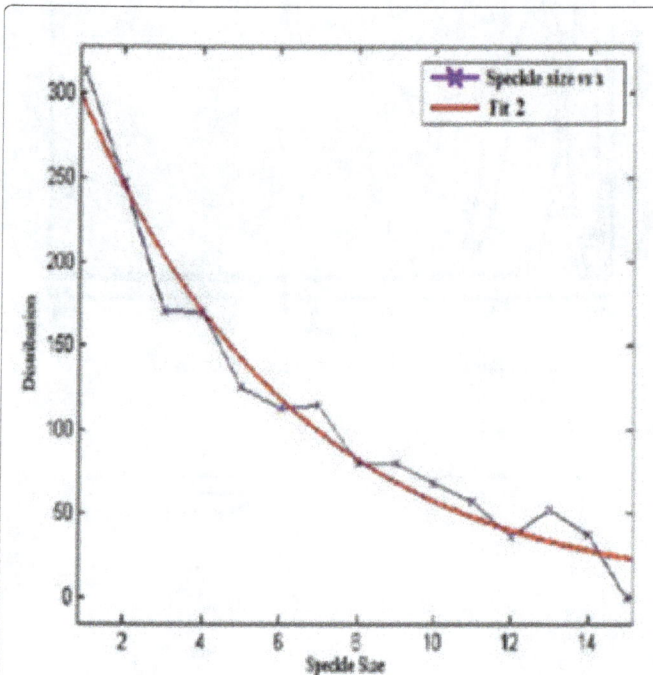

Figure 11: Speckle size distribution of a specklegram, fitted to an exponential function specimen 2.

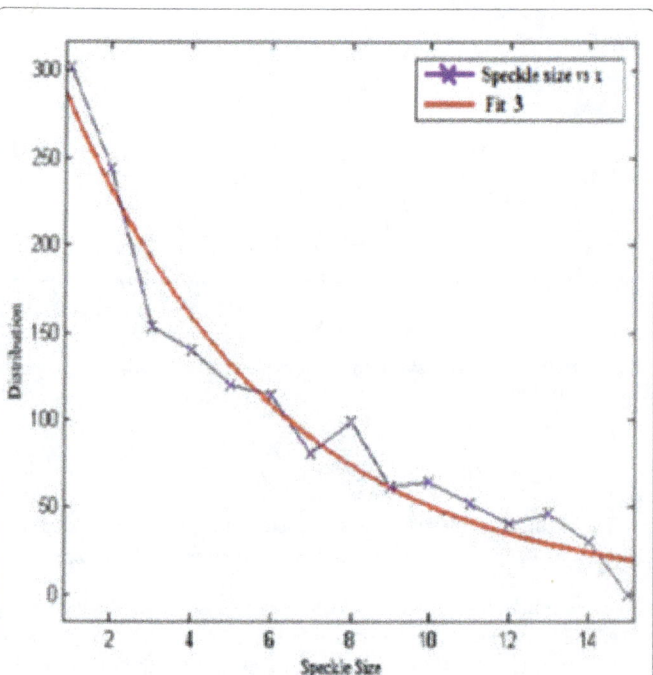

Figure 12: Speckle size distribution of a specklegram, fitted to an exponential function specimen 3.

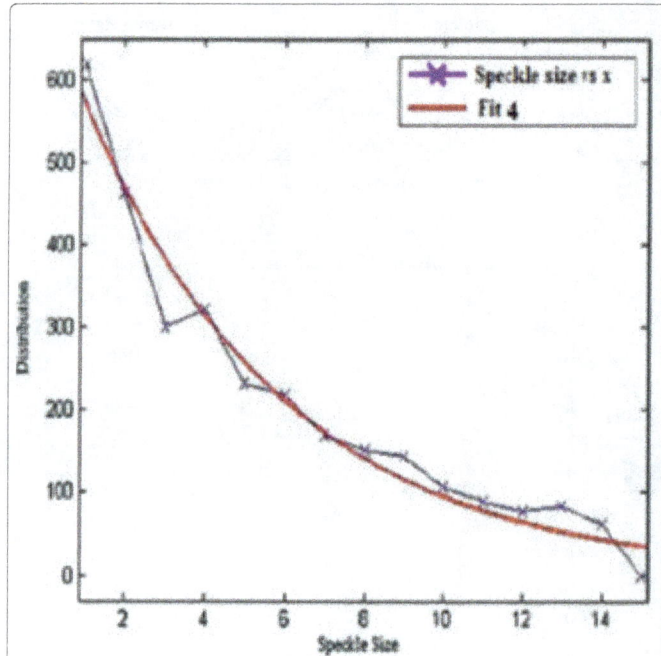

Figure 13: Speckle size distribution of a specklegram, fitted to an exponential function specimen 4.

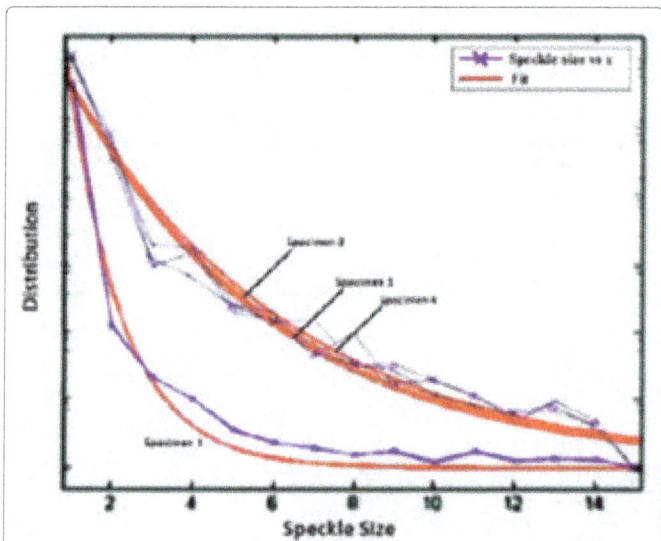

Figure 14: Comparison of fitting curves for the Speckle size distribution for all the specimens.

specklegram is normally distributed at random. The fitting shows that it is identical with Rayleigh distribution and is in agreement with the theoretical results. The contrast ratio of the specklegram depends on the roughness of the surface from which specular reflection occurs. The materials can also be characterized using the maximum speckle size that is obtained using the required wavelength.

Acknowledgement

All authors acknowledge the Department of Science and Technology (DST) and University Grant Commission (UGC) for the financial support. V P N is thankful to Kerala University and International School of Photonics.

References

1. Cloud G (1998) Optical methods of engineering analysis. 0-521-45087-X Cambridge University Press New York.

2. Parks VJ (1980) The range of speckle metrology. Exp Mech 20: 181191.

3. Gregory DA (1978) Topological speckle and structural inspection in Speckle Metrology RK Erf Ed Academic New York. 183223.

4. Nikita A (1998) Fomin Speckle Photography for Fluid Mechanics Measurements Experimental Fluid Mechanics 61-84.

5. Lehmann P (1999) Surface-roughness measurement based on the intensity correlation function of scattered light under Speckle-pattern illumination. Applied optics 38: 1144-1152.

6. Fujii H, Asakura T (1974) Effect of surface roughness on the statistical distribution of image speckle intensity. Optics Communications 11: 35-38.

7. Bahr, Alfred J (1982) Microwave nondestructive testing methods CRC Press 1.

8. Kadono H, Toyooka S (1991) Statistical interferometry based on the statistics of speckle phase. Optics letters 16: 883-885.

9. Pedersen Hans M (1976) Theory of speckle dependence on surface roughness. JOSA 66: 1204-1210.

10. Robert Erf (2012) Speckle metrology. Elsevier.

11. Joseph WG (2007) Speckle phenomena in optics: theory and applications. Roberts and Company Publishers.

12. Dainty JC, Wolf E (1976) Progress in optics North-Holland Amsterdam 4.

13. Kadono H, Takai N, Asakura T (1986) Statistical properties of the speckle phase in the diffraction region. JOSA A 3: 1080-1089.

14. McKinney JD, Webster MA, Webb KJ, Weiner AM (2000) Characterization and imaging in optically scattering media by use of laser speckle and a variable-coherence source. Optics Letters 25: 4-6.

15. Fujii H, Asakura T, Shindo Y (1976) Measurement of surface roughness properties by means of laser speckle techniques. Opt Commun 16: 68.

16. Toh SL, Sang SL, Tay CJ (1998) Surface-roughness study using laser speckle method. Opt Laser Eng 29: 217.

17. Leonard LC, Toal V (1998) Roughness measurement of metallic surfaces based on the laser speckle contrast method. Opt Lasers Eng 30: 433.

18. Fujii H, Asakura T, Shindo Y (1976) Measurements of surface roughness properties by using image speckle contrast. J Opt Soc Amer 66: 1217.

19. Persson U (1992) Real-time measurement of surface roughness on ground surfaces using speckle-contrast technique. Opt Laser Eng 17: 61-67.

24

Optimization of Aircraft Pitch Trim Rate of Movement Using Model Based Approach and Improving the Software Algorithm

Rathinakumar V*, Manju Nanda and Jayanthi J

Department of Aerospace Electronics and System Division, CSIR-National Aerospace Lab, Bangalore, India

Abstract

In safety critical systems such as aerospace, it becomes more important since the non-performance of the system as per the requirement may lead to a catastrophe. Also, the work-around to modify the design as per the requirements, generate code, obtain safety clearance from the authorized agency before porting to the target is very time consuming and a cumbersome approach.

In this paper, we propose a model-based approach to improve the performance of the software algorithm and optimize the pitch trim movement before porting the code to the target. The effectiveness of the approach is demonstrated with a case study of aerospace domain. The approach encompasses the aircraft sub-system dynamics and the software which operates the sub-system. The analysis of the functionality with performance provides a high level of confidence in the software that is to be ported on to the target. The test crew can provide feedback on the overall functionality and performance of the software at the model-level. The proposed approach not only increases the efficacy of the process but also provides higher safety assurance earlier in the process.

Pitch-trim is a critical sub-system of the aircraft which is modeled and the improved software algorithm is incorporated into the model for analyzing the overall functionality and performance of the sub-system. Based on the model simulation and analysis result, the changes in the algorithm were made and finally ported onto the target. The performance and functionality of the pitch-trim sub-system on the aircraft was as per the simulation analysis results indicating the correctness of the model and the proposed approach.

Keywords: Safety critical system; Software algorithm performance; Pitch trim analysis; V&V

Introduction

In aeronautics and automotive industries, verification and validation procedure (V&V) is an important part of safety critical system. V&V phase performs verification and validation at various activities such as unit-level, integration-level, hardware-software integration level, system level and finally at the aircraft level. Conventional V&V approach amounts to a lot of time. In order to reduce this time and make the V&V phase effective novel techniques are being adopted. Formal methods, model-based engineering, and requirements-based V&V are few such techniques being adopted to make the process effective.

Modelling and simulation are very important tools of systems engineering that have now become a central activity in all disciplines of engineering and science. Not only do they help us gain a better understanding of the functioning of the real world, they also play a key role during the design and analysis of new system design. Model-based approach enable in analyzing the system behaviour with respect to the project requirements earlier in the process. Modelling and simulation also help in analyzing systems accurately under varying operating conditions. Simulink is one of the popular modelling software capable of simulating, and analyzing dynamic systems [1,2].

Model based development provides a development process from requirements to code, ensuring that the implemented systems are complete and behave as expected. The problem of effectively modeling and analyzing software algorithm to realize functional and non-functional requirements is an important research topic. The significant benefits of such work include detecting and removing defects earlier, reducing development time and cost while improving the software algorithm [3-5].

This paper proposes an approach to improve the engineering process by analyzing the performance of the system at the model-level rather than at target level. In order to validate the system performance a very realistic system model needs to be designed. The case study of the aircraft pitch trim system is used as an example to demonstrate the approach. Aircraft pitch trim simulation model using Simulink allows the dynamic analysis of pitch trim for various test environments as per the requirements. The tests which were conducted at system level can now be tested at model level. This modification in the process improves the overall quality of the software as the design flaws can be detected earlier in the phase. Model-based approach has the capability to simulate scenarios as per the user requirement. The simulation model provides results to analyse software algorithm for its functionality and performance [6].

Related Work

Efficient methods are required for improving the software algorithm to allow early performance evaluation of completeness and correctness. Law et al. [1] present a specific level model based approach to improve the performance of software algorithm. This approach/

*****Corresponding author:** Rathinakumar V, Department of Aerospace Electronics and System Division, CSIR-National Aerospace Lab, Bangalore, India
E-mail: rathina2020@gmail.com

analysis is used to improving the performance of software algorithm before implementing into the target. Importance of this method is to check functional and non functional activities of the model in very early stage. Matlab simulink tool help library is very useful to generate the Simscape model. Help menu provides lot of examples related to dynamic system model. For example, to model a small dynamic system of car, Figure 1 represents a simple model of a car suspension. It consists of a spring and damper connected to a body (represented as a mass), which is agitated by a force. The model parameters, such as the stiffness of the spring, the mass of the body, or the force profile, and view the resulting changes to the velocity and position of the body [3] can be varied.

The approach is executed by modelling the mechanical design in Matlab Simulink tool and the source code is imported to the model. Comparing simulation results to original model, source code and check, verification cases and procedure (VCP) documents. Matlab Simulink Simscape tool provides complicated Aircraft dynamics systems [5].

Importance of the model-based approach is to reduce gap between source codes to target analysis. Model-based approach is an evolutionary step in the software field that changes the focus of software development from code to models. The model can be used to study the performance impact of different modes and/or configuration alternatives under different workloads, leading to advice for improving the system and software algorithm. Performance evaluation of a model may be done by simulating the model and collecting results [6-9].

Flight control systems on aircraft have evolved from mechanisms that warped the wing to change its lift characteristics, to the three-axis positive aerodynamic controls (elevator, rudder, ailerons) used on almost every airplane flying today [1]. Aircraft trim systems range from simple metal tabs that can only be adjusted on the ground to complex systems of in-flight adjustable tabs. Ground-adjustable trim tabs are thin metal tabs on the trailing edge of an elevator, to achieve trim in one specific mode of flight, usually cruise. In any other mode, such as climb, slow flight, or descent, the aircraft will be out of trim. Simple in-flight adjustable pitch-trim devices are generally cable operated and work in one of several ways. One method is to vary the angle of a small tab attached to the trailing edge of the elevator. Changing the angle of the trim tab changes the aerodynamic neutral point of the elevator [10].

Pitch trim movement is control the aircraft nose down/up

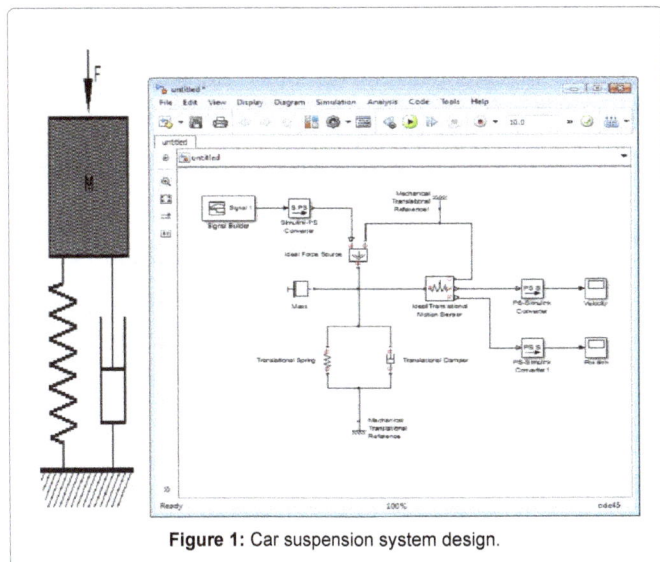

Figure 1: Car suspension system design.

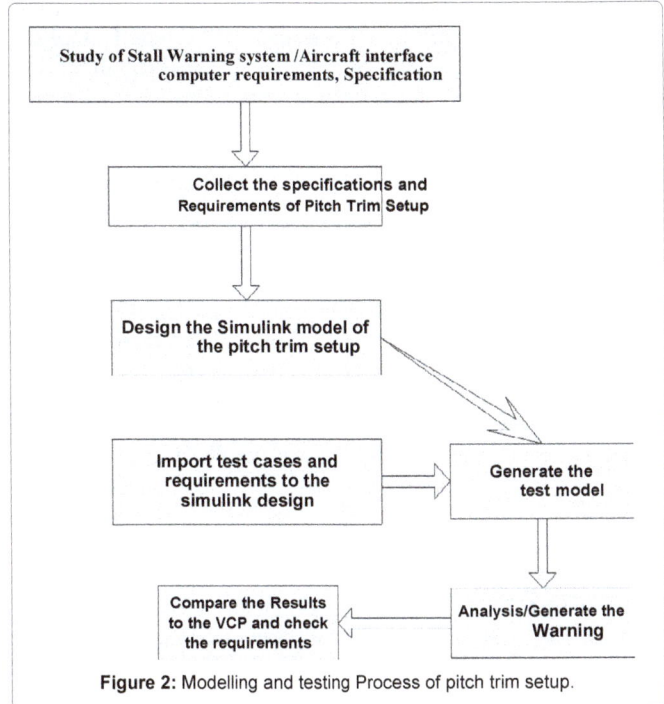

Figure 2: Modelling and testing Process of pitch trim setup.

movement. For the pitch up command aircraft goes nose down and vice versa for pitch down command. This commands generated by pilot/co-pilot or automatic flight control system. Every aircraft pitch trim movements have some limitations, once it's crossing that level, it will alert the warning signal to the pilot.

Modelling and Validating the Pitch Trim setup

Proposed approach

The proposed approach follows a sequential well-defined pattern consisting of the requirements capture, model the eletro-mechnaical system, embed the command and monitor algorithm, and generate the test scenarios to demonstrate the functionality and performance. This approach is shown in Figure 2. The approach starts with understanding the system, and collecting the requirement specification for the system. After the requirements are captured, the system is modeled. The model consists of electrical-mechanical components with the algorithm implemented in the model. The model is validated for its correct and complete implementation against the aircraft data. The functionality and performance of the system is validated by translating the embedded test scenarios to model level to generate the test model. The outputs of the results are compared to expected output.

Pitch trim set-up specification and the functionality requirements are implemented based on the SWS/AIC system requirements. Specification of servomotor, applied voltage, length of connecting wires, length of linkages, pitch trim tab dimensions, specification of potentiometer with applied voltage, types of joints with dimension, servomotor, maximum speed, and maximum torque values are used to model the pitch trim setup. The specifications are implemented into the model using Matlab 2013a [10-12].

Validating the approach

The model correctness is validated at two levels: Pitch trim set-up validation, functionality and performance validation.

Trim set-up model validation: The first level validation is carried out for trim set-up which is an electro-mechanical model. Once set-up is modelled, voltage supply is given and the movement of the trim surface is checked from one extreme to other. The time for complete movement, voltage corresponding to each degree is captured and verified against the actual time and movement captured on the aircraft [12-14].

Functionality and performance analysis of the model: The pitch trim command and monitor algorithm is embedded into the trim model. Pitch trim functionality and performance of trim algorithm is analyzed with respect to the requirements and validated against the Hardware-Software Integration Verification Cases & Procedures (HSI-VCP) [15]. HSI VCP test scenarios are provided to the model and the output of the model were compared to the test cases for the correct and complete implementation of the model. The results are analyzed for validating the end-end correctness of trim functionality

Analysis

Figure 3 shows block diagram for implementation and analysis approach. The analyses of the models help in validating the correct implementation of the model with respect to the referenced document and data. The output of the analysis is saved in the Matlab workspace for each and every test scenario. Further the output data is plotted for further analyses [16-20]. The pitch-trim setup of SARAS is based on the project specification and requirements. The trim-tab setup is modeled in Matlab 2013a. The set-up model is validated for its correct implementation by proven aircraft trim functionality data. The algorithm is checked against the trim functionality requirements.

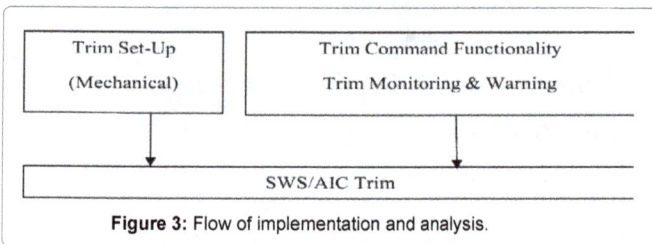

Figure 3: Flow of implementation and analysis.

The simulink model of Pitch trim setup has three main subsystems:

a. Pitch Trim Setup emulating the aircraft pitch-trim,

b. Pitch Trim command logic, and

c. Pitch Trim Warning Logic (1)

Top level model

The Pitch Trim top level model consists of the pilot/co-pilot inputs, pitch trim command logic, and pitch trim mechanical setup, and monitoring/warning logics. Figure 4 shows the Top Model. The inputs are given in the form of manual switch discrete manner (1-ON and 0-Off). Outputs of the simulation are shown in display and scope. Outputs also in the form of discrete (1-warning and 0-no warning) [3-7].

Pitch trim command logic model

The pitch trim input command is given by pilot either co-pilot. The command is given in the discrete form. Figure 5 has four input port (pilot pitch up and pitch down, co-pilot pitch up and pitch down). With four inputs 16 combination of inputs are possible. Out of 16 inputs some combinations result in pitch up command and some combinations results in pitch down command are true. K-Map technique is used to reduce the complexity. In this technique we avoid some invalid combination of inputs. The pitch trim command logic is modeled based on the pilot and co-pilot pitch up/pitch down command

Pitch trim mechanical model

The mechanical Pitch Trim Setup is modeled by three subsystem servomotor mechanism, crank-lever mechanism, and linkages mechanism. Servomotor will activate when pilot/co-pilot gives a command. Servomotor output torque is the input of the next crank lever mechanism subsystem. Connecting rods are moving because of applied torque from servomotor. At the end of the connecting rods the trim surface is connected. When connecting rods are activate then the trim surface will move up and down based on the pilot command. A joint between trim tab and connecting rod, the potentiometer is connected and this potentiometer excitation voltage is 11.3V. The

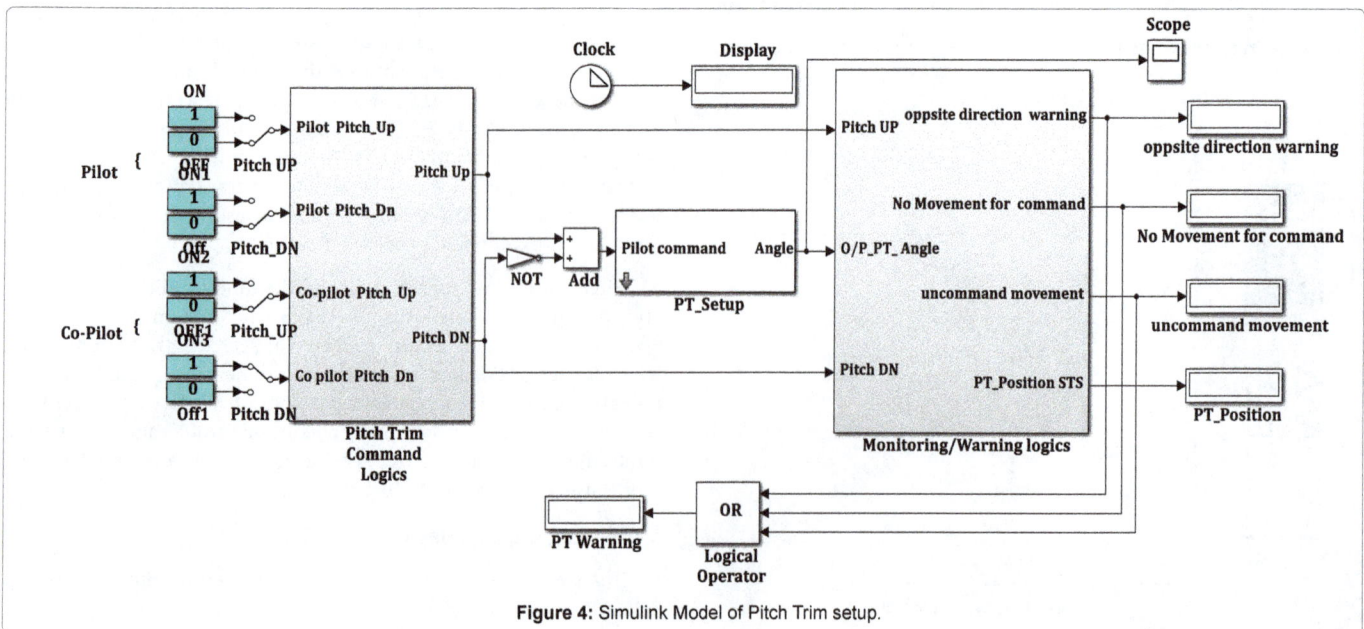

Figure 4: Simulink Model of Pitch Trim setup.

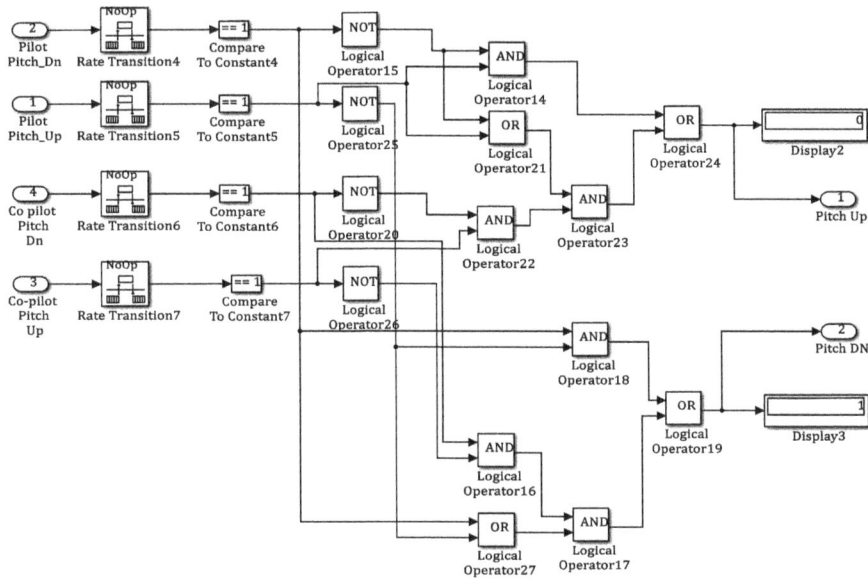

Figure 5: Pitch Trim Command logic.

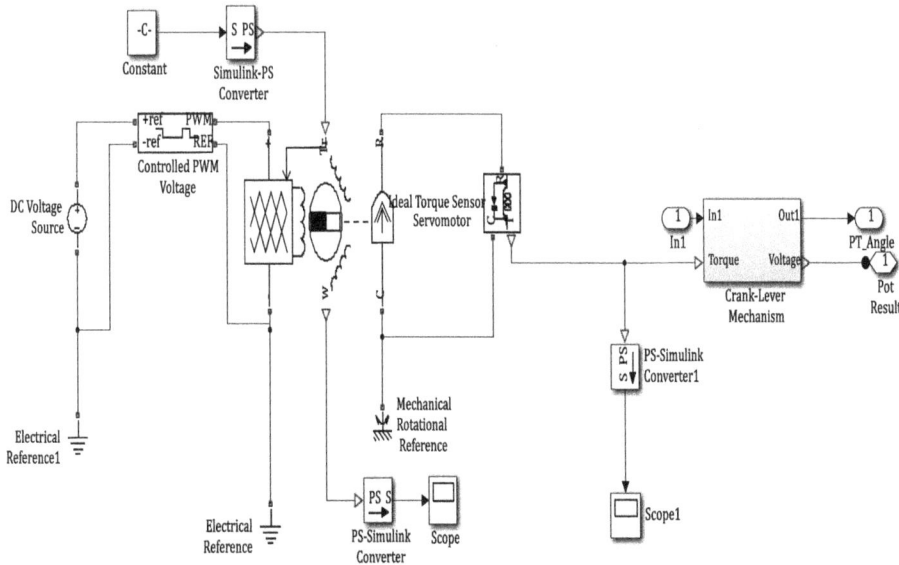

Figure 6: Servomotor Mechanism.

excitation voltage (of 28 V) is measured for all the trim tab position for various time steps [7,8].

Servomotor mechanism: Figure 6 consists of servomotor block, controlled PWM voltage block, electrical reference, ideal torque sensor block and DC voltage source. Once command is activated the servomotor starts to rotate and produces torque. This torque is measured by using ideal torque sensor block and it passes through the crank lever mechanism. For the pilot pitch trim up/down command for, the crank lever mechanism is connected as input port in1.

Crank-lever mechanism: The output of the linkages mechanism is connected to the potentiometer. Potentiometer measures the voltage based on the trim surface movement. It's shown in Figure 7.

Linkages mechanism: Dimensions of connecting rods, connecting

wheels, types of joints and trim-tab dimension are measured in the actual experimental setup. Figure 8 shows the linkage mechanism, where the solid blocks represent the connecting rods.

This subsystem models all the mechanical part of the pitch trim setup. The model has been designed as per the actual setup. The requirements and specification of servomotor, potentiometer, voltage supply, surface dimension and types of joints are got from Aircraft design and requirements documents

Pitch trim warning/monitoring algorithm: The pitch trim surface position is provided as an analog input to the system by means of feedback signal. When pilot gives the pitch up command, the surface will move in clockwise direction and for pitch down command the surface will move in anti-clockwise direction. Based on the pilot

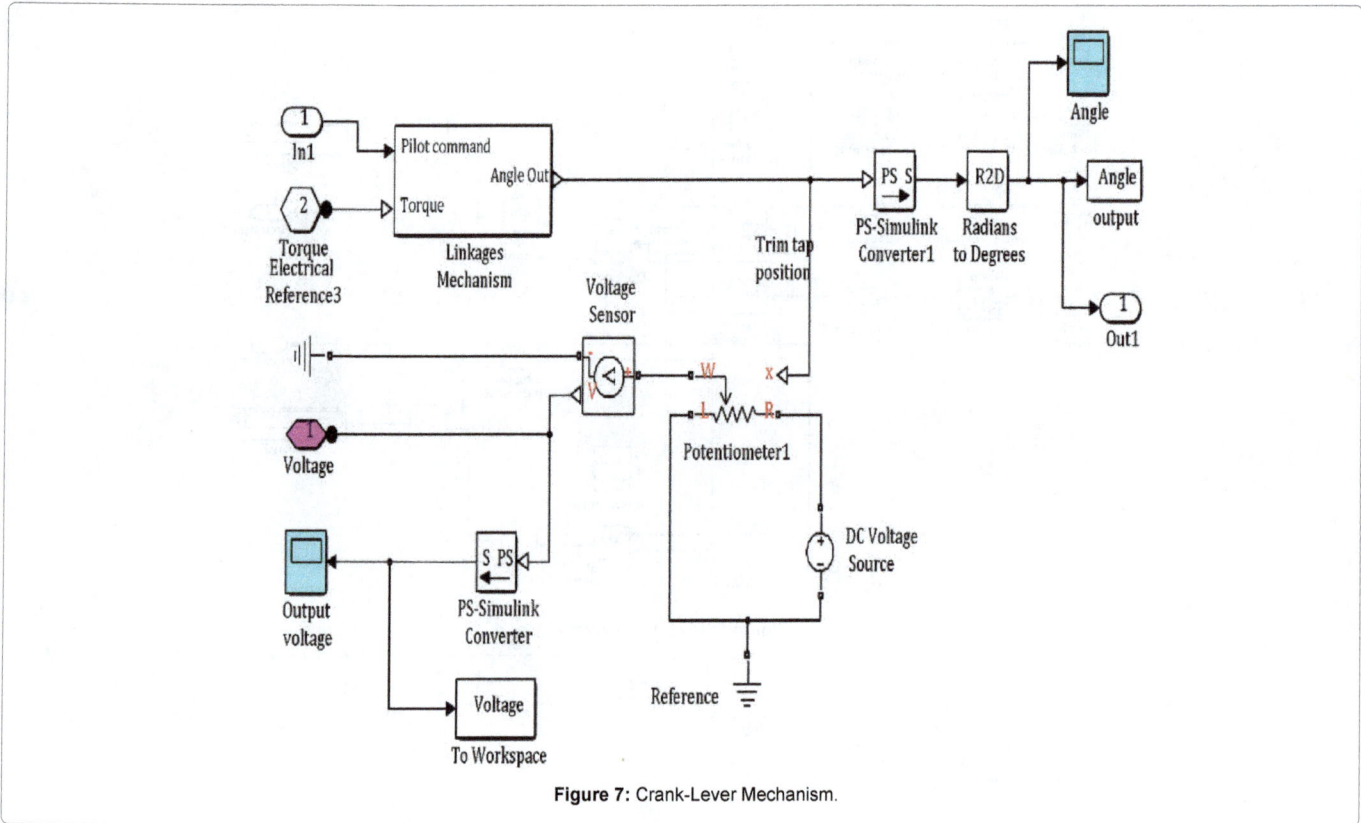

Figure 7: Crank-Lever Mechanism.

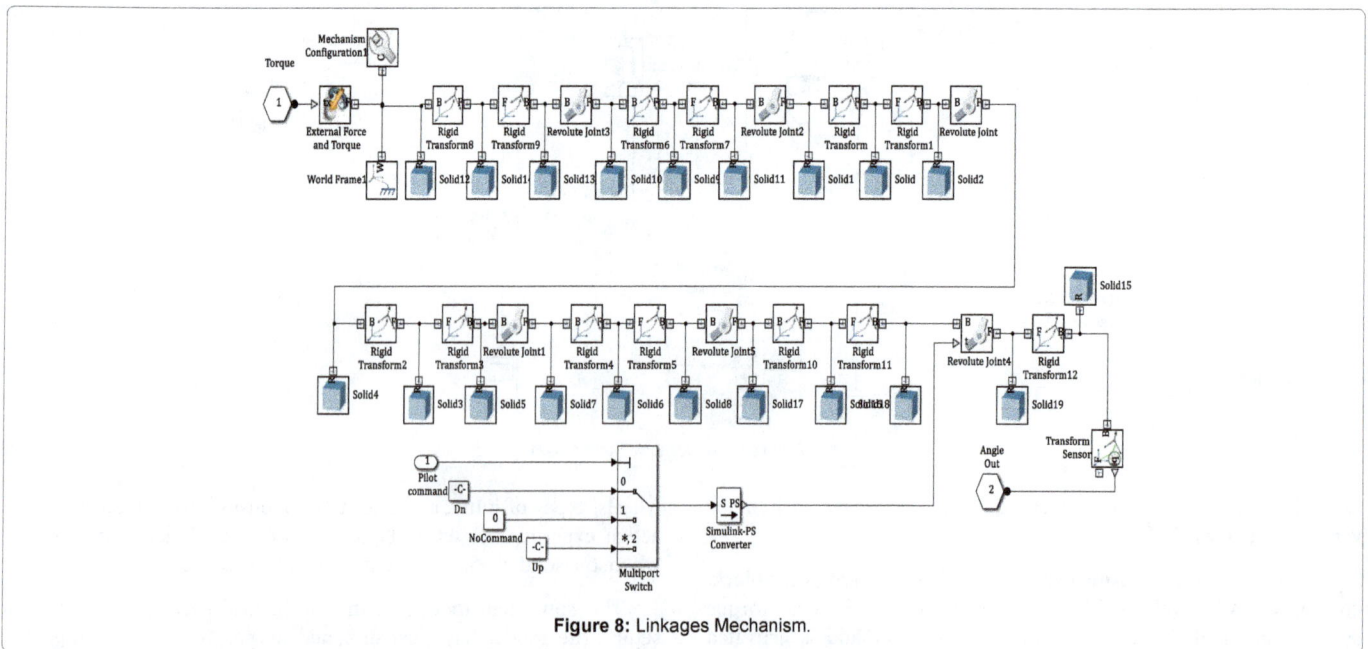

Figure 8: Linkages Mechanism.

command, motor provides the PWM command and the surface moves. The monitoring functionality monitors the surface movement as per the command and in case there is a problem then the warning is announced after the persistence time. Figure 9 shows the monitoring and warning logics of the pitch trim setup. Output of this subsystem is the final to save in the matlab workspace. Those outputs are used for further analysis and validation purposes [21-24].

Result Analysis

The excitation voltage from potentiometer and the corresponding trim tab positions are saved in the Matlab workspace. The total trim-tab movement is from 0 to 23 degrees and the corresponding measured voltage value is 0.06805 V to 7.85244 V. Trim-tab moves linearly with respect to time and the measured voltage value varies linearly from 0.06805 V to 7.85244 V. These results are compared with the original

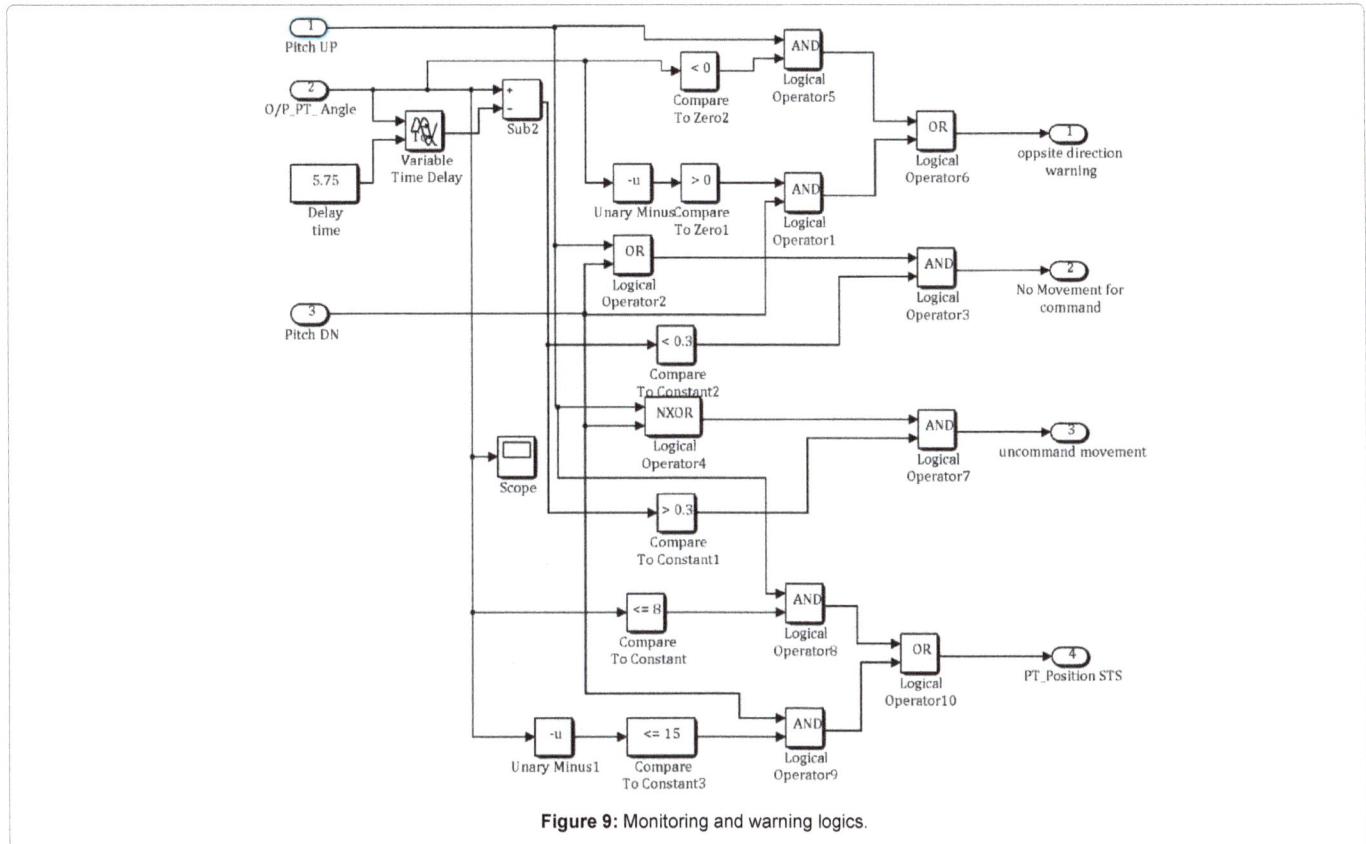

Figure 9: Monitoring and warning logics.

Degrees	Voltage(Model)	Voltage (Lab Set up)	Difference
0DEG (-15deg)	0.06805V	0.4751V	0.40345
23DEG (-8deg)	7.85244V	7.827V	0.0252
Total Voltage Swing			
Model	7.85244-0.06805=	7.78438V	
Lab set up	7.827-0.4751=	7.3519V	
Voltage Swing Difference: 7.78438 V-7.3519 V=0.43248(5.5%)			

Table 1: Result analysis.

laboratory setup. In the Experimental setup, the Trim tab also moves 23 degrees i.e. from 8 degree to -15 degree. The measured voltage varies from 0.4751 V to 7.827 V corresponding to -15 and 8 degree.

The percentage of error obtained with the model based approach for the trim surface and algorithm validation is about 5.5% corresponding to 0.43248 V (Table 1).

The surface movement for persistence time 500 msec, 400 msec, 300 msec and 200 msec is analyzed and the results are shown in below (Figure 10-13). The analysis is required to fix the persistence for providing the warning. Figure 10 shows the pitch trim warning after a persistence time of 500 msec. Trim warning is provided for trim runaway as seen in the Figure 10, with 28 V supply corresponding to 100% PWM, the surface is moves from 0 degree to 0.348369 degree with a pitch down command in 500 msec duration Trim surface movement is greater than 0.3 degree in 500 msec, and as per the current algorithm pitch trim warning will announced to the test crew. Figure 14 is representing the pitch trim surface movement at various persistence times. Meaning of the red line is corresponding to the persistence time 500 msec and this point, surface exceed limit and generate the pitch trim warning.

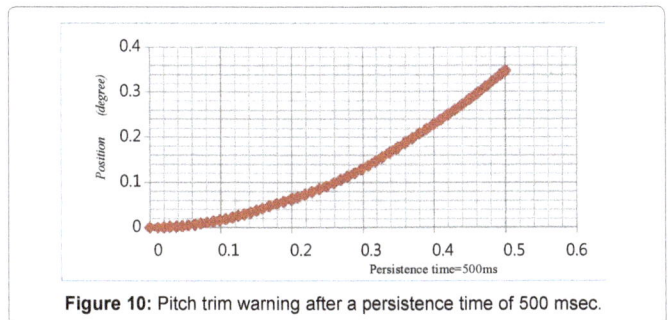

Figure 10: Pitch trim warning after a persistence time of 500 msec.

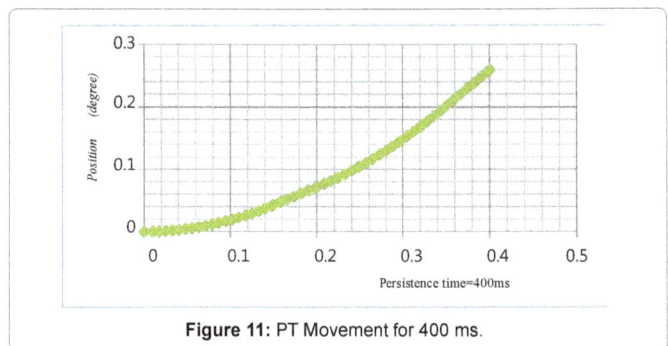

Figure 11: PT Movement for 400 ms.

PWM output analysis

Table 2 is representing the different pitch trim rate for different applied PWM voltage. All the results of PWM output are shown in Table 2. From the PWM analysis the rate of pitch trim movement is good for 62.5%. It was accepted by the wing commander from NAL.

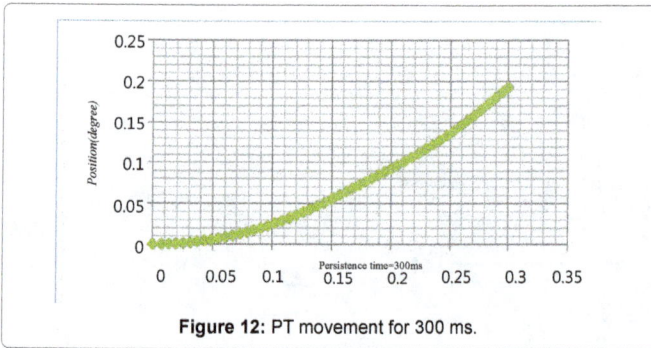

Figure 12: PT movement for 300 ms.

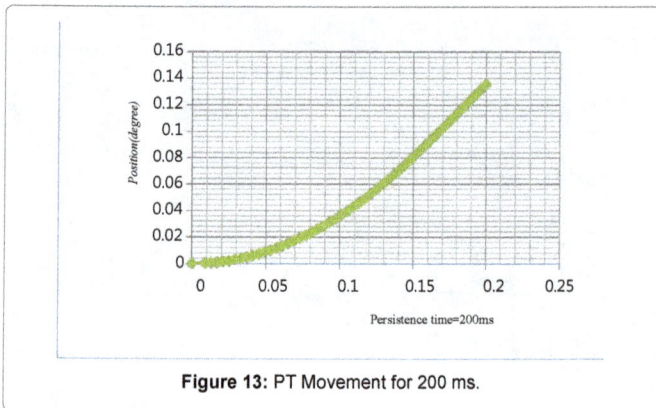

Figure 13: PT Movement for 200 ms.

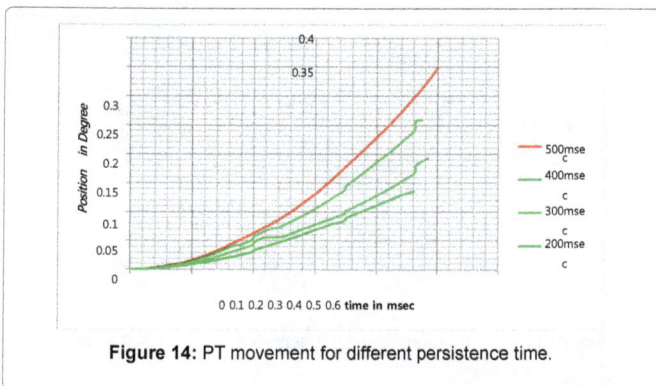

Figure 14: PT movement for different persistence time.

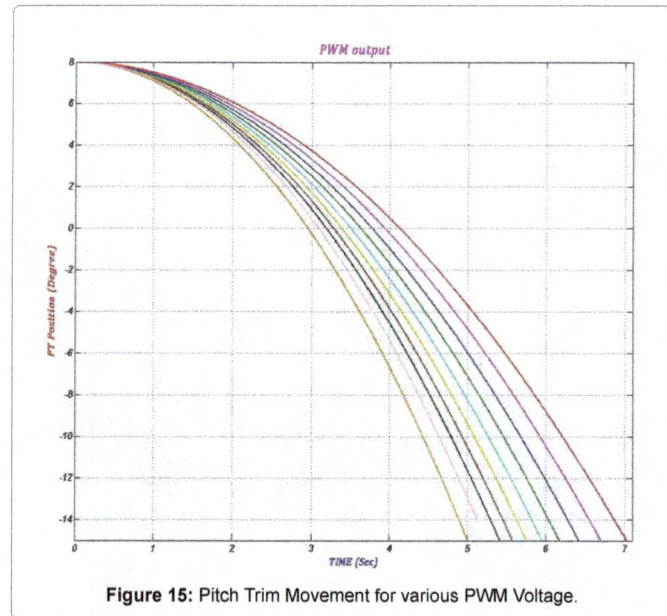

Figure 15: Pitch Trim Movement for various PWM Voltage.

The rate of movement at 62.5% PWM is 0.25 d/s. Figure 15 is shown the variation of pitch trim movement from 100% PWM to 50%. PWM of 62.5% is giving the Pitch trim smooth movement not to speed and not a too slow. So, the optimized value of rate is considered in this PWM and the rate is 0.25 degree/sec.

Conclusion

Model-Based approach can be used to analyze critical sub-system functionality and performance. The case study of pitch-trim set-up and the detailed analysis proves the effectiveness of this approach. The correctness of the model set-up with the modified algorithm is validated against the data from the aircraft pitch trim functionality test. The results of the analysis is tabulated, plotted and compared with experimental results. The percentage of error obtained is 5.5% proving the model correctness 94.5% which is similar to the original experimental setup in the laboratory. The persistence time of 500 msec is the optimum time for generating the trim warning as the other persistence time does not provide the surface movement for warning generation.

The model-based approach to analyze the performance can be incorporated in the engineering process help the engineers in detecting the design flaws and inconsistency earlier in the process.

References

1. Averill ML, David Kelton W. Simulation Modelling and Analysis 3/e McGraw. Hill 3rd edn: 0070592926.

2. Saxena V, Pratap A (2014) A Framework for performance estimation of Object-Oriented Databases. International Journal of computer Science and Information technology and security 4: 39-45.

3. Mathworks Inc (2015) Simcape.

4. Coopera K, Daia L, Dengb Y (2005) Performance modelling and analysis of software architecture: an aspect-oriented UML based approach. science of computer programming 57: 89-108.

5. Daia L, Cooper K (2006) Modeling and performance analysis for security aspects. science of computer programming 61: 58-71.

6. Torabi A, Ahari AA, Karsaz A, Hosssin Kazemi S, Intelligent Pitch controller identification and design. Proceeding of the 2013 International Conference on System control Signal processing and informatics.

% PWN	Total Swing (degree)	Time taken for total swing (sec)	Rate of Movement (d/s)	Rate for movement for 100 m sec
100	~ =23	5	4.6	0.46
95	~ =23	5.12	4.49	0.449
90	~ =23	5.26	4.37	0.437
85	~ =23	5.41	4.25	0.425
80	~ =23	5.57	4.13	0.413
75	~ =23	5.75	4	0.4
70	~ =23	5.95	3.87	0.387
65	~ =23	6.17	3.73	0.373
60	~ =23	6.42	3.58	0.358
55	~ =23	6.7	3.43	0.343
50	~ =23	7.02	3.28	0.328

Table 2: PWM output with rate of movement.

7. Sampaio RCB, Becker M (2009) Mechatronic Servo System Applied to a Simulated- based Autothrottle module. Proceeding COBEM 2009 20th International Congress of Mechanical Engineering November 15-20.

8. Zhang J, Cheng BHC, Model Based Development of Dynamically Adaptive software. In International Conference on Software Engineering (ICSE06) Received ACM SIGSOFT Distinguished paper award 371-380.

9. Engels G, Kuster JM, Heckel R, Lohmann M (2003) Model Based Verification and Validation of Properties. Electronic Notes in Theoretical Computer Science 82: 133-150.

10. Stylianos SPP (2006) Robust High Performance Servo Controller Design technique Using Matlab Simulink. Proceeding of the 5th WSEAS Int. Conference on System Science and Simulation in Engineering Tenerife Canary Islands Spain, 16-18: 125-129.

11. Murphy B, hayurst C, Friedman J, Mohtadi C, Anderson R, Mosterman P Verification and Validation integrated within processes using Model Based Design. The Math Works Inc.

12. Sargent RG (2011) Validation and Verification of Simulation Models. Proceeding of the WinteSimulationConference 183-198.

13. Foster H, Uchitel S, Magee J, Jeff Kramer (2006) LTSA-WS: A Tool for Model- Based Verification of Web Service Compositions and Choreography. Proceedings of the 28th international conference on Softwareengineering: 771-774.

14. Zhang L, Glab M, Ballmann N, Teich J (2013) Bridging Algorithm and ESL Design: Matlab Simulinik Model Transformation and Validation. Forum on Specification & Design Languages (FDL): 1-8.

15. Meenakshi B, Bhatnagar A, Roy S (2006) Tool for Translating Simulink Models into Input Language of a Model Checker. ICFEM LNCS 4260 Springer-Verlag Berlin Heidelberg: 606-620.

16. Rus I, Neu H, Munch J (2003) A Systematic Methodology for developing Discrete Event Simulation Models of Software Development Processes. Proceedings of the 4th International Workshop on Software Process Simulation and Modeling (ProSim): 3-4.

17. Martens A, Koziolek H, Becker S, Reussner R (2010) Automatically improve software Architecture models for performance reliability and cost using evolutionary Algorithms: 105-116.

18. Denford M, Neill TO, Leaney J (2003) Architecture Based Design of Computer Based System. International Conference and Workshop on the Engineering of Computer-Based Systems (ECBS'03): 1-8.

19. Busse LJ, Crimmins TR, Fienup JR (1995) A Model Based Approach to Improve the performance of the Geometric filtering Speckle reduction Algorithm. IEEE Ultrasonic's Symposium 2: 1353 -1356.

20. Broy M, Kirstan S, Krcmar H, Schtz B, Zimmermann J (2012) What is the benefit of a model-based design of embedded software systems in the car industry. Emerging Technologies for the Evolution and Maintenance of Software Models IGI Global 343-369.

21. Feiler PH (2010) Model Based Validation of Safety Critical Embedded System.

22. Musgrove J, Cukic B, Cortellessa V, Proactive Model Based Performance Analysis and Security Tradeoffs in a complex System. IEEE 15th International Symposium on High Assurance System Engineering.

23. Ould-Ahmed-Vall EIM, Woodlee J, Yount C, Doshi KA, Seth Abraham (2007) Using Model Trees for Computer Architecture Performance Analysis of Software Applications.

24. Joshi A, Heimdahl MPE (2015) Model-Based Safety Analysis of Simulink Model Using SCA|DE Design verifier.

25

Developing the Framework for Integrating Autonomous Unmanned Aircraft Systems into Cloud Seeding Activities

DeFelice TP[1]* and Duncan Axisa[2]

[1]*Sykesville, Maryland, MD, USA*
[2]*Research Applications Laboratory, National Center for Atmospheric Research, Boulder, CO, USA*

Abstract

This paper introduces an engineering approach to develop autonomous unmanned aircraft systems technology for integration in future weather modification (cloud seeding) programs with the goal to improve operational efficiency and evaluation accuracy. It builds upon the process already established in a previous paper by Axisa and DeFelice who constructed a framework underlying the development of new technologies for use in cloud seeding activities, identifying their potential benefits and limitations and providing initial guidance.

Keywords: Unmanned Aircraft; Autonomous UAS; Cloud seeding Operations; UAS and Cloud Seeding

Introduction

This paper addresses an approach to develop new sensing technologies and their integration in unmanned aircraft systems (UAS) for use in real-time guidance of airborne cloud seeding activities to increase precipitation efficiency (i.e., precipitation enhancement) based on an engineering and scientific process-based framework described in Axisa and DeFelice [1]. The new sensor suite will optimally provide *'in-situ'*, real-time temporal and spatial sensitivities to overcome the predictability issues related to, or sparseness of, environmental parameters needed to identify conditions suitable for airborne cloud seeding and to test the implementation of the seeding.

Precipitation enhancement projects have been conducted primarily in regions where orographic clouds (those developed by the lifting of moist air as it flows over elevated topography) are common in the cold season, or where warmer-season cumuliform clouds are generated by vigorous convection, since the mid-1940s [2] based on the scientific principles of the precipitation process [3-4]. Simply stated, cloud seeding is conducted on cloud systems or portions of clouds that are naturally inefficient at converting their moisture into precipitation, hence, cloud seeding makes clouds more efficient precipitators. Operational cloud seeding projects have been conducted since the first tests of both dry ice [2] and silver iodide, AgI [5,6], as cloud seeding materials [7-10]. Cloud seeding for enhancing winter snowpack in western mountainous areas is considered highly successful since the mid-1980s [11]. The results of mixed phase convective cloud seeding have been inconclusive. The seeding of isolated individual clouds has led to definite, mostly positive changes in the precipitation amounts [7,12-14]. Woodley and Rosenfeld [14] developed and tested a method for the objective evaluation of short-term, nonrandomized operational convective cloud seeding projects on a floating-target-area basis. Their results indicated that rainfall was increased downwind of the seeding activity, primarily as the seeded clouds moved out of the target and into downwind areas. Downwind or extra-area effects are further discussed by DeFelice et al. [15]. Cloud seeding evaluations are used to gauge operational efficiency of the seeding operation, and when successful a benefit/cost ratio greater than 200/1 can be achieved [16].

Airborne cloud seeding with AgI is conducted from near or at cloud base with strong updrafts or at cloud top depending on conditions, whereas dry ice is typically used via aircraft just inside or above cloud top depending on conditions. This type of seeding often referred to as glaciogenic seeding, is applied in clouds that contain high concentrations of super-cooled liquid water and relatively warm temperatures. Glaciogenic seeding, whether seeding near cloud base or cloud top requires maximum updrafts at cloud base, or cloud tops growing vertically above the freezing altitude, and to occur early in the convective cloud's lifetime. Cloud seeding may also be conducted in clouds too warm for AgI and dry ice, known as warm cloud seeding or hygroscopic seeding. Mather et al. [17] have carried out successful hygroscopic seeding in South Africa; Silverman et al. [12] report statistical significance and substantial increases in radar-estimated rainfall (ranging from 30% to 60%) from the seeded clouds using hygroscopic seeding techniques; Rosenfeld et al. [18] report a broadening of the drop size distribution following the seeding of continental convective clouds with hygroscopic salt powder, indicating that the salt material was acting to accelerate the warm rain process. Hygroscopic seeding usually occurs just below cloud base, in clouds that are microphysically continental (i.e. contain high concentrations of small drops due to the absence of large hygroscopic aerosols), in the area of maximum updraft to ensure that the cloud ingested the seeding agent, and early in the convective cloud's lifetime.

Background

The western hemisphere is in the midst of a significant drought [20], with millions in danger of starvation. Hence the need to develop the science and technology that improve the appropriate systems used to monitor and manage atmospheric water should remain at the forefront of current research. Better technology, appropriately designed and implemented to improve the efficiency of cloud seeding and independently verify such, will yield more water returned to the surface in the form of precipitation. More precipitation will help resolve the direct and indirect issues related to drought.

*Corresponding author: DeFelice TP, Sykesville, Maryland, MD, USA
E-mail: clddoc1@gmail.com

Cloud seeding technologies may be effectively applied [9,10,21,22] to facilitate the water and energy cycles [15], which are key to dealing with many present and potential future scientific, environmental, public concerns, and socio-economic issues. This review paper builds upon the basic context and initial guidance for weather modification operators that might integrate UAS technology in future cloud seeding operations provided recently by Axisa and DeFelice [1]. Axisa and DeFelice [1] provide a first look at the integration of UAS for cloud seeding operations and research. They define cloud seeding operations and research with the greatest need for advanced technology and technique development into three functional components: (a) cloud seeding activity monitoring and simulation, (b) seeding agent delivery and dispersion, and (c) cloud seeding evaluation technology, techniques and protocols. Instrumented UAS technology was determined to be at the operational or near operational readiness level and therefore suitable for integration in cloud seeding operations. They identify the primary issues with UAS integration, noting such were most likely related to government policy, technology advancements, and operational considerations. They formulate a conceptual configuration for operations and for evaluation of the operational use of UAS for modern cloud seeding operations (Table 1). The use of UAS in operational cloud seeding operations also have benefits that go beyond overcoming some of the operational safety concerns, but they also provide:

1. A cost effective means to evaluate cloud seeding using near real-time cloud system relevant measurements for evaluating the operations.

2. A cost effective means to advance our understanding of the relevant science and engineering aspects, and

3. A framework for application to other science and engineering areas. For example, using UAS in tandem to gain the ability for cost effective, concurrent, real-time Eulerian and Lagrangian analyses of seeding processes throughout cloud life cycles on sub-cloud scales. The latter would benefit other disciplines including climate change, forecasting extreme and severe weather, flooding, drought and the impacts of such events on society, the economy and more.

Axisa and DeFelice [1] provide a more detailed list of benefits.

The configuration of a UAS for cloud seeding under either mode in Table 1 would be the ultimate goal. In practice, one would start with a much simpler version of the configuration in Table 1, guided by the modern-day cloud seeding operational capabilities [9,10,21,22].

Goal and Objectives

As we develop and assess an autonomous UAS platform for cloud seeding operations our first goal would be to (1) develop simple, calibrated and well-validated lightweight payloads that measure meteorological state parameters, wind, turbulence and aerosol-cloud microphysical properties in conditions that are conducive to seeding, and (2) develop algorithms that use *in-situ* real-time sensor data to guide the platform towards suitable targets to implement the seeding.

In its simplest form, the UAS could be guided by weather radar or satellite data products to navigate to regions of suitable convection. The UAS payload would consist of lightweight sensors designed to provide 'real-time' *in situ*-based measurements that support operational flight guidance of the UAS. The flight guidance system would navigate the UAS autonomously to areas of suitable temperature, relative humidity, updraft velocity, aerosol size distribution and droplet size distribution to implement optimal seeding. Optimal seeding means that seeding starts and proceeds at a rate that will yield maximum conversion of cloud water to precipitation that falls in the intended location on the ground, or target area. The latter ability to have the precipitation fall in an intended area on the ground is known as targeting. The data collected from the payload sensors, and seeding apparatus, during an entire flight would be collected and downloaded for use by others to improve and validate model parameterizations especially when applied to simulating seeding agent dispersion

Large datasets collected during airborne cloud seeding experiments already exist [e.g. 17, 23-36] and provide valuable sources of data to develop and constrain the algorithms that guide the UAS. These data can be mined, analyzed and features extracted to locate representative time-series of key sensors from research aircraft flying at or below cloud base (e.g. sensors that measure updraft velocity, aerosol size distribution and droplet size distribution). Similar analysis would also be conducted on weather radar and satellite data in regions that are known to be suitable for seeding in order to establish representative radar [37] and satellite signatures [38] for the corresponding periods and locations.

Data collected in previous campaigns can be aggregated following the data assimilation process and passed through a UAS simulator for evaluation. The simulator can implement software in the loop (SIL) technology that has the ability to simulate the UAS flight characteristics, with navigation driven by sensor data collected from previous campaigns. Radar and satellite algorithms can be implemented to now-cast convection that may be suitable for targeting the seeding. The simulations can be compared to actual flight paths flown on previous cloud seeding missions to understand differences in behaviors between manned operations and that performed by the UAS in the simulation. This analysis can serve as guidance for improving the simulation software.

Function	Capability
Sensing	Atmospheric profiles surface to flight-level: air temperature, dewpoint temperature, wind field, turbulence, static pressure, spectral irradiance, supercooled liquid water content (SLW) Atmospheric constituents (aerosols, cloud, precipitation, trace gases, total water content) Surface characteristics (temperature, moisture content, spectral reflectance, soil moisture, soil temperature profiles). Ancillary, auxiliary (e.g. GPS, platform velocity, acceleration, attitude, pitch, roll, video) [e.g. AgI, dry ice (DI), hygroscopic agent dispenser].
Sensor Coverage	Omni-slight skew toward forward hemisphere; [Sub-UAS point^ (AgI; DI)].
Data Processing	Able to process hundreds of Terabytes of data per second; functional tools, decision support; calibration/validation; archive; [Seed start and stop, GPS locations, amount AgI/DI dispensed].
Software	Algorithms to yield required information: Capability Maturity Model Integration, level III (CMMI III). Command, Control, Communications, Computers, Intelligence, Surveillance, and Reconnaissance (C4ISR); data logging; data processing; [Algorithms to yield required information (e.g. seeding decision), control operations (e.g. ignite squib-burn AgI solution/flare or other, flight path, sensing); data logging; data processing].

^Sub-UAS Point is defined as the "point of intersection with the earth's surface-geoid of a plumb line from the UAS to the center of the Earth" (i.e., intersects surface at a 90 degree angle).

Table 1: Conceptual functional configuration of a system to identify and monitor cloud seeding opportunities and [Additional configuration included to conduct cloud seeding operations]. Adapted from Axisa and DeFelice [1]).

Once the simulations have been refined then at least two field campaigns involving the lightweight UAS can be conducted. The first campaign would evaluate the sensors for performance within their operational limits. The second campaign would test the sensor and algorithm equipped UAS as a unit for the first time during an intensive observation period (IOP). The location of the IOP would be in an area where cloud seeding has a high potential of success, and may even be piggy-backed on an existing operational program. The IOP will help establish the viability of UAS as a weather modification research platform with possible cloud seeding applications. Specifically, it would achieve a range of technological objectives, including:

1. Assimilate data from previous rainfall enhancement field campaigns to define the key sensor parameters required for optimizing and evaluating cloud seeding operations and determining a suitable sensor payload for these parameters.

2. Develop software in the loop (SIL) based simulator that tests the performance of cloud targeting algorithms.

3. Evaluate the use of UAS in seeding of convective clouds with real time guidance from radar, satellite and an *in-situ* measurement based systems.

An approach to develop and assess an autonomous UAS platform that utilizes *in-situ* real time data to sense, target and implement seeding might proceed generally as provided in the following sections for each technical objective.

Approach for developing the sensor payload

The first step in this development process is to determine which physical quantities best describe clouds that are amenable to seeding. Once these operational and research-like quantities have been established, their threshold values for seeding need to be determined by analyzing large datasets collected during airborne cloud seeding experiments. This would require developing the capability to directly measure those values inside and around clouds. The sensor payload would be designed with the capability to measure properties (i.e., temperature, relative humidity, wind, updraft velocity, aerosol size distribution and droplet size distribution) within threshold values of operational quantities to determine when to seed, and research quantities for subsequent operations verification or evaluation. For example, determining thresholds based on analysis of measured drop size distribution and their relationship to the production of rain. A broad drop size distribution with a tail of large drops might not be suitable for hygroscopic seeding especially if large hygroscopic aerosol particles are present below cloud base. This would necessitate the design of a system consisting of two UAS, one flying below cloud base (at the cloud formation level) while another flying in the vicinity of the optimal seeding location inside the candidate cloud (at a flight level above the cloud base). Each of the UAS would have a similar sensor payload, each transmitting data to the ground control station. These data would need to be simultaneously processed to determine aerosol and drop size distribution parameters for concurrent periods, and then compared against the thresholds to determine cloud seedability.

The two UAS, UAS1 (high cloud base/spotter) and UAS2 (cloud formation level/seeder), would each have on-board data processing systems, and each be controlled by a ground control station that controls the actions of each UAS. UAS1 and UAS2 should be equipped with a lightweight payload that measures basic thermodynamic properties (i.e., pressure, temperature and relative humidity), wind velocity, aerosol size distribution, and drop size distribution. UAS2

would also be equipped with seeding apparatus (i.e., dry ice dispenser, acetone generator, hygroscopic flares and/or glaciogenic flares).

An instrument that measures 3D wind velocity could be the Rain Dynamics multi-angle inertial probe (MIP), and one that measures drop size distributions could be the Droplet Measurement Technologies (DMT) backscatter cloud probe (BCP), [41], and one that measures aerosol size distribution could be the Hendix Scientific printed optical particle spectrometer (POPS),[42]. (These instruments are merely listed as examples and other instruments that perform a similar function might be suitable.) We have considered the concept of UAS1 sampling cloud drop-derived aerosol residuals through a counter-flow virtual impactor, and to have these residual particles collected on scanning electron microscope (SEM) stubs, for example, for analysis of their chemical composition. The aerosol chemistry, especially the residual aerosols that form the drops, are important in understanding the aerosol-cloud interactions inherent to the cloud and its formation [e.g. 40] as well as in evaluating the impact of the seeding within the cloud [e.g., 18,19]. This sampling technique could be very important in determining whether a seeding response is present in the cloud being modified and would also be useful in evaluating the seeding operation and assessing any environmental impacts during post operational assessments. However, this technique is dismissed as immediately feasible due to technological limitations of obtaining the simplest chemical composition of a single aerosol residual, given the stopping distance of a drizzle drop may be too large for a UAS inlet (and inlet counter-flow rate), for example.

The data processing system onboard the UAS would feed information to a more powerful computational platform at the ground station. Our autonomous control module will interface with a back trajectory module that uses wind measurements from the two MIPs. Both UAS1 and UAS2 processing systems will calculate back trajectories but the final position of UAS2 will be adjusted to be relative and downstream of UAS1. By positioning the aircraft in this formation, the position of UAS2 would be ideal for cloud seeding in the case when UAS1 is sampling seedable clouds. Hence, UAS1, which needs to spot the conditions for seeding, would use the environmental information and the trajectory results to arrive at the location where conditions are conducive for seeding, and similarly but also taking into account location of UAS1, UAS2 would use that information to implement the seeding.

The existing datasets collected from research aircraft on past campaigns would be processed so that their output will be similar to that produced by the UAS payload (i.e. temperature sensor, relative humidity sensor, MIP, BCP and POPS). These data would then be analyzed to develop and constrain the algorithms that guide the UAS, to finalize and test sensor payload algorithms; to perform the data analyses; and to develop the radar and satellite algorithms. The results of these activities would be used in the next phases of development, including aerodynamically optimizing the sensor payload weight and location on each platform. In cloud seeding operations, small UAS might be capable of carrying some seeding material in the form of ejectable or burn-in-place flares. However the seeding material to be carried will depend on the payload capability and the type of seeding to be performed. In this paper, we will assume that UAS1 and UAS2 stay in the warm parts of the cloud system, so unlike for the glaciogenic seeding application, there would not be a need for de-icing. At the end of the simulations our sensor and seeding payloads will be aerodynamically optimized on UAS1 and UAS2 and ready for the next phases of development. Small UAS, despite their payload limitations, have operated successfully in the vicinity of thunderstorms as part of an observational campaign [43].

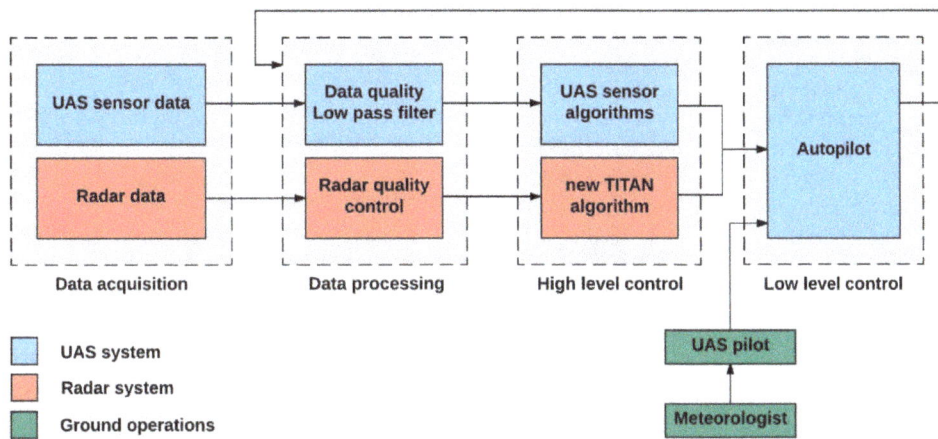

Figure 1: Autonomous UAS control routine for cloud seeding operations. Sensor and radar data equipped with cloud seeding algorithms provide seeding actions to the UAS. A UAS pilot and meteorologist have the option to modify or interrupt the actions taken by the UAS.

Approach for developing an algorithm that uses sensor data to guide the platform towards suitable targets to implement the seeding

In this section the approach shifts toward transferring the verified algorithm to the on-board data system computer and interfacing it with the actual sensors using standard software development lifecycle principles. The development proceeds on the hypothesis that weather radar software such as TITAN (Thunderstorm Identification, Tracking, Analysis, and Now-casting, [37]) can be modified to not only now-cast the location of convection, with real-time radar echo data input about the cloud environment, but also with sensor data input from the UAS. The combination of radar and sensor data would improve the ability to forecast optimal seeding conditions.

Once operational the new TITAN algorithm would provide the UAS with the coordinates of the ideal region to sample and the UAS, equipped with the sensor payload, would proceed with *in-situ* sampling as they navigate to those new locations. TITAN continuously updates optimal seeding coordinates throughout the flight, but once the UAS reach the ideal location, seeding begins and ends once the sensors indicate favorable and then unfavorable seeding conditions, respectively. The seeding cycle continues until the UAS must return for fueling or there is an unsafe situation. The algorithm controlling the seeding simply ingests the location coordinates and time stamps from TITAN and the time stamps from the sensors, along with the environmental (e.g., 3D winds, temperature, relative humidity, pressure), aerosol and drop microphysical properties (e.g., drop concentration, drop size distribution, effective drop size) from the sensor payload. The data are quality controlled using a simple test (e.g. range test), and processed in real time (e.g. passed through low pass filter) to provide updraft velocity, droplet size and corresponding droplet concentration, and total aerosol concentrations in the fine, accumulation and coarse mode of the size distribution. These values are then passed through a series of if/then statements which essentially encompass the threshold criteria to indicate seedability. If the thresholds are met, then seeding occurs via UAS2, and UAS1 continues to make measurements in formation. The thresholds are not exclusively used for seeding, but also for establishing natural variability and addressing Lagrangian analyses, among others. They may in some circumstances be relatable to control cases in the event seeding occurred in a nearby cloud. The latter will be approached following the proven design of a

software module that simulates software in the loop (SIL) technology, based on the autonomous control schematic shown in Figure 1.

We realize the scientific complexity of this task and more importantly the difficulty in operationally getting such a routine to work consistently and accurately and our approach adds the critical first step of starting with trial and error flights in a simulated environment. The simulations will include placing a cloud in the "new TITAN" with a set of assimilated observable parameters, then running the simulation to see if the UAS finds that target cloud. If it finds the area of maximum threshold condition, then it starts seeding there, until it then finds the position of the minimum threshold condition where it stops seeding. This is repeated for different clouds and environmental conditions until the "new TITAN" updates with new seeding coordinates. The simulator implementing SIL technology simulates the UAS flight characteristics, with navigation driven by sensor data collected from the previous campaigns. The SIL simulation and algorithm performance of the targeting will be evaluated by running an ensemble of simulation scenarios. Our approach using the simulations would follow the guidance of Axisa and DeFelice [1] and once the updates to the "new TITAN" software result in near perfectly accurate simulation results in selecting seedable areas and their locations, we would take UAS1 and UAS2 into the field for experimentation.

Approach for field testing the algorithms that use sensor data to guide the platform towards suitable targets to implement the seeding

Once the guidance algorithm is interfaced with the sensors and working adequately, our approach shifts toward using a UAS with the lightweight payload to locate regions of seedability, and a UAS to monitor the results following a scheme first mentioned in Axisa and DeFelice [1]. The approach would include deployment during at least two field campaigns, in an area conducive for effective cloud seeding, and preferably with active seeding programs underway, with a sole objective of testing how well the updated TITAN software performs. The system would not and should not yet be intended to be operational in any way, but meant to simply work out any challenges implementing the system.

UAS1 and UAS2, as previously defined, would be programmed to fly in a manner that exploits the advantages of flying in formation. For example, UAS1 would profile the planetary boundary layer to determine the thermodynamic and aerosol microphysical properties. It would then climb to the 5°C isotherm while UAS2 would profile downwind of UAS1 and up to the cloud formation level. Both UAS would fly in formation while approaching the cloud and profile up and down (in a saw tooth type pattern) through the top of the boundary layer while keeping a safe minimum separation. Once near the cloud each UAS would assume their position and commence their seeding mission profile where UAS1 penetrates the cloud and UAS2 samples the cloud updraft and the cloud formation level. Once seeding stops more sampling could resume which may involve a series of cloud penetrations and sampling of aerosols below cloud base (while maintaining separation).

Besides the technological challenges that must be overcome or adequately worked around, the societal and regulatory issues remain and must be respected. Axisa and DeFelice [1] highlight the latter, but the most immediate issue in this research and development case lies with aviation regulatory limitations [39]. In the United States, the Federal Aviation Administration (FAA) is the regulatory entity for air safety from the ground up, whether manned or unmanned, and irrespective of the altitude at which the aircraft is operating. While Axisa and DeFelice [1] provide details, suffice it to say that in the United States [39] the regulatory agency does not provide for UAS to be used for weather modification operations, and certainly not without a certificate of waiver or authorization (COA). The COA would at least allow an operator to use a defined block of airspace and includes special provisions unique to the proposed operation, such as, requiring flight under Visual Flight Rules (VFR) only, and/or only during daylight hours. An example of UAS operations with a COA in a cloud environment is the Verification of the Origins of Rotation in Tornadoes Experiment, or Vortex2, field campaign where a lightweight UAS measured meteorological state parameters and wind along gust fronts associated with super-cell thunderstorms [43].

Concluding Remarks

We have developed a concept for autonomous cloud seeding using UAS. We introduced an engineering approach to develop autonomous unmanned aircraft systems technology for integration in future weather modification (cloud seeding) programs with the goal to improve operational efficiency and evaluation accuracy. The broader impacts and benefits lie within evolving improved technology and automation of cloud seeding operations while lowering the operational footprint in order that we can optimize the effectiveness and efficiency of cloud seeding programs. The proposed technology could have an impact on the future of rainfall enhancement operations in arid and semi-arid regions of the world especially in those countries with limited infrastructure. The sensor package and algorithms can also be used on manned project aircraft to guide the seeding. The data collected from each seeding mission can be used in real-time to improve model parameterizations, and improve processing throughput while maximizing quality by acting as input into coupled models. The latter will also facilitate the development, improvement and/or validation of weather modification-relevant operational and evaluation models, and decision support tools.

Acknowledgement

The authors wish to thank the anonymous reviewers for their constructive review comments. NCAR is funded by the National Science Foundation.

References

1. Axisa D, DeFelice TP (2016) Modern and prospective technologies for weather modification activities-A look at integrating unmanned aircraft systems. Atmospheric Research 178: 114-124.

2. Schaefer VJ (1946) The production of ice crystals in a cloud of super-cooled water droplets. Science 104: 457- 459.

3. Bergeron T (1935) On the physics of clouds and precipitation. Proc 5th Assembly IUGG IUGG (Lisbon 1933): 156-178.

4. Findeisen (1938) Die kolloidmeteorologischen vorgange bei der niederschlagsbildung. Meteor Z 55: 121-131.

5. Vonnegutt B (1947) The nucleation of ice formation by silver iodide. JWM. J Appl Phys 18: 590-592.

6. Vonnegutt B (1981) Misconception about cloud seeding with dry ice. JWM. J Appl Phys 18: 593-595.

7. Dennis AS (1980) Weather modification by cloud seeding Academic Press New York.

8. Marwitz J (1986) A comparison of winter orographic storms over the San Juan Mountains and the Sierra Nevada. Precipitation enhancement A scientific challenge RR Braham Jr ed Meteorol Monogr AMS Boston 21:109-113.

9. Keyes CG, Bomar GW, DeFelice TP, Griffith DA, Langerud DW (2016) (3rdedn) ASCE Manuals and Reporsts on Engineering Practice No 81. ASCE Reston VA USA 220.

10. ANSI/ASCE/EWRI (2017) ASCE standard practice for the design and operation of precipitation enhancement projects (42–16) per: DeFelice TP (ed) ASCE/EWRI reston VA USA (In Public Ballotting for ASCE & ANSI).

11. Elliott RD (1986) Review of wintertime orographic cloud seeding. Precipitation enhancement-A scientific challenge RR Braham Jr Ed Meteor Monogr AMS Boston 2143: 87-103.

12. Silverman BA, Rosenfeld D, Sukarnjanaset W, Talumassawatdi R (1999) The Thailand warm cloud seeding experiment 2 Results of the statistical evaluation. Prepr 7th WMO Scientific Conference on Wea Modif WMO Geneva: 9-12.

13. Woodley WL (1999) The Thailand cold-cloud seeding experiment: 2 Results of the statistical evaluation. Prepr 7th WMO Scientific Conf Wea Modif WMO Geneva: 25-28.

14. Woodley WL, Rosenfeld D (2004) The development and testing of a new method to evaluate the operational cloud-seeding programs in Texas. J Appl Meteorol 43: 249-263.

15. DeFelice TP, Golden J, Griffith D, Woodley W, Rosenfeld D et al. (2014) Extra area effects of cloud seeding-an updated assessment. Atmos Res 135: 193-203.

16. Axisa D (2004) The Southern Ogallala Aquifer Rainfall (SOAR) Program: A new precipitation enhancement program in West Texas and South-eastern New Mexico. The J Weather Modification 36: 25-32.

17. Mather GK, Terblanche DE, Steffens FE, Fletcher L (1997) Results of the South African cloud-seeding experiments using hygroscopic flares. J Appl Meteor 36: 1433-1447.

18. Rosenfeld D, Axisa D, Woodley WL, Lahav R (2010) A quest for effective hygroscopic cloud seeding. J of Applied Meteorology and Climatology 49:1548-1562.

19. Warburton JA, Young LG, Stone RH (1995) Assessment of seeding effects in snowpack augmentation programs: Ice nucleation and scavenging of seeding aerosols. J Appl Meteorol 34: 121-130.

20. McNutt M (2014) The drought you can't see. Science 345: 1543-1643.

21. ANSI/ASCE/EWRI (2013) American National Standards Institute American Society Civil Engineers & Environmental Water Resources Institute, Standard practice guidelines for the design and operation of super-cooled fog dispersal programs. DeFelice TP (ed.) ASCE Reston, VA, USA.

22. ANSI/ASCE/EWRI (2015) ASCE Standard Practice for the Design and Operation of Hail Suppression Projects. Langerud D (ed) ASCE Reston VA USA 39-15: 62 .

23. Bigg EK (1997) An independent evaluation of a South African hygroscopic cloud seeding experiment 1991-1995 Atmos Res 43: 111-127.

24. Cooper WA, Bruintjes RT, Mather GK (1997) Calculations pertaining to hygroscopic seeding with flares. J Appl Meteor 36: 1449-1469.

25. Bruintjes RT, D Breed, MJ Dixon, BG Brown, V Salazar et al. (1999) Program for the augmentation of rainfall in Coahuila (PARC): Overview and preliminary results. Pre-prints 14th Conference on Planned and Inadvertent Weather Modification Everett WA.

26. Bruintjes RT (1999) A review of cloud seeding experiments to enhance precipitation and some new prospects. Bull Amer Meteor Soc 80: 805-820.

27. Silverman BA (2000) An independent statistical evaluation of the South African hygroscopic flare seeding experiment. J Appl Meteor 39: 1373-1378.

28. Bruintjes RT, Breed D, Salazar V, Dixon M, Kane T et al (2001) Overview and results from the Mexican hygroscopic seeding experiment. Preprints AMS Symposium on planned and inadvertent weather modification. Albuquerque NM.

29. Caro D, Wobrock W, Flossmann AI (2002) A numerical study on the impact of hygroscopic seeding on the development of cloud particle spectra. J Appl Meteorol 41: 333-50.

30. Silverman BA (2003) A critical assessment of hygroscopic seeding of convective clouds for rainfall enhancement. Bull Amer Meteor Soc 84: 1219-1230.

31. Kucera PA, Axisa D, Burger RP, Collins DR, Li R, et al. (2010) Features of the weather modification assessment project in the southwest Region of Saudi Arabia. J Wea Mod 42: 78-103.

32. Rosenfeld D, Axisa D, Woodley WL, Lahav R (2010) A quest for effective hygroscopic cloud seeding. J of Applied Meteorology and Climatology 49:1548-1562.

33. Kulkarni JR (2012) The cloud aerosol interaction and precipitation enhancement experiment (CAIPEEX): overview and preliminary results Curr Sci 102: 413-425.

34. Posfai M, Axisa D, Tompa E, Freney R, Bruintjes P, et al. (2013) Interactions of mineral dust with pollution and clouds: An individual-particle TEM study of atmospheric aerosol from Saudi Arabia Atmos. Res 122: 347-361.

35. Semeniuk TA, Bruintjes RT, Salazar V, Breed DW, Jensen TL, et al. (2014) Individual aerosol particles in ambient and updraft conditions below convective cloud bases in the Oman mountain region. J Geophys Res 119: 2511-2528.

36. Tas E, Teller A, Altaratz O, Axisa D, Bruintjes R, et al. (2015) The relative dispersion of cloud droplets: its robustness with respect to key cloud properties Atmos Chem Phys 15: 2009-2017.

37. Dixon M, Weiner G (1993) Titan: thunderstorm identification tracking analysis and how casting a radar-based methodology. J Atmos Ocean Technol 10: 785-797.

38. Woodley WL, Rosenfeld D, Strautins A (2000) Identification of a seeding signature in Texas using multi-spectral satellite imagery. The J Weather Modification 32: 37-52.

39. FAA (2016) Unmanned aircraft systems.

40. DeFelice TP, Cheng RG (1998) On the phenomenon of nuclei enhancement during the evaporative stage of cloud. Atmospheric Research 47: 15-40.

41. Beswick K, Baumgardner D, Gallagher M, Volz-Thomas A, Nedelec P (2014) The backscatter cloud probe a compact low-profile autonomous optical spectrometer Atmos Meas Tech 7: 1443-1457.

42. Gao RS, Telg H, McLaughlin RJ, Ciciora SJ, Watts LA, et al. (2015) A light-weight high-sensitivity particle spectrometer for PM-2.5 aerosol measurements. Aerosol Science and Technology: 88-99.

43. Elston JS, Roadman J, Stachura M, Argrow B, Houston A, et al. (2011) The tempest unmanned aircraft system for in situ observations of tornadic supercells: Design and Vortex-2 flight results. J Field Robotics 28: 461-483.

26

Enhancing Flight Data Monitoring and Analysis can Increase Flight Safety

PadmanabanS* and Mahendran SME

Aeronautical Engineering, Hindustan University, Chennai,Tamilnadu, India

Abstract

The scope of the project is to Enhancing Flight Data Monitoring and Analysis can increase Flight Safety assists an operator to identify, quantify, assess and address operational risks. This analysis can be effectively used to support a range of airworthiness and operational safety tasks. The scope of this project is to de-code the recorded avionics parameter of interest based on the OEM's recommendation by using the logical extraction of data from the data frame of the recorder based on ARINC standard and Air born software standard. This project involves different processes from Data down loading from the DFDR, Raw data extraction, Optimum Parameter configuration, Logical Event configuration, Logical calculation of various flight scenarios, Comparison with FCOM, Flight Health monitoring, Exceedance Analysis based on regulatory guidance, Statistical analysis of various avionics parameter's impact on flight safety. The recommendation and solution found will be represented by various graphs and chart. Graphs of the checked parameters to show their evolution during cruise, take-off and landing phases of a same flight and an analysis of the validity of parameters based on graphs and corresponding tables, A check on the chronological structure of the complete recording, based on the aircraft flight history. It will be used for identifying and defining the risk index, and the inclusion and exclusion of the necessary maintenance programs based on the OEM.

Keywords: Flight data analysis; Exceedance; Airworthiness; Flight; Safety

Introduction

Over the past several years, airlines have initiated or participated in a number of safety data programs. Each involves collection of voluntary safety reports or the monitoring of flight data. These initiatives grew from recognition that mitigating safety risks requires monitoring a variety of data streams – reports, observations, and flight data. They have spawned technologies within air carriers, including Airline Safety Action Programs (ASAP), Line Operational Safety Audits (LOSA), improved analysis of training and checking data through the Advanced Qualification Program (AQP), and Flight Operational Quality Assurance (FOQA) programs. This paper will discuss the functions that can be served by flight data analysis [1-6].

The aviation community is under constant pressure to achieve safety improvement. Operational Flight Data Monitoring(OFDM) offers an efficient solution to this challenge. OFDM is to some extent aquality assurance process but also has a vital Safety Management dimension. It involves the downloading and analysis of aircraft flight recorder data on a regular and routine basis. It is widely used by aircraft operators throughout the world to inform and facilitate corrective actions in a range of operational areas by offering the ability to track and evaluate flight operations trends, identify risk precursors, and take the appropriate remedial action [7-11]. The potential of OFDM programmes has been materially enhanced by the rapid expansion in the number of data parameters which can be captured using digital recorders now routinely carried on aircraft.

To ensure the highest levels of safety each flight crewmember must carefully monitor the aircraft's flight path and systems, as well as actively cross -check the actions of each other. Effective crew monitoring and cross-checking can literally be the last line of defense; when a crewmember can catch an error or unsafe act, this detection may break the chain of events leading to an accident scenario. Conversely, when this layer of defense is absent the error may go undetected, leading to adverse safety consequences [12-15].

Methodology

Systematic flight data monitoring

The systematic approach of the FDM system allows an operator to compare their Standard Operating Procedures (SOPs) with those actually achieved in everyday line flights [16].

The analysis will be done for the recommended parameter and the methodology involves the process as follows:

- Parameter Configuration

- Validation Of Raw data

- Raw data extraction

- Phase configuration

- Exceedence monitoring

- Parameter grouping

- Parameter analysis

Parameter configuration is the process where the information about the parameter like analog, discrete and documentary will be defined based on the data frame layout with specific information like which sample per second, most significant bit and least significant bit to locate the parameter. After the configuration of parameter it is mandatory to audit for verifying and validation of the parameter

---

*Corresponding author: Padmanaban S,Aeronautical Engineering, Hindustan University, Chennai, Tamilnadu, India
E-mail: padmanabanpillai@gmail.com

configuration based on the data frame layout to ensure that the configured parameter values are matching to the recommended values [17-19]. The raw data extraction is decoding of the recorded parameter based on the parameter configuration and decoding law. During the raw data extraction flight phases will be defined based on the configuration of different phases of flight like taxi out, takeoff, climb, top of climb, cruise, top of descent, descent, approach, final approach, landing, touchdown, taxi in. Exceedence monitoring of the parameter will be based on the parameter limit range defined based on the safety standard and SOP's. Limit range defines the safe range of operation and it is further classified in to three categories as below:

- Green – Safe range

- Yellow – Acceptable range of operation

- Red – Danger range

Parameter analysis is based on the excedence or event definition, the event logic will be configured in to the software, based on the exceedence definition the exceedence range will be monitored and predicted as and when it exceeds the limit range. The analysis will result the total number of exceedence occurred and also it will help to predict the problem occurred in the system. The parameter analysis will help to find the problem occurred due to pilot behavior like human error.The trend monitoring of the parameter will predict the exact nature of the error and malfunction of the system or component this will improve the safety measure and help us to define the performance indicator or index value [20].

The Analysis will be based on the exceedence limit defined for the parameter,Safe range is mentioned as exceedence limit'1'and marked as Green, which is in the boundary of safe range of operation limit and it should within range of less than 5 percent of the recommended value and this percentage will vary depend on the parameter nature and performance and also it is defined in the SOP's of the operator. Acceptable level range is mentioned as exceedance limit '2' and marked as Yellow, which is in the boundary of acceptable level of safety but little away from the safe range still this is acceptable for minimum occurrence but this has more tendency to become risk if it is not considered properly and it should be within range deviation not more than 10 percent of the exceedence limit value and this percentage will vary depends on the parameter nature and performance and also it is defined in the SOP's of the operator.Unacceptable or Danger zone or level range which is mentioned as exceedence limit '3' and marked as red, which is away from the boundary of acceptable level of safety and away from the acceptable range still this is acceptable for minimum occurrence but this has more tendency to become risk if it is not considered properly and it should be within range deviation is more than 10 percent of the exceedence limit value and this percentage will vary depends on the parameter nature and performance and also it is defined in the SOP's of the operator.Trend monitoring of this analysis will be taken for the defined period like quarter-early, Half-early, and yearly and sometime based on the annual statistical analysis will predict the problem naturebased on the occurrence and place of occurrence which will be used to find the exact problem and allow taking the decision on recommendation and this will have the direct impact on the Sop's.

Six sigma methodology

Six sigma is a structured, data -driven approach to eliminating defects. The primary objective of the Six Sigma methodology is the implementation of a data based strategy that focuses on variation reduction and process improvement through the application of Six Sigma improvement projects. DMAIC – Define, Measure, Analyze, Improve, and Control – is the method used to engage in process improvement [21]. It was asserted that Six Sigma methods might be effectively used in FOQA programs, especially for addressing very infrequently occurring events.A disciplined quality approach to improving safety is needed in the airline industry. Airlines would benefit by increasingly embracing and employing quality principles in designing, implementing, and managing safety programs, including FOQA. Six Sigma is one quality-based program that may be used to increase the effectiveness of FOQA, particularly for process improvement initiatives. Whether an airline employs Six Sigma or various other methods in its safety improvement efforts, quality in airline safety must be the goal.

Six sigma techniques applied for FOQA

The parameters recorded during flight allow for a FOQA air carrier to monitor adherence to standard flight protocols. Each parameter can be monitored for variance based on set tolerance thresholds as determined by the air carrier upon appropriate validation. For example, a target value of 165 knots could be established for a certain phase of flight, with a maximum allowable variation of ±10 knots. Any exceedance (which in Six Sigma terms can be considered a 'defect') of these limits is flagged as an 'event', which is differentiated by severity levels. Therefore, a recorded parameter of 172 knots might be considered a level 1 severity event, while an exceedance of 180 knots could be considered a severity 3.

When excessive numbers of severity 1 and 2 events are detected by the FDM software, airline managers might elect to re-evaluate the tolerances since they might be too strict. However, when a severity 3 is detected, it usually points to a potentially dangerous violation of standard procedures; thus, they usually warrant close examination. If an airline continues to detect excessive numbers of severity 3 or other events after adjusting severity thresholds, the potential for an incident or accident may be indicated. FOQA's proactive nature means that it functions by concentrating on level 1 and 2 events, proactively implementing remedial action and standardizing the operations in order to avoid level 3 events from occurring. In the commercial air transportation is already highly standardized and level 3 events are rare, but they do occur. Examples of level 3 events are tail strikes during takeoff, and overshooting or undershooting runways during final approachdue to energy mismanagement. The rarity of these events makes it problematic to utilize rate based methods that depend on events that have already occurred in order to estimate the chances of any future occurrences. To illustrate, for an air carrier operating thousands of flights per month, FOQA trend data will be increasingly abundant with commonly occurring events such as speed or pitch violations. As data is collected and analyzed, the distribution will eventually become normalized, allowing for proper predictive statistics. However, for extremely rare events such as tail strikes, the distribution will not likely be normal, but rather highly skewed due to the extended amount of time without any occurrence. There will not be enough data to support proper predictive statistics.

Software Implementation

Flight data monitoring and analysis FDMA tool is used for the analysis of the Avionics parameter analysis for monitoring flight safety This incorporates the programming method used by the data acquisition system location of parameters, number of bits used to encode parameters, type and method of encoding the functions used to convert the recorded value into the actual physical value. For each parameter, the conversion function is checked with the calibration of the measuring and processing channel, Data acquisition systems output a binary file sequenced in four-second frames, depending on the FDR's The entire set of recorded data are copied for analysis and then converted into engineering units using decoding software which is programmed according to data frame layout documents here calibration is made because conversion functions provided by OEM's are only theoretical therefore differ from the ones of the actual aircraft. Calibration checks demonstrate if conversion equations identified are appropriate. These equations should convert recorded binary words into parameters expressed in engineering units. If conversions are shown to be inappropriate, acquisition channel elements or conversion equations should be adjusted. The processes are as follows shown in Figure 1.

- Data Frame
- Data Frame Structure
- Data conversion
- Algorithm
- DFDR recorded parameters decoding law

A. Data process flow

Raw data extraction will be done based on parameter configuration and flight slicing will be based on the phase configuration and exceedance analysis will be done based on the exceedance configuration, the data process flows as per the below chart shown in Figure 2.

B. Monitoring performance can be improved

As an industry, we seem to have accepted the axiom that, "Humans are not good monitors". While it may be true that humans are not naturally good monitors, we firmly believe that crew monitoring performance can be significantly improved through policy changes and crewmember training.Traditional CRM courses have generally improved the ability of crewmembers to challenge others when a situation appears unsafe or unwise; however, many of these courses provide little or no explicit guidance on how to improve monitoring. "First, we must change our approach to monitoring. Instructors must insist that the non-flying crewmember monitors the flier effectively. A system that grades monitoring must be established. Good monitoring skills are not inherent in a pilot as they progress in their careers. Therefore, effective monitoring techniques must be trained and rewarded.

Systematic Flight Data Monitoring

The systematic approach of the FDM system allows an operator to compare their Standard Operating Procedures (SOPs) with those actually achieved in everyday line flights.

- Identify areas of operational risk and quantify current safety margins.
- Identify and quantify changing operational risks by highlighting when non-standard, unusual or unsafe circumstances occur.
- To use the FDM information on the frequency of occurrence, combined with an estimation of the level of severity, to assess the risks and to determine which are or may become unacceptable if the discovered trend continues.
- To put in place appropriate risk mitigation to provide remedial action once an unacceptable risk, either actually present or

predicted by trending, has been identified.

- Confirm the effectiveness of any remedial action by continued monitoring.
- FDM is a closed loop system enhances the systematic approach to fulfill the problem statement.
- A feedback loop that should be part of a Safety Management System.
- (SMS), will allow timely corrective action to be taken where safety may be compromised by significant deviation from SOPs.

Advantage of Proposed System over the Existing System

- Gives knowledge of actual operations rather than assumed.
- Gives a depth of knowledge beyond accidents and incidents.
- Setting up an FDM program gives insight into operations.
- Helping define the buffer between normal and unacceptable operations.
- Indicates potential as well as actual hazards.
- Provides risk-modeling information.
- Indicates trends as well as levels of risk.
- Can provide evidence of safety improvements.
- Feeds data to cost-benefit studies.
- Provides a continuous and independent audit of safety standards.
- Can help identify area where flight crew training can be further improved.

Conclusion

FDM has increased gradually over the last 30 years as analysis techniques and data recording technologies have improved. The processes used in the past have tended to be rather ad hoc, locally implemented and controlled by informal procedures with less than ideal 'check and balance' records after issues have been raised and acted upon. Having said that, despite this lack of established process, many significant safety issues have been raised and resolved. The systematic approach should provide a more quantitative risk picture to the organization to help it manage its risks and measure the success of its mitigation actions.Parameter Analysis for Monitoring Flight Safety covers the basic Parameter configuration rules and methodology with clear understanding of the parameter type and decoding procedures.

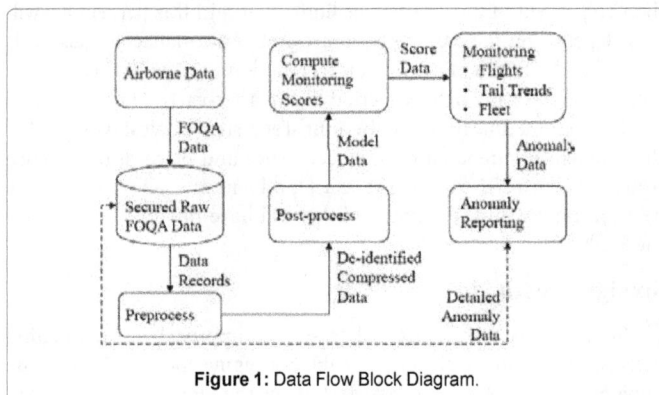

Figure 1: Data Flow Block Diagram.

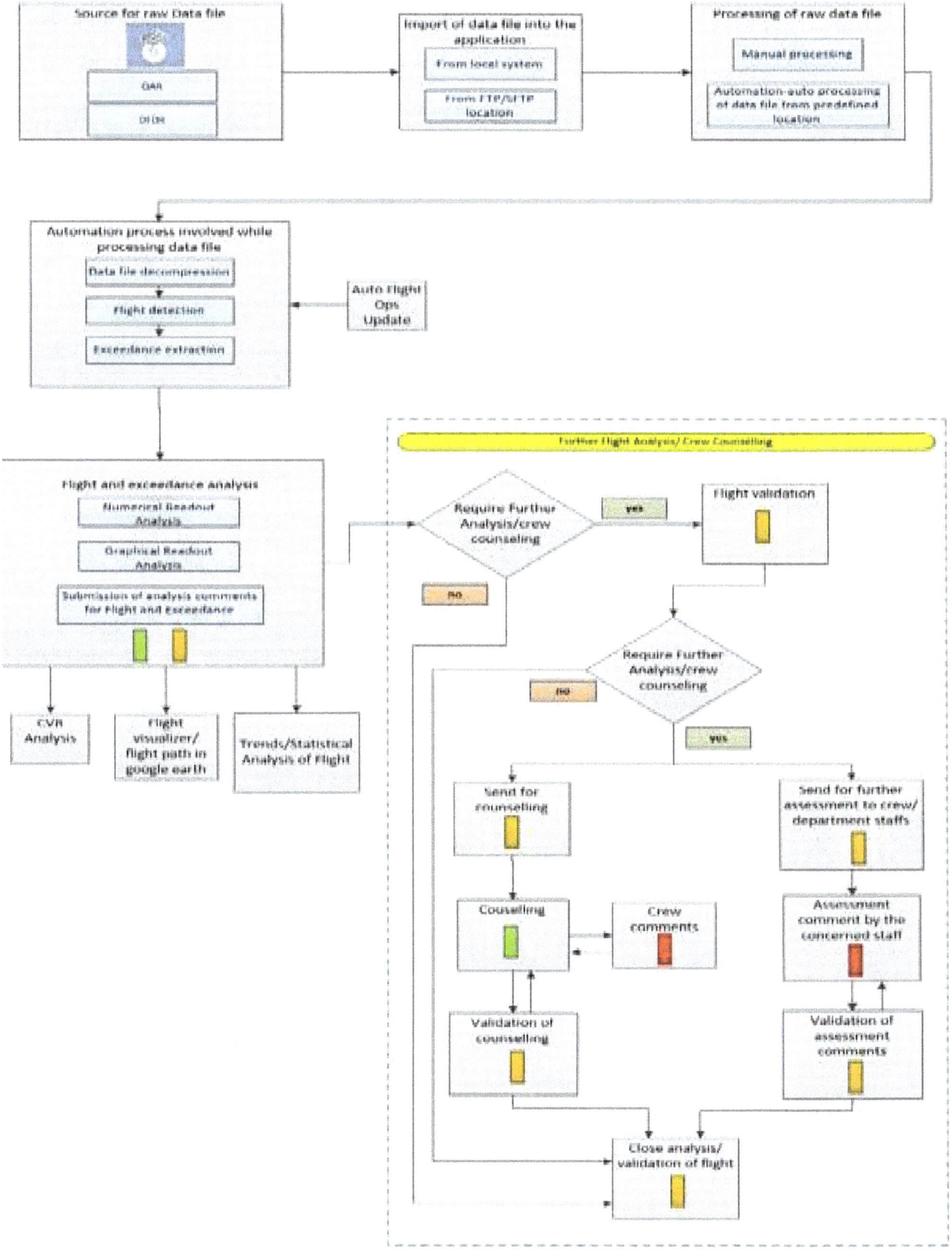

Figure 2: Data process flow.

And also explains the systematic approach for monitoring flight data, the detailed analysis with the program triggers, Configuration details and exceedance conditions for event monitoring and parameter analysis. Flight Operations Quality Assurance has been one of the most highly regarded and potentially effective airline safety initiatives to emerge in the past 20 years. It is a program based on quantifiable, objective data collected from the air carrier aircraft's data recording system. On some modern aircraft, over 2000 parameters each second are recorded. The FOQA system uses expert software to analyze the data from individual flights of interest, or aggregated data from multiple flights in order to examine trends that may affect safety. Unfortunately, with very few exceptions, the analysis of FOQA data has been limited to relatively simple statistical methods. It has been surmised that the application of more sophisticated quality and statistical methods may increase the effectiveness of the program and the air carrier's return on investment.

References

1. Ananda CM, Kumar R (2008) Configurable Flight Data Analysis For Trends And Statistics Analysis Of Avionics Systems An Embedded Perspective Of An Efficient Flight Data Analysis International Conference on Aerospace Science and Technology IISC Bangalore.

2. APMS (2012) International conference advances in production management systems.

3. (1999) Analysis of FDR Data for Accident/Incident Prevention is to be used as a guideline. The CAR Section-5 Air Safety Series 'F; on Monitoring of DFDR/DR/QAR / PMR Data For Accident/Incident Prevention.Air Safety Circular No. 2.

4. Operational Serviceability, and readout of flight data recorder system and cockpit voice redorder system -CAP 731.

5. CAA (2003) Flight data monitoring A guide to good practice. London.

6. Campbell NH (2007)The Evolution of Flight Data Analysis Regional Seminar Australia.

7. Dr. Tulinda Larsen, masFlight, Bethesda MD (2013) Cross-platform aviation analytics using big-data methods, Integrated Communications Navigation and Surveillance Conference (ICNS).

8. Gorinevsky D (2012) From Mitek Analytics LLC from NASA Ames Research Center, and Rodney Martin from NASA Ames Research Center Aircraft Anomaly Detection using Performance Models Trained on Fleet Data CIDU - Conference On Intelligent Data Understanding Boulder Co.

9. EASA (2007) European Aviation Safety Agency, Position Paper, safety management systems (SMS), Annex III ORO AOC 130.

10. Balakrishnan H (2011) Estimation of Aircraft Taxi-out Fuel Burn using Flight Data Recorder Archives Harshad Khadilkar Massachusetts Institute of Technology Cambridge MA 02139 USA.

11. EASA (2012) Guidance for National Aviation Authorities. European Authorities coordination group on Flight Data Monitoring (EAFDM) Guidance: Annex III -ORO.AOC.130.

12. FDS.Visualisation as a debrief tool.

13. (STATE) Flight Data Analysis Programme: A Guide to Good Practice CAA / DGCA.

14. CAR section 2 Series I Part-V the mandatory parameters to be recorded, listed in Appendix - 'K' of Air Pegasus Flight Safety Manual.

15. ICAO (2014) Operation of Aircraft. Office of the Director General Of Civil Aviation. Annex 6.

16. Yan J, Histon J, Ramana MV (2013) Flight Data Monitoring and Human Factors Risks identification A Review of Best Practices. University of Waterloo ON Canada QuESTFlight Operational Quality Assurance through Exploitation of Flight Data Recorders.

17. Manoel P (2010) RAZABONI, Instant flight data analysis.EMBRAER Air Safety Department.

18. Neil AH. The Evolution of Flight Data Analysis Campbell MO3806 .

19. (2013) CAA document on FDM CAP 739UK.

20. Yan J, Histon J (2013) Flight Data Monitoring and Human Factors Risks Identification: A Review of Best Practices CASI 60th Aeronautics Conference and AGM.

21. Charles D (1972) "Fire Detection System Performance in USAF Aircraft" AFAPL-TR-72-49Air Force Aero Propulsion Laboratory Wright Patterson AFB OH.

27

Can the Vertical Motions in the Eyewall of Tropical Cyclones Support Persistent UAV Flight?

Chung-How Poh, and Chung-Kiak Poh*

Aero-Persistence Research, 23 Halaman York, 10450 Penang, Malaysia

Abstract

Powered flights in the form of piloted aircraft or unmanned aerial vehicles (UAVs) have been flying into tropical cyclones to obtain vital atmospheric measurements. The flight durations of these aircraft typically last only between 12 and 36 hours. Convective vertical motion properties of tropical cyclones have previously been studied. This work investigates the possibility to achieve persistent flight by harnessing the generally pervasive updrafts in the eyewall of tropical cyclones. An endurance/persistence UAV capable of vertical take-off and landing (VTOL) is proposed and its flight characteristics are simulated using the RealFlight® simulator. Results suggest that the concept of persistent flight within the eyewall is promising and may be extendable to the rainband regions.

Keywords: Tropical cyclones; UAV; Persistent flight

Introduction

Tropical cyclone looks serene and elegant from the space above, but the sheer scale of the storm can generate winds in excess of 250 kmh⁻¹ and drive storm surge causing massive destruction to the surface below [1,2]. The main structural features of a tropical cyclone are the rain bands, the eye, and the eye wall [3]. The eye has a typical diameter of 32 to 64 km, though its formation mechanism is still not fully understood [3,4]. The eye is the calmest part of the storm with winds that usually do not exceed 24 kmh⁻¹ [3]. The eye wall region consists of a ring of tall thunderstorms that produce heavy rains and often the strongest winds [3].

The wind circulations of a matured tropical cyclone can be broadly divided into the primary and the secondary circulation [5]. The primary circulation refers to the tangential flow rotating about the central axis, and the secondary circulation refers to the "in-up-and-out circulation" (low and middle level inflow, upper-level outflow) [5]. Thus, the general air flow model of a tropical cyclone is air parcels spiraling inwards, upwards and outwards [5].

Despite advances in remote sensing, in-situ measurements with reconnaissance flights are still necessary to obtain accurate location of pressure center and wind speeds to aid reliable forecasting [6]. Of particular interest to forecasters and the public is the maximum sustained 10 m surface wind [7]. UAVs have recently been deployed for tropical cyclone missions. Two such well-known UAVs are the NASA Global Hawk and the Aerosonde, with flight endurances of 30 hours and 26 hours (minimal payload), respectively [8,9].

Comprehensive study on the vertical motions in intense tropical cyclones has been carried out by Jorgensen et al. [10]. It involved dataset from four mature hurricanes (Anita, David, Frederic and Allen) and a total of 115 penetrations from nine flight sortie at altitudes from 0.5 to 6.1 km [10]. A total of over 3000 updrafts recorded and the downdrafts total was nearly 2000. A convective updraft or downdraft was defined as having vertical velocity (w) measurements with at least one point > $|0.5|$ ms⁻¹ transversing over at least 500 m of horizontal distance [10]. Stronger cores were further distinguished from the drafts if the $|w| > 1$ ms⁻¹ [10]. Thus, a single draft may contain several cores [10].

Thermal soaring is a form of flight where the flying objects use only convection currents to stay aloft without any additional power source (motor power in the case of airplane, or flapping of wings in the case of birds) [11]. Cross-country flights have been undertaken by birds and man-made sailplanes [11]. The soaring birds gain height by circling in thermals with wing spread until desired height is reached. They then glide forward in a sinking phase to the next available column of thermal [11]. UAVs with autonomous thermal seeking capability have been demonstrated [12,13]. Whenever an updraft is rising faster than the sink rate of a sailplane, there will be a gain of altitude. The rate of climb of a sailplane, V_c with its sink rate, V_s in an ascending updraft of average velocity, $\bar{w}$ is [14]:

$$V_c = \bar{w} - V_s \qquad (1)$$

This work investigates the potential to develop persistent UAV by harnessing the updrafts in the eye wall of tropical cyclones. Viability of the concept was assessed based on the work published by Jorgensen et al. on the characteristics of vertical motions in hurricanes [10] and by simulating the performance of a small sailplane UAV. The UAV is intended to advance from an updraft core to the next using similar flight scheme as that of the cross-country flights. The VTOL capability of the UAV is demonstrated by simulation and the effect of payload is also evaluated. The long-term goal of the research is to develop an UAV capable of remaining airborne with the storm throughout its life-span to enable uninterrupted acquisition of dataset, particularly information relating to the 10 m surface wind.

Materials and Methods

Characteristics of vertical motions

A simple graphical representation of the updraft cores within ±5 km from the radius of maximum wind (RMW) was generated using

---

Corresponding author: Chung-Kiak Poh, Aero-Persistence Research, 23 Halaman York, 10450 Penang, Malaysia
E-mail: nanophotonics@yahoo.com

the results reported by Jorgensen et al. [10]. The primary aim was to establish the estimated mean distance between 2 adjacent cores in the eye wall. This information will allow one to assess whether a gliding UAV with a given flight performance has the required range to reach adjacent core, taking into account other possible environmental conditions.

Data analysis by Jorgensen et al. revealed that the size and strength of the drafts and cores were very much similar for all the four hurricanes considered. Four properties considered were the 1) intercepted length of the draft or core, D (loosely referred to as the "diameter"); 2) mean vertical velocity of each draft or core $\bar{w}$; 3) the maximum 1-s vertical velocity in the draft or core w_{max} and 4) total mass transport per unit length normal to the flight track. The distributions of these properties were found to be approximately linear on a log-normal plot.

The followings summarize the key atmospheric characteristics of updrafts and downdrafts in tropical cyclones as reported in [10] which will be needed for the conceptual design of the persistence UAV proposed in this work as well as to determine if such concept is viable:

• In both the eye wall and rain band regions updrafts dominated over downdrafts in terms of counts and mass transport.

• Approximately two to three times more updraft cores than downdraft cores were encountered, particularly at the lower altitudes.

• 99% of the updraft cores were < 7 km in diameter, and 99% of the downdraft cores were less than about 5.6 km.

• Their results suggested that relatively small area of these mature hurricanes was covered by significant vertical motions and that the eye wall region was dominated by updrafts. The dominance was presented as percentages of an annulus defined as ±5 km around the RMW covered by the cores.

• The inner halve was defined as the eyewall region and the outer region was defined as the rainband region.

• At altitude of 0.5 km, the percent of eyewall region covered by the updraft and downdraft cores were 54.7% and 22.7%, respectively.

• At altitude of 0.5 km, the top 10% levels of the diameter of the updraft cores in the eyewall and rainband regions have an average value of 3.2 km.

This work focused on the atmospheric conditions at the altitude of 0.5 km because it is of particular relevant to surface wind measurement. RMW of Hurricane Anita of 21 km [15] was used in this work to generate the graphical representation of the updraft core distribution at altitude 0.5 km. Given the similarity, it should typify the other three hurricanes or other tropical cyclone in general.

Simulation details

Simulation work was performed using the RealFlight[®][1] 6.5 simulator [16] running on a quad-core 2.2 GHz computer. The Spirit 100 (shown in Figure 1) with highly-efficient Selig 703 airfoil was used as the base platform to create the gliding UAV. The motorless Spirit 100 has a wing span of 2.53 m and flying weight of 1.7 kg. Modification of the Spirit 100 was done using the Accu-Model™ aircraft editor. Surface areas of the rudder and elevator were increased by adjusting the chord ratios, and the flaps were replaced by a pair of ailerons. Control surface areas of the original Spirit 100 and its modified variant (termed as Spirit 100-VT) are as shown in Figures 2(a) and 2(b), respectively. The following

[1]RealFlight is a registered trademark of Hobbico, Inc. used with permission.

Figure 1: The as-supplied simulation model of the Spirit 100.

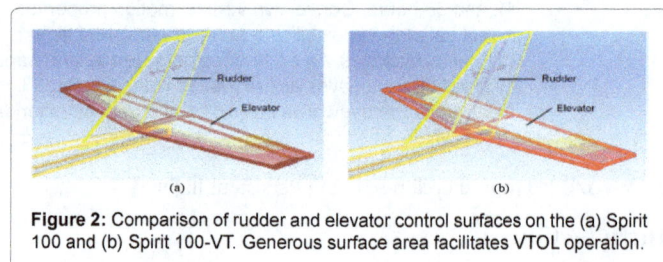

Figure 2: Comparison of rudder and elevator control surfaces on the (a) Spirit 100 and (b) Spirit 100-VT. Generous surface area facilitates VTOL operation.

components were added to the platform to enable powered flight: A 2204 W brushless motor with 121.72 N thrust, a pair of co-axial counter-rotating folding propellers (size: 432×174 mm), and a 6-cell 8000 mAh lithium polymer (Li-po) battery pack. Angular position dependent 3-axis (roll, pitch, yaw) gyro was added to the platform for enhanced flight stability. The completed model weighed in at 3.085 kg.

Gliding performance was characterized using the polar curve, with the motor turned off and the propellers folded. The flight power was acquired by setting the sailplane on a straight and level flight, adjust the throttle until the desired airspeed was reached with the variometer showing zero readout; if the airspeed, altitude and electrical power value remained invariant for 1 minute, equilibrium conditions were considered reached and the power value was recorded as the power requirement. Effect of payload on flight performance was also evaluated at 1, 3, and 5 kg. The center of mass of the aircraft was made to coincide with the center of lift throughout the simulation to ensure flight characteristics were not adversely affected.

Results and Discussions

Properties of updraft cores in the eyewall

Figure 3 shows a simple graphical representation of the updraft cores within the eyewall and rainband at altitude 0.5 km based on published data mentioned in Section 2.1. The diameter of cores in both regions were 3.2 km. The density of updraft in the rainband was included in Figure 3 as well even though the focus of this study was on the eyewall region. The average core-to-core distances in the eyewall and rainband were estimated to be 2.9 and 4.75 km, respectively.

VTOL capabilities

Normal flight and VTOL capabilities were successfully demonstrated with good control over the roll, pitch and yaw, even with the payload of 5 kg. The use of the counter-rotating propellers was essential as it reduced the propeller torque effect and the P-factor to an almost negligible level, and consequently, virtually no aileron deflection was needed to maintain roll-free vertical hovering. This allowed for smaller ailerons to be used, symmetrical roll-rate on both directions, and improved the hovering stability. Less aileron deflection

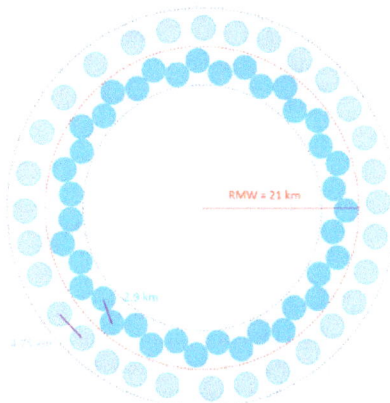

Figure 3: A simple graphical model showing a plausible distribution of updraft cores ±5 km from the radius of maximum wind (RMW) for Hurricane Anita at altitude 0.5 km.

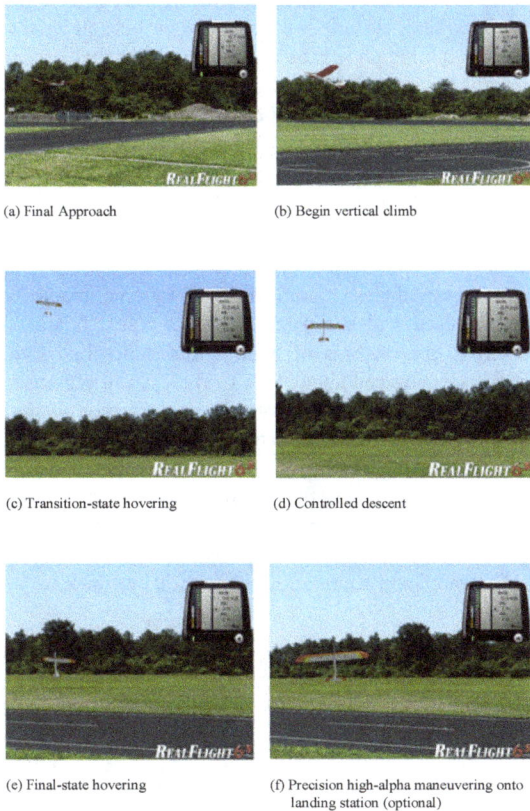

(a) Final Approach

(b) Begin vertical climb

(c) Transition-state hovering

(d) Controlled descent

(e) Final-state hovering

(f) Precision high-alpha maneuvering onto landing station (optional)

Figure 4: Vertical landing sequence.

also resulted in greater vertical acceleration, which can be used to lift heavier payload.

Figures 4(a) to (f) show the vertical landing sequence for the gliding UAV with 5 kg payload onboard. The landing sequence begun with the final approach and followed by the vertical climb and reduction in airspeed. As the nose pitched upward, the motor thrust was gradually increased until the platform came to a transition hovering (variometer showing 0 ms^{-1}). The platform was then brought into a constant rate of

descent (-0.9 ms^{-1}, in this case) and terminated with the final hovering. Optionally, the platform could be steered with precision onto a landing station, if necessary [Figure 4(f)]. The VTOL capability will allow the platform to be launched from ships, e.g. an unmanned marine vehicle (UMV).

Figure 5 shows the hovering power for payload ranging from 0 to 5 kg, and a linear fit has been applied to the plot with a R^2 value of 0.9974. The 3.085 kg platform itself consumed 343.68 W for hovering and a linear fitting gave 230 W kg^{-1} for any addition of payload.

Gliding performance

Gliding performance of the hybrid sailplane with varying payloads was evaluated using the polar curves, as shown in Figure 6, from which minimum sink rates and maximum glide ratios were obtained. The glide ratio refers to how far a glider can travel for a given altitude. A line is drawn from the origin tangent to the polar curve, and the speed indicated by the point of tangency is the speed to achieve maximum glide ratio [17]. The minimum sink rate is the speed at which the glider is descending at a slowest possible rate through the air [17]. Table 1 summarizes the maximum glide ratio and the minimum sink rate for different payloads, along with other crucial parameters such

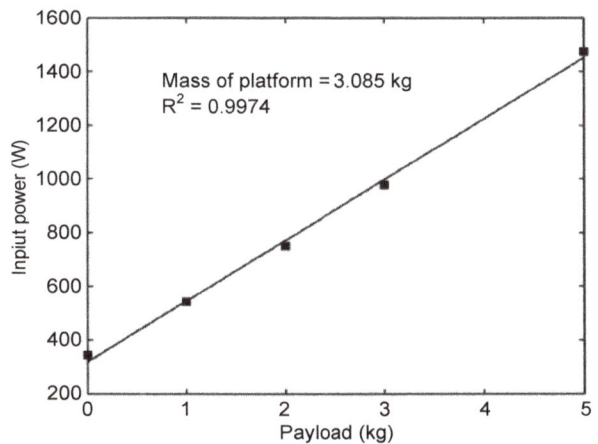

Figure 5: Variation of hovering power with payload.

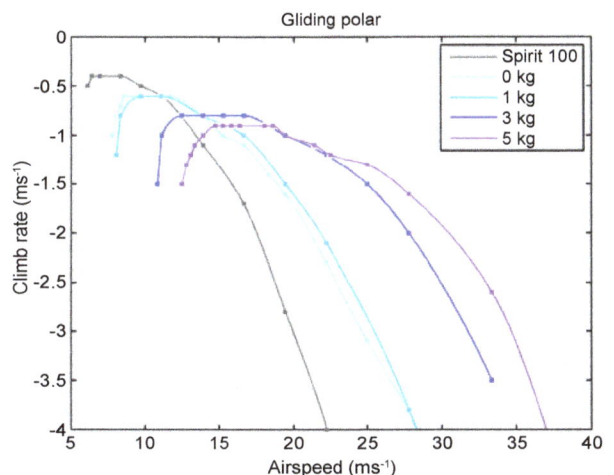

Figure 6: Polar curves for the original Spirit 100 and the Spirit 100-VT with different payloads.

Configuration of Platform	Flying Weight (kg)	Wing Loading (Nm^{-2})	Hovering Input Power (W)	Stall Speed (kmh^{-1})	Minimum Sink Rate (ms^{-1})	Best Glide Speed (kmh^{-1})	Maximum Glide Ratio
Spirit 100 (as supplied)	1.745	25.27	N/a	22	0.4	31	21.1
Spirit 100-VT (0 kg)	3.085	44.67	343.68	28	0.6	42	19.7
Spirit 100-VT (100 kg)	4.085	59.14	542.82	29	0.7	42	19.0
Spirit 100-VT (3 kg)	6.085	88.10	978.80	39	0.8	62	21.1
Spirit 100-VT (5 kg)	8.085	117.06	1474.86	49	0.9	67	20.7

Table 1: Key flight parameters for the simulated Spirit 100-VT and the effect of payload. Motorless Spirit 100 was included for comparison.

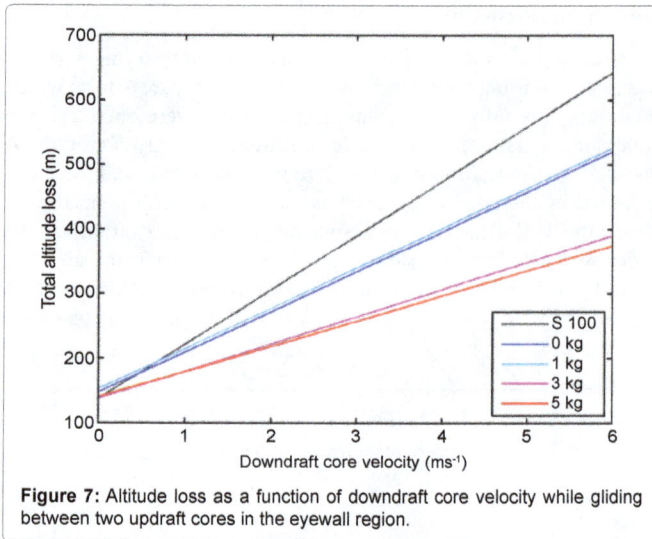

Figure 7: Altitude loss as a function of downdraft core velocity while gliding between two updraft cores in the eyewall region.

as the flying weights and wing loadings. It is interesting to note that despite relatively large variation in the payload, the minimum sink rate only varied between 0.6 and 0.9 ms^{-1}. There was little variation in the maximum glide ratio. A glide ratio of 21:1 which means that in smooth air the plane will sink 1 km for every 21 km forward. The stall speeds tabulated in Table 1 were also derived from the polar curves (Figure 6), and as expected, they increased with payload.

Sink rates (V$_s$) of 0.6 to 0.9 ms^{-1} found in this simulation study implied that average updraft velocity ($\overline{w}$) of 1 ms^{-1} or more is sufficient to keep the platform airborne, as indicated by Eq. 1. Obviously the thermalling conditions will be more easily met with lighter payloads or a wing with a higher lift-to-drag ratio.

Analysis of gliding flight in tropical cyclone

This section evaluates the potential of Spirit 100-VT to achieve persistent flight in the eyewall of tropical cyclone by analyzing its gliding performance and the characteristics of vertical motions. From Section 3.1, the mean distance between two updraft cores is 2.9 km. The maximum glide ratio of the Spirit 100-VT without payload was 19.7 (Table 1). Calculations based on its gliding performance indicated that it will lose 147.2 m in altitude while gliding to the next updraft core at its best glide speed of 11.67 ms^{-1} (42 kmh^{-1}). From Figure 3, the updraft cores are in close proximity to each other as the updrafts were dominant within the eyewall with the percent coverage of ~54% at 0.5 km altitude. Now, let's assuming there is a downdraft that exists along the gliding phase and occupying 1/4 of the path. With a top 10% average downdraft core value, w_{down} of 3.2 ms^{-1} [10], this will lead to a total altitude loss of 346.1 m with remaining height of 153.9 m. Though the lowest safe altitude (LSALT) for aviation is generally 304.8 m (1000 ft), ~150 m of safety buffer should be adequate for a small UAV. In case

the downdraft core value = $w_{\text{down max}}$ (6 ms^{-1}) [10], the total altitude loss was 520.1 m which exceeded the initial flight height of 500 m. To mitigate this problem, the initial height can be increased to 600 m or powered flight can be initiated if altitude drops below 50 m. The percent area of downdraft cores in the eyewall was only 22.7% [10] and therefore the probability of encountering downdraft would not be too significant.

Figure 7 shows the altitude loss as a function of downdraft core velocity for different payload configuration. S 100 refers to the motorless Spirit 100. As it turned out, the variant with higher wing loading suffered least altitude loss because their best glider speeds are higher while retaining glide ratio around 21:1. For that reason, it has been a common practice to equip cross-country sailplanes with water ballast to achieve higher cruising speeds [18]. These simulation results strongly suggest that persistent UAV flight in the eyewall of a tropical cyclone is feasible. Future work will extend the study to the rainbands as well as the early stages of cyclogenesis.

Conclusions

Persistent UAV flight within the eyewall region of tropical cyclone has been proposed and simulated. Previous research and analyses of direct flight observations of mature hurricanes had found that the updraft cores covered about 55% of the eyewall regions with $\overline{w}$ of 4.5 ms^{-1} at 0.5 km altitude. Given the vertical motion properties, simulation work suggested that persistent flight is indeed viable. Glide ratio of 20.7 and a wing loading of 117.06 Nm^{-2} resulted in an altitude loss of under 300 m during the gliding phase across a core-to-core distance of 2.9 km. VTOL capability of the persistent UAV had also been demonstrated via simulation. Persistent flight would usher in a new era of research and observations of tropical cyclones.

References

1. Cheung KKW, Chang LTC, Li Y (2014) Rainfall Prediction for Landfalling Tropical Cyclones: Perspectives of Mitigation. Typhoon Impact and Crisis Management. Series: Advances in Natural and Technological Hazards Research , 40: 175-201.

2. Chan JCL, Kepert JD (2010) Global Perspectives on Tropical Cyclones: From Science to Mitigation. World Scientific Publishing Co Pub Ltd, Singapore.

3. Tropical Cyclone Stucture.

4. Piñeros MF, Ritchie EA, Tyo JS (2010) Detecting Tropical Cyclone Genesis from Remotely Sensed Infrared Image Data. IEEE Geosci Remote Sens Lett 7: 826–830.

5. Smith RK (2006) Lectures on Tropical Cyclones.

6. Hurricane Hunters Association.

7. Franklin JL, Black ML, Valde K (2003) GPS Drop Windsonde Wind Profiles in Hurricanes and their Operational Implications. Wea Forecasting 18: 32-44.

8. HS3 Hurricane Mission. Hurricane and Severe Storm Sentinel.

9. Aerosonde.

10. Jorgensen DP, Zipser EJ, and LeMone MA (1985) Vertical Motions in Intense Hurricanes. J Atmos Sci 42: 839–856.

11. Ákos Z, Nagy M, Leven S, Vicsek T (2010) Thermal Soaring Flight of Birds and Unmanned Aerial Vehicles.

12. Daugherty SC, Langelaan JW (2014) Improving Autonomous Soaring via Energy State Estimation and Extremum Seeking Control. AIAA Guidance, Navigation, and Control Conference, National Harbor, Maryland. AIAA 2014-0260.

13. Allen MJ (2007) Guidance and Control of an Autonomous Soaring UAV. NASA Dryden Flight Research Center.

14. Shannon HD, Young GS, Yates MA, Fuller MR, Seegar WS (2002) Measurements of Thermal Updraft Intensity Over Complex Terrain Using American White Pelicans and a Simple Boundary-layer Forecast Model. Bound-Layer Meteor 104: 167-199.

15. Jorgensen DF (1984) Mesoscale and Convective-scale Characteristics of Mature Hurricanes. Part I: General Observations by Research Aircraft. J Atmos Sci 41: 1268-1286.

16. RealFlight.

17. Holtz, R (2012) Glider Pilot's Handbook of Aeronautical Knowledge. White Oak Communications, North Carolina, USA.

18. Flying with Water Ballast.

28

Architecture Level Safety Analyses for Safety-Critical Systems

Kushal KS*, Manju Nanda and Jayanthi J

Aerospace Electronics and Systems Division, CSIR-National Aerospace Laboratories, Bangalore, Karnataka, India

Abstract

The dependency of complex embedded Safety-Critical Systems across, Avionics and Aerospace domains, on their underlying software and hardware components has gradually increased with progression in time. Such application domain systems are developed based on a complex integrated architecture, which are modular in nature. Engineering practices assured with system safety standards to manage the failure, faulty and unsafe operational conditions are very much necessary. System safety analyses involves the analysis of complex software architecture of the system, a major aspect in leading to fatal consequences in the behavior of Safety-Critical Systems, provides high reliability and dependability factors during their development. In this paper, we propose an architecture fault modeling and the safety analyses approach that will aid in identifying and eliminating the design flaws. The formal foundations of SAE Architecture Analysis and Design Language (AADL) augmented with the Error Model Annex (EMV) are discussed. The fault propagation, failure behavior and the composite behavior of the design flaws/ failures are considered for architecture safety analysis. The illustration of the proposed approach is validated by implementing the Speed Control Unit of Power Boat Autopilot (PBA) system.

The Error Model Annex (EMV) guides with the pattern of consideration and inclusion of probable failure scenarios and propagation of fault conditions in the Speed Control Unit of Power Boat Autopilot (PBA). This helps in validating the system architecture with the detection of the error event in the model and its impact in the operational environment. This also provides an insight of the Certification impact that these exceptional conditions pose upon at various criticality levels, design assurance levels and its implications in verifying and validating the designs.

Keywords: Architecture analysis and Design language (AADL); Error annex (EMV); Fault-tree analysis (FTA); Primary events database (PED); Safety analyses; Safety-critical systems

Introduction

Systematic analyses of the architectural models modelled using the Model-Based Engineering (MBE) [1] practices, early and at every abstraction level imbibes a greater confidence in the integration of the system. The creation and analysis of architectural models of a system supports prediction and understanding the system's capabilities and its operational quality attributes. These attributes include performance, reliability, reusability, safety and security. All along the developmental lifecycle, the faults such as their failure modes and their propagation effects, at system-level can be predicted. Such issues remain un-noticed until system integration and testing. This proves to be a costly rework resulting in an unaccounted project time, cost and maintenance.

For safety critical advanced complex embedded systems, the system design and development is in compliance with the safety standards, and engineered with practices as specified by MIL-STD882 [2], SAE ARP-4761 [3] and DO-178B/C [4]. The process of development, management and controlling these systems in conformance with the safety practices, proves to have an impact on the system requirements, post system integration and test. With the evolution of the system, availability and reliability of these models are to be consistent and this poses a great challenge.

These safety practices include various availability and reliability prognosis with the help of system architectural models. Model-Based Engineering approaches for safety analyses address these issues and prove to provide consolidated information about the informal requirements and the architecture model of the system. The safety analyses performed on a system also takes into consideration, the physical environment of its deployment and functioning. Due to insufficient support of the formal languages trend is to make use of architecture description

languages such as Architecture Analysis and Design Language (AADL), and Society of Automotive Engineers (SAE) standard. AADL, a high-level architectural descriptive language, basically provides a platform for overall integration of various system recommended components via formal semantics and syntax. This component-based modelling language is extended with the introduction of sublanguages as Annexes. AADL is packaged with multiple Annex sublanguages such as Error Model Annex (EAnnex) and Behaviour Annex (BAnnex) as standards. The EAnnex standard is suitably augmented with safety semantics and ontology of fault propagation, supporting error annotations on the architectural models [5]. This thus enables the component error models and their interactions to be considered in context to the system architecture modeled using AADL.

This paper presents our contributions as a case study implementation (Speed Control Unit of Power-Boat Autopilot), to the standard approach for the illustration of its application. The paper is organized as follows: Firstly, we summarize the concept of Architecture Analysis and Design Language (SAE AADL) and Error Model Annex (EAnnex/EMV2). Next we provide an illustration of the architecture fault model specification for Speed Control Unit of a Power-Boat Autopilot (PBA). We also discuss the various safety analyses methods involved in MIL-STD882

---

*Corresponding author: Kushal KS, Aerospace Electronics and Systems Division, CSIR-National Aerospace Laboratories, Bangalore, Karnataka, India
E-mail: ksk261188@gmail.com

safety practice. Finally, we conclude the paper with the assessment of these safety analyses based on the architecture fault models.

Error Model Annex in Architecture Analysis and Design Language (AADL)

Architecture Analysis and Design Language (AADL), an SAE International standard, a unified framework providing extensive formal foundations for Model-Based Engineering (MBE) practices. These practices extend throughout the system design, integration and assurance with safety standards. AADL distinctly represents a system hardware and software components and their interactions via interfaces. Critical real-time computational factors such as performance, dependability, safety, security and data integrity can be rigorously analysed with AADL.

AADL also integrates custom analyses and specification techniques during the engineering process. This allows in the development and analysis of a single, unified system architectural model. AADL can be extended using the specialized language constructs that can be attached to the components of the architectural model defined by AADL. These components are reinforced with additional characteristics and requirements, referred to as Annex languages. The architectural model components are annotated with these properties and annex language clauses for functional and non-functional analyses. Error Model Annex (EMV), which is an extension of AADL aids in describing the failure conditions and fault propagations as error events, propagations, occurrence and their distribution properties. With the integration of these constructs in the AADL model/s, as shown in Figure 1 [6], the existing components are extended as current models liable for Safety Evaluation and Analyses. This can be done with the help of the algorithms in OSATE or by using other third party tools.

Error annex

The Error Model Annex (EAnnex) is a sublanguage of AADL This sublanguage extension includes the analyses of the runtime architectures. The EAnnex [7,8] annotates the hardware and the software component architectures with error states, error events, error transitions and error propagations that may affect the component interacting with each other. In Error Model Annex sub clause conditions can be specified under which the errors are propagated through designated component ports. Error Model Annex basically helps in defining the fault models, hazards, fault propagation, failure modes and effects, as well as specifying compositional fault behavior.

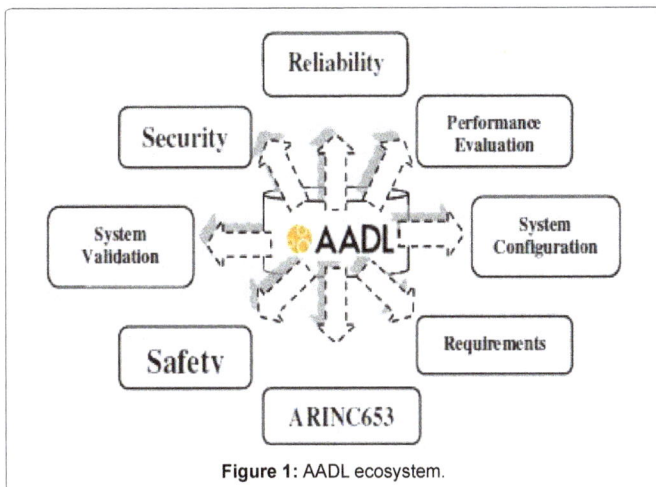

Figure 1: AADL ecosystem.

AADL Error Model Annex supports architectural fault modeling at three levels of abstraction [9].

1. Modeling of faults in systems and their implications on other dependent components of the physical environment of its operation through propagation of these faults. (Includes Hazard identification, fault impact analysis).

2. Modeling of faults occurring in a component of the system and analyzing the behavior of the same across various modes termed as failure modes and its effects on other components and its related propagations. It is also inclusive of the recovery strategies involved.

3. Compositional abstraction of system error behavior in terms of its subsystems.

Error Model Annex (EMV2) overlays major focus on the standards set of error types and error propagation, defined by AADL as a standard syntactic construct through the introduction of annex libraries. These annex libraries provides an overlook of the formally specified error propagation behaviors [10,11]. Some of the common error types being [9];

1. **Commission and omission errors**: Represents loss of message/command, failure to provide readings from a component.

2. **Timing errors**: Arrival rate, service too early or late, unsynchronized rate.

3. **Value errors**: Individual service item error or errors in a sequence of values

4. **Replication errors**: Replicates of states or services being communicated.

5. **Concurrency errors**: Accessing shared logical or physical resources.

Along with these the error model types can be referenced in the Error Model Annex sub clause. The constructs for the EMV2 are similar to the syntax and style as defined for AADL. An exceptional being that, any set of textual language constructs can be included within an annex, that includes Object Constraint Language (OCL) [12] or a temporal logic notation [13].

Implementation of proposed research: In this section we exhibit the architecture fault modeling in AADL, along with the extension of EMV2, at three levels of abstraction with a suitable case study, Speed Control Unit of Power-Boat Autopilot (PBA). This unit is a simplified speed control model, including a pilot interface unit for input of relevant Power-Boat Autopilot information, a speed sensor that sends speed data to the PBA, the PBA controller, a throttle actuator that responds to the commands specified by the PBA controller and a display unit. The type definitions defining the component, component names, their runtime category and interfaces are identified and defined. The speed sensor, pilot interface, throttle actuator and the display unit are modeled as devices, while the PBA control functions are represented as process, as shown in Figure 2. With all these we perform the safety analyses with the specification of the source of error and its propagation across the system and its components. This is carried out by defining the error states and their corresponding compositional fault behaviour. This is followed by the expansion of the fault logic with respect to its error behaviour related with each component of the system and its response to the failures.

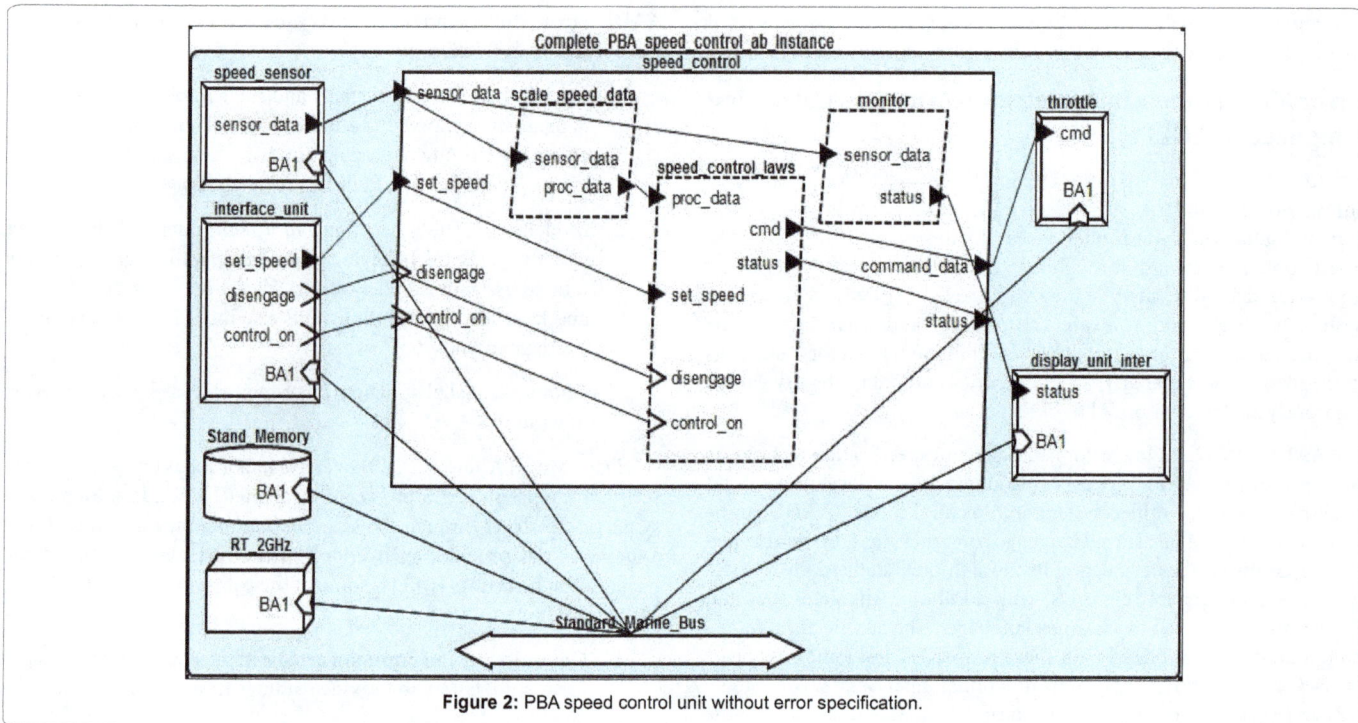

Figure 2: PBA speed control unit without error specification.

Specification of error source and propagation

The source of errors and their propagation with respect to each component of a system in PBA Speed Control Unit is defined, as shown in Figure 3. In the case study on_flow_src related to the device, pilot interface unit is sourcing the fault. The component error propagations are also defined with the error No-Value and No-Service.

The component error behaviour is also defined for the system components that correlate to the faults that are possible to occur. Here in this system the No-Value due to failure passes on from the pilot interface unit to the throttle actuator. The same is being conveyed to the display unit feature status. In addition to this fault, there occurs another propagation of error i.e. No-Service. This fault results in the failed state of the system. Here we can observe that the specification is automatically inherited by the instances of each component and their interactive neighbours. The error propagation paths inherent in such system architecture AADL models form a basis, as a need for the representation of Failure Mode and Effect Analysis (FMEA) and Common Cause Analysis (CCA).

Composite error behaviour

The error model annex library is associated to the state machine defined for the system component model using the declaration use behaviour, as shown in Figure 4. This maps the error state behaviour of the sub-components (both hardware as well as software components) onto the error states of the system itself. In this case study of Speed Control Unit of PBA, we have two error states defined for each component i.e. Failure and Failed. But here we have considered only the Failed state as the sub-component error state and the state Operational as the recovery state. We can see in this example that the system error behaviour is mapped from the sub-component behaviours defined as;

[throttle. Failed and display_ unit_ inter. Failed]-> Failed;

We assume that the system fails if either of the devices i.e. throttle actuator or the display unit behaves in the Failed state. While it tends

to recover from the Failed state and remains to be Operational even if the display unit fails, as the speed control unit mainly depends on the throttle command in maintaining and controlling the speed of the PBA.

[display_ unit_ inter. Failed]-> Operational;

This provides a scope for redundancy management for fault management capability of the system as well analyse for extensive solutions for reliability and availability analyses through various hierarchical levels of the system architecture. This methodology is not advisable for Markov Chains as the systems tends to grow quickly with their dependencies among various components within a system, as the number of components increases.

Component error behaviour

The modeller will have the flexibility of analysing the possible error behaviour that may correspond to individual components of a system. This also provides an insight into the component internal failures and the divergent factors that may result in failure mode, in turn having an impact on other components. The case study in this paper specifies that there might be multiple failure modes like Failure and Failed. In Failed mode the entire component is assumed to be redundant while in the Failure the component is working but having erroneous outputs/ output states, as shown in Figure 5.

The failure modes are represented using the error states with more likely coupled error behaviour of the sub-system/component. The consistency checker associated with the Error Model Annex abstracts the propagation specification to introduce unique and distinctive error types. While the modeling tool associated with the Error Model Annex validates for the organization of the component error behaviour along with the propagation specification specific to each of the component in the system architecture. The actual system architecture must include the safety system component/s that regulates the fault management and aids in safety analyses (Figure 6).

```
device interface
features
set_speed : out data port;
disengage : out event port;
control_on : out event port;
BA1 : requires bus access Marine.Standard;

flows
on_flow_src : flow source set_speed;

annex EMV2{**
use types ErrorModelLibrary;
use behavior ErrorModelLibrary::Simple;

error propagations
set_speed: out propagation{NoValue};
disengage: out propagation{NoService};
control_on: out propagation{NoService};

flows
fPath_Src: error source set_speed{NoValue};
end propagations;
```

Figure 3: Error source and propagation.

```
system implementation
Complete.PBA_speed_control_ab
subcomponents
speed_sensor : device sensor.speed;
throttle : device actuator.speed;
interface_unit : device interface.pilot;
speed_control : process control_ex.speed;
display_unit_inter : device display_unit;
RT_2GHz : processor Real_Time.two_GHz;
Standard_Marine_Bus : bus Marine.Standard;
Stand_Memory : memory RAM.Standard;

annex EMV2{**
use types ErrorModelLibrary;
use behavior ErrorModelLibrary::Simple;

composite error behavior
states
[throttle.Failed and
display_unit_inter.Failed]-> Failed;
[display_unit_inter.Failed]-> Operational;
end composite;
```

Figure 4: Composite error behaviour.

Safety Analyses

Safety [14] Analyses involves various analytical processes such as Consistency checks, Fault tree analysis (FTA), Failure modes and effect analysis (FMEA), Functional hazard assessment (FHA), and Common mode assessment (CMA) of the architectural model. The architecture model and its associated fault model is designed and developed in Open Source AADL Tool Environment (OSATE) [15]. It is an Eclipse based AADL modeling framework. There is also need the safety analysis tool such as OpenFTA [14] an Open-Source tool for FTA is integrated into Eclipse environment, to assist in generation of FTA and its relevant documents. While CMA, FMEA, FHA reports are generated as a built-in feature from OSATE.

Consistency checks

The consistency checks at the system integration level checks for the consistency in their functionality and the interfaces between various models/components, as shown in Table 1. This thereby strengthens the Virtual integration and analysis of the architecture model of the system. The consistency of various models deals with their integration feasibility while the consistency of the internal components in a model concentrates on the propagation capabilities, redundancies etc. With Error Model Annex the concept of consistency across the error models as specified checks for the consistency with respect to the component error behaviour along with the composite error behaviour of the system. It helps in defining the correctness of the error state as per the components specified in the architectural model. This may be proven with the substantial inclusion of Behaviour Annexes (B-Annex) [16] along with the Error Model Annex. The consistency report generated by the OSATE plugin for the case study is as shown below:

Consistency report

Warning! Complete_PBA_speed_control_ab_Instance:C13: component Complete_PBA_speed_control_ab_Instance does not define occurrence for and state Failed.

Complete_PBA_speed_control_ab_Instance:C13: component Complete_PBA_speed_control_ab_Instance has consistent probability values for state Operational.

Warning! Complete_PBA_speed_control_ab_Instance:C13: component Complete_PBA_speed_control_ab_Instance does not define occurrence for and state Failed.

Warning! Complete_PBA_speed_control_ab_Instance:C13: component Complete_PBA_speed_control_ab_Instance does not define occurrence for and state Failed.

Warning! Complete_PBA_speed_control_ab_Instance:C13: component Complete_PBA_speed_control_ab_Instance does not define occurrence for and state Failed.

Complete_PBA_speed_control_ab_Instance:C13: component Complete_PBA_speed_control_ab_Instance has has consistent probability values for state Failed.

Fault tree analysis (FTA)

A widely used safety and reliability analysis [17] feature in

```
device display_unit
features
status : in data port;
BA1 : requires bus access Marine.Standard;
flows
on_flow_snk : flow sink status;

annex EMV2{**
use types ErrorModelLibrary;
use behavior ErrorModelLibrary::Simple;

error propagations
status: in propagation{NoValue};
flows
fPath_Snk: error sink status{NoValue};
end propagations;

component error behavior
transitions
t0: Operational -[status{NoValue}]-> Failed;
end component;
```

Figure 5: Component error behaviour.

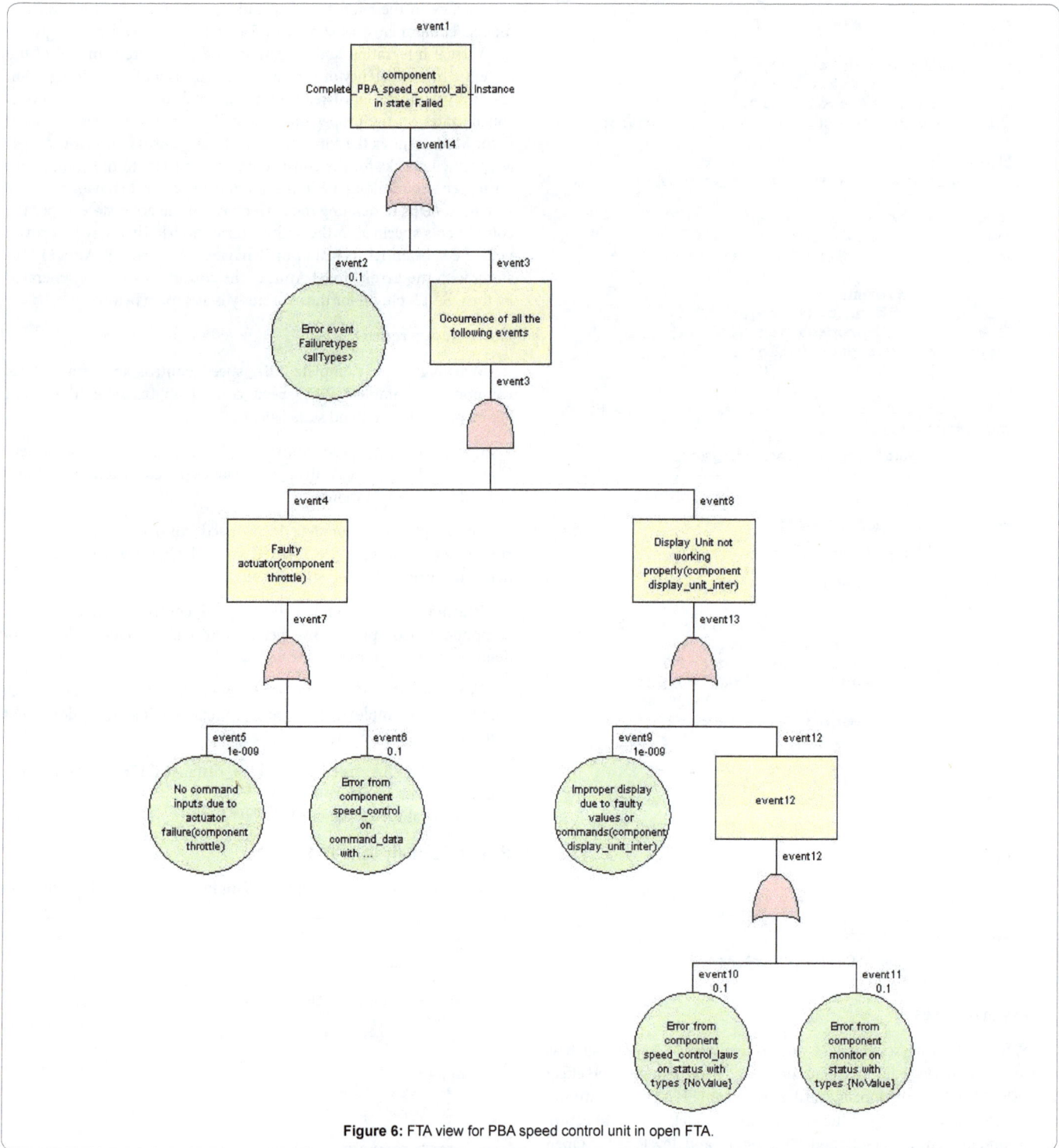

Figure 6: FTA view for PBA speed control unit in open FTA.

aerospace, medical electronics and industrial automation industries [18]. In this analysis the major focus is on the top level event (Minimal Cut-Set), from a set of combinations of basic events (Faults). It provides a hierarchical representation of the errors of the system (top-level event) from the basic events, related to components as specified in component error behaviour, in the form of a tree. OSATE depicts this composite error behaviour of the system from the underlying component error behaviours as a fault tree that represents specific

error state of the system. This is achieved in the form of two files from OSATE for the representation of the fault tree, one being the database of primary events (.ped), as shown in Figure 4, causing the top-level error event and the fault tree analysis file (.fta). These files are viewed using Open FTA, as shown in Figure 7.

The FTA analysis is in conformance with MIL-STD882 standard and the generated fault tree is validated, as shown in Figure 8.

Consistency Report
Warning! Complete_PBA_speed_control_ab_Instance:C13: component Complete_PBA_speed_control_ab_Instance does not define occurrence for and state Failed
Complete_PBA_speed_control_ab_Instance:C13: component Complete_PBA_speed_control_ab_Instance has consistent probability values for state Operational
Warning! Complete_PBA_speed_control_ab_Instance:C13: component Complete_PBA_speed_control_ab_Instance does not define occurrence for and state Failed
Warning! Complete_PBA_speed_control_ab_Instance:C13: component Complete_PBA_speed_control_ab_Instance does not define occurrence for and state Failed
Warning! Complete_PBA_speed_control_ab_Instance:C13: component Complete_PBA_speed_control_ab_Instance does not define occurrence for and state Failed
Complete_PBA_speed_control_ab_Instance:C13: component Complete_PBA_speed_control_ab_Instance has consistent probability values for state Failed

Table 1: Consistency report.

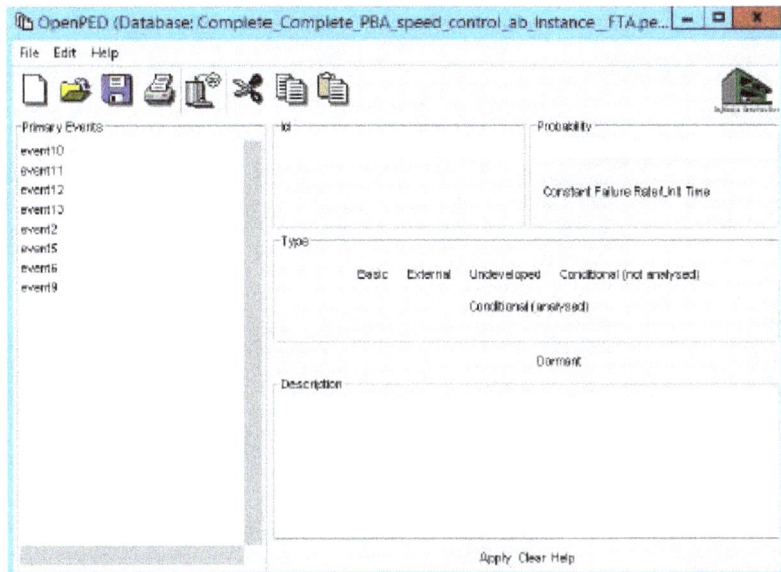

Figure 7: PED view in Open FTA.

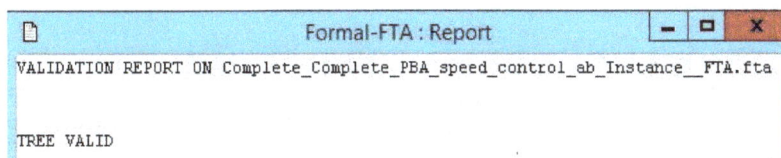

Figure 8: FTA validation report.

The artifacts related to FTA as specified by MIL-STD882 deals with error composites and error events. FTA is a top-down approach of analysis [19-21]. The Minimal Cut Set is evaluated in the OpenFTA tool and is as shown below in Figure 9.

Failure modes and effects analysis (FMEA) and functional hazard assessment (FHA)

Analysis of the failure modes associated with the system and the determination of its effects over the hierarchical evolution, performed systematically with a bottom-up approach is FMEA. With respect to the errors of the system, FMEA provides the information about the deficient component/models and their related effects. It also provides sufficient overview of the failing component such its phase of failure, severity/impact, etc. FMEA is based on the artifacts that include error propagation paths (error source, error path and the error sink). FHA provides with the possible list of error upon the synthesis of the architectural model of the system. The major artifacts from FHA comprise of the source of the error and the error events, as shown in Figure 10. The details of FHA are processed from the OSATE tool after

the model is instantiated and the relevant error information is suitably extracted from these architecture models. The report will be in the form an excel spreadsheet with the specification of the error event details.

Conclusion

In this paper, we have proposed a novel approach of safety analyses of Safety-Critical Systems using AADL and the related Error Model Annexes. In spite of the comprehensive activities involved in safety analyses, the needs for such approaches are proved to be very much necessary. This is achieved and projected with the implementation of a suitable case study, Speed-Control Unit of Power-Boat Autopilot. The employment of analysis techniques such as Fault-Tree Analysis (FTA), Functional Hazard Analysis (FHA) and Consistency of the model along with the conduction of qualitative and quantitative reliability analyses as part of these techniques can assess the system hazards and faults. The assessment covers the generation of suitable reports justifying the analyses. These methodologies or techniques provides grant for early identification and probability of the occurrence of potential problems. This also provides a perspective to explore additional architectural

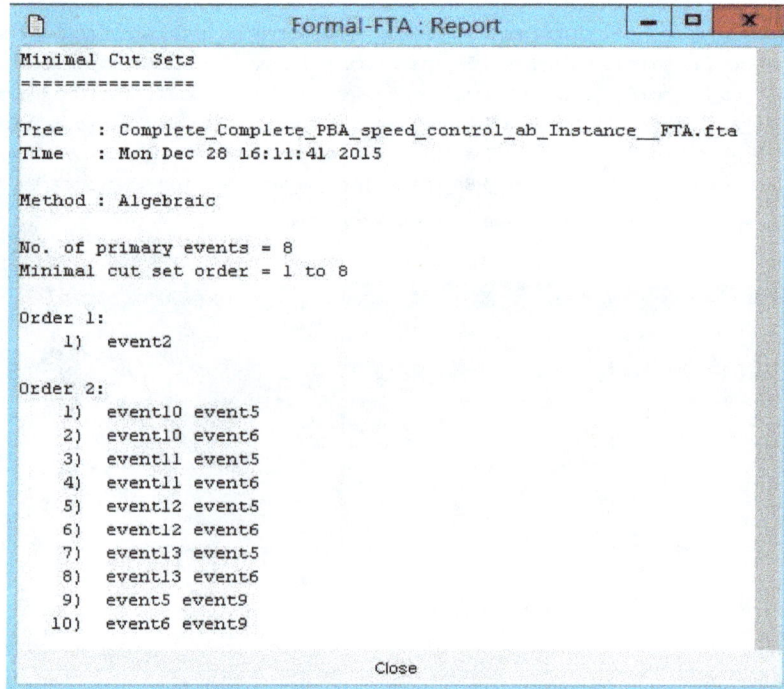

Figure 9: Minimal-cut set analysis report from Open FTA.

Figure 10: FHA report.

properties. Re-use and analysis of the evolved models, provided with suitable extensions with limited effort can be achieved with this approach. The overall effect induces a greater confidence over abstracted stages of development and safety analyses of these architectural models of the system. Also analysing the system based on the Safety-Critical Requirements, with the expectation of exceptional conditions, hazards expedites in the development of Safety System architecture models which will have an impact in certifying the same. This also avoids the unnecessary certification costs by understanding the change impact or the exceptional causes that impact during system engineering.

Acknowledgement

Our thanks to the Director, CSIR-NAL, Bengaluru, for supporting this work.

References

1. Feiler HP, Gluch DP (2012) Model-based engineering with AADL-An introduction to the SAE architecture analysis and design language. Addison-Wesley Pearson Education Inc.

2. Department of Defence (2015) MIL-STD882 (E) Standard practice - System safety.

3. SAE International (1996) Guidelines and methods for conducting the safety assessment process on civil airborne systems and equipment.

4. RTCA-I (2011) Software considerations in airborne systems and equipment certification.

5. Joshi A, Vestal S, Binns P (2007) Automatic generation of static fault trees from AADL Models, In: Proceedings of IEEE/IFIP Conference on Dependable Systems and Networks'-Workshop on Dependable Systems, Edinburgh, Scotland-UK.

6. Julien D, Peter F (2014) Architecture fault modeling with AADL error-model annex. In Proceedings of 40th Euromicro Conference on Software Engineering and Advanced Applications.

7. Hall B, Driscoll KR, Madl G (2013) Investigating system dependability modeling using AADL, NASA/CR-2013-217961, Honeywell International, Inc., Golden Valley, Minnesota.

8. Li Q, Gao Z, Luo X (2016) Error modeling and reliability analysis of airborne distributed software based on AADL. Advance Science Letters - American Scientific Publishers.

9. Delange J (2013) Safety evaluation with AADLv2. Software Engineering Institute Carnegie Mellon University.

10. Powell D (1992) Failure mode assumptions and assumption coverage. In Proceedings of Twenty-Second International Symposium on Fault-Tolerant Computing.

11. Walter CJ, Suri N (2003) The customizable fault/error model for dependable distributed systems. Theor. Comput. Sci 2: 1223-1251.

12. Jordi C, Martin G (2010) Object Constraint Language (OCL): A Definitive Guide.

13. Benammar M, Belala F (2010) How to make AADL Specification More precise. Int J Computer Applications.

14. Open-FTA (2015) Open-FTA - An advanced tool for fault tree analysis.

15. OSATE (2014) OSATE - An open-source tool platform to support AADL v2.

16. SAE International (2015) Annex behaviour language compliance and application program interface.

17. Li C, Yang H, Liu H (2016) An approach to modelling and analysing reliability of Breeze/ADL-based Software Architecture. Int J Automation and Computing. 1-10.

18. Xiang J, Yanoo K, Maeno Y, Tadano K (2011) Automatic synthesis of static fault trees from system models. In: 5th International Conference on Secure Software Integration and Reliability Improvement.

19. Liu Y, Shen G, Wang F, Si J, Wang Z (2016) Research on AADL model for qualitative safety analysis of embedded systems. Int J Multimedia and Ubiquitous Engineering 11: 153-170.

20. Grunske L, Han J (2012) A comparative study into architecture-based safety evaluation methodologies using AADL's error annex and failure propagation models. In: Proceedings of 11th IEEE High Assurance Systems Engineering Symposium 283-290.

21. Feiler PH, Gluch DP, McGregor JD (2015) An architecture-led safety analysis (ALSA) method. SEI Digital Library. Ada User Journal 36: 192-196.

29

Design and Analysis Aircraft Nose and Nose Landing Gear

Rajesh A and Abhay BT*

Department of Mechanical Engineering, New Horizon College of Engineering , Bangalore, India

Abstract

Tri-cycle arrangement landing gear is extensively used as it is simple; convenient both structurally as well as aerodynamically. Though it is advantageous over other configuration is has its own draw backs. Factors like its weight drag, sudden application of load, acoustics, fatigue etc tend to slow down its performance and life. Among main landing gear and nose landing gear; the former carries about 85% of total weight of aircraft and latter carries around 12-15% of weight. The nose landing gear is also a source of noise and its effect is prominent when compared to main landing gear. In this project the executive jet aircraft are studied thoroughly and a nose landing gear similar to those of executive jets is modeled using CATIA. The same geometry is imported to ANSYS ICEM and flow on the body is analyzed for different angle of attack. Pressure variation, temperature, density and velocity distribution around the body is noted and then Coefficient for Lift and Drag are plotted against angle of attack for obtained results. It is also important to check the strength and stiffness of designed landing gear. Hence using ANSYS APDL and Explicit; Static structural and Impact test has been carried out for designed geometry. Stress distribution and deformation was noted for two distinct materials such as steel and aluminum alloy and primary results of acoustics has been compared with the available data.

Keywords: Angle of attack; Deformation; Flow over body; Coefficient of lift; Coefficient of drag Impact landing; Nose landing gear; Stress distribution; Acoustics

Introduction

Landing gear [1] is one of the important parts of an aircraft; often referred [2] to as undercarriage. Landing gear is a structure which is installed on the aircraft for the purpose to support the weight of the aircraft while it is on the ground and also allows the aircraft for smooth maneuver such as takeoff and landing. Landing gear also provides mobility to the aircraft on ground or water. It is capable of reaching the largest local load on the plane. The landing gear's main function is to control the rate of compression/extension and to prevent further damage to itself as well as other parts of an aircraft when a load is applied. Loads can be either static or dynamic in both cases the structure must withstand applied loads and deformation and continue to do its purpose. Among the available configuration the most common type of landing gear arrangement is tri-cycle arrangement. Considering the nose landing gear configuration of a tri cycle the stress distribution and deformation [3] is noted for an executive jet aircraft for applied loads.

Nomenclature of Landing Gear

Landing gears are located on the under carriage of aircraft . The front landing is called as nose landing gear while main landing gear [4] is located on rear side of aircraft. Nose landing gear consists of actuators to retract and extension; drag brace to lock the pins metering pin extension; trunnion; rotation lock pins; aft braces; oleo cylinder; oleo piston; brake assembly; tires and wheels which is shown in Figure 1.

Design and Operational Requirement

Operational Requirement showed in Table 1

$$H_{CG} = \Delta H_{clear} + \frac{D_{prop}}{2} \tag{1}$$

$$C = s/b \tag{2}$$

$$= 1.2 + 1.9$$

$$= 3.1$$

$$b = \sqrt{s \times AR} \tag{3}$$

$$= \sqrt{60 \times 12}$$

$$= 26.883 \text{m}$$

$$\bar{C} = \frac{S}{b} \tag{4}$$

$$= \frac{60}{26.883}$$

$$= 2.236$$

$$V_R = 1.1 V_s \tag{5}$$

$$= 48.1 \text{ m/s}$$

$$D_{To} = \frac{1}{2} \rho v^2 S C_{DTo} \tag{6}$$

$$= 6267. 4 \text{ N}$$

Based on the design requirement and as per the need of the project the nose landing gear dimensions are altered and are designed in

Maximum Takeoff weight [5]	18000 Kg
Static load acting	17650 N
Impact load acting	67032 N

Table 1: Operational Requirement.

***Corresponding author:** Abhay BT, Department of Mechanical Engineering, New Horizon College of Engineering , Bangalore, India
E-mail: abhayt123@gmail.com

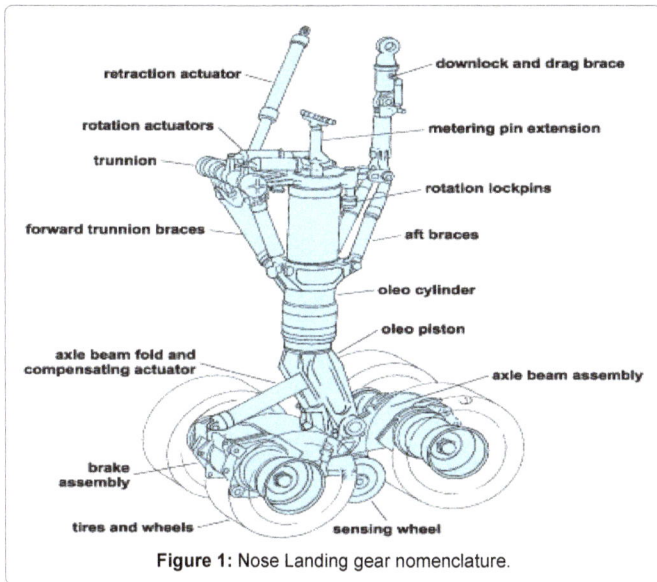

Figure 1: Nose Landing gear nomenclature.

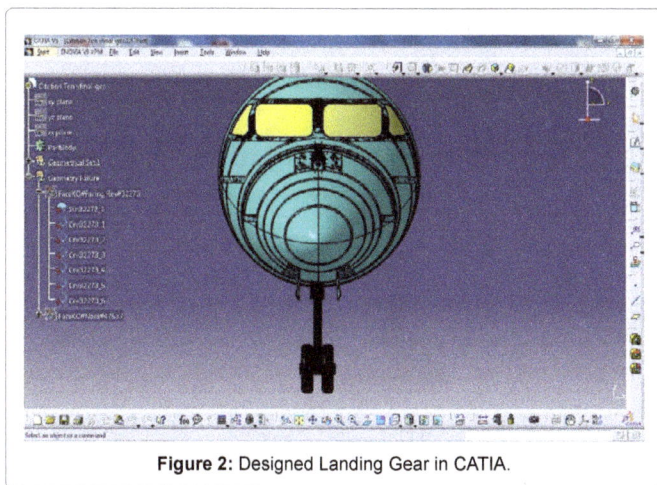

Figure 2: Designed Landing Gear in CATIA.

CATIA V5 R19. Since only nose landing gear is concern of this project nose landing gear is designed along with nose cone as shown in Figure 2.

In order to analyze the fluid flow, the geometry is spilt into number of elements. If the number of elements is high the accuracy of results is high [5]. TRIA surface mesh is used to mesh the geometry. Auto mesh is used to mesh the curved surface and edges.

The geometry is subjected to flow analysis. It is done by assigning flow properties to the geometry within the domain. The body was tested for five different conditions by varying its angle of attack from -5° to15° with the equal increments of 5° and by maintaining flow speed constant at 0.2 mach.

The domain properties are as such that the outlet from the geometry is 5 times when compared to inlet. The flow is assumed to be continuous which is also non buoyant. The static temperature assigned is 300 K. It is also found that there are 468848 nodes and the number of elements in the domain is 1599260. The meshed geometry shown in Figure 3.

The landing gear is checked for its stability and stiffness by assigning it two material properties Aluminum alloy and Structural

steel [6]. The material properties of steel and aluminum alloy are given in below Table 2.

Results and Discussion

A. Flow analysis-pressure contour

From Figure 4a; a flow analysis has been carried out for the landing gear assembly and nose cone. The geometry is placed in the domain and fluid is allowed to flow over aircraft portion with a Mach number of 0.2 with an angle of attack of -5°. Variation of pressure is obtained for a range of -10742.89 Pa to 3364.05 Pa. Pressure distributions at 0° angle of attack is explained in Figure 4b. The maximum pressure variation is 3333.33 Pa. There is an increase in pressure distribution with the increase in angle of attack. Similar analysis is carried for 5°, 10° and 15° angle of attack as given in Figures 4c-4e respectively. In all the cases it

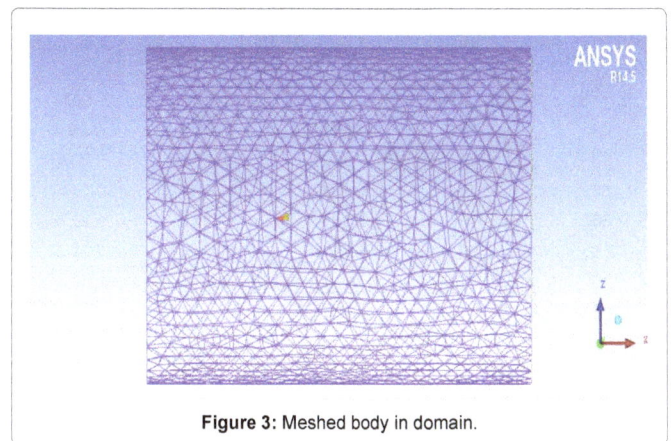

Figure 3: Meshed body in domain.

Properties	Aluminum 2024	Steel
Density (g/cm³)	2.27	7.85
Coefficient of Thermal expansion (10⁻⁶/°C)	22.8	0.12

Table 2: Material properties.

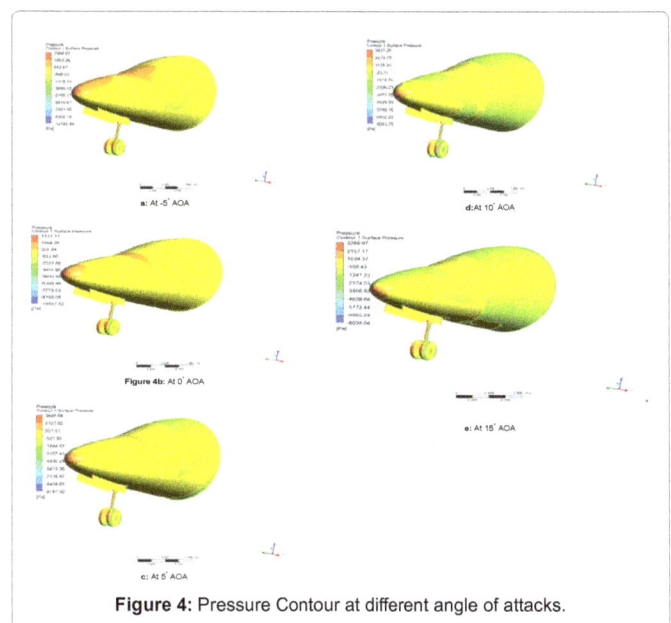

Figure 4: Pressure Contour at different angle of attacks.

is noticed that the maximum pressure acts on front portion of landing gear and nose cone.

From the pressure plots using dynamic equations, coefficient of lift and drag can be calculated and are tabulated in Table 3.

Using the values from Table 3 coefficient of lift and drag values are plotted against angle of attack.

Graph for C_L v/s Angle of attack is shown in Figure 5. It is known that with the increase in angle of attack lift coefficient increases and after reaching a certain angle it tends to decreases which is known as stall angle. From the Figure 5 at 15° angle of attack there is higher lift coefficient which would fall down if the angle of attack is increased further.

Figure 6 is a graph for C_D v/s Angle of attack. At an angle of attack -5° greater amount of drag is experienced by landing gear. With the increase in angle of attack from 0 to 15° drag reduces gradually to minimum value because thrust is more than drag.

V (m/s)	AOA (deg)	C_D	C_L	L/D
68	-5	1.0181	-0.083	-0.0815
68	0	0.9801	0.0094	0.0096
68	5	0.8989	0.0998	0.1111
68	10	0.8215	0.1814	0.2208
68	15	0.7969	0.2242	0.2813

Table 3: Lift and Drag Coeffiecient.

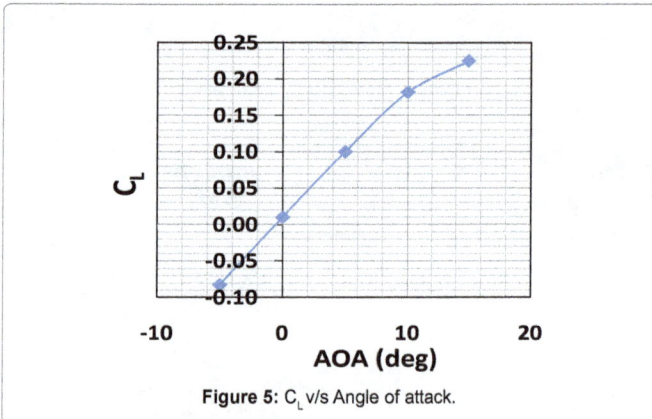

Figure 5: C_L v/s Angle of attack.

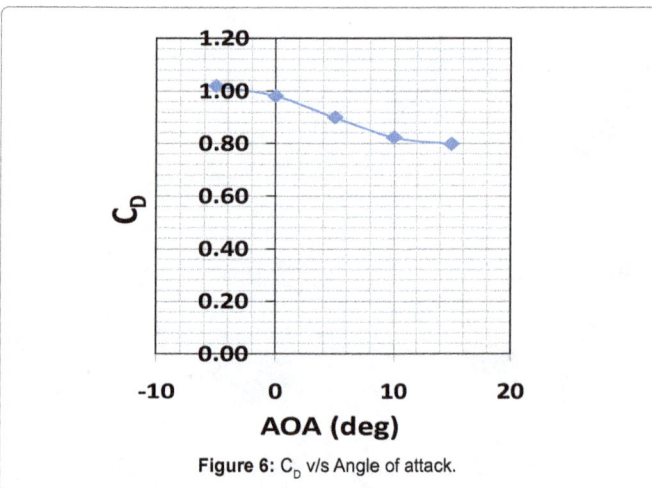

Figure 6: C_D v/s Angle of attack.

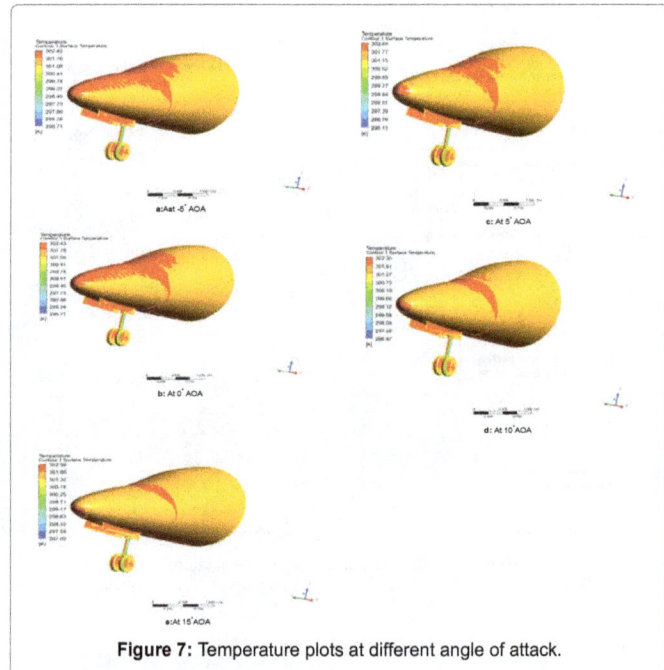

Figure 7: Temperature plots at different angle of attack.

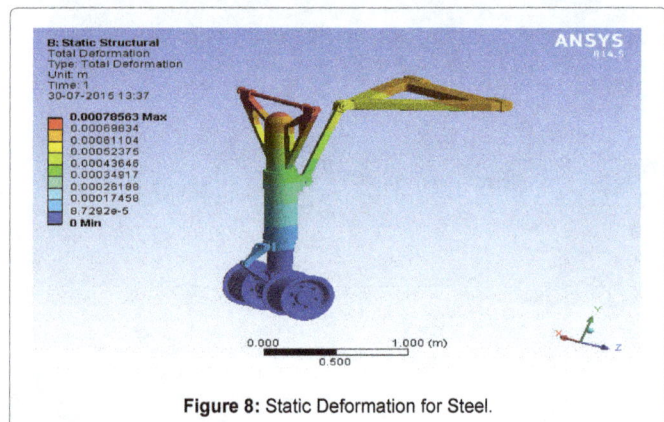

Figure 8: Static Deformation for Steel.

B. Flow analysis –temperature contour

Temperature plots for geometry subjected to flow analysis at mach 0.2 are given in Figure 7. Figure 7a-7e shows geometry at -5°, 0°, 5°, 10° and 15°. In all the cases there is a slight variation on temperature for changing angles and maximum temperature is found at the tip of nose cone and front portion of landing gear.

C. Mechanical analysis

The landing gear assembly has undergone a deformation, when a constant load is applied on the assembly which is at static condition. From Figure 8 maximum deformation of 0.78 mm which is indicated in red color region, and a minimum deformation of 0.000087 mm as shown in blue color region for applied loads. And landing gear assembly is suggested to be safe for a given material and load.

The landing gear assembly has undergone a deformation, when the aircrafts lands on hard runway. An impact load is induced on landing gear which is then transferred to axle and then to strut. From Figure 9 maximum deformations of 4.1 mm which is indicated in red color

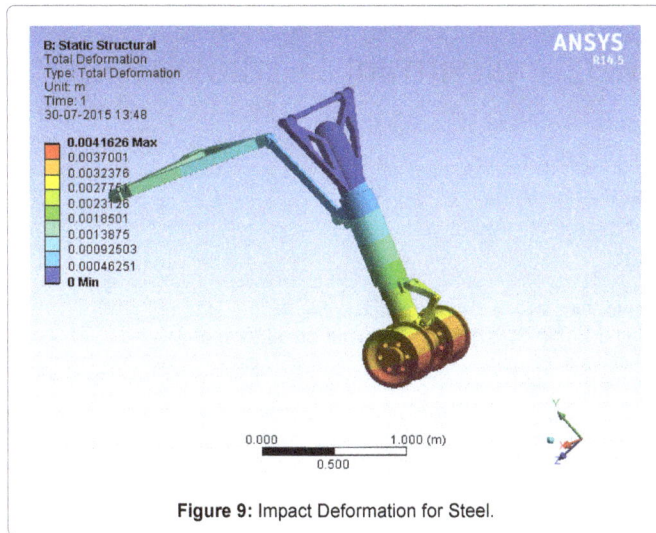

Figure 9: Impact Deformation for Steel.

Material	Max Von-Mesis Stress (M Pa)	Max Deformation (Mm)
Steel	15.9	0.78
Aluminum Alloy	12.9	2.1

Table 4: Static Analysis results.

Material	Max Von-Mesis stress (MPa)	Max Deformation (mm)
Steel	82.6	4.1
Aluminum Alloy	65.7	10.96

Table 5: Impact Analysis results.

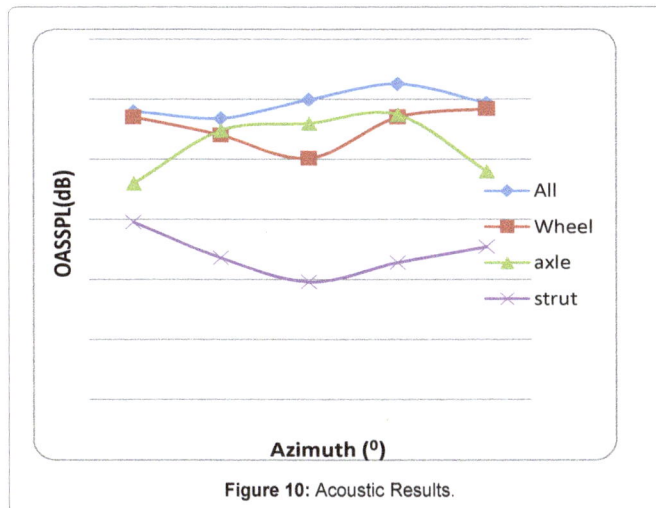

Figure 10: Acoustic Results.

region, and a minimum deformation of 0.46 mm as shown in blue color region for applied loads. And landing gear assembly is suggested to be safe for a given material and load.

Table 4 explains the results of static analysis. The maximum deformation occurs in case of aluminum alloy and a maximum stress distribution is for steel.

During impact test at a given loading condition a maximum deformation occurs for aluminum alloy and stress distribution is maximum in case of steel which is tabulated in Table 5.

D. Aero acoustic results

An initial approach has been carried out to study the noise generated [7] by landing gear by dwelling its surface. The obtained results were of noise in terms of frequency and azimuth angle. It was compared with the available data [8] which closely matched with the result which is given in Figure 10.

Conclusion

CFD analysis has been carried out to study flow around nose landing gear and nose cone, where the flow has been considered into 3 regions on nose landing gear and nose cone namely upper region flow, mid region flow and lower region flow. The upper region flow has its effect on strut while mid region flow effected axle and slightly on wheel. The effect of lower region flow was experienced on wheel.

Sudden application of load deforms the structure. The static deformation and stress distribution are tabulated in Table 4 and impact deformation and stress distribution are tabulated in Table 5.

Stress and deformation value which is obtained for applied loads on structure are found to be within design range and hence the nose landing gear is suggested to be safe.

A possible solution to reduce landing gear noise is streamlining the flow field through the use of fairings. However, streamlining of the entire landing gear structure is difficult due to its size and function (wheels must rotate, brakes require cooling) and a more complicated weight trade-off analysis becomes necessary to determine potential benefits.

References

1. Niu MCY (1989) Airframe structural design. Practical design information and data on Aircraft structures Lockheed Aeronautical systems company, Burbank, California, Conmilit Press ltd Chapter1.

2. Antonio A. Aircraft Classifications. Trani Associate Professor Department of Aeronautical Engineering Virginia Tech.

3. Ganorkar K, Deshbhratar V (2014) Design optimization of Landing Gear of an Aircraft – a review IOSR Journal of Mechanical and Civil Engineering (IOSR-JMCE) 01-04.

4. Megson THG. "Aircraft Structures for engineering students. (3rdedn).

5. Nguyen TD (2010) Finite Element Analysis of a Nose Gear During Landing. UNF Theses and Dissertations Paper 215.

6. Raymer DP. Aircraft Design. A conceptual approach, Conceptual research corporation Sylmar, California. Chapter 11.

7. Gaikwad AV, Sambhe RU, Ghawade PS (2013) Modeling and Analysis of aircraft landing gear : experimental approach, International Journal of Science and Research (IJSR) 2: 366-369.

8. Liu W. Numerical Investigation of Landing Gear Noise, Airbus Noise Technology Center, University of Southampton.

30

Stress Analysis of an Aircraft Fuselage with and without Portholes using CAD/CAE Process

Fayssal Hadjez* and Brahim Necib

University of Constantine, Faculty of Science and Technology, Department of Mechanical Engineering, Algeria

Abstract

The airline industry has been marked by numerous incidents. One of the first, who accompanied the start of operation of the first airliners with jet engines, was directly related to the portholes. Indeed, the banal form of the windows was the source of stress concentrations, which combined with the appearance of micro cracks, caused the explosion in flight of the unit. Since that time, all aircraft openings receive special attention in order to control and reduce their impact on the aircraft structure. In this paper we focus on the representation and quantification of stress concentrations at the windows of a regional jet flying at 40000 feet. To do this, we use a numerical method, similar to what is done at major aircraft manufacturers. The Patran/Nastran software will be used the finite element software to complete our goals.

Keywords: Aircraft; Bay; Modeling; Portholes; Fuselage; Pressurization; Loads

Introduction

There are some methods to realize the appearance of additional expenses at the window. The calculation by finite element is one of the main methods used [1], particularly by aircraft manufacturers in the interests of economy, speed and reliability [2]. This case will have to obtain internal loads and levels constraints from well-defined external loads [3-6]. To do this, Figure 1 shows the process of generating internal loads. Thus, the action plan is shown in the following figure.

To do this, we will use the calculation by finite element software Patran/Nastran, which is considered standard in the aerospace industry and is used in all major aerospace companies [7]. The launch into the heart of this paper will be done after an upgrade on the operation of this software. Geometric modeling is the most important step in the process of degeneration of internal forces, to the extent that it serves as support for the creation of the finite element mesh [8]. Thus, our body is composed of frames, stiffeners, coating and finally the floor. The stiffeners support larger primary times, resume compression-tension efforts, and are guarantor of the overall stiffness. Frames give the external shape of the fuselage and support, in large part, the circumferential stress upon pressurization. The coating shows a part of shear efforts then distributes to the stiffeners, which in turn transfer them to the frames. The floor, meanwhile, adds to the rigidity on the transverse and longitudinal planes [9-12]. Table 1 below summarizes all the geometric data used for modeling.

Description of the Problem

The issue is part of a structural analysis context. Indeed, aircraft are subjected to various loads and load cases during their flight cycles. Then there exists in each flight phase, one, or even several cases of loading. It is then necessary to situate our study, Phase flight on which you will base our analysis. In our case, we focus on the "cruise" phase of a commercial flight. During this phase of flight several loading cases may exist depending on the portion to be analyzed on the aircraft. In our case, we will analyze the fuselage. A major case of loading in the fuselage pressurization. This usually occurs in the cabin given its importance to the survival of the passengers, at high altitude. Indeed, taking the altitude the air is thinner and lower atmospheric pressure

gradually. It is then necessary to pressurize the cabin to the survival and passenger comfort. However this pressurization is not beneficial to the structure as it involves additional structural loads. The first airliner to experience the consequences of these additional expenses is the Dehavilland Comet 1. This aircraft was the first commercial aircraft to fly with jet engines, allowing pressurization of the unit. However, the first version of this unit possessed rectangular windows, which were the cause of many accidents. Indeed, the banal form of the windows was the stress concentration source combined with the appearance of micro cracks caused the explosion in flight of the unit.

The windows, like other openings in an aircraft when receive special attention. Among the openings, gates, baggage doors, cut-out for kerosene tanks, or openings for antennas, etc., are all sensitive parts on aircraft.

In this project we focus on the representation and quantification of stress concentrations at the windows of the aircraft flying at 40,000 feet

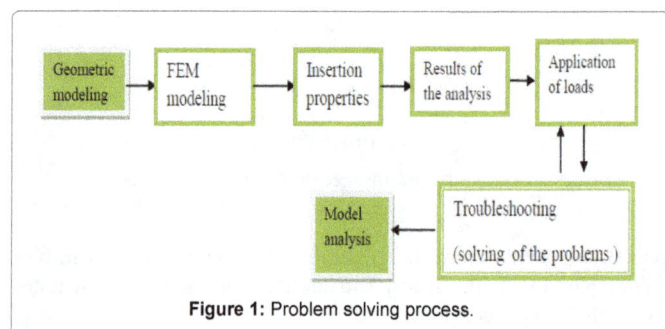

Figure 1: Problem solving process.

---

*Corresponding author: Fayssal Hadjez, Professor, University of Constantine, Faculty of Science and Technology, Department of Mechanical Engineering, Alegeria, E-mail: hadjez103@gmail.com

above sea level. It should be noted that during the initial presentation of the project, the subject referred to "openings" in their globalities. Given the impossibility of treating all types of possible openings for this project, I chose to restrict and deepen my study of the case of the portholes. We will analyze first place pressurization and now and then twisting the fuselage.

The objective is to achieve the most important two stages namely obtaining a finite element model of integrity, and the calculation of internal forces as shown in the examples in the Figures 2 and 3.

Methodology Employee

As stated previously, the modeling by finite element based on the geometry, to be defined. Due to the CQUAD4 surfaces elements are obtained and through the curves, CROD elements. The frames are made of soles, modeled by CROD elements, and their soul is modeled with CQUAD4 elements, their normal pointing backwards to ensure consistent results. The stiffeners are modeled by CROD elements, the coating of CQUAD4 elements. Normal of these elements should always point outwards from the cabin. The floor beams are made of soles and souls. The soles are then modeled by CROD elements and the core elements by CQUAD4 with their normal pointing backward or outside of the aircraft. As regards the properties of the elements, the coating is made of aluminum alloy Al 2024-T3 having a thickness of 0.06 ", the frames are Al 7475-T7351 (with a thickness of 0.06 " for CQUAD and an area of 0.2508 in^2 for ROD) along the floor (with a thickness of 0.1 " and for CQUAD 0.51in^2 area for ROD). Finally the stiffeners are made by 2024-T62 (0.1272 in^2 for the area) to those of the upper crown of

the fuselage and Al 7475-T7351 for those of the lower crown (0.1676 in^2 for the area). Regarding the portholes thereof are in fact methyl polymethanecrylate (PMMA or Perspex with 0.02" thick).

The fuselage will be subject to pressurization loads firstly, then torque secondly. We are part of a cruise context to 40,000 feet, and we assume the cabin altitude of 6,000 feet. Thus, a closed cylinder pressurization in induces the appearance of a circumferential stress (hoop stress σH) and longitudinal (σL) as shown in Figure 4.

To represent the loading on the finite element model, we have to manually apply the longitudinal stress, through MPC, because our body is not closed. This σL stress is 3624 psi and strength associated 65578,359 lbs. An internal pressurization 9.06 psi is then applied to all the plates of the covering. The results speak for themselves. Regarding the stiffeners, we note that height openings in a baie without openings, they do not include any load, their presence in this area is therefore virtually useless. By against the stiffeners located immediately above and below the openings experimenting stress concentration factors of up to 1.56. For managers we can say that the plates are at the openings, the top of the fuselage and about at the floor, undergo an increase in internal stress with a stress concentration factor of up to 1.29. In addition, it appears that top plates immediately resumes lower fewer loads (Figure 5).

Geometric modeling

Firstly revisit the structure of a fuselage. Airliner fuselages are very similar. The type of the most commonly used structure is the semi-monocoque, that is to say, an enhanced cell. The semi-monocoque structure is effective for its weight compared to its rigidity. It allows among other things, to tolerate a crack avoiding the whole structure is affected by the redistribution of loads in other members. Thus, partitions (bulkheads), frames, stiffeners (stringers) and spars are regularly used to give support to the structure. Figure 6 highlights the set of these components.

The rails placed lengthwise, bear the brunt of the primary bending loads (bending moment). Stiffeners show the compression tension forces along the rails, and are guarantees the overall stiffness. Management sets the external form to the fuselage and largely repeats the circumferential stress during pressurization. The coating bears part of shear efforts and distributes them to the stringers which in turn transfer them to the frames. The floor in turn adds to the rigidity of the transverse and longitudinal planes.

Finally the transverse walls are of "plug" at the front and rear of the pressurized section and are located at places subject to larger stresses (location engine, wing roots of the wings, landing gear, etc.) [5]. for this study we model almost all the components of a conventional fuselage except for longitudinal and transverse bulkheads (because we have no large loads on the fuselage). The choice not to model the beams from the fact that all the geometric information needed to model the structure come from aircraft models, and that they do not use rails for their regional aircraft. In fact every time efforts are then taken up by the stringers.

We define the word 'bay' all items between two frames. The purpose of this study is to find the distribution of stresses at the windows, so we better model several bay, to obtain a faithful model that either disturbed by the boundary conditions. A choice of seven (7) bays seems so sensible. Thus Table 1 summarizes all the geometric data used for modeling. On Patran we obtain a suitable cross-section of the fuselage, as well as our seven bays desired (Figure 7).

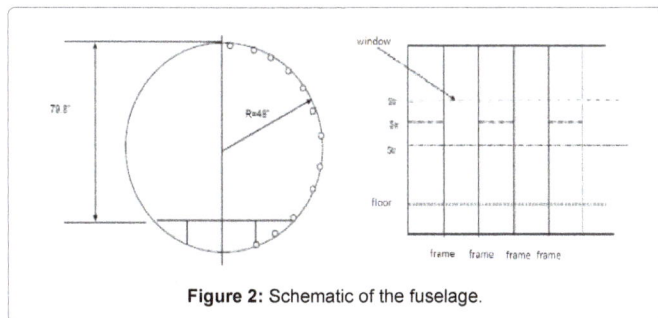

Figure 2: Schematic of the fuselage.

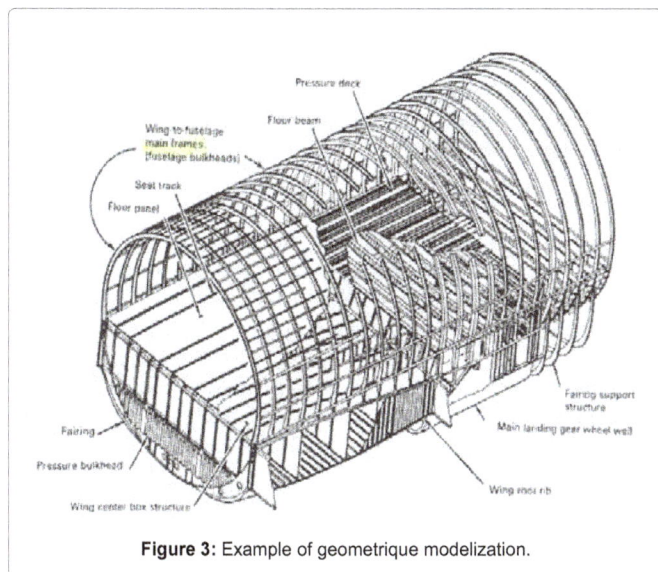

Figure 3: Example of geometrique modelization.

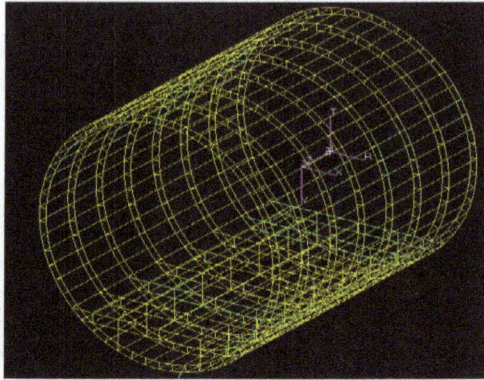

Figure 4: The geometrique model of the fuselage.

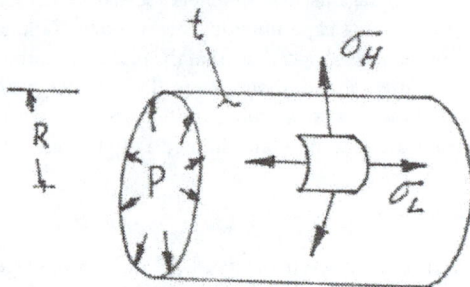

Figure 5: Resulting stress due to internal pressure.

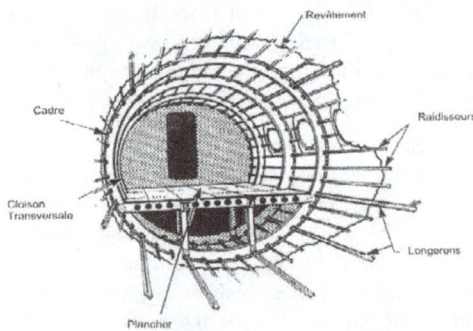

Figure 6: Components of the fuselage.

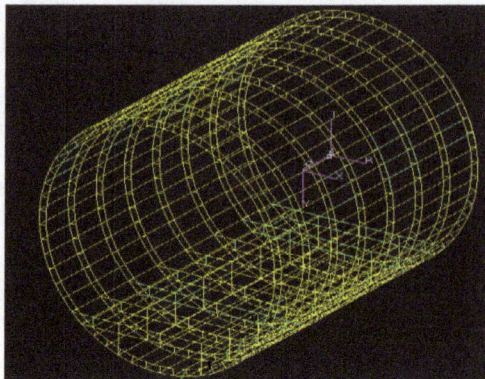

Figure 7: Geometric model of the fuselage.

Objet	Value
Diameter	96 in
Distance between the floor and the fuselage	79,8 in
Frame location	16 in
Overall length	112 in
Number of window frames	1
Number of stringers	54

Table 1: Geometric Data

Finite element modeling

Now the geometric model is completed, it is time to create the finite element model; this model is based on the geometric structure to generate the information necessary for the resolution of the problem. To do this, we have several types of elements used in the construction of a global finite element model of an airplane. Table 2 describes these types of items.

The frames are made of soles, modeled by CROD elements, and their soul is modeled with CQUAD4 elements, their normal pointing backwards to ensure consistent results. The stiffeners are modeled by CROD elements, the coating of CQUAD4 elements. Normal of these elements should always point outwards from the cabin. The floor beams are made of soles and souls. The soles are then modeled by CROD elements and soul by CQUAD4 elements with their normal pointing backward or out of the plane. The next step would be the mesh. However, before creating the mesh, we must control the way of subdividing the geometric elements like surfaces and curves. For this, we use the 'Mesh Seed'. Since our geometric model is detailed enough, we just need to do a subdivision for all components.

We can now mesh the model: mesh curves CROD elements and elements CQUAD surfaces. We note that for the representation of portholes we remove CROD elements stiffeners#10, 11, and 12 on either side of the fuselage. We obtain the finite element model (Figure 8).

Setting properties

The definition of property covers of all the elements that we just created. Indeed, it is essential to combine these elements of geometric and material properties. The coating is made of aluminum alloy Al 2024-T3 with a thickness of 0.06 '', the frames are Al 7475-T7351 (with a thickness of 0.06 '' for CQUAD4 and an area of 0.2508 in2 for CROD) as well as the floor (with a thickness of 0.1 '' and for CQUAD4 0.51 in 2 area for CROD). Finally the stiffeners are in 2024-T62 (0.1272 in 2 for the area) to those of the upper crown of the fuselage and Al 7475 T7351 for those of the lower crown (0.1676 in 2 for the area). Regarding the portholes thereof are in fact methyl polymethanecrylate (PMMA or Perspex with 0.02 ''thick).

Definition of loadings

As previously announced the fuselage will be subject to pressurization loads first, then when and torsion second. Indeed, we are part of a cruise flight at 40,000 feet context, and we assume the cabin altitude of 6,000 feet. And atmospheric pressure to 40,000 foot being of 2.72 psi and the 6000 foot being of 11.78 psi, then we obtain an internal pressurization 9.06psi cabin. This case of loading in a closed cylinder induces the appearance of a circumferential stress (hoop stress σH) and longitudinal (σL) as shown in Figure 9.

The longitudinal stress is calculated according to the formula [12]

$$\sigma L = \frac{\rho * r}{2t}$$

[12]

	Spring Elements	Line Elements	Surface Elements	Solid Elements	Multiple Constraint Elements (MPC)	
Physical Behavior	Simple Spring	Rod (axial load only), Beam	Shear, Membrane, Plate	Brick, Tetrahedron	Rigid Body Element	Interpolation Constraint Element
MSC/NASTRAN Element Name	CELAS1	CROD CBAR	CQUAD4 CTRIA3	CHEXA CTETRA	RBE1 RBE2	RBE3
Associated property entry	PELAS	PROD PBAR	PSHELL	PSOLID	None	None
Example of utilization	Connection between control surfaces and aircraft	Simple representation of fuselage stringer	Representation of fuselage and wing skin	Honeycomb of composite	Representation of engine	Load distribution

Table 2: Types of elements used to FEM Global.

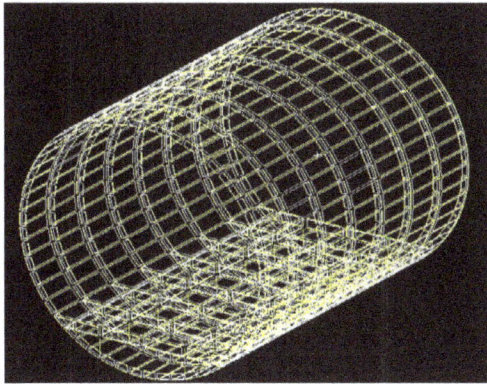

Figure 8: Finite element model of the fuselage.

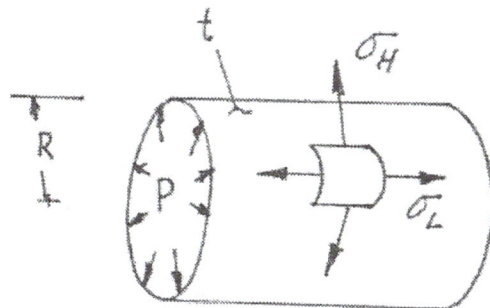

Figure 9: Resulting stress due to internal pressure.

And the longitudinal force exerted on the surface is expressed. With t being the thickness of the coating and r is the radius of the cylinder, it can be deduced that; σL=3624 psi and FL=65578,359 lb.

Things are slightly different for the circumferential stress. Indeed if the body was not made up of reinforcements (frames and stiffeners) equation of this constraint would be

$$\sigma L = \frac{\rho * r}{2t} \qquad [12]$$

So we get σ = 6565.83 psi instead of 7248 psi without reinforcement. To represent as closely loading into Patran, then we must apply the longitudinal force on the nodes of both ends of the fuselage, and the internal cabin pressure is 9.06 psi. For not applying the longitudinal stress at each node, we create a RBE3 element connecting each end of

the fuselage, taking as the central node depending on each end of the fuselage, and the independent nodes are all other nodes to devices both ends. Finally, we can apply the force dependent node for transmission to turn to independent nodes. We choose to allow all the translations and rotations of the dependent node, while the rotations are blocked for independent nodes because our focus righteous in their translations (all in the global coordinate system). As regards the internal pressure, it is applied to each of the coating panel (Figure 10).

Furthermore, for the purposes of the moment and the torsion, we define a moment of 5,000,000 in lbs, corresponding to a lateral gust for example, and a torsion 1,000,000 in. lbs. which corresponds for example to forces from the rudder when it is pressed.

We also need to apply boundary conditions to 'fix' the model. So we choose a node located at the junction between the floor and an extreme environment, where we block all translations (relative to the global benchmark). Then on vis-à-vis node that shares the same framework, we block the translations in Y and Z (relative to the global benchmark). Finally, we are left with only one degree of freedom: the model can still move in rotation on the Y axis (relative to the global benchmark).

To solve this, we then choose to block the translation of Y node in

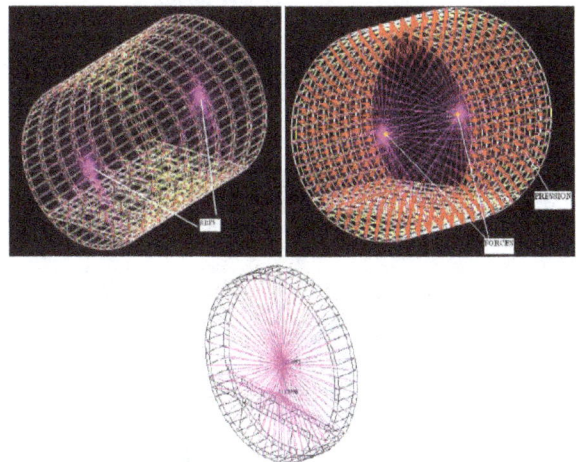

Figure 10: Creation of MPC (Multiple Constraint Elements) and expenses application.

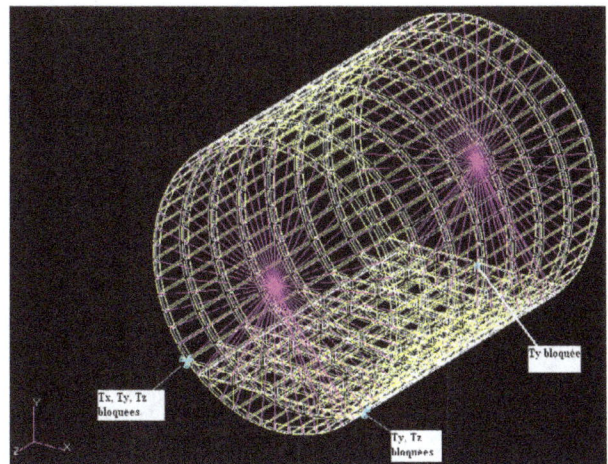

Figure 11: Application of boundary conditions.

the middle of the floor on the two end frames (Figure 11). The model is thus 'fixed'.

We note that a finite element model 'without porthole is also created in order to make comparisons, and better assess the effect of openings on the fuselage. This model is created from the modification of the one with windows, changing the properties of the surfaces of the windows that become identical to those of the coating. Then adding stiffeners which had been removed at the windows, taking care of their applied the same properties as the upper stiffeners.

Results

General comments in such an analysis (static), the Nastran solver generates several files including a file whose extension F06. This is the file that will allow us to get detail results of the internal efforts. But before that, we can get an overall view of the effect of external load on the structure. For this, we need only read a Patran result files and then display the results in the form of color variation (color plot). By posting Von Mises stress, we get the equivalent stresses (Figures 12 and 13).

We note that the application of SPC has effects on the surrounding structure, since the stress distribution is not symmetric with respect to the normal plan Z. Thus for better correlation of the results will be analyzed rather berries plants, that is to say those far boundary conditions.

In analyzing the results of the model with portholes, we can make some general remarks:

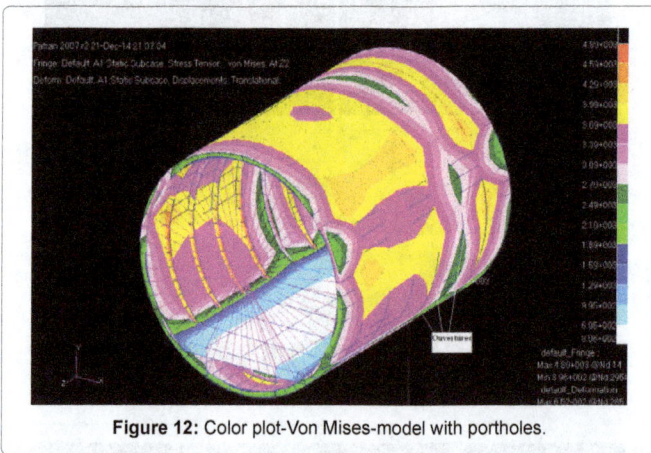

Figure 12: Color plot-Von Mises-model with portholes.

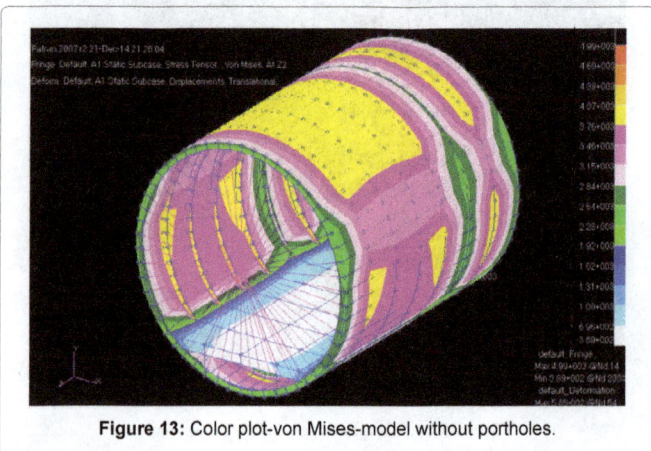

Figure 13: Color plot-von Mises-model without portholes.

1) The distribution of equivalent stresses is not very uniform

2) The internal forces are small at the openings (about 3390 psi) as opposed to the rest of the coating (about 3690 psi)

3) The upper and lower surfaces of the bays where the openings are topics found to stress concentrations (approximately 3990 psi)

4) The adjacent frames are subject to stress concentrations (approximately 4590 psi)

5) The resulting σx=2300 psi average (which represent the longitudinal stresses) and σy=3260 psi average (which represents the circumferential stresses) are lower constraints previously calculated in the methodology section (Figures 13).

This can be explained by the presence of the floor, which includes part of these constraints (Figures 14 and 15).

Unfortunately this color distribution does not allow us to visualize the effects on the stiffeners. For this, we must look directly internal loads stiffeners result in the fo6 file.

The stiffeners

The internal loads in the ROD elements are represented in several ways. For our analysis we will use the 'end loads', which are made by a summation of "grid points balance force', the balance of all the loads acting on a node in a given direction. This is basically free body diagram of a node; the sum of forces acting on a node must always be zero, giving a steady state to the node between the external loads and internal forces. And to realize the distribution of internal loads, we will seek end loads stiffeners bay 3 (adjacent to the central bay), and the bay 4 for configurations with and without openings (Figures 16 and 17).

The origin (0) represents the top of the fuselage, while the end (20) is the floor level. This reference system will be the same for all analyzes.

By analyzing these graphs and specifically the case with openings, we note that the stiffeners at the openings in the berry 3 do not include any load; their presence in this area is therefore virtually useless. By against the stiffeners located immediately above and below the openings experimenting stress concentration factors of up to 1.56. A strengthening structure at this level is then highly desirable. The frames are modeled by CQUAD elements and to know the internal efforts, we will seek, always in the f06 file, in the category 'strengths in quadrilateral element'. This gives us the forces Fx, Fy and Fxy [12] (Figure 18).

$$Fx = \frac{F21+F34}{\Delta x} = \frac{F12+F43}{\Delta y} \quad Fy = \frac{F41+F32}{\Delta x} = \frac{F14+F34}{\Delta y} \quad [12]$$

Once obtained these strengths, we calculate the maximum principal stress in order to have a comparison tool. This constraint is obtained as follows:

$$\partial max = \frac{Fx+Fy}{2} + \sqrt{\left(\frac{Fx+Fy}{2}\right)^2 + Fxy^2} \quad [12]$$

In calculating the bending stresses it is usually assumed that the elementary beam theory is sufficiently accurate resulting in a bending distribution given by the equation $f_z = \dfrac{M_{xx}Y}{I_{xx}} + \dfrac{M_{yy}X}{I_{yy}}$ [12]

In order to agree with the bending theory the fuselage shear distribution over the frames should be in accordance to the shear equation $q = \dfrac{V_y A\bar{y}}{I_{xx}} + \dfrac{V_x A\bar{x}}{I_{yy}}$ [12]

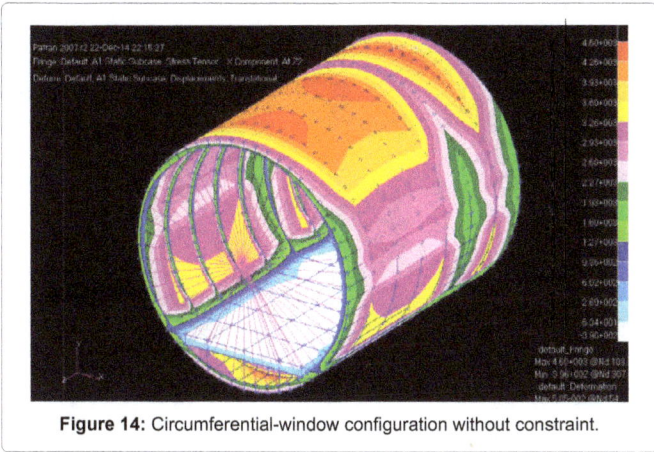

Figure 14: Circumferential-window configuration without constraint.

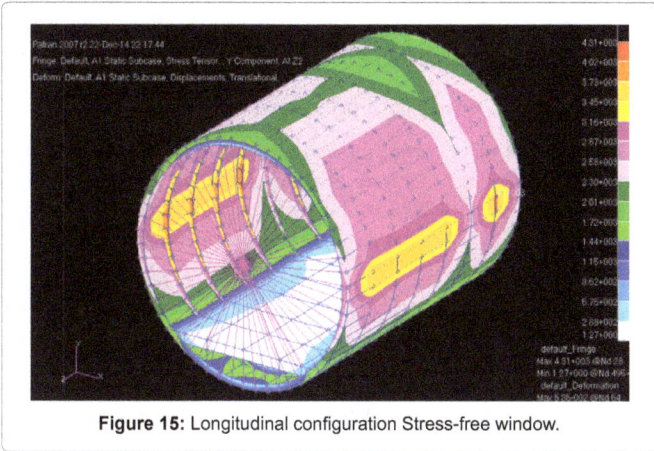

Figure 15: Longitudinal configuration Stress-free window.

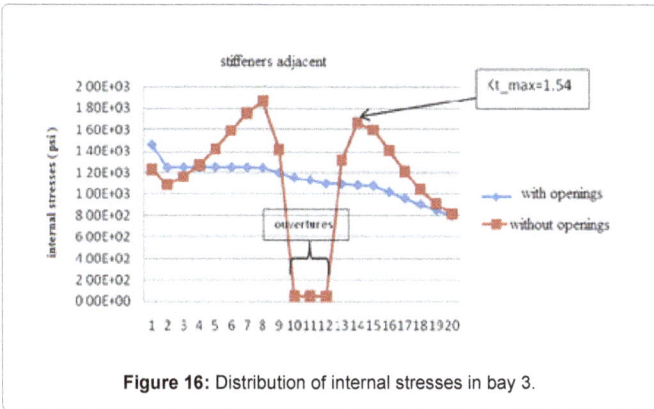

Figure 16: Distribution of internal stresses in bay 3.

It is common practice to use the simplified beam theory in calculating the stresses in the skin and stringers of a fuselage structure. If the fuselage is pressurized, the stresses in the skin due to this internal pressure must be added to the stresses which resist the flight loads (ie $\sigma_{sk} = \sigma_b + \sigma_L$.) (Figure 9) illustrates a distributed stringer type of fuselage section.

Up to the point of buckling of the curved sheet between the skin stringers, all the material in the beam section can be considered fully effective and the bending stresses can be computed by the general flexure formula $\sigma_b = \dfrac{M_y z}{I_y}$ [12], where I_y is the centroid moment of inertia of the section.

We analyze the frames 4 and 5 by comparing the configurations with and without openings (Figures 19 and 20).

Compared to these two graphs and specifically the case with openings, we can say that the plates are at the openings, the top of the fuselage and about at the floor, undergo an increase in internal stress

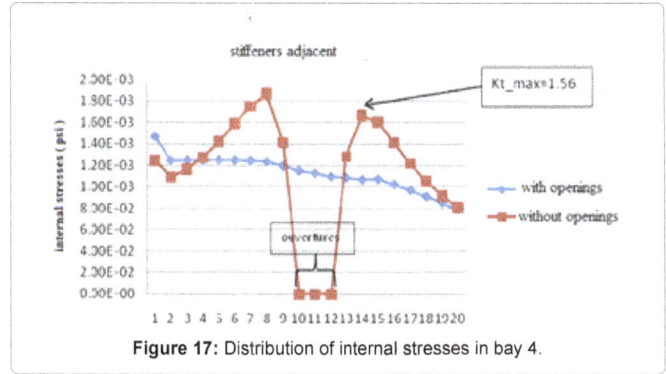

Figure 17: Distribution of internal stresses in bay 4.

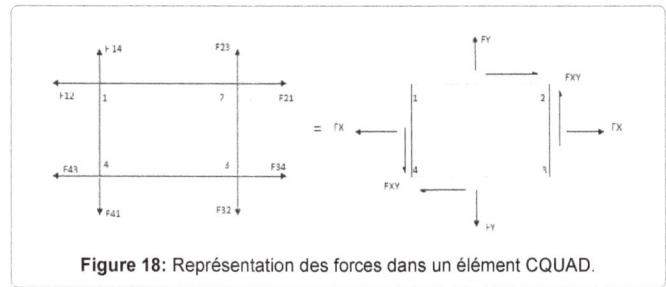

Figure 18: Représentation des forces dans un élément CQUAD.

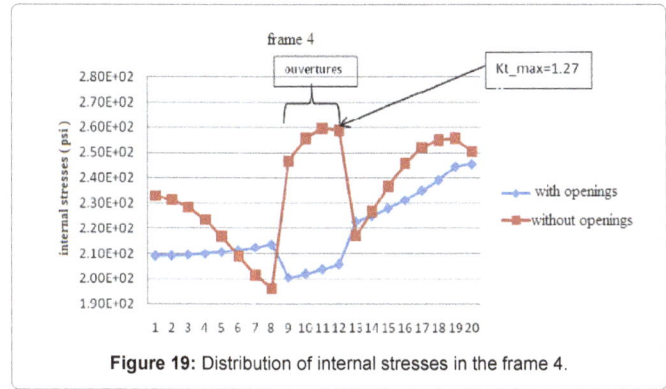

Figure 19: Distribution of internal stresses in the frame 4.

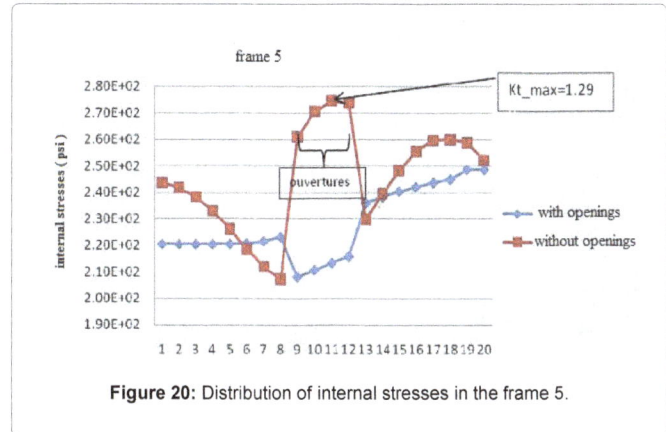

Figure 20: Distribution of internal stresses in the frame 5.

with a concentration factor stress up to 1.29. In addition, it appears that top plates and immediately resumes lower charges.

The skin

Compared to managers, the coating is composed of CQUAD elements. Thus the extraction method of internal forces will be the same. We analyze the plates of the bay 3 and 4 for the two configurations. Unlike following graphs, the latter two are a little different. This is explained by the fact that the bay 4 has possessed of openings unlike the bay 3. Thus, these openings due to their properties show fewer charges than the rest of the coating. Thus, we note the adjacent bay increased the internal stresses at the openings with a maximum stress concentration factor of 1.42. At the bay 4 additional charges

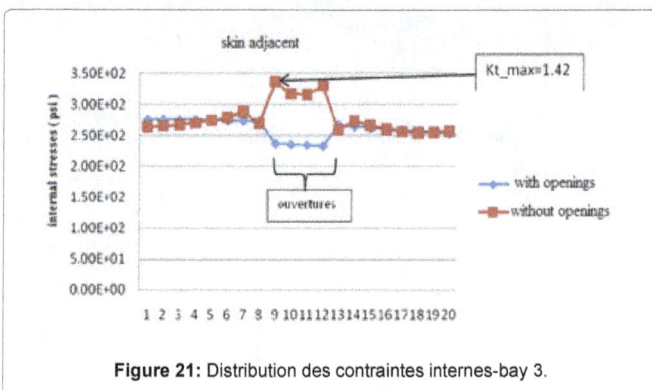

Figure 21: Distribution des contraintes internes-bay 3.

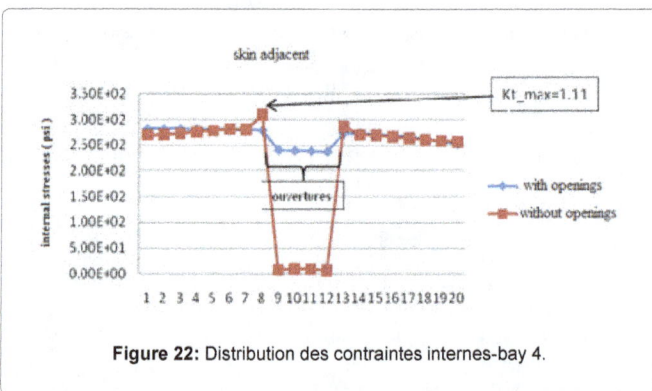

Figure 22: Distribution des contraintes internes-bay 4.

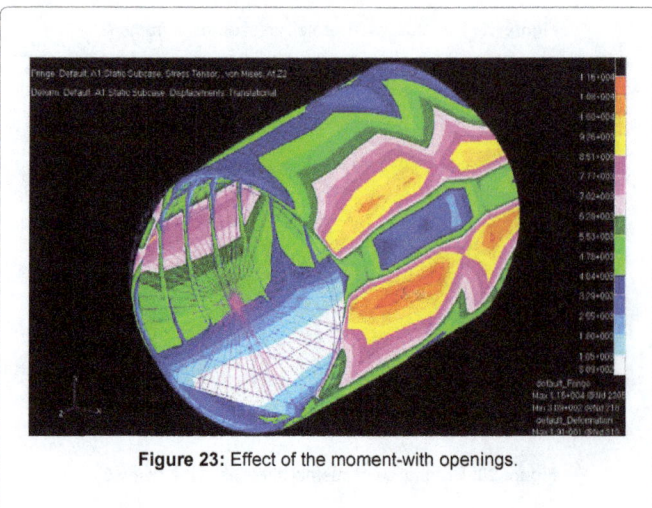

Figure 23: Effect of the moment-with openings.

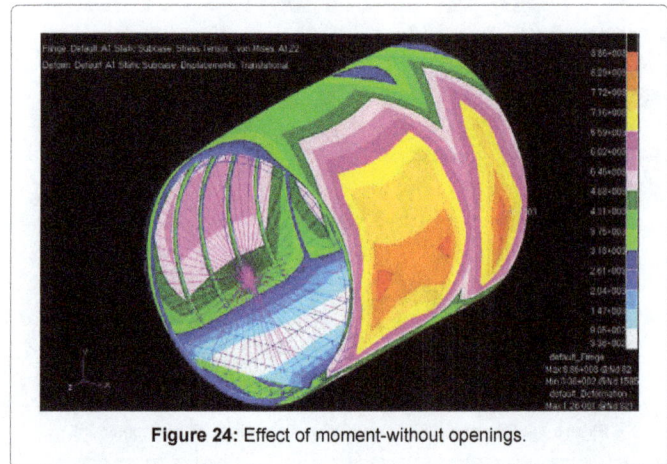

Figure 24: Effect of moment-without openings.

are transferred immediately to the plates of the upper and the lower coating (stress concentration factor of up to 1.11) (Figures 21 and 22).

Effects of the moment with pressurization

(Figure 23 and 24) From these graphs, we clearly distinguish the effect of the openings increases of the internal stresses, up to about 1,14E4 psi instead of 8.29E3 psi below the openings (with a moment of 5000000 lbs.in). There has also reduced of the internal stress at the openings.

Effects of torsion with pressurization

(Figure 25 and 26) The openings effect when the body is subjected to pressure and torsion is less obvious to notice. Indeed the major changes taking place at the management level as shown by the figures. The openings level executives note that the plates show more charges if there are no openings.

Critical Discussion

Previous results are quite consistent, in that we realize the overall effect of the openings of the surrounding structure. Indeed, we can point to the views of different graphs, the new internal load distribution, and also the occurrence of stress concentration. The model constructed by finite element seems to be quite faithful to the only global representations of these effects. However, this model has some weaknesses, such an asymmetry with respect to the XY plane. Given that the fuselage has a symmetrical distribution of these elements (windows, frames, charges ...) we would have expected the same way to a symmetrical distribution of internal stresses. However we see from the above figures that it's not really the case, contrasting the consistency of results. After investigation we realize that the application of travel restrictions (SPC) greatly influences the stress distribution. Made in the characterization made of the SPC is not really realistic. In flight, the fuselage is supported by the wings that generate lift, so it is normally at the junction between the fuselage and the wings that we would have the force displacement. But this is beyond the scope of our study, we decided to keep the boundary conditions as have already been established and defined above. Another weakness of the model is that when analyzing the results, we reported a flaw in the model with the configuration without openings. Indeed, as shown in (Figure 27), we notice at the openings, an abnormal decrease of the internal forces to the configuration without opening. A thorough investigation will probably solve this problem.

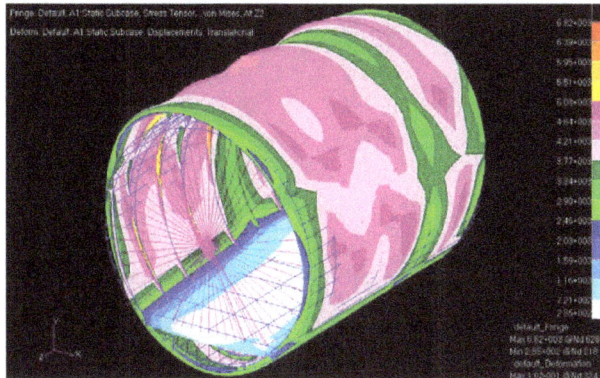

Figure 25: Effect of torsion-with openings.

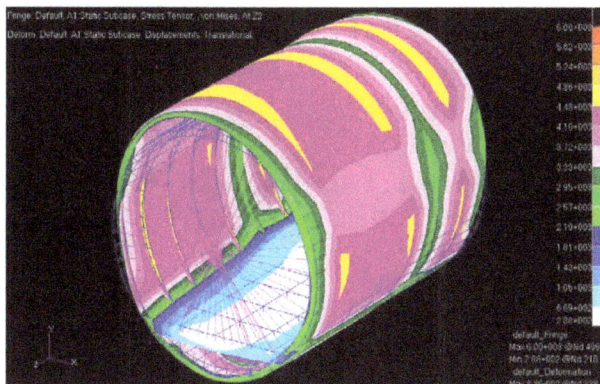

Figure 26: Effect of torsion-without openings.

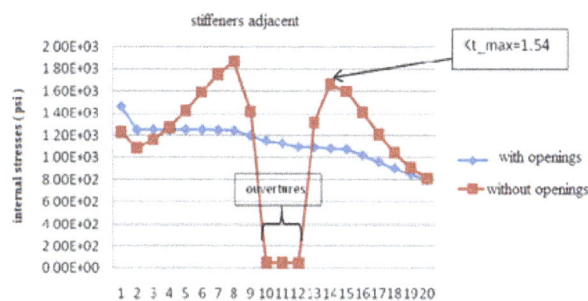

Figure 27: Distribution of internal stress-Bay 3.

The largest model of limitation lies in the fact that illustrates the overall effects and non-local. Indeed, the model is not refined enough to highlight precisely the value of internal efforts. We can only simulate the magnitude of the consequences of the openings in the fuselage structure.

Conclusion

Related to the aviation industry, the aim of the project was to realize the scale, and try to quantify stress concentrations at the openings, through a numerical method: finite elements. So from a geometric model, properties for elements and different load cases, 30 related to real situations in flight, we were able to visualize the phenomenon of stress concentrations. Thus we can realize the significant effect of openings in the structure of an aircraft. Indeed, these stress concentrations can be fatal if not pay attention. It is therefore essential to develop the structure accordingly to the most likely places to reduce the risk of damage. We tried to explain and comment on the phenomenon through this report, however, this study is only a first step to try to get closer to reality. In the future, it would be interesting to conduct a refining model to quantify precisely the stress concentrations. Or, to modify the boundary conditions to define those that best represent the reality for a given body structure. Finally, it could also be interested in the influence of the shape of the openings on the distribution of stress concentrations and thus devise solutions to reduce these concentrations and model them.

References

1. Bruhn EF, Analysis and Design of Flight Vehicle Structures. Jacobs Publishing Inc.

2. C Ledermann, C Hanske, J Wenzel, P Ermanni, R Kelm (2005) Associative parametric CAE methods in the aircraft pre-design. Aerospace Science and Technology 7: 641-651.

3. Christopher J, Sergio ERS, Pitter C, Use of Cad for weight estimation aircraft conceptual design. 24th International congress of the aeronautical sciences

4. Howe D (2000) Aircraft Conceptual Design Synthesis. Professional Engineering Pub Ltd, UK.

5. Brandt SA, Stiles RJ, Bertin J, Whitford R (1997) Introduction to Aeronautics: A Design Perspective. AIAA Education Series 1997

6. Raymer DP (1999) Aircraft Design: A Conceptual Approach, third edition AIAA Education Series.

7. Lloyd RJ, James FM III (2003) Aircraft Design Projects for engineering students.

8. Hürlimanna F, Kelmb R, Dugasb M, Oltmannb K, Kress G(2011) Mass estimation of transport aircraft wingbox structures with a CAD/CAE-based multidisciplinary process. Aerospace Science and Technology 15: 323-333.

9. Niu MC (1999) Airframe Stress and Analysis and Sizing. Conmilit Press Ltd.

10. Ramberg W, Osgood W (1943) Description of Stress-Strain Curves by Three Parameters. NACA Technical Note 902

11. P Kuhn, J Peterson, LR Levin (1952) A Summary of Diagonal Tension: Part 1-Method of Analysis. NACA Technical Note 2661

12. Nicolai LM (1975) Fundamentals of Aircraft Design, METS, Inc., 6520 Kingsland Court, San Jose, CA, 95120.

31

Generating Cost Efficiency Charts: A Comparison between B737, A319 and A321

Harasani W*

Aeronautical Engineering Department, King Abdul Aziz University, Jeddah, Saudi Arabia

Abstract

With a help of a local airline, the aim of this paper is to construct a cost efficiency charts, for three given aircraft to operate in three independent sectors. These chart are generated by an Excel code, the efficiency charts would be useful for airlines and fleet planners in their decision making process.

Keywords: Cost efficiency charts; Fleet planning; Direct operation cost

Introduction

The intent of this paper is to calculate the cost efficiency for three aircraft namely B737-500, A319, and A320 for prescribed sectors. The cost efficiency is to be calculated using the flight data provided by an airline and the output was assessed for two possible flight scenarios. The cost efficiency charts would be useful for airlines and fleet planners in the discussion making, and in-depth rout analysis. The two routs considered here are Jeddah to Medina (JED to MED) and Jeddah to Riyadh (JED to RUH). All trips would be assumed as round trip flights. Table 1 shows the selected engine type for the study

Methodology

The first step in all scenarios was to determine the flight utilization. This was done by stating values for the time to climb and descend, and the associated climb and descent speed, along with various associated distances.

The take-off weight of each of the aircraft is calculated by the weight fraction method as identified in Figure 1, [1]. Since the empty weight was known for each aircraft, the total weight and fuel weight is easily determined.

The first step would be to calculate the payload weight Wpl from equation 1

$$Wpl = number\ of\ passengers \times (174 + 40) \tag{1}$$

The second step would be to calculate the fuel weight Wf from equation 2

$$Wfl = (1 - Mff)Wto \times Wfuel\ res \tag{2}$$

Where M_{ff} if the fuel fraction calculated, the fuel fraction ware calculated in each phase of flight, startup, taxi, take off, climb, decent and landing

The W_{to} is the assumed takeoff weight for the aircraft at this flight, while the fuel fraction in cruise is calculated from equation 3

$$Rcr = \left(\frac{Vcr}{Cj}\right)\left(\frac{L}{D}\right)In\left(\frac{W4}{W5}\right) \tag{3}$$

Where

$\left(\frac{L}{D}\right)$ Lift to drag ratio

C_j specific fuel concumtion in lb/lb/hr

V_{cr} aircraft crusing speed in kts

R_{cr} cruise range in n.m

The thread step would be to calculate the operating weight empty Woe equation 4

$$Woe = Wto + Wf + Wpl \tag{4}$$

The forth step would be to calculated the empty weight from equation 5

$$Wetent = Woe + Wtfo + Wcrew \tag{5}$$

Where

W_{crew} from mission specification

W_{tfo} trapped fuel and oil

W_{tfo}=5% W_{to}

The empty weight then is compared from the empty weight allowable from Ref1 equation 6

$$We = inv.Log\left\{\frac{LogWto - A}{B}\right\} \tag{6}$$

Where A, and B are constant from page 47, [1], [Table 2]

If the error between W_{eten}t calculated and W_e is less than 5% then the W_{etent} is acceptable and W_p, and W_{to}, are acceptable if not change the W_{to}

Note: all dimensions are in Ib

Figure 1 summarize the process in which all weights are calculated.

Aircraft Type	Engine Type
B737-500	CFM56-3B1R
A 319	CFM56-5A4
A320	CFM56-5B4/P

Table 1: Aircrafts specification.

*Corresponding author: Harasani W, Associate Professor, Aeronautical Engineering Department, King Abdul Aziz University, Jeddah, Saudi Arabia
E-mail: wharasani@kau.edu.sa

Step 1. From mission specification

Wpl =PAX Number × (175+40) = ... lb

Step 2. Wto guss from similar airplanes

A/C Type	Wpl (lb)	Wto (lb)	Vcr (kts)	Range (n.m.)

Step 3. Determine Wf

Mff = (W1/Wto) Σ (Wi+1/Wi)
Mff = (W9/W8)(W8/W7)(W7/6)(W6/W5)(W5/4)
(W4/W3)(W3/W2)(W2/W1)(W1/Wto)
Wfused = (1-Mff)Wto
Wf = (1-Mff)Wto+Wf res

phase 1: (W1/Wto) Start up from table   2.1
phase 2: (W2/W1) Taxi   from table   2.1
phase 3: (W3/W2) Take-off from table   2.1
phase 4: (W4/W3) Climb to crise table   2.1
phase 5: (W5/W4) Cruise from eq.2.10
    Rcr = (Vcr/Cj) (L/D)cr ln (W4/W5)
        (L/D)cr ~ 16   from table 2.2
        Cj   ~ 0.5 lb/lb/hr   from table 2.2
        Vcr   in kts
        Rcr = design range -172* =   n.m
phase 6: (W6/W5) Loiter from eq. 2.12 p15
    Eltr = (1/Cj)ltr (L/D)ltr ln(W5/W6)
        (L/D)ltr ~ 18   from table 2.2
        Cj   ~ 0.6 lb/lb/hr from table 2.2
        Eltr is in Hours !
phase 7: (W7/W6) Descent from table 2.1
phase 8: (W8/W7) Fly to Alternative and Desc.
    R = (V/Cj) (L/D) ln(W7/W8)
        (L/D) ~ 10 **
        Cj   ~ 0.9 **
phase 9: (W9/W8) landing,taxi,shut-down
        from table 2.1

Step 6. Allowable value for We
            from figure 2.9
            or from eq. 2.16
We = inv.Log { (Log Wto - A) / B }
    A = 0.0833 from page 47 table 2.15
    B = 1.0393 from page 47 table 2.15

Step 4. Calculate Airplane Operating Weight Empty
        Woe = Wto - Wf - Wpl

Step 5. Calculate We
        Wetent = Woe - Wtfo - Wcrew
Wcrew   from mission specification
Wtfo trapped unusable fuel and oil
    Wfto = 5% Wto
    Wtfo = 0.005 × Wto

No   withen error? er=Wetent-We

Yes

Print: We, Wto, Wf. All in (lb)

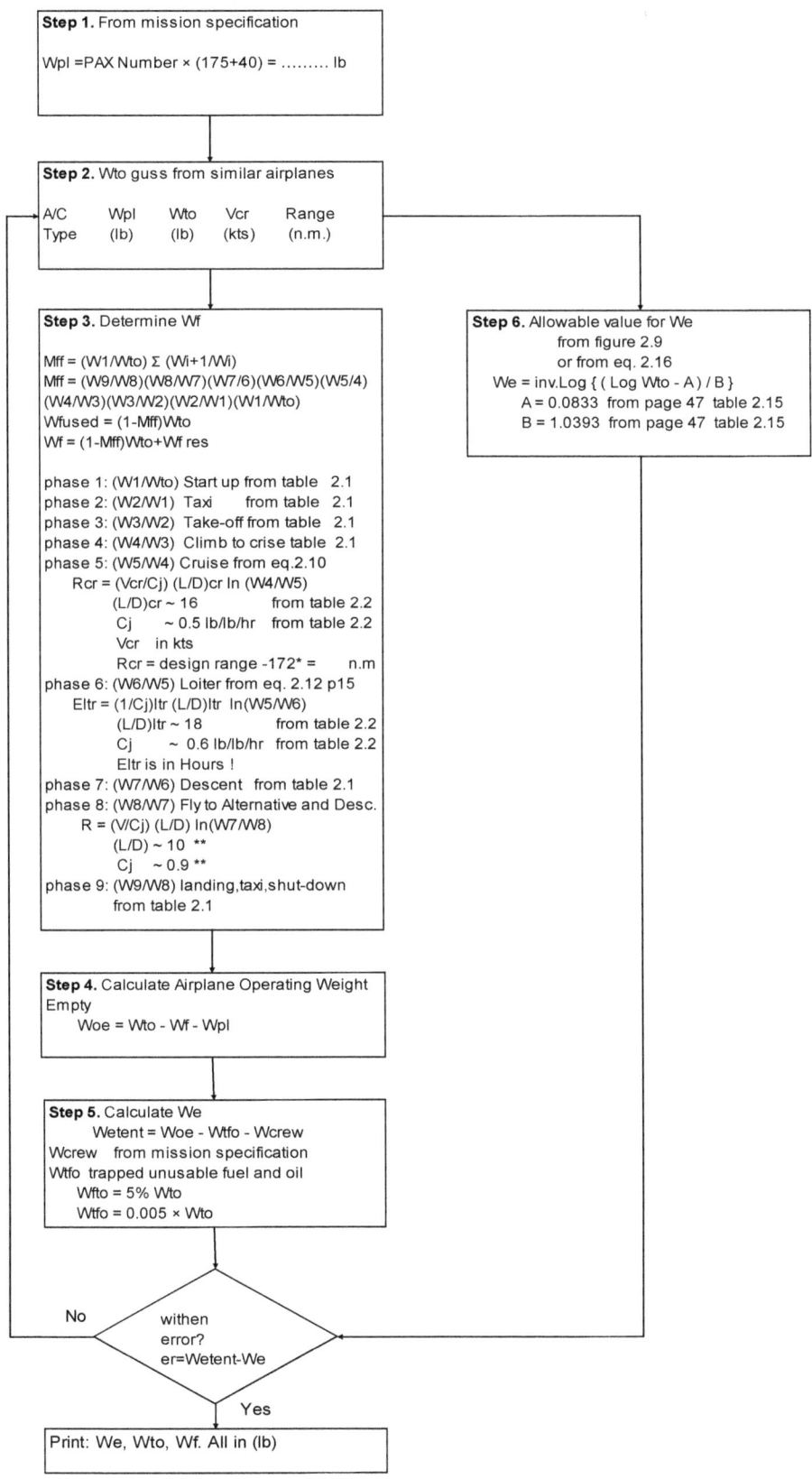

Figure 1: Calculating the fuel and takeoff weight through the weight fraction method Ref [1].

Next the actual Direct Operating Cost DOC, components is calculated. The DOC is calculated by using the method used in [2], starting with the DOC of flight. Within the DOC of flight the cost of the crew, fuel and oil, and insurance all can be directly determined. All values for the crew cost were assumed based on the values from the local airline and [3]. As were the input values for the fuel and oil cost and insurance. [Table 2].

$$DOC=DOCfly+DOCmaint+DOCdepr+DOClnr+DOCfin \qquad (7)$$

Where:

DOCfly is the direct operating cost of flying in $/n.m.

DOCmaint is the direct operating cost of maintenance in $/n.m.

DOCdepr is the dirct operating cost of depreciation in $/n.m.

DOClnr is the direct operating cost of landing fees, navigation fees, and taxes in $/n.m.

DOCfin is the direct operating cost of finance in $/n.m.

n.m. nautical miles

The DOCfly is given by

$$DOCfly=Ccrew+Cpol+Cins \qquad (8)$$

Where

Ccrew is crew cost given by

$$Ccrew=SUM\ [(n_{cj})\ \{(1+K_j)/V_{bl}\}\ (SAL_j/AH_j)+(TEF_{j}/V_{bl})] \qquad (9)$$

nc_j is the number of crew member of each type (i.e. captain, and co-pilot)

V_{bl} is the airplane block speed in n.m/hr.

SAL_j is the annual salary paid to crew members of each type

AH_j is the number of flight hours per year of each type

TEF_j is the travel expense factor

K_j factor which accounts for items such as vacation pay, cost of training

Cpol is the fuel and oil cost per nautical mile given by

$$Cpol=1.05\ (Wf/R)\ (FP/FD) \qquad (10)$$

Wf is the fuel weight in lb

R range in n.m

FP is the price of fuel in $/gallon

FD is the fuel density in lbs/gallon

Cins is the airframe insurance cost in $/n.m given by

$$Cins=(fins)\ (AMP)/\{(Uann)\ (V_{bl})\} \qquad (11)$$

fins is the annual hull insurance rate in $/$/year

AMP is the airplane market price

Uann is the annual hour utilization

The DOCmaint is given by

$$DOCmaint=Clab/ap+Clap/eng+Cmat/ap+Cmat/eng+Camb \quad (12)$$

Where

Clab/ap is the labour cost of airframe and systems in $/n.m

$$Clab/ap=1.03\ (MHRa)\ (R/V_{bl}) \qquad (13)$$

MHRa is number of airframe and systems maintenance hours needed per block hours

Clap/eng is the labour cost of engines in $/n.m

$$Clap/eng=1.03\ (1.3)\ Ne\ (MHRe)\ (R/V_{bl}) \qquad (14)$$

Ne number of engines

MHRe is the number of engines maintenance hours needed per block hours

Cmat/ap is the cost of maintenance materials for the airframe and systems $/n.m

Cmat/eng is the cost of maintenance materials for the engines $/n.m

Camb is the applied maintenance burden in $/n.m.

The DOCdepr is given by

$$DOCdepr=Cdap +Cdeng +Cdav +Cdapsp +Cdengsp \qquad (15)$$

Where

Cdap is the cost of airplane depreciation without engines in $/n.m

Cdeng is the cost of engine depreciation in $/n.m

Cdav is the cost of depreciation of avionics systems in $/n.m

Cdapsp is the cost of the depreciation of airplane spare part in $/n.m

Cdengsp is the cost of the depreciation of engine spare part in $/n.m

The DOClnr is given by

$$DOClnr=Clf +C\ nf+Crf \qquad (16)$$

Where

Clf the direct operating cost due to landing fees in ($/n.m) are calculated by

$$Clf=(Caplf)/\{(V_{bl})\ (t)\} \qquad (17)$$

Where

Caplf is the landing fees per landing given by

$$Caplf=0.002Wto\ \$/lbs \qquad (18)$$

Parameter	value
Annual salary paid for one pilot [$/year]	100000
Annual salary paid for for one co-pilot [$/year]	80000
Cost of maintenance materials for airplane [$/n.m]	404
Cost of maintenance materials for engine [$/n.m]	217
Annual hull insurance rate [$/$/year]	0.015
Maintenance manhours per flight hours [hrs/hr]	5.86
Number of flight hours / year	750
Fuel density FD [lbs/gallon]	8
Fuel price FP [$/gallon]	1.4
L / D	15
engine maintenace labor rate [$/hr]	12

Table 2: Some data given Ref [3].

W_{to} is the airplane takeoff weight in lbs

Cnf the navigation fees in \$/n.m

$$Cnf=(Capnf)/\{(V_{bl})\,(t)\} \tag{19}$$

Where

Capnf is the navigation fees charged per airplane per flight

Crt is the direct cost of registry taxies in (\$/n.m) are calculated by

$$Crt=(frt)\,DOC \tag{20}$$

Where frt is a factor suggested from [3]

$$Frt=0.00 +(10^{-8})\,W_{to} \tag{21}$$

Where

W_{to} takeoff weight in lbs

The DOCfin is given by

$$DOCfin=0.07\,DOC \tag{22}$$

In order to calculate the cost per aircraft per trip and the cost per seat mile, it is calculated as follows

$$Cost\ per\ aircraft\ per\ trip=DOC\ [\$/n.m] \times Distance\ [n.m] \tag{23}$$

$$Cost\ per\ seat\ mile=DOC\ [\$/n.m] \div Number\ of\ seats \tag{24}$$

More details are available in [2].

The the DOC of maintenance was calculated, mostly based on the values from a local airline and values founded in [2]. This was also true for the DOC of the depreciation, as well as the DOC of the landing and navigation fees. Once all of these components were calculated, the total direct operation cost could be calculated by just adding these values together, for each of the flight scenarios. The unite of DOC is dollars per nautical miles

After calculating DOC, and with the known distance and the seats for each aircraft at each sector, the cost efficiency chart could be generated and determined (Table 3).

Results

Figures 2 and 3 shows the efficiency of each aircraft at a given sector. Different aircraft types are not only compared with their trip cost but also with their seat mile cost, the lower the two parameters for the given aircraft the better, the aircraft is said to be more efficient if both parameters are low.

Figure 2, shows the cost efficiency chart results for the sector Jeddah to Riyadh, Figure 3 shows the cost efficiency chart results for the sector Jeddah to Medina.

Conclusion

- In the cost efficiency chart the best-performed aircraft in this

Aircraft Type	DOC in \$/n.m	
	JED - MED	JED - RUH
B 737-500	15.3	15.0
A 319	16.6	16.3
A 321	21.0	20.5

Table 3: Direct operating cost in \$ per n.m.

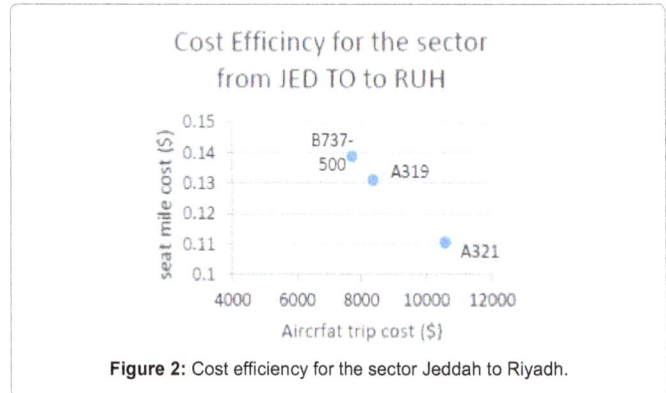

Figure 2: Cost efficiency for the sector Jeddah to Riyadh.

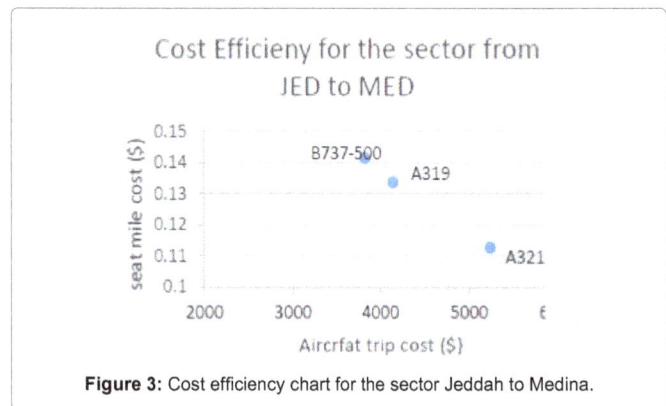

Figure 3: Cost efficiency chart for the sector Jeddah to Medina.

sector would be the lowest seat mile cost and the lowest aircraft trip cost.

- B737-500 has the highest seat mile cost but the lowest trip cost for each sector.

- A321 has the lowest seat mile cost but has the highest trip cost.

- Since it is preferential to have the lowest seat mile cost and the lowest trip cost the A319 performed better than the B737-500 and A321.

References

1. Roskam J (1990) Aircraft Design Part I Roskam Aviation and Engineering Corporation Ottawa Kansas U.S.A.

2. Roskam J (1990) Aircraft Design Part VII Airplane cost estimate Roskam Aviation and Engineering Corporation Ottawa Kansas U.S.A.

3. Local Airline

32

Casimir-Like Macroscopic Propulsion and Environmental-Energy Conversion

Scott Smith W*

Spokane, Washington, USA

Abstract

According to Quantum Electrodynamics (QED), a hidden source of great momentum and energy, called the EM Quantum Vacuum, fills the Universe. A Quantum light-sail will use refraction to induce asymmetric boundary conditions in the isotropic EM radiation pressure of the Quantum Vacuum. In terms of Newton's First Law, these asymmetric radiation-pressures act as outside-forces that push harder on one side of a light-sail than on its opposite side. This radiation-pressure will provide the motive force for a new class of macroscopic prime-movers, a kind of massless propulsion; a new environmental-energy conversion device.

Keywords: Propellp physics; Solar sails; Light sails; Interstellar propulsion; Space drive; Reactionless propulsion; Quantum optics; Environmental energy conversion; Massless propulsion

Introduction: A New Prime-Mover

According to Quantum Electrodynamics, QED, the EM Quantum Vacuum fills the Universe [1]. If it could be seen, it would look like the tiny flashes of static-snow that are seen on televisions in-between channels. It is believed to consist of photons that momentarily manifest then promptly disappear. NASA's Breakthrough Physics Propulsion Program report says that it might be possible to induce Casimir-like propulsive forces by developing new means of imposing asymmetric boundary conditions on the EM Quantum Vacuum [2] ractical approach to accomplishing that is explored in this paper.

Just as sails on ships and vanes on windmills capture the cost-free energy and momentum of the wind, so also, Quantum Sails will be driven by inducing asymmetries in the energy and momentum of the isotropic radiation pressure of the EM Quantum Vacuum. This radiation-pressure will provide the motive force for a new class of macroscopic prime-movers, a kind of mass-less propulsion, a new environmental-energy conversion device. All of this is reconciled with the Laws of Thermodynamics and the Laws of Motion.

Objectives

The purpose of this paper is not to prove that any of this is certain; rather, a main point is to establish that the proposed experiment is worthwhile. It is also intended to recruit individuals who will help join, assemble, fund and equip a research team to design and produce the prototype Quantum-Sail that is described in this paper. To accomplish these two objectives, it is necessary to overcome a large number of unwarranted theoretical objections to this sort of proposal. Rather than being some sort of new discovery, this paper is a novel perspective on settled science. Although the proposed device is simple, its merit is not apparent if the science that it is based upon is not first re-articulated from this new perspective.

Deriving the Radiation-Pressure of the EM Quantum Vacuum

According to Quantum Electrodynamics, the Quantum-Vacuum Zero-Point Energy field is what is left in space even if all matter and heat were removed. It is believed to consist of particle-pairs that pop into existence then vanish a short time later. Some of these particles are photons. Its photons continuously appear in every possible wavelength. Then they vanish after traveling about half of a wave-length.

The EM Quantum Vacuum Radiation Pressure can be derived from energy values that can be experimentally detected. These values and their relationships are described by Planck's Black Body Radiation Spectrum Formula. Each energy value is represented by u. Planck's Constant is ħ.

$$u = \hbar\omega \, (n+1/2); \{ n=0, 1, 2, 3 \ldots n, n+1\} \tag{1}$$

The existence of the Zero Point Energy Field is most clearly seen in the instance where n=0; this leaves Eq. 2.

$$u_{zpe} = \hbar\omega/2 \tag{2}$$

The radiation that is attributable to emissions from atoms is given by n, which is always a whole number. However, actual readings of Black Body Radiation reveals the extra term: 1/2. So in addition to the part of the Black Body Radiation field which originates in orbital transitions inside of atoms, another field is present. It would be detected by itself when n equals Zero. Zero-Point Energy, u_{zpe}, is the energy that comprises the EM Quantum Vacuum.

Planck's constant is the energy per cycle, the energy per single photon. Planck's constant is represented by the symbol ħ or h; ω is light frequency in radians per second; f is frequency in cycles per second.

$$(\text{Eq. 2}) = \hbar \, \omega/2 = [(h/2\pi) \, (2\pi \, f)/2] = u_{zpe} = h \, f \, /2 \tag{3}$$

A single photon is one wavelength long; $1/\lambda^3$ gives the total number of photons of one wavelength that can fit into one cubic meter. (Eq. 4) gives the total energy of quanta of that one wavelength that is in a cubic meter of space.

$$(h/2) * (1/\lambda^3) = u_{zpe} \, (\lambda^3) = h/2 \, \lambda^3 \tag{4}$$

*Corresponding author: Scott Smith W, Spokane, Washington, USA
E-mail: scott712@hotmail.com

$$c = f \lambda \tag{5}$$

$$h/2 \lambda^3 d \lambda = u_{zpe} (f) df = (h f^3/2 c^3) df \tag{6}$$

Integrating u_{zpe} (f) df sums the energies of all photons of every frequency that can simultaneously occupy one cubic meter in a given range of frequencies.

$$U_{zpe} (f) = \int (f^3/2 c^3) df = U_{zpe} (f) = h f^4/8 c^3 \tag{7}$$

$$(h f^4/8 c^3) * c/c = (h c f^4/8 c^4) = U_{zpe} (\lambda) = h c/8 \lambda^4 \tag{8}$$

Ludwig Boltzmann proved that the isotropic radiation-pressure Pr, that is acting uniformly on a surface is equal to the energy-density above the surface, divided by three.

$$U_{zpe} (\lambda)/3 = Pr = h c/24 \lambda^4 \tag{9}$$

In practice, one can justify not counting wavelengths that are larger than λ specified since one only needs to specify wavelengths that significantly impact the material that has been chosen. The energy density of progressively longer wavelengths quickly becomes insignificant. This is because every two-fold increase in wavelength is accompanied by a sixteen-fold decrease in energy-density. This is due to the λ^{-4} term. Likewise, consideration of ever-smaller wavelengths also converges rapidly on irrelevance for the present purposes; this is because matter rapidly becomes ever-more transparent as ever-smaller wavelengths are considered.

Using the EM Quantum-Vacuum to Push, a First Quantum Sail Experiment

Henrik Casimir is credited with being the first person to propose using the Radiation Pressure of the EM Quantum Flux to move a physical object [3] described what would occur if EM Zero-Point Energy really exists. In 1948, he proposed a thought experiment. He described two electrically-neutral, electrically-conducting, parallel plates. At least one plate could move freely. They would be separated from each other by a very small distance. He pointed out that these plates would prevent the EM Quantum Flux from forming photons with wavelengths that were too long to form between the plates.

In Figure 1, the smaller, purple waves represent all of the wavelengths that can form both inside-of and outside-of the space between the plates. Since the smaller waves are equal and opposite to each other, they exert no net forces on the plates; however, the longer red wavelengths can only form outside of the space between

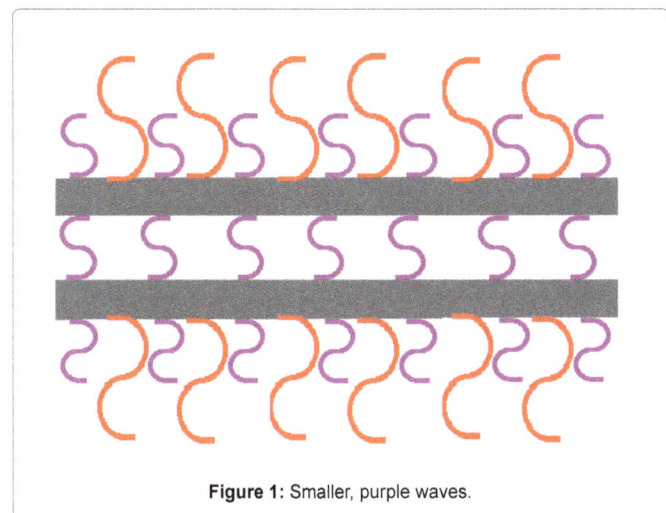

Figure 1: Smaller, purple waves.

the two plates since they are too long to fit between the plates. The inward-directed Radiation Pressure, outside of the plates, consists of all possible wavelengths. This is because all-possible wavelengths can form outside of the confines of the space between the Plates. The longer (red) wavelengths move the plates inward since there are no long, red wavelengths inside the cavity to counteract them by pushing outwardly.

In 1996, Steve Lamoreaux [4] then at the University of Washington in Seattle, experimentally confirmed a version of this gedanken. Many additional independent experiments have consistently verified this phenomenon. The force that moves the "plates" together is called the Casimir Force. These plates are a kind of quantum sail since they are driven by an induced asymmetry in the radiation pressure of the EM Quantum Vacuum.

The Quantum-Flux Has Already Done Work

Suppose Casimir's original Radiation Pressure Theory correctly models a real physical process. This means that the plates are not exerting any significant forces on each other. Instead, the plates are effortlessly altering the EM Quantum Vacuum Density by simply being there. They are then moved because of the radiation-pressure differences on opposite sides of each plate. They are moved by the energy and momentum of the Radiation Pressure of the EM Quantum Flux. In other words, no human is providing this energy. This is energy that would have been there, in the environment, whether or not anyone chose to use it. Harold Puthoff, Director of the Institute for Advanced Studies at Austin, Texas, has pointed out that this energy is left behind as heat if the plates collide. In other words, Zero-Point energy has already been used to propel real objects. Again, these plates constitute a kind of Quantum Sail since they are moved by induced asymmetries in the radiation pressures in the EM Quantum Vacuum.

Casimir Plate Movement and the Four Laws

If Casimir Plates did not move, that would violate the Zeroeth Law. This is because the freely-moveable Plate is located directly between a high energy-density radiation-pressure region and a low-energy radiation-pressure region; therefore, it would have to move.

The First Law is not violated. This is because the energy that is entering the system as a high-energy flux of photons is equal to the work and low-energy photon flux that is leaving system as heat.

The Second Law is not violated. This is because an already-existing high energy-density, low entropy photon flux is crossing the system boundary, performing work and shedding high-entropy, low energy-density heat which then exits across the System Boundary. Therefore Entropy is increasing and energy density is decreasing, just as they should do.

The Third Law concerns the usual inability of energy to transfer in the absence of two differing Thermal Reservoirs. It is not the machine that is exceptional in this case. Rather, the Quantum Flux itself is defined as the energy that remains in otherwise empty space, when all heat energy has been removed. This is one reason why it is called Zero-Point Energy (ZPE.) Of course it is also still present at all non-zero temperatures.

ZPE literally has no thermal potential because differences in energy potential cannot flow together. This is because each generation of particles vanishes before they can redistribute themselves. Each generation of particles is replaced by a new generation of particles that appears in the same general distribution as the previous generation of particles appeared. Casimir's Plates move because they store multiple generations of energy that is collected as the photons collide more with one side of each plate than with the opposite side.

Though the Quantum Flux is non-thermal, its spectrum is, highly energetic, especially at wavelengths below 50 nm. In other words, being at Zero-Temperature does not mean that its energy potential is Zero. Again, one is using the low-entropy, high-energy density Radiation Pressure of very-intense, very-small wavelengths to do work. That work is dissipated as low-energy, high-entropy wavelengths of infrared radiation.

Basically, the Third Law comes into play more abstractly: High and low frequencies always exist in the ZPE Spectrum at all temperatures. Even at Absolute Zero Degrees, the high energy, low-entropy wavelengths function as the high-energy reservoir and the cool heat-sink of space still serves as the low-energy heat-reservoir, as usual. So a high energy reservoir and a low energy reservoir are both still present. So, in principle, the Third Law is still being observed. This adaptation of the Third Law is no more extraordinary than when the Third Law is adapted to cover systems that require non-thermal differences in electrical potential or hydraulic pressure.

Many object that using some of this energy before it winked-out would change the amount of energy remaining in the Universe. This depends entirely on where one draws the system-boundaries of the Universe. On the one hand, if one defines the Quantum Flux mechanism as part of the Universe then, perhaps, energy is not really being created and destroyed. Instead, it might be alternating between a hidden state and a manifest state. This is the essence of Paul Dirac's Theory of a vast Sea of Particles that alternate between a positive energy level and a negative energy level. Any energy that left the Quantum-Flux would have begun in the Universe and then remained in the Universe.

On the other hand, if one posits that the Quantum Flux is not part of the Universe, then he is admitting that energy can enter and leave the Universe. Therefore, one would have no basis for assuming that matter and energy could not temporarily accumulate in one Universe while another Universe is temporarily depleted. Perhaps this energy would be passed back and forth more or less equally over time. These musings are no more- or less- fanciful than insisting that one even can know if the energy-balance of the universe has to match our small-minded, pathetically-uninformed expectations. In other words a quantum sail is not required to return an equal amount of energy to the Quantum field any more than one would require a coal-fired steam plant to return an equal quantity of coal to the ground.

A Patented, Potentially-Practical Casimir Device

The Casimir Effect is very real; it currently presents severe problems in Microscopic Electromechanical Machines, MEMs. It causes sticktion which means that the small clearances that are between small parts form unintended Casimir Cavities. In other words, Casimir Forces tend to jam very-small parts together so that they cannot function properly. So already, the Casimir Effect is a force to be reckoned-with. However, one man has found a potentially-practical method of utilizing the effect.

Fabrizio Pinto has obtained a patent for reciprocating Casimir Plates [5]. Semiconductor plates would alternate between a conducting and non-conducting state to turn the Casimir Effect off and on. Unfortunately such an approach may only produce small forces, and motions that are less than a micrometer in length. It may very well prove to be useful for microscopic machines; but it may be impossible to use it to create useful macroscopic forces. This is due to the difficulty of maintaining large parallel surfaces at nano-scale separations. It also would not be useful for propellantless propulsion.

Asymmetric Boundary Conditions on Two Sides of a Single Plate

Instead of using a pair of plates, a single isolated plate will impose asymmetric boundary conditions with the EM Quantum Vacuum on its two sides. These asymmetric conditions will cause a single plate to experience different radiation pressures on its two sides; this will cause a net force to act on the plate.

This Quantum sail could be easily distributed over large areas. Also, many layers of sails could be stacked; they would be separated by a material that allows the EM Quantum-Flux to manifest within it. In this way, macroscopic combined-forces might be obtained, even if individual layers contribute extremely small forces. Such an arrangement would be useful both for propulsion and for environmental-energy conversion.

Conclusions of NASA's Breakthrough Physics Propulsion Program (BPPP)

Can the two sides of a single plate interact asymmetrically with the radiation pressure of the EM Quantum Vacuum and experience a net force? Propellant-less Propulsion is the sort of concept that will cause many readers to quickly put this paper down. That is because, on its face, it seems like a hopelessly absurd thing to attempt. NASA's Breakthrough Physics Propulsion Program (BPPP) evaluated possible avenues of novel research. It hoped to identify new insights and methods that might enable science to surpass current expectations of what is considered possible. The BPPP authors examined many proposals; electric rockets, nuclear rockets, laser powered sails; interstellar ramjets. No mass-based propulsion methods appear adequate to take anyone, even just to the second-nearest Star, in a single lifetime. Therefore, the BPPP was particularly interested in identifying avenues of research that might lead to Propellant-less propulsion. Propellant-less Propulsion would not need on-board reaction mass. This is important because the reaction-mass requirement of rockets is the major part of what makes them fuel-hungry, large, expensive and slow.

NASA's Breakthrough Physics Propulsion Program final report mentioned the possibility of developing a Quantum light-diode sail [2]; it would let more EM Quantum Vacuum radiation pass through the diode from one side than from the opposite side. This would cause the isotropic EM Quantum-Vacuum radiation-pressure to exert a net force toward the less-transparent, more-reflective side of the Quantum-Diode. In other words, instead of using Newton's Third Law which requires reaction-mass, one would use the radiation-pressure of the Quantum-Vacuum as an outside-force, in accordance with Newton's First Law.

It sounds theoretically impossible to derive a net force from equal and opposite influxes. It is tempting to dismiss this entire notion by simply invoking Conservation of Momentum; however, this general objection is easily refuted by a simple example, a mechanical diode for rubber balls: Two teams of astronauts throw equal and opposite numbers of rubber balls at a wall that is floating in Space between them. At this point, most people insist that one should already know that the wall will not be moved in either direction. They will argue that all of the forces that are applied to the balls are equal and opposite; therefore, they say that whatever one side of the system accomplishes, the opposite side of the system must undo it. In other words, they are saying, since the applied momentum sums to zero, then so must the final momentum sum to zero. This much is true, but it still does not mean that the wall is not moved.

What if the wall itself responds differently to the two influxes of balls? What if the wall itself introduces asymmetric reaction-forces? If the wall is made of little doors that only open in one direction, then half of the balls will impart a full-measure of momentum to the wall as they bounce off of the one side where the doors stay closed; but the remaining balls will only impart a small amount of their momentum to the wall, in the opposite direction, as they push the doors open and continue onward. In other words, a door experiences a stronger collision-force from a rubber ball if that door remains closed than from a rubber ball that pushes on through the door. This happens because the collision-force that a given rubber ball exerts on a door must equal the reaction force that door applies to the rubber ball. On one side, that reaction-force is limited to whatever small amount of force it takes to push a door open. In effect, the balls on the other side encounter a solid wall.

It is quite remarkable that two equal and opposite influxes can indeed impact the same wall asymmetrically. This does not prove that one could in-practice make a light-diode; but it does prove that this sort of thing is not impossible from the standpoint of Momentum Conservation. In this example, the doors experienced different reaction forces on their two sides. Similarly, in the next example, a two-sided mirror is approached by equal and opposite influxes of EM Quantum-Vacuum radiation-pressure. Again, the reaction-forces are different on each of the two sides. In this case, a difference in pressure arises on opposite sides of the two-sided mirror because one side is covered with a highly-refractive material and the opposite side is bare or is covered with a low-refraction material.

The Defining Problem

This paper proposes a realistic, simple solution to a profoundly difficult problem: The EM Quantum Flux Radiation-Pressure is equally energetic in every direction. Put another way, this radiation-pressure normally pushes equally hard on opposite sides of most objects. All of these forces would normally be totally used-up, just counteracting each-other. Since these forces are normally equal and opposite among themselves, a single, isolated two-sided mirror floating in Space would normally remain stationary (Figure 2).

The two yellow rays represent the light of the EM Quantum-Vacuum. In particular, they represent the fact that its light rays strike any given point on the mirror at mostly non-perpendicular angles.

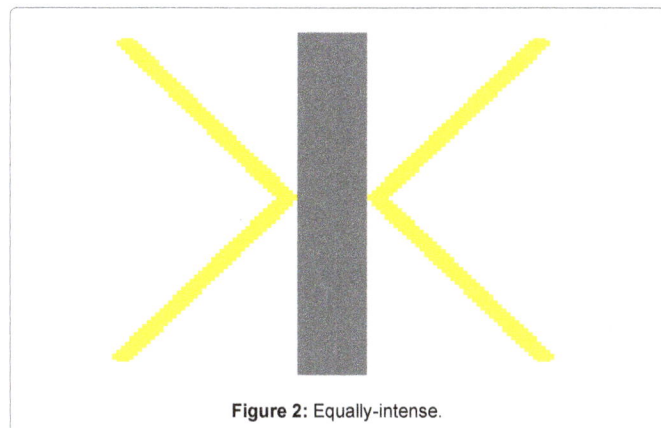

Figure 3: Two beams that are equally-intense.

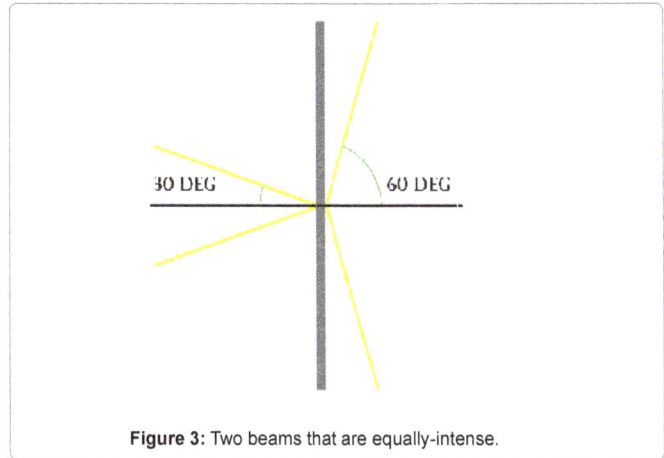

Figure 4: Two equal and opposite beams of light remain equally-intense but become non-opposite.

They also represent the fact that, usually, the light rays of the EM Quantum-Vacuum would strike a two sided mirror equally on each side. Therefore, this floating, two-sided mirror would not be pushed in either direction by the light of the EM Quantum-Vacuum.

A Simple Solution

The problem in Figure 2 is not merely that the rays of light are equally-intense. Rather, the difficulty lies in the fact that they are equally-intense and opposite.

Figure 3 illustrates one way to obtain a net force from two beams that are equally-intense. Two equally-intense beams of light are striking a two-sided mirror that is floating freely in Space. The beam on the left strikes the mirror at a 30 degree; this angle is more-perpendicular to the plate than the beam on the right which is striking at a 60 degree angle. The force of the light striking each surface is given by F cos θ; F cos 30 is greater than F cos 60; therefore, the beam on the left hits the plate harder than the beam on the right. Consequently, the plate will be pushed to the right, even though the beams are equally-intense. This could be done with two equally-intense lasers.

sIn Figure 4, two equal and opposite beams of light remain equally-intense but become non-opposite. One light-beam approaches the blue refractive-material at a sixty-degree angle that is equal and opposite to the light-beam that is approaching the bare side of the mirror on the right, at sixty degrees; but then the beam on the left is refracted. The light on the left strikes the mirror at a thirty degree angle which is steeper than the light that strikes the right side of the mirror. Therefore, a net force to the right acts on the two-sided mirror, just as in Figure

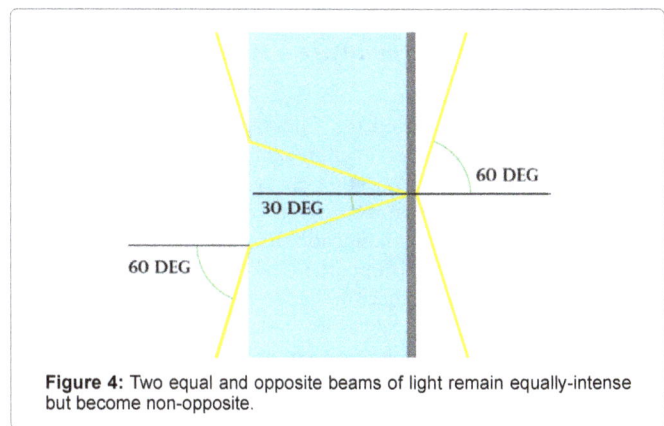

Figure 2: Equally-intense.

3. Many experiments, for many years, proved that, under these circumstances, light does in-fact exert a stronger force on a mirror that is in a refractive material, than the same light exerts on the same mirror when no refractive material present [6-10]. The forces that act on the refractive material will be discussed later in this paper.

As with the rubber-ball diode, the applied momenta start out equal and opposite to each other, yet the mirror is moved. In the case of the ball-diode, the two equal and opposite influxes of balls applied asymmetric collision forces to the doors because the doors applied asymmetric reaction-forces to the balls. With the refractive mirror system, the refractive material and the light exert refraction forces on each-other; thus, the path of the light is altered on the refractive side of the mirror and not altered on the bare side of the mirror, as in Figure 4. This is true for every equal and opposite pair of light rays that approach the system. It will also be true of every photon of the EM Quantum Vacuum that is refracted by the refractive material, from every possible angle.

Reduced Radiation-Pressure in a Refractive Medium

So far, the refractive mirror setup has had light entering a refractive surface that is parallel to the mirror. This arrangement, shown again in Figure 5, yields a result where the force on the mirror increases. The yellow lines represent the momentum vectors of the light beam. The red arrows represent momentum that is perpendicular to the mirror. The purple arrows represent the momentum component of the light that is parallel to the mirror. It is important to note that changes in these vector lengths are a consequence of the fact that the angle of the ray of light changes, relative to the mirror. The red perpendicular momentum vectors verify this change in momentum; the red perpendicular-momentum vector is longer, after the light enters the refractive material.

Other times the force on the mirror will decrease inside a refractive medium. Figure 6 demonstrates how one can obtain this result with an experimental setup that is similar to the one in Figure 5.

This time, the light enters the refractive material through a surface that is perpendicular to the mirror. Therefore, in this case, the refraction is causing the light to strike the mirror inside the refractive material at a shallower angle than if the refractive material were not present.

Therefore, this time, the force on the mirror is less when the refractive material is present. This can be verified by comparing the momentum component vectors. In Figure 5 the red perpendicular vector became longer as the light entered the refractive material. In contrast, in Figure 6 the red perpendicular momentum vector becomes shorter as the light enters the refractive material. The system in Figure 6 is nothing more than a trivial variation of the system in Figure 5.

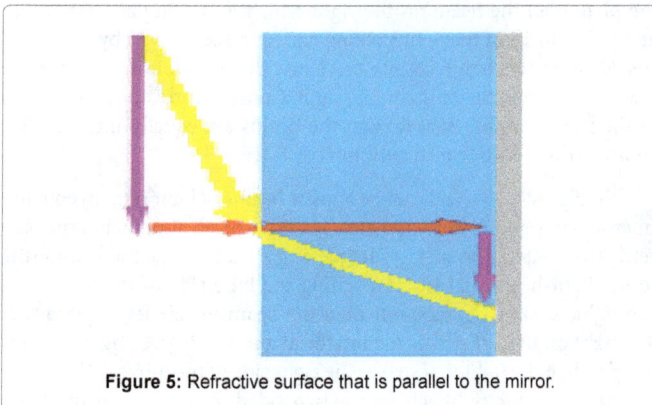

Figure 5: Refractive surface that is parallel to the mirror.

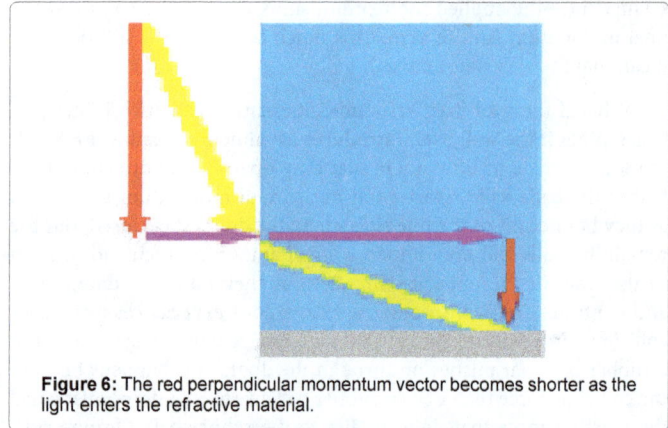

Figure 6: The red perpendicular momentum vector becomes shorter as the light enters the refractive material.

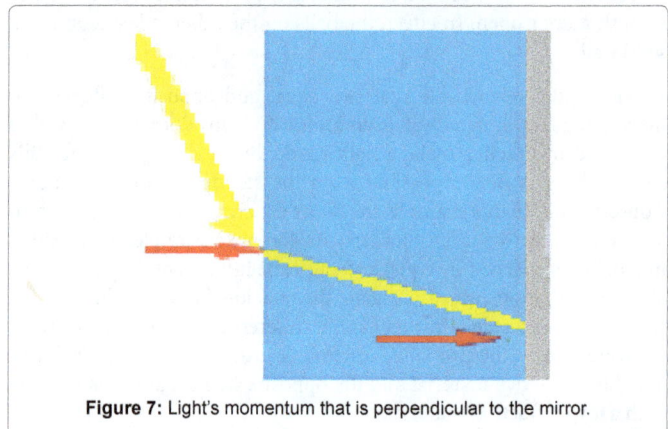

Figure 7: Light's momentum that is perpendicular to the mirror.

Comparing Figures 5 and 6 also reveals how momentum is transferred between the component vectors as the light refracts.

Momentum in a Refractive Medium

The mirror in these arrangements will definitely experience a net force; [6] but, does the refractive-force that acts on the refractive-material exert a force that is equal and opposite to the net force that acts on the mirror, so as to render the entire system motionless? The answer to this question has long-been obscured by the century-old Minkowski-Abraham discussion concerning [7] the momentum of light inside a refractive-medium.

According to Minkowski, light acquires extra total momentum as it enters a more refractive medium. It is said to increase by a factor that is equal to the refractive index of the medium [8]. According to Abraham the total momentum will decrease by a factor that is equal to the inverse of the refractive index of the medium [9]. Each of these hypotheses is seemingly-supported by an abundance of experimental observations. These two viewpoints are examined in the following discussion.

A refractive material covers the back side of the mirror in Figure 7; it is represented by the blue area. The yellow rays represent the average of all of the rays that are approaching each mirror.

In Figure 7, the red arrows represent the component of the light's momentum that is perpendicular to the mirror. The red arrows in Figure 7 show what would happen if momentum were conserved in the expected way. These two rays are shown as though they had the same perpendicular vector-momentum both before- and after- the ray enters the refractive material. The perpendicular momentum outside of the refractive material is represented as being equal to the perpendicular momentum that is inside the refractive material.

Figure 8 shows what really happens. The magnitude of the perpendicular momentum vector is actually greater, inside the refractive material, even though it is still the same ray of light as it was outside of the more-refractive medium.

Figure 9 consists of a two sided mirror. The back side of the mirror is covered with the refractive material. The front side is bare. Equal and opposite light influxes approach both sides of the entire system. Why doesn't Figure 9 outright-prove that one can use refraction to get a net force from equal and opposite influxes of light? Minkowski took this dilemma at face value and concluded that the beam of light somehow actually acquires extra total momentum as it enters the refractive medium. No causal mechanism has been established; this is just as inexplicable today as when Minkowski first made this famous conjecture. This so-called extra total-momentum is called Canonical Momentum. He took the position that this situation would still be consistent with conservation of momentum, as long as the light gains or loses the same amount of extra momentum as it enters and exits the refractive medium. In other words, he was adding extra momentum into his math for the light that was inside the more-refractive medium. Therefore, to salvage his approach to the problem while still Conserving Momentum, he had to subtract this extra momentum out of his figures for when the light exited the refractive medium. Again, no causative mechanism is proposed. This would also mean that any extra-momentum forces that are acting on the mirror because of this Canonical momentum would have to be counteracted by refractive

forces that operate in the opposite direction, as the light enters and exits the refractive medium.

In Figure 10, the green arrows indicate the direction that the refraction forces would have to act if Minkowski was correct. On the one hand, if Minkowski is correct, then there would be no net force acting on the entire sail system (Figure 11). On the other hand, if there is no equal and opposite refraction force to counteract the net reflection force that acts on the mirror then Minkowski is mistaken; then, there will be a net-force acting on the entire sail system as is shown in Figure 12.

There Are No Refractive Counter-Forces

As light enters the more-refractive material, it slows down. It is as though it is exerting deceleration reaction forces on the refractive medium. These forces are analogous to a car that imparts a forward-directed force to the roadway as it decelerates upon entering a school zone. This forward-directed force is directed into the slow-speed zone.

Surprisingly, it turns out that a force also acts back toward the refractive medium as the light exits the refractive medium [11]. This exit force has been experimentally demonstrated. A laser was shined through a glass fiber and a recoil force was noted; it bent the fiber as the light exited out the end of the fiber [12]. Loudon and Baxter also

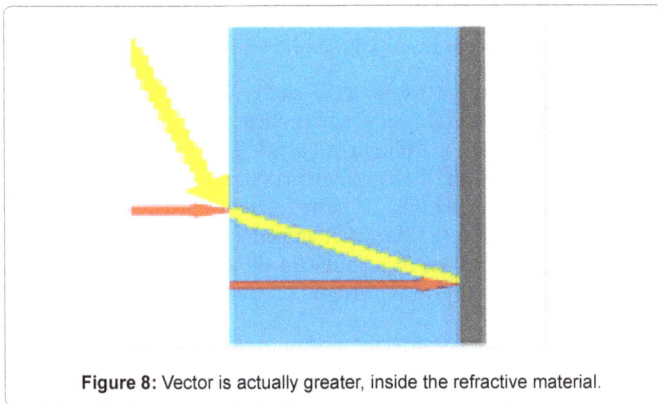

Figure 8: Vector is actually greater, inside the refractive material.

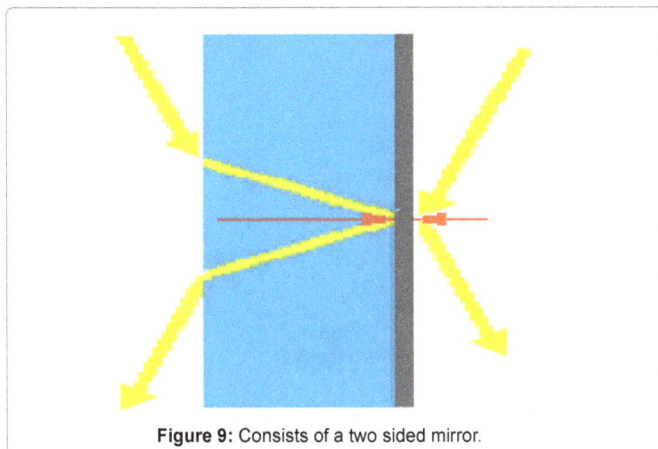

Figure 9: Consists of a two sided mirror.

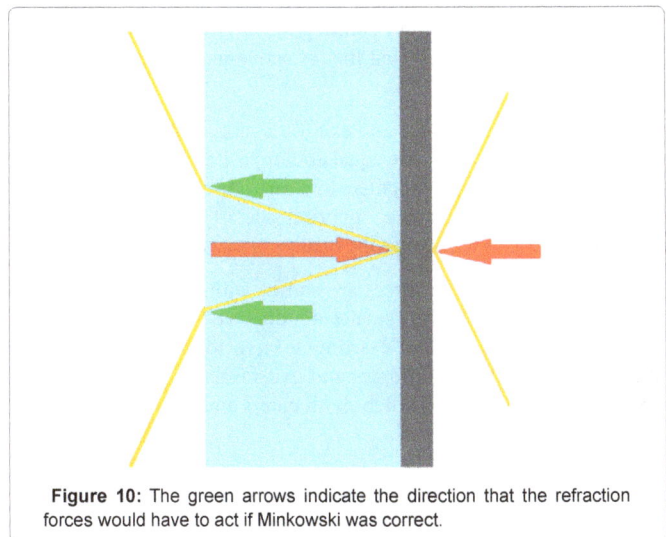

Figure 10: The green arrows indicate the direction that the refraction forces would have to act if Minkowski was correct.

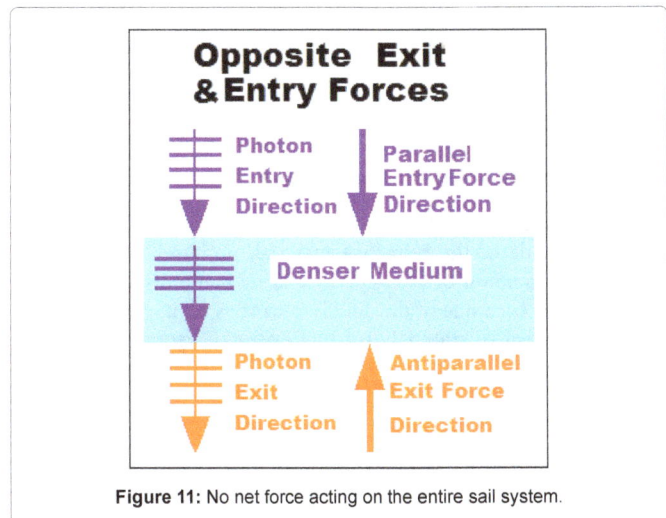

Figure 11: No net force acting on the entire sail system.

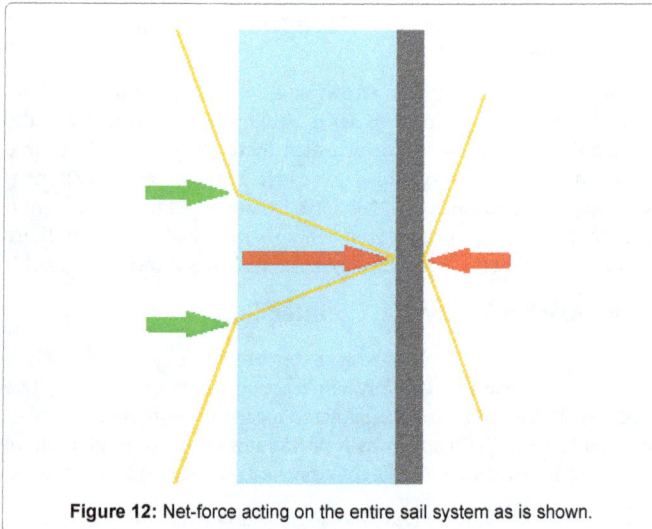

Figure 12: Net-force acting on the entire sail system as is shown.

claimed this result in their paper on Poynting's work [13]. Analogously, this time, the car in the example is exerting backwards-directed forces on the roadway, as it pushes against the road to accelerate while leaving the school-zone. In other words, the refraction-forces that act on the refractive material, are always pointing toward the refractive-material, both when the light is entering the refractive-material and when it is exiting the refractive-material.

In Figure 11, these entry and exit forces are equal and opposite-to each-other; however, they are equal and opposite only because, in this particular instance, the light enters into one side of the more refractive material and exits from the opposite side of the more refractive medium; but this is not always the case.

The actual refraction forces are shown with the green arrows in Figure 12. Both refractive forces are still directed toward the refractive medium; but in this case, "exerting a force toward the refractive-material" means the entry-forces and exit-forces are both exerted in the same direction since the light both enters and exits from the same side of the refractive material.

This time, the refraction forces do not counteract each-other. Neither do they counteract the net reflection force that is acting on the mirror that is implied by the longer red arrow. All three forces are directed toward the back of the mirror. Therefore, a net-force will act on the entire sail system, including the refractive material as well as the mirror.

The green arrows show that both the entry- and exit-forces act in the same direction as the net reflection-force.

Redistribution of Energy and Momentum

Extra energy and extra momentum are positively not magically appearing inside of the refractive material, not temporarily, not at all. The same amount of total-energy and momentum-amplitude are present on the bare side of the double-mirror as on its refractive side, inside the refractive-material; but that energy is reorganized by the refractive-forces. The total momentum magnitude does not change as the light enters the refractive material; instead, its direction of application changes.

Suppose a car is coasting in a circle, its speed is a steady ten miles per hour, but its velocity constantly changes. Its velocity changes

only because its motion changes direction. Even so, its speed and its momentum-amplitude remain the same but are continually expressed in new directions. Its kinetic energy also changes direction, but it also retains the same kinetic-energy amplitude in each new direction. In Figure 13, energy and momentum are constantly transferring between the horizontal vectors and the vertical vectors. So also can the parallel vectors and the perpendicular vectors of the light exchange momentum and energy when refraction forces cause the light path to change direction.

The momentum of the car was altered by the forces that acted between the tires and the road; likewise, the momentum of the turning light is altered by the refraction forces; therefore, its momentum and energy change direction but they do not change amplitude. The light is just as energetic now as it was before it changed direction. Each photon takes its energy and translates that energy to a different orientation as it decelerates into the refractive material. The photons and the refractive material exert forces on each other, forces that are directed perpendicularly toward one side of the mirror.

There are two ways to turn a vehicle: One can turn some or all of the wheels in the manner that one turns an automobile's front wheels; alternatively, one can make some wheels travel faster or slower than the other wheels like the way a Segway turns. In Figure 14, each wheel of these Segways is programmed to individually rotate slower when it crosses over onto the grass. The outside wheels had to slow down first since they were first to enter the grass. These outside wheels make the light-brown tracks in the grass. They start going slow, sooner than the inside wheels. Meanwhile, the still-faster inside wheels travel further in the same amount of time. The dark brown tracks show the longer distance that is traveled by the opposite wheel on each Segway, in the same amount of time. This has the effect of turning the vehicle since the faster wheel is pivoting around the slower wheel.

Likewise, the refractive material exerts forces on the refracting light; first, it slows one side of a light particle; an instant later, it slows the opposite side of the same light particle as it finishes turning the particle. In other words, the fast side of a photon pivots around its slow side as it is crossing over into the more refractive material. The photon exerts a force toward the refractive material. It slows down because the refractive-material exerts an outward-directed reaction force on it. Really, each photon is pivoting around this reaction force.

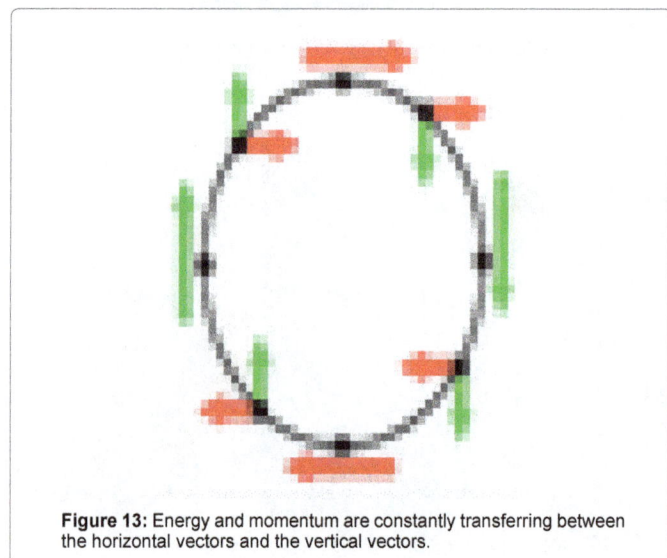

Figure 13: Energy and momentum are constantly transferring between the horizontal vectors and the vertical vectors.

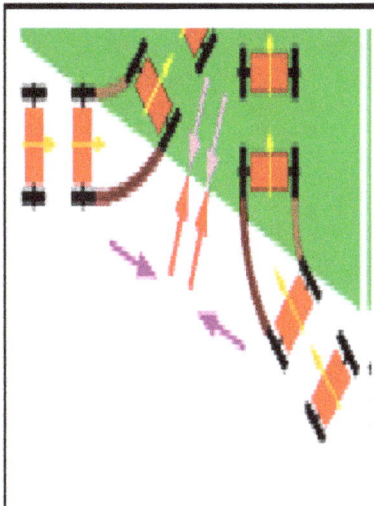

Figure 14: Each wheel of these Segways is programmed to individually rotate slower when it crosses over onto the grass.

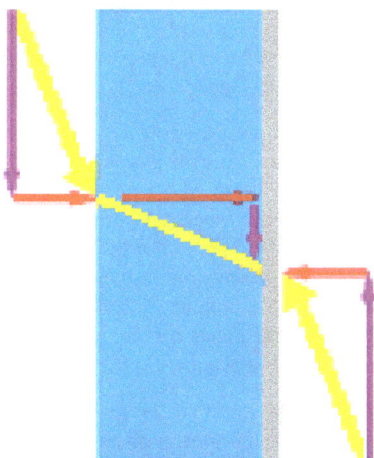

Figure 15: The average-ray (yellow) that is approaching the refractive material on the back of the sail.

beam turns. The direction of its total momentum magnitude has also turned as the light resumes traveling in a straight line, but in a new direction. It is now acting more-perpendicularly to the mirror inside the refractive-material than is the light that is acting in the opposite direction on the bare side of the mirror; therefore, the reflection force that is acting on the refractive side of the mirror is stronger than the reflection-force that acts on the bare side of the double-mirror. This can be verified by comparing the momentum-component vectors that act on both sides of the double-mirror, in Figure 15. The red perpendicular component, inside the refractive medium, is longer than the red perpendicular component that is acting in the opposite direction, on the bare side of the mirror. Again, it is critical to recognize that the vector lengths are a function of the changes in the angle of the yellow light-vectors, relative to the double-mirror.

As the light enters the refractive medium, the short red perpendicular momentum vector outside of the more-refractive material becomes longer, inside the refractive medium; but the parallel

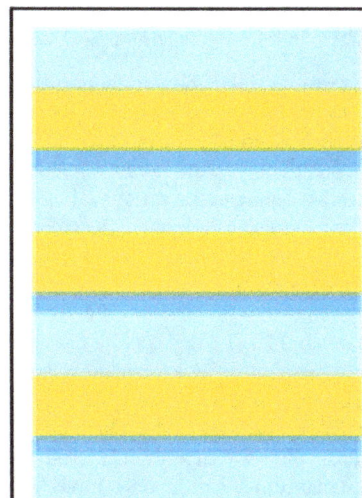

Figure 16: The material with the lowest refractive index is light blue.

In Figure 15, the average-ray (yellow) that is approaching the refractive material on the back of the sail, is equal and opposite to the corresponding average ray that is approaching the bare mirror that is on the front of the sail; this can be verified by observing their short red perpendicular component vectors; they are equal and opposite.

As the light enters into the refractive material, the parallel momentum of the light decreases; it decreases by the same amount that the perpendicular momentum increases. This can be verified by comparing the parallel momentum-component vectors of the light before and after it enters the refractive-material: The purple parallel component vector is longer outside the refractive-material but shorter inside the refractive material. However, the red perpendicular component vector was lengthened as it entered the refractive material. This implies that momentum and energy are being caused to flow less in the parallel direction and more in the perpendicular direction as the light is bent by refraction.

As the light passes into the more-refractive material, the entire

Figure 17: It shows the pressures that can be expected by exploiting a given wavelength.

momentum vector (purple) is longer outside of the refractive-material and shorter, inside the refractive-material.

Moving Forward

The project can be broken-down into three phases. Phase One would consist of testing the Refractive Mirror Concept using lasers to make certain that equal and opposite light beams really can simultaneously produce a net force in a refractive-mirror system. Confirming that premise will provide a reason to persist in solving any problems that arise uniquely at smaller wavelengths, using the distinctively ephemeral photons of the EM Quantum Vacuum.

The second phase of research would involve trying to measure Quantum Vacuum EM radiation forces that are acting on a single-sail system. It appears likely that net pressures of two- or three-hundred Pascal are within reach of a reasonably modest research effort. This does not sound like much pressure or energy; but such a sail would easily lift thousands of times its own weight, since it could be as small as a few-hundred nanometers in thickness; however, it would have to be mounted on a heavier substrate. (The sail cannot be exposed to the air when it is supposed to be exploiting VUV wavelengths since these wavelengths do not form in ionizable mediums such as air.)

A third phase of research will involve stacking many layers of sails to form a meta-material that is thousands of sails thick. There is no particular constraint on how broad an area might be covered by each sail. In Figure 16, the material with the lowest refractive index is light blue. It is the material that separates the various sails. The material that has the highest refractive index is dark blue. The gold color is gold or some other highly reflective metal. Individual gold/dark-blue pairs comprise individual sails. Generally speaking, the metal-layers must be at least three times the thickness of the skin-effect for the wavelengths of interest; it can be much thicker if this is more convenient. The light blue, low-refraction material must have a thickness that is at least two times the wavelengths of interest; but again, it can be as much thicker as is convenient. The dark blue highly refractive material must be less than a quarter of the smallest wavelength of interest. This is because the photons of the Quantum-Vacuum only travel about half of a wavelength before vanishing.

Smaller wave-lengths are much more powerful but the metal is more transparent to smaller wave-lengths so it may require much modelling, trial and error to determine what wavelengths will prove to be optimum with which materials.

A prototype might resemble many layers of Mylar. It might consist of thin aluminum sheets that are coated with plastic on one side and a highly refractive substance on the opposite side. Many yards of this could be wound up and cut off a large flat roll, already stacked. It may be useful to use a vitreous metal sheet to eliminate the crystal structure of the metal to enhance its smoothness.

It is necessary to recruit investors, and experts in photonics and in optical films to design and perform these experiments. It will require an optical lab with lasers and equipment for measuring very small forces. Please contact the author for participation information.

Eq. 10 is based on Eq. 9; Eq. 10 gives the net pressure Pr that acts upon a two-sided refractive-mirror. The net pressure is obtained by subtracting the unaltered pressure that acts on the mirror's bare side from the increased pressure of the refractive side of the mirror. (n_1 is the refractive index of the material that coats one side of the light-sail. n_1 is the index of the most refractive material. n_2 is the refractive index of the material that separates the different sails).

Eq. 9=h c/24 λ^4

Net Pressure=Pr=(n_1-n_2) h c/24 λ^4 (10)

Eq. 10 gives the ideal net pressure assuming that the refractive index n_1 is two. It uses the following equation.

(n_1 is the most refractive material. n_2 is the refractive index of the material that separates the different sails)

Pressure=Pr=(n_1-n_2) h c/24 λ^4 (11)

Figure 17 is based on Eq. 10; it shows the pressures that can be expected by exploiting a given wavelength and all wave-lengths that are greater than that wavelength. For example, exploiting 50 nm wavelengths would yield greater than atmospheric pressure if it was just one hundred layers deep.

Summary

The ball-diode proves that equal and opposite influxes can, in principle, be tapped to produce a net force, and to extract energy from equal and opposite influxes. Casimir's original interpretation of the Casimir phenomenon was that it was driven by asymmetries in the radiation-pressure of the EM Quantum-Vacuum. Casimir's plates can be considered a kind of experimentally-confirmed Quantum light-sail. Therefore it is reasonable to try to develop other methods of inducing useful asymmetries in the EM Quantum-Flux.

For many years, various experiments have revealed that light can exert more pressure on a mirror if it is covered with a refractive material than if it is not covered with a refractive-material. However, this idea has acquired extra-baggage in the form of incorrect assumptions concerning the direction of the refractive forces that act on the refractive material. According to these incorrect assumptions, there could be no net force acting on the entire system. According to the experimentally-confirmed facts concerning refractive forces, a net force should act on the entire proposed system.

The simplest explanation is that the momentum-magnitude and energy intensity of a beam of light remains constant as the beam is refracted and changes direction. Therefore, it may be possible to use refraction to cause the isotropic radiation-pressure of the EM Quantum Vacuum to exert a net force on a passive Light Sail.

The ability to lay down many layers of mirrors and refractive-materials means that even very modest forces may potentially be combined to produce practical net forces. This may have potential for energy production and vehicular propulsion of all kinds, using currently available equipment and techniques.

Although every separate aspect of this experiment has been experimentally verified, seemingly, no one has simultaneously measured the refraction forces and the reflection-forces to determine if a net force can be generated by the refractive mirror system that has just been discussed. Again, the point of this paper was not to prove that this idea must work; rather, the point is to stimulate discussion and action on the proposal and to justify performing the proposed experiments.

References

1. Milonni PW, Goggin ME, Cook J (1988) A Radiation pressure from the vacuum: Physical interpretation of the Casimir Force. Theoretical Division Los Alamos National Laboratory Los Alamos New Mexico Rank J Seiler Research

Laboratory United States Air Force Academy Colorado Springs Colorado. Physical Review A 38: 1621.

2. Millis MG, Davis EW (2009) Frontiers of Propulsion Science. Progress in Astronautics and Aeronautics Reston Virginia NASA Glen Research Center Cleveland Ohio Institute for advanced Studies at Austin Texas 227: 153-156.

3. Casimir HBG (1948) Proc Kon Ned Akad Wetensch 51: 793.

4. Lamoreaux SK (1997) Demonstration of the Casimir Force in the 0.6 to 6µm Range. Phys Rev Lett 78.

5. US Patent Number 8 317:137

6. Jones RV, Richards JCS (1954) The Pressure of Radiation in a Refracting Medium. Proc Roy Soc A 221: 480.

7. Pfeifer RNC, Nieminen TA, Heckenberg NR, Rubinsztein Dunlop H (2007) Colloquium: Momentum of an electromagnetic wave in dielectric media. Reviews of Modern Physics 79: 1197.

8. Minkowski H (1908) Die Grundgleichungen für die elektromagnetischen Vorgänge in bewegten Körpern. Nachrichten von der Gesellschaft der Wissenschaften zu Göttingen, Mathematisch-Physikalische Klasse 1908: 53-111.

9. Abraham M (1909) Zur Elektrodynamik bewegter Körper Rendiconti del Circolo Matematico di Palermo 28: 1-28.

10. Jones RV, Leslie B (1978) The Measurement of Optical Radiation Pressure in Dispersive Media. Proc R Soc Lond A 360: 347.

11. Campbell GK, Leanhardt AE, Mun J, Boyd M, Streed EW, et al. (2008) Photon recoil momentum in dispersive media. Phys Rev Lett 94: 170403.

12. She W, Yu J, Feng R iv, She W. State Key Laboratory of Optoelectronic Materials and Technologies. Sun Yat-Sen University, Guangzhou, China.

13. Loudon R, Baxter C (2012) Contributions of John Henry Poynting to the understanding of radiation pressure. Proc R Soc A 468: 1825-1838.

Permissions

List of Contributors

Madhunuri Raju
Department of Aeronautical Engineering in MLR Institute of Technology, Hyderabad, India

Abene Abderrahmane
University of Valenciennes and Hainaut Cambrai, ISTV Mount Huy, F-59304 Valenciennes, France

Sabatini R, Zammit-Mangion D and Jia H
Department of Aerospace Engineering, Cranfield University, Cranfield, Bedford, UK

Richardson MA
Defence Academy of the United Kingdom, Shrivenham, Swindon, Cranfield University, UK

Cantiello M, Toscano M and Fiorini P
Italian Ministry of Defense, Air Staff, Rome, Italy

Djavareshkian MH and Faghihi AR
Faculty of Engineering, Department of Mechanical Engineering, Ferdowsi University of Mashad, Mashad, Iran

Roberto Sabatini, Anish Kaharkar and Celia Bartel
Department of Aerospace Engineering, Cranfield University, Cranfield, Bedford, MK43 0AL, UK

Tesheen Shaid
Leopoldo Rodríguez Salazar, Aerotech Systems Ltd, Milton Keynes, Buckinghamshire, MK11 1BY, UK

Kostopoulos V, Kotzakolios T and Vlachos DE
Applied Mechanics Laboratory, Department of Mechanical Engineering and Aeronautics, University of Patras, Patras University Campus, GR-26500 Patras, Greece

Bras M, Vale J and Lau F
Instituto Superior Técnico, Lisbon, Portugal

Suleman A
University of Victoria, Victoria BC, Canada

Roberto Sabatini
Department of Aerospace Engineering, Cranfield University, Cranfield, Bedford MK43 0AL, UK

Mark A Richardson
Defence Academy of the UK, Cranfield University, Shrivenham, Swindon SN6 8LA, UK

Ermanno Roviaro
Electro-Optics R&D Laboratories, SELEX-ELSAG, Cogoleto (Genova) 16016, Italy

Arvind Prabhakar and Ayush Ohri
Department of Mechanical and Manufacturing Engineering, Manipal Institute of Technology, Karnataka, India

Ian R McAndrew
Embry Riddle Aeronautical University, Brandon England, UK

Kenneth Witcher
Embry Riddle Aeronautical University, Las Vegas Nevada USA

Abene Abderrahmane
Laboratory of Aerodynamics, University of Valenciennes, France

Juntao Chang, Shibin Cao, Junlong Zhang, Jianfeng Z and Wen Bao
Academy of Fundamental and Interdisciplinary Sciences, Harbin Institute of Technology, Harbin 150001, China

Wenqing S and Zhixin Li
Beijing Research Institute of Mechanical and Electrical Engineering, 100074, Beijing, China

Ntantis E
School of Engineering, Cranfield University, Bedford, UK

Menail Y
University of Badji Mokhtar, Sidi Ammar, LR3MI, BP 12, 23000, Annaba, Algeria

Abderrahim EL Mahi
University of Maine, LAUM, CNRS UMR 6613, Avenue Olivier Messiaen, 72085 Le Mans Cedex 9, France

Assarar M
University of Reims Champagne-Ardenne, LISM, EA 4695, IUT de Troyes, 9 rue de Québec, 10026 Troyes Cedex, France

Jokela Jorma
Laurea Simulated Hospital, Laurea University of Applied Sciences, Hyvinkää, Finland

Laapotti Heli
Päijät-Häme Social and Health Care Group, Centre for Prehospital Care and Emergency Medicine Lahti, Finland

Engblom Janne and Harkke Ville
Turku School of Economics, University of Turku, Finland

Rukshan Navaratne, Marco Tessaro, Weiqun Gu, Vishal Sethi and Pericles Pilidis
Department of Power and Propulsion, Cranfield University, Bedfordshire, UK

Roberto Sabatini and David Zammit-Mangion
Department of Aerospace Engineering, Cranfield University, Bedfordshire, UK

Ahmad Sedaghat and Mohammad Amin Aghahosaini
Department of Mechanical Engineering, Isfahan University of Technology, Isfahan, 84156-83111, I. R. of Iran

Rossett M
Rowan University, USA

Mahendra Kumar Trivedi, Rama Mohan Tallapragada, Alice Branton, Dahryn Trivedi and Gopal Nayak
Trivedi Global Inc, 10624 S Eastern Avenue Suite A-969, Henderson, NV 89052, USA

Omprakash Latiyal and Snehasis Jana
Trivedi Science Research Laboratory Pvt. Ltd , Hall-A, Chinar Mega Mall, Chinar Fortune City, Hoshangabad Rd., Bhopal-462026, Madhya Pradesh, India

Harasani W
Faculty of Engineering, King Abdul Aziz University, Saudi Arabia

Hong Z
AVIC Commercial Aircraft Engine Co., Ltd., R & D Center, 3998 S. Lianhua Road, Shanghai Minhang, 201108, China

Cao G and Chen WR
AVIC Commercial Aircraft Engine Manufacturing Co., Ltd., 77 Hongyin Road, Shanghai Lingang, 201306, China

Shagaiya Daniel Y
Department of Mathematical Sciences, Kaduna State University, Kaduna, Nigeria

Retheesh R, Samuel B, Radhakrishnan P, Nampoori VPN and Mujeeb A
International School of Photonics CUSAT, Kochi 682022, India

Rathinakumar V, Manju Nanda and Jayan-thi J
Department of Aerospace Electronics and System Division, CSIR-National Aerospace Lab, Bangalore, India

DeFelice TP
Sykesville, Maryland, MD, USA

Duncan Axisa
Research Applications Laboratory, National Center for Atmospheric Research, Boulder, CO, USA

PadmanabanS and Mahendran SME
Aeronautical Engineering, Hindustan University, Chennai,Tamilnadu, India

Chung-How Poh and Chung-Kiak Poh
Aero-Persistence Research, 23 Halaman York, 10450 Penang, Malaysia

Kushal KS, Manju Nanda and Jayanthi J
Aerospace Electronics and Systems Division, CSIR-National Aerospace Laboratories, Bangalore, Karnataka, India

Rajesh A and Abhay BT
Department of Mechanical Engineering, New Horizon College of Engineering , Bangalore, India

Fayssal Hadjez and Brahim Necib
University of Constantine, Faculty of Science and Technology, Department of Mechanical Engineering, Algeria

Harasani W
Aeronautical Engineering Department, King Abdul Aziz University, Jeddah, Saudi Arabia

Scott Smith W
Spokane, Washington, USA

Index

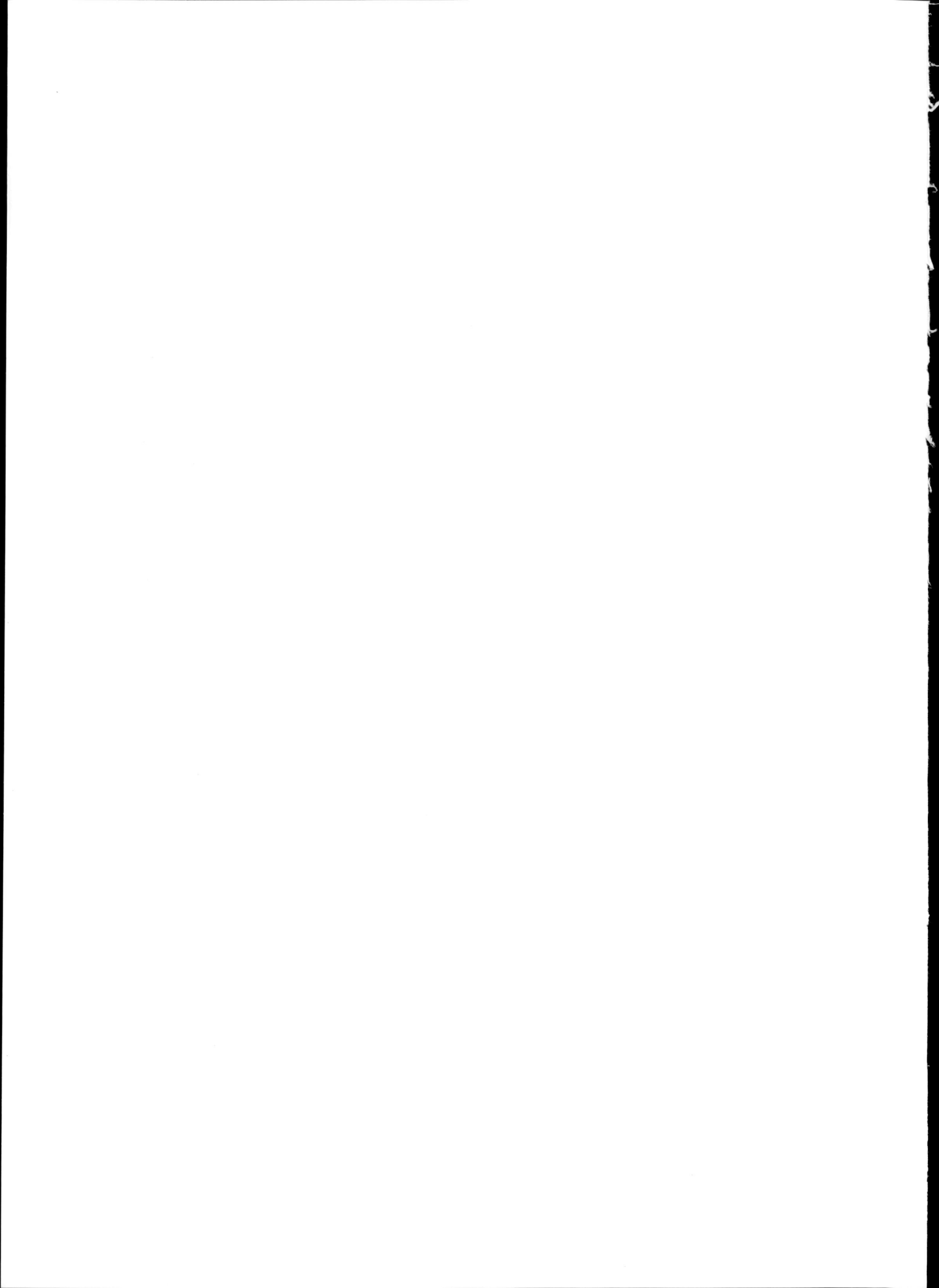